Philadelphia
02. 17. 05

To Peter

with apreciation for the friendship, support and scientific cooperation

[signature]

AF701112

Pore-forming Peptides and Protein Toxins

Cellular and molecular mechanisms of toxin action
Edited by Philip Lazarovici
The Hebrew University of Jerusalem, Israel

A series of books on various aspects of toxin research, giving a broader emphasis on the mechanism of action, structure–function relationship, the use of toxins as research tools and their therapeutic applications.

Volume 1
Toxins and Signal Transduction
Y. Gutman and P. Lazarovici

Volume 2
Secretory Systems and Toxins
M. Linial, A. Grasso and P. Lazarovici

Volume 3
Site-selective Neurotoxicity
D. Lester, W. Slikker Jr. and P. Lazarovici

Volume 4
Chimeric Toxins
H. Lorberboum-Galski and P. Lazarovici

Volume 5
Pore-forming Peptides and Protein Toxins
G. Menestrina, M. DallaSerra and P. Lazarovici

Pore-forming Peptides and Protein Toxins

Edited by
Gianfranco Menestrina,
Mauro Dalla Serra
and Philip Lazarovici

Taylor & Francis
Taylor & Francis Group
LONDON AND NEW YORK

First published 2003
by Taylor & Francis
11 New Fetter Lane, London EC4P 4EE

Simultaneously published in the USA and Canada
by Taylor & Francis Inc,
29 West 35th Street, New York, NY 10001

Taylor & Francis is an imprint of the Taylor & Francis Group

Typeset in Goudy by
Newgen Imaging Systems (P) Ltd, Chennai, India
Printed and bound in Great Britain by
TJ International Ltd, Padstow, Cornwall

British Library Cataloguing in Publication Data
A catalogue record for this book is available from the British Library

Library of Congress Cataloging in Publication Data
A catalog record for this book has been requested

ISBN 0–415–29852–0

Contents

Figures

Tables

Contributors

Gregor Anderluh
University of Ljubljana
Biotechnical Faculty
Department of Biology, Ljubljana
Slovenia

David Andreu
Department of Experimental and Health Sciences
Universitat Pompeu Fabra
Barcelona, Spain

Rustam Azimov
Department of Psychiatry
UCLA School of Medicine and Neuropsychiatric Institute and West LAVeterans Administration
Medical Center
Los Angeles, California

Rushana Azimova
Department of Psychiatry
UCLA School of Medicine and Neuropsychiatric Institute and West LA
Veterans Administration Medical Center
Los Angeles, California

L. Baba Moussa
Département de Biochimie et de Biologie Moléculaire
Faculté des Sciences et Techniques
Université Nationale du Bénin
Cotonou, Benin

Alessandro Ballio
Dipartimento di Scienze Biochimiche A. Rossi Fanelli
Università di Roma La Sapienza
00185 Roma, Italy

Roland Benz
Lehrstuhl für Biotechnologie
Theodor-Boveri-Institut (Biozentrum) der Universität Würzburg
Am Hubland, D-97074 Würzburg
Germany

Katalin Blasko
Institute of Biophysics
Semmelweiss University of Medicine
Budapest, 1444, Hungary

Argante Bozzi
Dipartimento di Scienze e Tecnologie Biomediche
Università dell'Aquila
Via Vetoio, Loc. Coppito
I-67100 L'Aquila, Italy

Joseph G. Brand
Monell Chemical Senses Center
Philadelphia, PA 19104, USA

Alejandra Bravo
Instituto de Biotecnología
Departamento de Microbiología Molecular. UNAM
Apdo postal 510-3, Cuernavaca
Morelos 62250, México

Stephen Bronner
Laboratoire de Physiopathologie et d'Antibiologie Bactériennes des Infections Emergentes et Nosocomiales
UPRES EA-1318
Institut de Bactériologie de la Faculté de Médecine de Strasbourg (ULP) – Hôpitaux Universitaires de Strasbourg
3, rue Koeberlé F-67000 Strasbourg
France

Armando Carpaneto
CNR – Istituto di Biofisica
Sezione di Genova
via De Marini 6, 16149 Genova, Italy

Didier A. Colin
Laboratoire de Physiopathologie et d'Antibiologie Bactériennes des Infections Emergentes et Nosocomiales
UPRES EA-1318
Institut de Bactériologie de la Faculté de Médecine de Strasbourg (ULP) – Hôpitaux Universitaires de Strasbourg
3, rue Koeberlé F-67000 Strasbourg
France

Manuela Coraiola
ITC – Centro di Fisica degli Stati Aggregati e CNR – Istituto di Biofisica
Sezione di Trento, via Sommarive 18
38050 Povo (Trento), Italy

Mauro Dalla Serra
ITC – Centro di Fisica degli Stati Aggregati e CNR – Istituto di Biofisica
Sezione di Trento, via Sommarive 18
38050 Povo (Trento), Italy

Lucantonio Debellis
Dipartimento di Fisiologia Generale e Ambientale, Università di Bari
via Amendola 165/A
70126, Bari, Italy

Vincenzo Fogliano
Dipartimento di Scienza degli Alimenti
Università di Napoli Federico II
80138 Napoli, Italy

Franco Gambale
CNR – Istituto di Biofisica
Sezione di Genova, via De Marini 6
16149 Genova, Italy

S. Geis
Institut für Pharmakologie und Toxikologie, Justus-Liebig-Universität
Giessen, Frankfurter Str. 107
D-35392 Giessen, Germany

Doron Gerber
Department of Biological Chemistry
The Weizmann Institute of Science
Rehovot, 76100 Israel

Antonio Di Giulio
Dipartimento di Scienze e Tecnologie Biomediche
Università dell'Aquila,Via Vetoio
Loc, Coppito
I-67100 L'Aquila, Italy

Isabel Gómez
Instituto de Biotecnología
Departamento de Microbiología Molecular
UNAM, Apdo postal 510-3
Cuernavaca, Morelos 62250, México

Alain Gravet
Laboratoire de Physiopathologie et d'Antibiologie Bactériennes des Infections Emergentes et Nosocomiales
UPRES EA-1318
Institut de Bactériologie de la Faculté de Médecine de Strasbourg (ULP) – Hôpitaux Universitaires de Strasbourg, 3, rue
Koeberlé F-67000 Strasbourg, France

Leopoldo Güereca
Instituto de Biotecnología
Departamento de Microbiología Molecular
UNAM, Apdo postal 510-3, Cuernavaca
Morelos 62250, México

Yutaka Hirakura
Department of Molecular Neurobiology
Advanced Research Center for Human Sciences, Waseda University, Tokyo

Bruce L. Kagan
Department of Psychiatry
UCLA School of Medicine and Neuropsychiatric Institute and West LA Veterans Administration Medical Center
Los Angeles, California

Yuri A. Kaulin
Monell Chemical Senses Center
Philadelphia, PA 19104, USA

Paul K. Kienker
Department of Physiology & Biophysics
Albert Einstein College of Medicine
Bronx, New York

Paavo K. J. Kinnunen
Helsinki Biophysics & Biomembrane Group, Department of Medical Chemistry
Institute of Biomedicine
University of Helsinki
Finland

M. Langewische
Institut für Pharmakologie und Toxikologie, Justus-Liebig-Universität
Giessen, Frankfurter Str. 107
D-35392 Giessen, Germany

Philip Lazarovici
Department of Pharmacology and Experimental Therapeutics
School of Pharmacy, Faculty of Medicine
The Hebrew University of Jerusalem
Jerusalem, 91120, Israel

Meng-Chin Lin
Department of Psychiatry
UCLA School of Medicine and Neuropsychiatric Institute and West LA Veterans Administration Medical Center
Los Angeles, California

Dietmar Linder
Biochemisches Institut, Justus-Liebig-Universität Giessen, Friedrichstr. 24
D-35392 Giessen, Germany

Monica Linder
Biochemisches Institut, Justus-Liebig-Universität Giessen, Friedrichstr. 24
D-35392 Giessen, Germany

Frieder Lutz
Institut für Pharmakologie und Toxikologie, Justus-Liebig-Universität Giessen, Frankfurter Str. 107,
D-35392 Giessen, Germany

Peter Maček
Department of Biology
University of Ljubljana, Biotechnical Faculty
Ljubljana, Slovenia

Valery V. Malev
Institute of Cytology, Russian Academy of Sciences
St Petersburg 194064, Russian Federation
Budapest, 14444, Hungary

Gianfranco Menestrina
ITC – Centro di Fisica degli Stati Aggregati e CNR – Istituto di Biofisica
Sezione di Trento, via Sommarive 18
38050 Povo (Trento), Italy

Cesare Montecucco
Dipartimento di Scienze Biomediche Sperimentali, Università di Padova, via
G. Colombo, 35121 Padova, Italy

Henri Monteil
Laboratoire de Physiopathologie et d'Antibiologie Bactériennes des Infections Emergentes et Nosocomiales
UPRES EA-1318
Institut de Bactériologie de la Faculté de Médecine de Strasbourg (ULP) – Hôpitaux Universitaires de Strasbourg
3, rue Koeberlé F-67000 Strasbourg
France

T. Nakada
Institut für Pharmakologie und Toxikologie, Justus-Liebig-Universität
Giessen, Frankfurter Str. 107
D-35392 Giessen, Germany

María E. Nuñez-Valdez
Instituto de Biotecnología, Departamento de Microbiología Molecular
UNAM, Apdo postal 510-3, Cuernavaca
Morelos 62250, México

Emanuele Papini
Dipartimento di Scienze Biomediche Sperimentali, Università di Padova, via G. Colombo, 35121 Padova, Italy

Rigoberto V. Pérez
Instituto de Biotecnología, Departamento de Microbiología Molecular. UNAM
Apdo postal 510-3, Cuernavaca
Morelos 62250, México

Gilles Prévost
Laboratoire de Physiopathologie et d'Antibiologie Bactériennes des Infections Emergentes et Nosocomiales
UPRES EA-1318
Institut de Bactériologie de la Faculté de Médecine de Strasbourg (ULP) – Hôpitaux Universitaires de Strasbourg, 3, rue
Koeberlé F-67000 Strasbourg, France

Andrea C. Rinaldi
Cattedra di Chimica Biologica
Dipartimento di Scienze Mediche Internistiche
Università di Cagliari
I-09042 Monserrato (CA), Italy

Luis Rivas
Centro de Investigaciones Biológicas (C.S.I.C)
Madrid, Spain

Anna Rufo
Dipartimento di Scienze e Tecnologie Biomediche, Università dell'Aquila
Via Vetoio, Loc, Coppito
I-67100 L'Aquila, Italy

Jorge Sánchez
Instituto de Biotecnología, Departamento de Microbiología Molecular. UNAM
Apdo postal 510-3, Cuernavaca
Morelos 62250, México

Yechiel Shai
Department of Biological Chemistry
The Weizmann Institute of Science
Rehovot, 76100 Israel

Stephen L. Slatin
Department of Physiology & Biophysics
Albert Einstein College of Medicine
Bronx, New York

Annette Sliwinski-Korell
Institut für Pharmakologie und Toxikologie, Justus-Liebig-Universität
Giessen, Frankfurter Str. 107
D-35392 Giessen, Germany

Mario Soberón
Instituto de Biotecnología
Departamento de Microbiología Molecular
UNAM, Apdo postal 510-3, Cuernavaca
Morelos 62250, México

Jon Y. Takemoto
Department of Biology, Utah State University, Logan, UT 84322 USA

Francesco Tombola
Centro CNR Biomembrane e
Dipartimento di Scienze Biomediche
Università di Padova, via G, Colombo 3
35121 Padova, Italy

Sandra Werner
ITC – Centro di Fisica degli Stati Aggregati e CNR – Istituto di Biofisica
Sezione di Trento, via Sommarive 18
38050 Povo (Trento), Italy

Hong Xia Zhao
Helsinki Biophysics & Biomembrane Group, Department of Medical Chemistry
Institute of Biomedicine
University of Helsinki, Finland

Mario Zoratti
Centro CNR Biomembrane e Dipartimento di Scienze Biomediche
Università di Padova, via G, Colombo 3
35121 Padova, Italy

Preface to the series

Pathogenic bacteria, and poisonous animals and plants have been known to mankind for centuries. These organisms produce toxins that act by a variety of mechanisms to immobilize or kill their prey. Recently, toxin research has rapidly expanded as a result of the powerful and productive contributions of recombinant DNA, monoclonal antibodies, microinjection, crystallography and patch clamp techniques. The number of toxins isolated and identified has increased, and more profound insights into their structure, mode of action and role in disease have been achieved. The stage is now set to re-examine our previous concepts about toxin action in the light of current findings and to trace new pathways for the future. Accordingly, the purpose of this series is to fill the need for a comprehensive, contemporary work at the cellular and molecular levels of toxin action. Although emphasis will be placed on particular achievements, the new data will be integrated with previous investigations. Stimulating critical evaluations and current views and suggestions for new lines of research have been encouraged. Because of the huge number of toxins now known, a certain degree of selection was necessary, of course, of a subjective nature.

The aim of this series is to provide a multidisciplinary approach orientated toward an understanding of the basic principles and cellular molecular mechanisms of the action of toxins and their potential use as research tools. For this reason, each chapter provides a description of a normal physiological cellular structure and function, the interference of toxins with this process, and the use of particular toxins in research. Similarly, the structure of each book in the series was determined partly on scientific, and partly on pedagogic grounds. The first chapter(s) comprise mainly a review of the general principles of the topic discussed in the book. The chapters that follow present specific reviews of the progress that has been made in different areas of this topic.

We planned five books in the series: Toxins and Signal Transduction, the first volume, presents selected mechanisms by which toxins affect molecular processes which transduce extracellular signals into intracellular messages regulating cell function. Secretory Systems and Toxins, the second volume, provides an updated state-of-the-art treatment of vesicle-mediated secretion with special emphasis on the specific action and recognition of the secretory organelle proteins and glycolipids by tetanus, botulinum and α-latrotoxin neurotoxins. The third volume, Site-Selective Neurotoxicity, presents different neurotoxicological aspects with a unique mechanistic perspective of neurotoxicity. Chimeric Toxins: Mechanisms of Action and Therapeutic Applications, the fourth volume, focuses on toxins affecting protein synthesis, their structure, genetic engineering, mechanism of action and therapeutic application in medicine. The fifth volume, Pore Forming Peptides and Protein Toxins, presents natural and synthetic peptides and toxins forming pores and ionic channels that cause cell membrane collapse and cell death.

This book series includes contributions by many leading researchers in the field. While each research group has chosen a particular toxin, or cellular or molecular system, collating all efforts into a single series will hopefully provide a unique source of information. Toxin research requires skill, special safety precautions, hardwork, and patience. I expect that this field of research will continue to reveal new cellular and molecular processes and provide new, selective research tools and prototypical compounds for drug development. If this series supports this effort in some small way, our work will be rewarded.

This undertaking has been made much easier by the excellent cooperation of the coeditors, Prof. Yehuda Gutman, Dr Michal Linial, Dr Alfonso Grasso, Dr Haya Lorberboum-Galski, Dr William Slikker, Dr David Lester, and Dr Gianfranco Menestrina. I would like to thank all authors for their commitment, time, and scholarship. We would also like to express our gratitude to Harwood Academic Publishers, for their encouragement, advice and practical assistance during the production of this book series.

Philip Lazarovici
Series Editor

Preface

Confronting competitors is a very tough, but vital task for all living organisms. An important part of the genome is devoted to encoding proteins employed in this competition. Such proteins include many offensive weapons, like toxins, meant to physically eliminate the competitors. Not surprisingly, the cell membrane is often the first target for such weapons. In fact, integrity of the cell membrane is crucial for appropriate cell function and, thereby, for life itself. Similar weapons, collectively called cytolysins or membrane-damaging toxins, form a large family of proteins and polypeptides of bacterial, animal and plant origin which share the capability of altering cell homeostasis. Among them, an increasingly large number are those that have been found to act as pore formers. These molecules inflict a lethal blow to the attacked cells simply by punching an aqueous channel, usually relatively large and poorly selective, through their membranes.

The pore concept was first elaborated to explain the lytic damage imposed by complement to bacteria and other non-self cells, but it was soon realised it could extend to a wider range of lytic agents. Early in 1979, classifications of cytolysins of different origin clearly indicated that many of these could form discrete lesions in the cellular membrane. However, the first direct proof of a pore-forming action had to wait until a few years later when studies were conducted on α-toxin from *Staphylococcus aureus* and aerolysin from *Aeromonas hydrophila*. Since then, pore formation has emerged as a major and widespread mechanism for toxin-mediated cell attack in the plant, animal and microbial world, and a number of books and symposia have been devoted to this topic. So far, hundreds of natural molecules, of peptide or protein nature, have been described to work in this way.

Pore-forming toxins (PFT) are topics of interest not only for people studying their functional role, like physicians or microbiologists, but also for a variety of other scientists, ranging from biochemists to nanoscientists, as it will become clear in the following. PFT are involved in important pathologies or envenomations. For example, large families of leucotoxins have been identified, aimed at contrasting the immune response of the host. They pertain either to Gram positive bacteria, like the cholesterol binding toxins and the α-toxin/bicomponent-toxins family of *Staphylococcus* spp. (reviewed in Chapter 1), or to Gram-negative bacteria, like the RTX superfamily. RTX (for Repeat in ToXins) are characterised by the presence of an octapeptyde repeat forming a variable number of Ca^{++} binding loops (reviewed in Chapter 2). Virtually all pathogenic bacteria, whose genome have been sequenced to date, possess at least one, but usually more, cytolysin of the PFT kind. Often, these toxins, although the parent by the mechanism of action, are completely unrelated by the structure and do not form known families. This is the case, for example, of *Pseudomonas aeruginosa* cytotoxin (Chapter 3) or the vacuolating toxin of *Helicobacter pylori* (Chapter 4), two toxins involved in serious human pathologies: the infection of airways of cystic fibrosis patients and gastric ulcer, respectively.

PFT are not limited to the bacterial world. In fact, plants and animals have adopted similar molecules as tools to contrast microbial infections or as part of their venoms. An example is provided by the family of cytolytic toxins found in the tentacles of sea anemones (*Actiniaria*) described in Chapter 8. Notwithstanding their origin, PFT have some peculiar properties that make them extraordinary tools and study systems. They exist, in fact, in two possible stable structures, a water-soluble and a membrane-inserted form. The polypeptide, once secreted by the producing organism, adopts a temporary water-soluble fold, suitable for diffusion in the organic fluids. When it reaches the target membrane, however, passing through different intermediate states, it ultimately inserts into the lipid and forms a stable pore. While, in most cases, the insertion of proteins in membranes is a complex process that requires the assistance of a number of proteins, besides the one to be inserted, in the case of PFT, insertion is a completely spontaneous, or self-assisted, process. This is apparently achieved through the interaction with specific lipids, but, more importantly, by an aggregation mechanism implying multiple copies of the same or related toxin components (see Chapter 1). For these reasons PFT are unique models with which to study mechanisms, whose importance certainly transcends that of the toxins themselves, for example, the conformational transitions leading to aggregation (protein–protein interaction), those leading to membrane insertion (protein–lipid interaction), and those leading to opening and closing of the pores (gating mechanisms).

Structural studies are certainly paramount and the most informative technique has been, also in this case, X-ray crystallography. Monomeric structures of PFT have recently become available in an increasing number of cases. To derive the structure of peptide PFT and some related smaller molecules however, NMR spectroscopy has proven at least as important, with the additional advantage of operating in a softer, native-like condition. The membrane-bound structure is often the most interesting, but certainly also the most elusive, for the well-known problems posed by membrane proteins: the structure is too disordered for the X-ray technique, but too ordered for the NMR approach. *S. aureus* α-toxin is the one exception, having been crystallised, and structurally solved, in the aggregated pore-forming form, albeit in the presence of a detergent and not in a lipid membrane. For all other membrane-bound protein PFT, the techniques used have lower resolution, and are those classically employed for intrinsic proteins, that is, electron and atomic force microscopy, as well as spectroscopic techniques such as flulorescence, infrared, circular dichroism and mass. With smaller peptides, additional structural data in the lipid environment could be obtained by low-angle X-ray and neutron scattering, by NMR and by theoretical calculations (e.g. molecular dynamics), as also described in Chapter 9.

Comparing the results in the two environments should provide clues about how, and through which intermediate states, the change from water-soluble monomer to membrane-inserted aggregate takes place. Interestingly, at least within protein PFT, and especially in the transmembrane region, the predominant secondary structure observed till now, is the β-sheet, resembling bacterial and mitochondrial porins. The existence of a common ancestor with porins was postulated. Nontheless, PFT forming of predominantly α-helical structures do also exist (called α-PFT as opposed to the previous β-PFT). Examples of α-PFT are colicins, lytic proteins used as lethal arms in the competition between different bacterial strains living on the same host (see Chapter 7). Pore formation by α-PFT implies the insertion in the lipid film of a bundle of helices, which would otherwise remain hidden in the hydrophobic core of the protein. The opening of the pore occurs together with the translocation of a large portion of the protein through the membrane. As explained in Chapter 7, this is in complete analogy with the mechanism of action of diphtheria toxin (DT).

DT is a toxin of the A/B type which are characterised by the presence of two protomers: the B subunit (responsible for cell binding and membrane translocation) and the A subunit (responsible for an enzymatic intracellular activity). In the case of DT, the A and B subunits belong to the same polypeptide chain. Interestingly, even DT is able to open ion channels into cell and model membranes, thus reinforcing its analogy with colicins and PFT. Such translocation mechanism can easily be used to import inside the cell foreign proteins that have been fused to the colicin, suggesting an important biotechnological application, as a transfection tool, for these molecules.

Due to their peculiar self-assembling properties, PFT may also find other interesting applications. They may be used as they are, in the wild form, for applications with either a commercial interest, like the insecticidal toxins of *Bacillus thuringiensis* and *B. sphaericus* widely used in agriculture (see Chapters 5 and 6), or with a basic science interest, for example, the controlled poration of cells in studies of intracellular mechanisms. Pore-forming peptides have also a strong potential as tools in dissecting cellular functions, like the uses of pardaxin described in Chapter 10, or as new-concept antibiotics and antifungals, for example, the metabolites of *P. syringae* described in Chapters 12 and 13. Other applications involve the engineering of PFT to produce new molecules with peculiar properties. Examples range from the production of antitumoral toxins to constituting the basic elements for the construction of sophisticated biosensors. In the case of smaller PFT, chimeric molecules (as those presented in Chapter 11) and even *de novo* designed and constructed peptides are now common strategies.

Finally, a number of peptides deriving, as secondary products, from the spontaneous and, in some cases, misconducted degradation of endogenous proteins, assume the character of pore-forming substances. These include the amyloid peptides characterising more than 20 human amyloid diseases. Amyloidosis (reviewed in Chapter 14) include pathologies that are assuming a greater importance in developed countries, for the risks posed by the ageing of the population (Alzheimer disease) and by the increasingly industrialised food production (Kreutzfeld-Jackob or prion disease).

The Editors

Glossary

Aβ The amyloid-beta protein of Alzheimer's disease, containing 39–43 amino acids.

A/B toxin A bacterial secreted toxin formed by two domains, one of which (A) embodies the intracytosolic-acting enzymatic activity, while the other (B) is responsible for binding to the target cell and translocating the active portion into it. B usually forms pores in cell and artificial membranes. Example: diphtheria toxin.

α-Conotoxin A family of small peptide neurotoxins from the cone snail family that antagonise the nicotinic acetylcholine receptor. α-Conotoxin IMI, from *Conus imperialis*, referred to in the text, has 12 amino acid residues and contains two disulfide bonds.

Alpha-helical peptide A peptide the secondary structure of which is exclusively or mainly formed by an α-helix, that is, the right-handed helical folding of a polypeptide such that amide nitrogens share their hydrogen atoms with the carbonyl oxygens of the fourth amide bonds towards the C-terminal end of the polymer.

Amphipathic/amphiphilic molecules Molecules having both polar and non-polar groups, for example, a detergent or a lipid molecule. A molecule with a hydrophilic head and a hydrophobic tail. That is, a molecule that has one end which attracts water and one end which repels water.

Amphipathic helix An alpha helix that is primarily hydrophobic on one side and hydrophilic on the other. Its lowest energy orientation at a lipid/water interface is with its hydrophobic side facing the lipid and its hydrophilic side facing the water.

Amphiphiles Molecules possessing both a water loving polar part (hydrophilic) and a water hating non-polar part (hydrophobic).

Amyloid Deposits containing proteins which aggregate into long fibrils in a β-sheet conformation and exhibit green birefrigence on staining with Congo red.

Amyloidosis Diseases characterised by amyloid deposition.

Antigens Substances capable of inducing a specific immune response and of reacting with the products of that response, that is, specific antibodies or specifically sensitised T-lymphocytes.

Apamin An 18-residue peptide toxin with two disulfide bonds, found in the venom of the honeybee, *Apis mellifera*, that blocks certain potassium channels.

Apoptosis Programmed cell death; it is the suicidal process of the cell involving among other features, PS expression at the outer leaflet of the plasma membrane and nuclear DNA fragmentation. Apoptotic cells are phagocytised by macrophages avoiding release of internal cellular material into the media, in contrast with necrosis, the non-programmed death with massive disruption of plasma membrane and release of intracellular material.

Asolectin Lipid mixture from soybean.

ATR dichroic ratio The ratio between two IR spectra taken with polarised light at 0° and 90°, respectively.

ATR-FTIR Attenuated Total Reflectance configuration used in Fourier Transformed Infra-Red spectroscopy, particularly suited for protein secondary structure investigations.

Bicomponent toxins Toxins constituted by two proteins that synergically attack some mammalian cells after being synthesised and secreted as separate, inactive, components.

Biocontrol The agricultural use of living things, such as parasites, diseases and predators, to control or eliminate others, such as weeds and pests, rather than by using chemicals.

Bursting A type of gating kinetics displayed by channels, consisting of relatively long periods in which the channel opens and closes with a high (typically in the KHz range) frequency (a burst) and of intervening periods of channel inactivation, lasting much longer than intraburst closures.

Ca^{2+}-activated Cl^- channels A class of anion-selective channels activated by Ca^{2+} influx or Ca^{2+} release from intracellular stores. The CLCA gene family (four known members in humans) has recently been found to encode channels of this type. Tissue-specific expression patterns.

Channel lifetime The time spent by the ionic channel in the open or closed state.

Charybdotoxin A 37-residue peptide, originally isolated from the venom of the scorpion *Leiurus quinquestriatus hebraeus*, containing three disulfide bonds. It is a potent blocker of certain potassium channels.

Cnidocytes (Nematocytes) of Cnidarians. Stinging cnydocysts contain an inverted tubule which can be abruptly ejected on mechanical or chemical stimulation.

Colicin A family of proteins, generally made by particular strains of *E. coli*, that are toxic to other strains of *E. coli*. Colicins have three functional domains with the lethal activity residing in the C-terminal domain, which is either an enzyme that acts cytoplasmically, such as a DNAse or an RNAse, or a pore former that permeabilises the inner membrane of the target bacterium.

Contact angle Angle formed between the liquid and solid surfaces at the solid–liquid–vapor boundary. It is related to the critical micelle concentration (CMC) of amphiphiles.

Corpeptin 22 amino acid residues lipodepsipeptide produced by *Pseudomonas corrugata*.

Critical micelle concentration (CMC) Concentration at which micelles in solution are in thermodynamic equilibrium with the constituent monomers. It is a characterising parameter of detergents and surfactants.

Diphtheria toxin A 535-residue protein produced by *Corynebacterium diphtheriae* that is the causative agent of diphtheria. The lethal activity resides in the N-terminal enzymatic domain, whereas the central T-domain is responsible for transferring the enzymatic domain into the cytoplasm of target cells.

Domains Spatially distinct portions of a protein. Example of the domain organisation of α-toxin: the cap domain, containing the pore entrance; the rim, which anchors the toxin to the membrane surface; the stem, which is the β-barrel forming the pore; the amino latch, connecting the α-toxin monomers.

Dominant negative An inactive mutated protein that inhibits the phenotype of the wild type protein when mixed together.

Fuscopeptin Cyclic lipodepsipeptides containing 19 amino acid residues produced by the bacterium *Pseudomonas fuscovaginae*.

GABA γ-aminobutiric acid, inhibitory neurotransmitter in central nervous system of vertebrates.

Gamma-haemolysin Previously described as a toxin with a wide range of haemolytic activity (Smith and Price, 1938), was found to be composed of two synergistic proteins. The locus encoding gamma-haemolysin consists of three genes *hlgA*, *hlgC* and *hlgB*. Gene products HlgA and HlgC are proteins related to LukS-PV, which can generate two different leucotoxins upon combining with HlgB.

Gating mode The reversible transitions of a channel between two conformational states (e.g. open and closed), describable by a set of two kinetic constants. A channel can display more than one gating (or kinetic) mode, as in the case of the commonly encountered bursting behaviour (see above).

6-His tag Six consecutive histidine residues introduced into a protein, often along with a small number of other residues, at one or both ends of it. six-His tags are typically engineered into proteins in order to facilitate their purification on nickel affinity columns, but they also turn out to block certain ion channels, a property exploited in some of the work discussed.

Hyalophora cecropia The American silk moth. It was extensively used for the isolation of antibiotic peptides, mainly cecropins, in its haemolymph, due to the big size of its caterpillar.

Hybrid toxins Chimeras of domain I of one toxin and domains II and III of another toxin, engineered by molecular biology.

Hydrophobic moment A measure of the amphipathicity of an α-helix. Each amino acid side chain is assigned a value, positive or negative, according to its hydrophobicity. These values are used to weight vectors for each residue as they are displayed around the helix. The summation of the vectors is the hydrophobic moment, μ_M.

Hydrophobicity scale A ranking of amino acid residues or side chains according to some quantitative measure of their hydrophobicity, for example the free energy of transfer of the side chain from dioxane or an alcohol to water. Mean hydrophobicity, H is the sum of the hydrophobicity values divided by the number of residues.

IAPP Islet amyloid polypeptide – a 37-residue peptide hormone from beta cells in the pancreas which deposits in type II diabetes mellitus.

Immunity protein A small protein that protects a colicin-producing cell from its own colicin. Immunity proteins of enzymatic colicins bind tightly to their cognate colicin and inactivate their enzymatic activity. Immunity proteins of pore-forming colicins are membrane proteins that interact with their cognate colicin only in its membrane-bound state.

Innate immunity A system of immune recognition which provides broad, but relatively non-specific host defenses that lack the properties of antigenic specificity and immunologic memory that characterise acquired immunity; also known as non-specific immunity.

Integrins Large family of transmembrane proteins involved in the adhesion of cells to extracellular matrix.

Irreversible binding step After receptor binding the toxin binds the membrane with an extremely high affinity. This process involves a structural rearrangement of part of the toxin and is considered irreversible.

Lactotrophs Neurosecretory cells from pituitary gland producing the hormone prolactin.

Leishmania A parasitic protozoa belonging to the Trypanosomatidae family. It is the causative agent of Leishmaniasis, a disease with a broad spectra of clinical manifestation. It is transmitted by the sandflies, and its life cycle has two major forms, the flagellated promastigote, living inside the insect, transmitted by the bite, and the amastigote, an aflagellated form that dwells inside the mammalian macrophage.

Lethal dose 50 (LD_{50}) The concentration at which half of the organisms are killed.

Leucotoxins Membrane-damaging toxins directed against leucocytes.

Lipid bilayer A double layer of amphiphilic molecules, arranged such that either the non-polar ends are on the inside screened by the polar ends or the polar ends are on the inside screened by the non-polar ends, depending on whether the solvent is polar or non-polar.

Lipid monolayer A layer of amphiphilic lipid molecules on the surface of a solvent, arranged such that the ends attracted to the solvent are in contact with it and the other ends point into the air.

Liposome A closed lipid bilayer of various size and lipid composition surrounding an aqueous interior; may be used to encapsulate exogenous materials and to mimic biological membranes.

m region A portion of the toxin VacA of *Helicobacter pylori* (see below), comprising approximately 250 residues in the central region of the p58 domain of the toxin. Approximately 80% of the total variations between VacA isoforms are localised in this section. The various m sequences have been grouped into two families of alleles, m1 and m2.

Madin–Darby Canine Kidney (MDCK) cells Two related cell lines (I and II) derived from dog renal epithelium, capable of forming polarised cellular monolayers (artificial epithelia) in vitro, when grown on porous filters. Epithelia formed by MDCK-I cells can reach trans-epithelial resistances (TER) as high as $12{,}000\,\Omega/cm^2$, while MDCK-II epithelia typically have TER values $<100\,\Omega/cm^2$ range.

Micelle A spherical structure in which all the hydrophobic portions of the amphipathic molecules are inwardly directed, leaving the hydrophilic portions in contact with the surrounding aqueous phase. The converse arrangement will be found if the major phase is hydrophobic.

Minimal inhibitory concentration (MIC) The lowest concentration of an antibiotic that prevents growth of a microorganism.

Molten globule A state in which a protein's structure 'melts' and the hydrogen bonds are broken.

MTS Methanethiosulfonate reagents are a class of chemicals of the form CH_3SO_2S–R consisting of the MTS functional group attached to some R group. They react specifically with sulfhydryl groups in their ionized S^- state. When reacted with protein containing a reduced cysteine residue, a disulfide bond linking the protein and the R group is formed.

Nitric oxide synthase isoform 2 (NOS2) It is an enzyme of the macrophage that is induced by TNF-α, interferon γ, LPS and other inflammatory stimuli. It catalyses the conversion of L-arginine into NO (nitric oxide, a molecule responsible for a wide variety of biological effects, including killing of intracellular pathogens.

Nosocomial infection An infection acquired during hospitalisation. It is often caused by strong antibiotic resistant bacterial strains.

Oxyntic cells Gastric acid-secreting cell, also called 'parietal' in mammalian, located on the wall of the tubular gastric gland. Oxyntic cells are found almost exclusively in the fundus and/or corpus of gastric mucosa and possess a distinctive apical invagination, the secretory canaliculi, greatly expanded during acid secretion.

Panton–Valentine leucocidin Two synergistic proteins called S and F (LukS-PV and LukF-PV) on the basis of their slow or fast elution in cation-exchange carboxymethyl cellulose chromatography.

Penknife model The pore-forming domain of the toxin binds the surface of the membrane. This triggers two helices to insert into the membrane, much like the opening of a blade. The rest of the protein remains on the surface of the membrane with little structural changes.

Planar lipid bilayer A membrane formed across a small hole in a solid support, such as Teflon or polycarbonate, separating two aqueous solutions. Planar bilayers, which can range from tens of microns up to several millimetres in diameter, can serve as a useful substrate for studying channel-forming proteins. They are used as an in vitro voltage clamp system for examining the single channel behaviour of isolated proteins and peptides.

Plasmodium The causative agent for malaria. It is a protozoa transmitted by the mosquito *Anopheles*. Its life cycle comprises: the sporozoite, the form in the salivary glands of the mosquito, transmitted by biting and that infects liver cells; inside those cells it transforms into the merozoite that lives inside the erythrocyte; the merozoite can develop the sexual forms of the parasite which, once they have been ingested with the blood meal, form the ookynete inside the gut of the mosquito. The cycle is closed when the ookynete crosses the gut epithelia and migrates and transforms into the merozoite at the salivary glands.

Polymorphonuclear (PMNs) leucocytes A class of leucocytes characterised by the presence of granules in their cytoplasm. These cells are active in allergic immune reactions such as arthritic inflammation and rashes. This class includes basophils, eosinophils and neutrophils; they are classified according to the pattern/colour they exhibit when they are stained with certain dyes.

Pore-forming domain The primary sequence of many toxins encodes for domain that has a distinct structure in solution. When this structure binds the membrane it transforms into a pore or a channel within the membrane.

Prion (PrP) Proteins which cause transmissible and inheritable diseases through alterations in protein conformation.

Protein translocation The transfer of a protein, or a part of a protein, from the aqueous phase on one side of a lipid membrane to that on the other side.

Pseudomonas syringae* pv. *syringae Widespread, Gram-negative, fluorescent plant bacterium that is both an epiphyte and opportunistic pathogen. It exists mainly on aerial plant surfaces and produces cyclic lipodepsipeptides, syringomycins and syringopeptins. Certain strains are implicated in necrotic plant diseases especially of stone fruits and grasses.

Pseudomycin A small antifungal cyclic lipodepsipeptide of the syringomycin family.

Reversible binding step Binding of the toxin to a specific receptor is a high-affinity although reversible process.

RGD Sequence of arginine, glycine and aspartate found in some extracellular matrix proteins and recognised by some transmembrane proteins (integrins) that bind these proteins.

RTX-toxins Proteins secreted from Gram-negative bacteria, that contain a variable number repeats of a glycine rich nonapeptides with the consensus (L/I/F)XGGXG(N/D) DX, where X is an arbitrary amino acid. The name comes from Repeats in ToXin. All of these proteins bind Ca^{2+}.

SAA Serum amyloid A, an acute phase reactant whose concentration rises dramatically in the serum during infection or inflammation.

Sea anemones (lat. Actiniaria), an anthozoan group belonging to Cnidarians. Mostly sedentary, radially symmetrical animals, characterised by cnidocytes, coelenteron and tentacles around an oral aperture.

Ion selectivity It is the ability of an ion channel to discriminate between ion a and b. It is measured by the permeability ratio, P_a/P_b.

Sterols Important components of natural membranes. They normally show a stabilising effect. Cholesterol is the main sterol component of mammalian cell membranes, whereas stigmasterol is a plant sterol and ergosterol a yeast and fungi sterol.

Streptavidin/biotin Streptavidin, a 66-kD protein produced by *Streptomyces avidinii* that binds ($K_d=10^{-15}$ M) to biotin, a small vitamin (MW 244). Streptavidin, like the egg white protein avidin, is a tetramer with four biotin-binding sites/tetramer.

Stretch-activated Cl^- channels A class of anion-selective channels whose open probability increases with increasing membrane tension. Presumed to be involved in cell volume regulation and to be at least in part encoded by members of the CLC gene family.

Superantigens Microbial antigens that have in common an extremely potent activating effect on T cells that bear a specific variable region.

Surface epithelial cells Mucus-secreting cells which cover the surface of the areas of the stomach in direct contact with the luminal contents and also project part way down the gastric cripts. These cells are also responsible for the alkaline secretion, first line of defense against luminal acid and pepsinogen.

Syringomycins A group of small (approximately 1,200 Da) cyclic lipodepsipeptides that have antifungal activities and are produced by *Pseudomonas syringae* pv. *syringae* and related pseudomonads. Several forms occur that vary in the length of lipid acyl chain.

Syringopeptins Cyclic lipodepsipeptides produced by *Pseudomonas syringae* pv. *syringae* and related pseudomonads. They are larger and more hydrophobic than the syringomycins and possess the ability to cause necroses in plant tissues and also to inhibit certain Gram-positive bacteria.

Syringostatin A small antifungal cyclic lipodepsipeptide of the syringomycin family.

Syringotoxin A small antifungal cyclic lipodepsipeptide of the syringomycin family.

TEA Tetraethylammonium, an organic cation, $N(CH_3)_4^+$, that blocks certain potassium channels.

TMS Transmembrane segment, any continuous stretch of protein that spans a membrane, no matter what its secondary structure.

Tolaasin 18 amino acid residues lipodepsipeptide produced by *P. tolaasi*.

Tonoplasts A semipermeable membrane surrounding the plant cell vacuole.

Toxin oligomerisation The toxins oligomerise as they bind and insert into the membrane. It is believed that this process is important for the formation of an active pore.

Umbrella model The pore-forming domain of the toxin binds the surface of the membrane. This triggers a drastic structural rearrangement in which the helices spread on the surface while two of them form a hairpin and insert into the membrane. This resembles the ribs and handle of an umbrella.

VacA Vacuolating cytotoxic produced by *Helicobacter pylori*, originally identified by its ability to induce intracellular vacuoles in the presence of weak bases.

Vacuolation Pathological generation in the cell cytoplasm of unusually large membrane vesicles, derived by alteration of cell organelles (endosomes, lysosomes, Golgi complex, endoplasmic reticulum) depending on the injury. They are in the ~5–10 μm diameter range and are normally visible by normal light microscopy.

Vacuole A large, fluid-filled vesicle found in plant cells that provides both storage and space-filling functions.

V-ATPase Vacuolar-type ATPase: a proton pumping enzymatic complex, homologous to the mitochondrial ATPase responsible for ATP synthesis, necessary for the acidification of endosomes and lysosomes in eukaryotic cells. Mostly localised in the membrane of these organelles, it is specifically inhibited by nanomolar concentrations of Bafilomicin A1.

Abbreviations

1-Eru-Gly	1-erucioyl-glycerol (1-*cis*-11-docosenoyl-glycerol)
AAD	antibiotic-associated diarrhoea
agr	accessory gene regulator
ANS	8-anilino-1-naphtalenesulphonate
CA	cecropin A
CB	cecropin B
CD	cecropin D
Chol	cholesterol
CL	cardiolipin
CP	cecropin P1
DLPG	1,2-dilaureoyl-*sn*-glycero-3-phosphoglycerol
DOPE	1,2-dioleoyl-*sn*-glycero-3-phosphoethanolamine
DOPG	1,2-dioleyl-*sn*-glycero-3-phosphoglycerol
DOPS	1,2-dioleoyl-*sn*-glycero-3-phosphoserine
DPhPC	1,2-diphytanoyl-*sn*-glycero-3-phosphocholine
DPPG	1,2-dipalmitoyl-*sn*-glycero-3-phosphoglycerol
EAP	eucaryotic antibiotic peptide
FDA	food and drug administration
FT-IR	Fourier transformed infra red
GABA	γ aminobutyric acid
GM1	monosialoganglioside-G_{M1}
HRBC	human red blood cells
i.v. LD_{50}	intravenous 50% lethal dose
IPC	inositol phosphoceramide
LD_{50}	lethal dose 50
LPS	lipopolysaccharide
LUV	large unilamellar vesicles
$M(IP)_2C$	inositol-phospho MIPC
M	melittin
MA	magainin 2
MIC	minimal inhibitory concentration
MIPC	mannosyl IPC
NMR	nuclear magnetic resonance
NOE	nuclear overhauser enhancement
NOS2	nitric oxide synthase isoform 2
PA	1,2-diacyl-*sn*-glycero-3-phosphatidic acid

PC	1,2-diacyl-*sn*-glycero-3-phosphocholine
PC-Cho	1,2-diacyl-*sn*-glycero-3-phosphocholine-cholesterol
PE	1,2-diacyl-*sn*-glycero-3-phosphoethanolamine
PEG	polyethyleneglycol
PG	1,2-diacyl-*sn*-glycerol-3-phosphoglycerol
PGLa	peptidyl-glycine-leucine-carboxyamide
PI	phosphatidylinositol
PMNs	polymorphonuclear cells
POPC	1-palmitoyl-2-oleyl-*sn*-glycero-3-phosphocholine
POPE	1-palmitoyl-2-oleoyl-*sn*-glycero-3-phosphoethanolamine
POPS	1-palmitoyl-2-oleoyl-*sn*-glycero-3-phosphoserine
PS	1,2-diacyl-*sn*-glycerol-3-phosphoserine
PVL	Panton–Valentine leucocidin
QSAR	quantitative structure–activity relationships
RRBC	rabbit red blood cells
RT-PCR	reverse transcription followed by polymerisation chain reaction
SA	stearoylamine
SAR	structure–activity relationship
SM	sphingomyelin
SUV	small unilamellar vesicles

Part 1

Pore-forming proteins

1 Staphylococcal bicomponent leucotoxins, mechanism of action, impact on cells and contribution to virulence

Gilles Prévost, Gianfranco Menestrina, Didier A. Colin, Sandra Werner, Stephen Bronner, Mauro Dalla Serra, Lamin Baba Moussa, Manuela Coraiola, Alain Gravet and Henri Monteil

Introduction

Staphylococcus aureus is one of the most frequently isolated bacterium in hospital routine, from infections that may affect every organ and tissue. Having developed resistance to most families of antimicrobials, it is now responsible for 5–15% of nosocomial infections, depending on hospital sites and services (Kloos and Bannerman, 1999). The pathogenicity of this bacterium is caused by a series of adhesion factors (Foster and Höök, 1998) and toxins that can be classified into three major categories: proteases, superantigens and pore-forming toxins (Alouf and Freer, 1999).

Till 50 years ago, furuncles d'ue to *S. aureus* were a cause of mortality because of frequent septic complications with possible pulmonary localisation (Elbaze and Orton, 1988). They were assumed to result from the lack of hygiene, although a recurrent form of this disease had already been observed. A leucocidal activity of *S. aureus* in a pleural infection experimental model was known as early as 1894 (Van der Velde, 1894). Only later however, were some *S. aureus* isolates, originated from furuncles, associated with a strong leucocidal activity of their culture supernatants, which was attributed to the presence of a leucocidin (Panton and Valentine, 1932).

Genetics of the subfamily of bicomponent leucotoxins

After the synthesis and secretion of the toxin was optimised (Gladstone and Fildes, 1940; Gladstone and Glencross, 1960), Woodin (1960) was the first to purify and characterise the Panton–Valentine leucocidin (PVL) as two synergistic proteins. These were called S and F (LukS-PV and LukF-PV) on the basis of their slow or fast elution in cation-exchange carboxy-methyl cellulose chromatography. In the 1960s–70s, the same author (Woodin, 1972) was able to establish that the toxin disrupted the cell permeability barrier at several levels, with various consequences on the metabolism and metabolic equilibrium. Later, two laboratories (Guyonnet and Plommet, 1970; Taylor and Bernheimer, 1974) concurrently demonstrated that staphylococcal gamma-haemolysin, previously described as a toxin with a wide range of haemolytic activity (Smith and Price, 1938), was also composed of two synergistic proteins. Even if two different *S. aureus* toxins were identified as being constituted in each of two distinct proteins, it was only 20 years later that they were shown to have common antigenic epitopes (Supersac *et al.*, 1993; Prévost *et al.*, 1995c). As PVL and gamma-haemolysin can both permeabilise human polymorphonuclear cells (PMNs), and may be simultaneously produced by bacteria, some confusion originated between the names and true identities of these proteins. This was complicated by the fact that the locus encoding gamma-haemolysin consists of three genes *hlgA*, *hlgC* and *hlgB*. Gene products HlgA and HlgC are proteins related to LukS-PV, which can generate two different leucotoxins upon combining with HlgB, which is related to LukF-PV (Cooney *et al.*, 1993; Kamio *et al.*, 1993). To date, four loci encoding bicomponent leucotoxins have been characterised in *S. aureus*: PVL (Prévost *et al.*, 1995c), gamma-haemolysin, LukE + LukD (Gravet *et al.*, 1998), and LukM + LukF′-PV (Kaneko *et al.*, 1997b). Another such leucotoxin, LukS-I + LukF-I is encoded by *Staphylococcus intermedius* (Prévost *et al.*, 1995a).

Except *hlgA*, all other genes are tandemly cotranscribed, the gene encoding for a protein similar to LukS-PV (class S component) being located one base (a thymine, T) upstream of the gene encoding a protein similar to LukF-PV (class F component). Due to their genetic location, and because they bind first to target cells, in spelling the leucotoxins, class S proteins are conventionally mentioned in the first place. With these conventions, and the historical names given to PVL and gamma-haemolysin, the different toxins have been collected in the subfamily of bicomponent, pore-forming leucotoxins (Prévost, 1999). Nucleotide sequences have enough differences to avoid homologous recombination in normal conditions of culture, in fact, no sequence longer than 14 bases is shared by any of these genes (Prévost *et al.*, 1994). Consistently, we have characterised some strains that simultaneously, and constantly, produce PVL, gamma-haemolysin and LukE + LukD, without loss of any of these toxins with time (Gravet *et al.*, 1998; Bronner *et al.*, 2000).

The genetic occurrence or acquisition of PVL locus has been attributed to the lysogenic conversion by a bacteriophage, ΦPVL (Van der Vijver *et al.*, 1972; Kaneko *et al.*, 1997a). In fact, gene sequences typical for some of the functions of a temperate bacteriophage have been found in the vicinity of the PVL locus. Recently, related sequences have also been found in different strains (ΦSLT) that led to infective particles promoting toxin production after lysogenisation in their host strain (Narita *et al.*, 2001). Nonetheless, a horizontal transmission of the PVL locus should be a very rare event, since neither the loss of this character in working strains, nor a natural spreading among different isolates, were ever observed. Moreover, patients with chronic infection by a PVL-producing strain were never observed to carry a second strain producing this toxin. Considering the phylum of these pore-forming toxins, it cannot be excluded that they originated from bacteriophage(s) transmission, reinforcing the concept that bacteria may have sequestered exogenous genetic material for their own interaction with privileged hosts. On the other hand, upon evolving, these genes became potent virulence factors, and *S. aureus* seems to have markedly used them by specialising different toxin variants.

Physicochemical properties of bicomponent leucotoxins

These proteins are generally positively charged, and are secreted in a water-soluble form before conversion into a transmembrane, pore-forming oligomer. Secretion occurs via signal peptides of 25–28 amino acids, conforming to the classical sequence of positively charged amino acids followed by hydrophobic residues that are cleaved through the secretion machinery of Gram-positive bacteria (Lee *et al.*, 2000). Class S proteins are from 277 to 286 amino acid long, from 30,688 to 32,239 Da in molecular mass, and have a pI ranging from 8.9 to 9.8 (Prévost, 1999). Class F proteins are from 296 to 301 residues, from 33,727 to 34,386 Da, with pI between 7.5 and 9.0. Sequence homologies are very important inside the two classes of proteins. Identities are up to 55–70% for class S and 70–80% for F proteins, but only 18–25% between the two classes (Gouaux *et al.*, 1997b; Prévost *et al.*, 2001). Additional homologies exist between the two classes of proteins and other pore-forming toxins such as staphylococcal α-toxin (Gray and Kehoe, 1984), β-toxin of *Clostridium perfringens* (Cpb, (Hunter *et al.*, 1993)), and two cytotoxins of *Bacillus cereus*, hemolysin II (HlyII (Baida *et al.*, 1999)) and cytotoxin K (CytK, (Lund *et al.*, 2000)). If aligned, all these proteins show 13 strictly conserved residues (Menestrina *et al.*, 2001) and additional homologies that concern both polar and non-polar residues (Figure 1.2A). A phylogenetic tree (Figure 1.1) suggests the existence of a common ancestor, whereby Cpb is the most divergent toxin, and S and F leucotoxin proteins seem to have been specialised more recently.

From the beginning, investigators have been looking out for a satisfying culture medium for toxin production. The Yeast Extract – Casaminoacids – Pyruvate (YCP) medium was found to be a good compromise (Gravet *et al.*, 1998) even if HlgA constantly appeared as a low-produced component. Recently, using a semi-quantitative RT-PCR test for messenger RNAs, Bronner *et al.* (2000) could confirm that HlgC + HlgB, LukS-PV + LukF-PV and LukE + LukD are produced during the last exponential growth phase and are positively regulated by the *sar–agr* regulation complex. Expression was more efficient in YCP than in Heart-infusion medium. Cation-exchange and hydrophobic interactions chromatography were found suitable for purification to homogeneity (Finck-Barbançon *et al.*, 1991; Prévost *et al.*, 1995a).

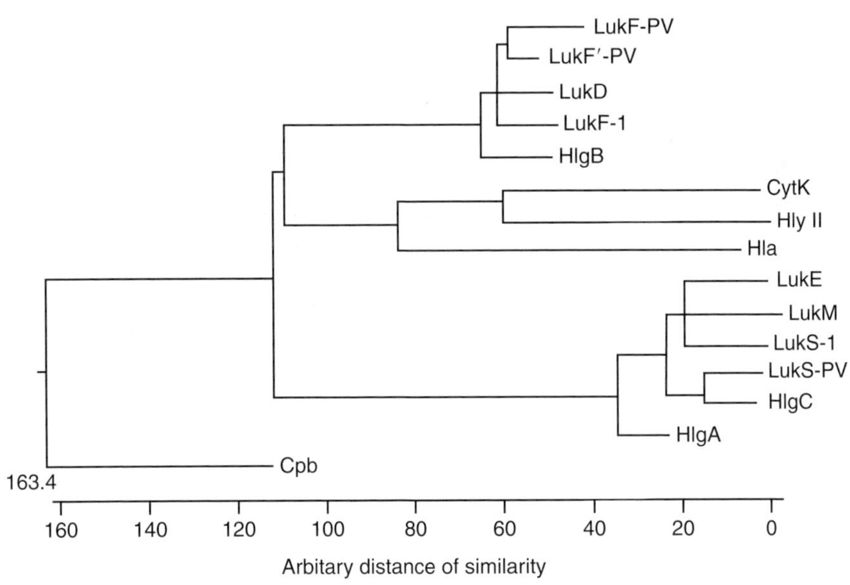

Figure 1.1 Phylogenetic tree based on the peptide sequences of the related bicomponent and single chain pore-forming toxins: Panton–Valentine leucocidin, LukS-PV + LukF-PV (EMBO-Genbank X72700); gamma-haemolysin, HlgA, HlgC + HlgB (L01055), LukE + LukD (Y13225), LukM + LukF′-PV (D42144, D83591), LukS-I + LukF-I from *S. intermedius* (X79188-1), staphylococcal α-toxin (M90536); CytK toxin from *B. cereus* (AJ277962); HlyII haemolysin from *B. cereus* (U94743); β-toxin from *C. perfringens* (L13198). The tree was made using Clustal software (http://www2.ebi.ac.uk/clustalw/).

Basic interests of leucotoxins

Bicomponent leucotoxins are intriguing toxins because of many specific features. (i) Several members can be genetically maintained and expressed by a given isolate, thus originating more leucotoxins by combination of class S and class F proteins. (ii) Their association with some precise clinical syndromes stimulates an interest in understanding their contribution to these pathologies. (iii) They constitute a model of association of two, non-equivalent proteins, which convert into an oligomeric transmembrane pore via several steps (Lesieur *et al.*, 1997; Gouaux, 1997a): binding, oligomerisation, transient activation of targeted cells, insertion of β-hairpins and arrangement into a β-barrel forming the functional pore.

Structural features of leucotoxins

While the structure of a related pore, the α-toxin heptamer, was known since 1996 (PDB: 7AHL, Song *et al.*, 1996; Song and Gouaux, 1998), that of a secreted leucotoxin monomer was determined only recently. HlgB (PDB: 1LKF, Olson *et al.*, 1999) and LukF-PV (PDB: 1PVL, Pédelacq *et al.*, 1999) were solved almost simultaneously by multiwavelength anomalous dispersion (MAD), and were found to be highly comparable (Figures 1.2A and B) mutually proving the pertinence of the characterisation. The two structures have the shape of an ellipsoid, $75 \times 34 \times 25$ Å, where three domains can be identified, two of which quite similar in the α-toxin protomer (Figure 1.2C). Such two domains form the core of the protein: a β-sandwich of two six-stranded antiparallel β-sheet (representing 54% of the sequence of LukF-PV), and a Rim domain formed by an antiparallel four-stranded open-face β-sandwich topped by two helical segments and two omega-loops. The β-sandwich corresponds to the Cap domain of the mushroom-shaped heptameric structure of α-toxin (Figure 1.2D). The Rim should become anchored to the membrane surface, as suggested by several studies on α-toxin (Valeva *et al.*, 1997; Vécsey-Semjén *et al.*, 1997). The two domains share one β-strand, residues 237–256 of LukF-PV (Figure 1.2A).

In the three-dimensional structure of HlgB (Olson *et al.*, 1999), a conserved pocket was identified, defined by Trp197 and Arg198, which binds phosphatidylcholine similarly to α-toxin (Song *et al.*, 1996). The minor differences existing between HlgB and LukF-PV structures are essentially located at the connecting loops between β-strands of the Rim and at the COOH-terminus (Figures 1.2A and B). Such discrepancies could either reflect technical differences in the acquisition of structural data, or specific constraints due to sequence differences. Comparing the core of HlgB, or LukF-PV with that of the α-toxin protomer, a rotation of 11–15° of the Rim domain of the monomer is required for complete superimposition (Figures 1.2A, B and D). It is not yet known whether this fundamental difference is consequent to the assembly of the monomers into the oligomer.

Another major difference, which indicates the idea that the different structures represent the endpoints of conformational changes occurring between monomers and oligomer, is in the folding of the third domain, the Stem. This portion is configured as three antiparallel β-strands stacked to the β-sandwich in LukF-PV and HlgB, whereas it is arranged as a β-hairpin, which represents the sole contribution of each monomer to the β-barrel forming the pore, in the α-toxin heptamer (Figure 1.2). Actually, the event promoting the stretching of this structure necessary for its insertion into the membrane is not yet understood. The existence of an intermediate heptameric α-toxin prepore, with the Stem in the folded configuration, was indicated by the detection of the movement of some amino acids of the Stem domain from a polar to an apolar environment (Valeva *et al.*, 1997).

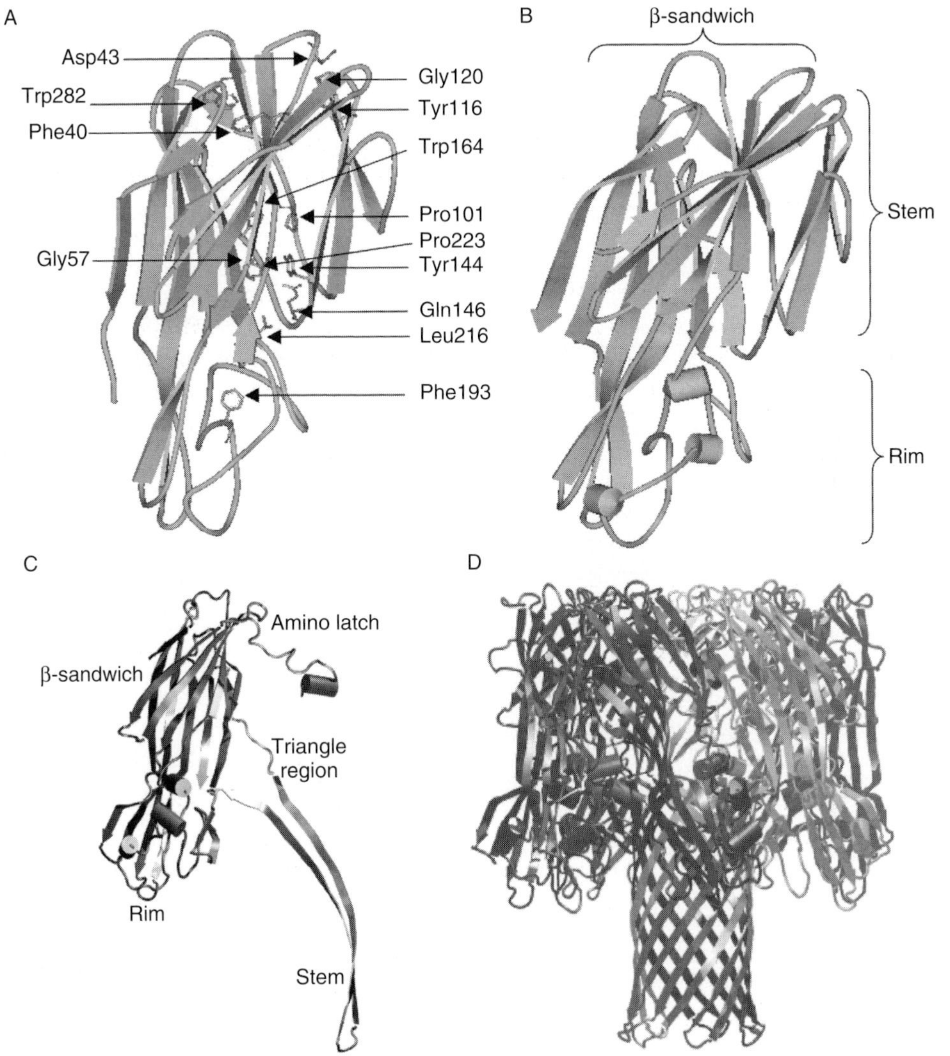

Figure 1.2 Comparison of the three-dimensional structures of leucotoxin water-soluble monomers with that of α-toxin protomer and heptameric pore. (A) LukF-PV (Pédelacq *et al.*, 1999, PDB code:1PVL) with the 13 residues conserved among all the related toxins highlighted by ball-and-stick representation. (B) HlgB (Olson *et al.*, 1999; PDB code:1LKF). (C) α-toxin protomer extracted from the heptameric pore (Song *et al.*, 1996, code PDB: 7AHL), different colours have been used for the different domains. (D) Side view of the α-toxin heptamer (Song *et al.*, 1996, code PDB: 7AHL), different colours have been used for the different monomers. Figures 2A and B were obtained with WebLab Viewer Pro (MSI Ltd), Figures 2C and D were obtained with VMD-Visual Molecular Dynamics (Humphrey, 1996). (See Colour Plate I.)

```
LukF-PV        109 QVQQTVGYSYGGDINI----SNG-LSGGGNGSKSFSETIN 143
LukF-PV-like   109 QVQQTVGYSYGGDINI----ING-LSGGGNGSKSFSETIN 143
LukD           109 QVQQTLGYSYGGDINI----SNG-LSGGLNGSKSFSETIN 143
LukF-I         109 QVQNTLGYSFGGDISI----SKG-LSGGLNGSESFSETIN 143
HlgB           109 QVQNTLGYTFGGDISI----SNG-LSGGLNGNTAFSETIN 143
CytK           115 TVTSSVSYQLGGSIKASVTPSGPSGESGATGQVTWSDSVS 154
HlyII          103 KVTSEVGYTLGGSVKVGVNDKGPNADASITGSFAWKESVS 142
Hla            111 EYMSTLTYGFNGNVTG---DDTGKIGGLIGANVSIGHTLK 147
LukE           109 DVGQTLGYNIGGNFQ-----SAP--SIGGNGSFNYSKTIS 141
LukM           104 DVGQTLGYNIGGKFQ-----SVP--SIGGNGSFNYSKSIK 136
LukS-I         105 DVSQTLGYNVGGNFQ-----SAP--LLGGKGEFNYSKKIS 137
LukS-PV        104 NVSQTLGYNIGGNFN-----SGP--STGGNGSFNYSKTIS 136
HlgC           106 NVSQTLGYNIGGNFQ-----SAP--SLGGNGSFNYSKSIS 138
HlgA           104 DVSQKLGYNIGGNFQ-----SAP--SIGGSGSFNYSKTIS 136
Cpb            129 TVSNTMGYKIGGSIEI--EENKP--KASIESEYAESSTIE 164
                                 15 16 17  turn  17 16 15
```

Figure 1.3 Alignment of the Stem of LukF-PV and the related proteins in this family of pore-forming toxins with that of α-toxin (Song *et al.*, 1996; Pédelacq *et al.*, 1999). The predicted turn/loop region of each toxin, or toxin component, is shaded in dark grey. The length of the strands forming the stem is variable between 15 and 17 residues (as indicated by the scale below the sequences). Strand residues that should face the side of the membrane are in bold over lightly shaded background.

Unfortunately, because of the reduced number of binding sites onto human PMNs (Gauduchon *et al.*, 2001) similar studies with leucotoxins did not achieve the necessary sensitivity. Leucotoxins might form hexamers, instead of heptamers, at least in erythrocytes and biomembranes (Ferreras *et al.*, 1996, 1998; Sugawara *et al.*, 1997). In a hexameric model (Pédelacq *et al.*, 1999), β-hairpins of the 12-stranded β-barrel would be more close to the vertical axis than those of α-toxin, thus achieving a 5 nm transmembrane length even if composed by shorter strands (Figure 1.3).

Finally, the different location of the amino latch in the α-toxin pore and in the leucotoxin monomers could be of great importance (Figures 1.2A, B and C). In α-toxin, this part is not folded with the rest of the protomer, but interacts with the adjacent protomer (Song *et al.*, 1996). In HlgB or LukF-PV, instead, it is folded against the β-sandwich and constitutes the exterior of that structure. Leucotoxin hexamers were modelled assuming that the N-terminus did not move (Pédelacq *et al.*, 1999). Among arguments that favour this hypothesis are: (i) when recombinant proteins are expressed fused to the glutathione-S transferase (GST) and then processed with the PreScission protease (Amersham, Pharmacia) an 8-residue stretch remains attached to the N-terminus after cleavage, which however does not disturb the specific activity of the leucotoxins (Baba Moussa *et al.*, 1999b); (ii) even proteins retaining the N-terminal fused GST maintain significant activity onto human PMNs (Prévost, unpublished result); (iii) the F components remain functional even introducing two cysteine substitutions that generate a disulphide bond between the first two N-terminal β-strands (Prévost, unpublished result).

Cell preference and binding of the bicomponent leucotoxins

Despite some variations within the subfamily, human monocytes, macrophages and polymorphonuclear cells can be considered the common targets of the staphylococcal leucotoxins. Gladstone and Van Heyningen (1957) and Woodin (1960) first reported that exclusive targets of the PVL are human and rabbit (but not mouse, rat or bovine (Szmigielski *et al.*, 1998)) monocytes, macrophages and polymorphonuclear cells. In fact, every other combination of an S protein with an F protein may lyse those cells (Prévost, 1999), even if LukE + LukD is particular because cells of only one-fifth of donors appear to be sensitive (Gravet *et al.*, 1998). Moreover, other couples may have a broader spectrum of susceptible cells like HlgC + HlgB, which shows haemolytic activity on rabbit erythrocytes, and HlgA + HlgB which lyses both rabbit and human erythrocytes (Prévost *et al.*, 1995a,c), the latter more potently than α-toxin. HlgA + HlgB may also target some T lymphocytes, as Jurkat or RAJI cells (Colin *et al.*, 1997; Ferreras *et al.*, 1998; Meunier 2000). Phenotype determination of these T lymphocytes did not evidence a clear distribution among sensitive cells. In fact, 85% of CD4/CD29+ cells, 50% of CD4/CD45 + cells, 25% and 45% of CD11b or CD18 cells, respectively, were sensitive to HlgA + HlgB. Some cell lines, such as U937, HL60 or THP1, were not sensitive enough to serve as a model. Considering the limited panel of cells tested, and some typical associations between staphylococcal infections and strains producing leucotoxins that will be exposed later, one could suspect that other specialised cells are sensitive to these toxins.

PMNs become sensitive to the PVL after the differentiation of the metamyelocyte (Meunier *et al.*, 1995), but no correlation was yet obtained between the appearance of a specific receptor. Staphylococcal α-toxin is able to permeabilise even synthetic bilayers (Freer *et al.*, 1968; Bashford *et al.*, 1996; Menestrina and Vécsey-Semjén, 1999), yet it is not active on human PMNs and only poorly on human red blood cells (HRBC). Its cell spectrum includes T lymphocytes (Jonas *et al.*, 1994; Bhakdi *et al.*, 1996), fibroblasts (Walev *et al.*, 1994), platelets (Bhakdi *et al.*, 1988), epithelial cells such as keratinocytes (Walev *et al.*, 1993), gastric cells (Thibodeau *et al.*, 1994), chromaffin cells (Anhert-Hilger *et al.*, 1985) and endothelial lung cells (Seeger *et al.*, 1990; Bhakdi *et al.*, 1996).

In the past, S proteins (HlgC or LukS-PV) were reported to bind to GM1 (Noda *et al.*, 1980; Ozawa *et al.*, 1994) and to vitronectin (Katsumi *et al.*, 1999). However, it is doubtful that these might represent a specific binding. In fact, GM1 is at least as abundant in other human blood cells as in the targeted PMNs, and the leucotoxins can bind synthetic membranes completely devoid of GM1 (Meunier *et al.*, 1997). On the other hand, the binding of LukS-PV to vitronectin, a circulating serum protein and not a membrane receptor, was not very efficient. A similar observation was previously made for α-toxin and another circulating protein, LDL (Bhakdi *et al.*, 1983). The most significant implication of this finding was that the leucotoxins might not be able to target cells if they are expressed in the blood. Indeed, infectious diseases associated with leucotoxins do not usually affect sites where all blood components are present, but rather soft tissues (such as hair follicles, intestines and lungs), where the primary targets may be specialised host defence cells. Furthermore, blood complement seems to inactivate PVL and γ-haemolysins: in fact, fresh human serum seems to inactivate PVL or γ-haemolysin, while, after heating at 56°C for 30 min, PVL retains a significant biological activity. Regarding their binding to target cells, S and F proteins are not equivalent. It was shown (Colin *et al.*, 1994) that biological activity could be retrieved when PMNs were treated with LukS-PV, washed and then treated with LukF-PV, but not when this sequence was inverted (Figure 1.4). The preliminary binding of S proteins was also observed on other cells (Meunier *et al.*, 1997) including rabbit and human RBC (Ferreras *et al.*, 1998). Inspite of this, other authors proposed that HlgB has to bind first onto HRBC to ensure haemolysis by HlgA or HlgC (Ozawa *et al.*, 1995). This observation, contradictory to ours, was never confirmed. In agreement with our proposal, when Thr28 of HlgA (or the

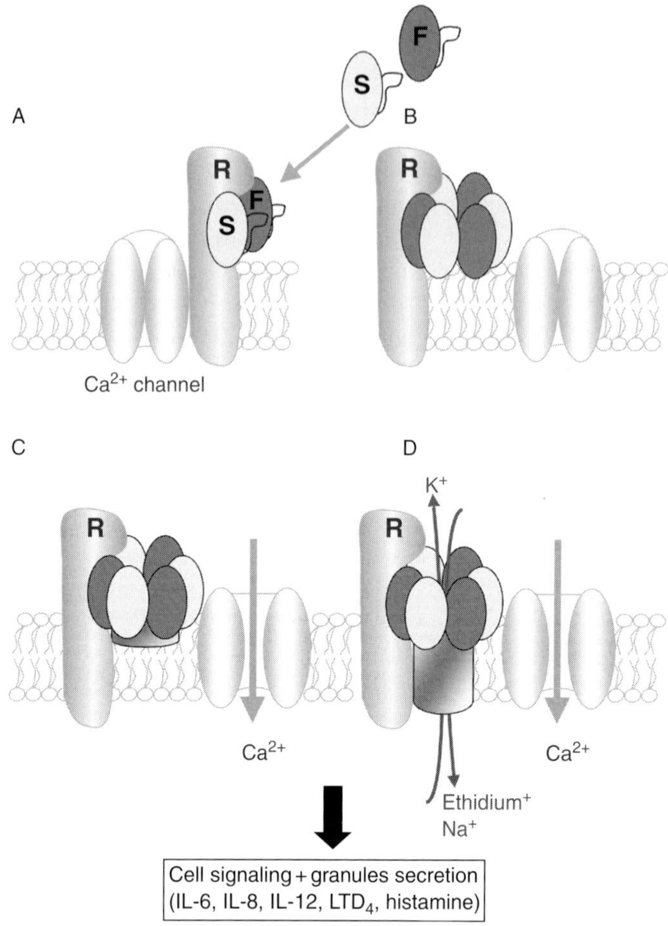

Figure 1.4 Scheme of action of the staphylococcal bicomponent leucotoxins. (A) Sequential binding: the S proteins bind to a high affinity cell receptor (R), followed by F proteins. (B) Prepore: oligomerisation leads to the formation of a pre-pore. (C) Ca^{2+} channel opening: oligomerisation and/or an additional conformational change induces the rapid opening of calcium channels. (D) Pore formation: insertion of the β-barrel leads to the formation of the pore. This pore is essentially permeant to monovalent cations, including ethidium. Ca^{2+} activation and pore-formation leads to signal transduction inside the cell and to the release of inflammation mediators. (See Colour Plate II.)

corresponding Thr30 of HlgC) were substituted by Asp, the mutants were still able to bind target cells and compete with the wt proteins, but the subsequent binding of HlgB was abolished, with a complete loss of biological activity (Meunier *et al.*, 1997). This was observed with human PMNs, erythrocytes, and PC-Cho small unilamellar vesicles, thus excluding that HlgB binds differently on different membrane models. Thr28 of HlgA aligns with His35 of α-toxin, a crucial residue for monomer–monomer interactions (Jursch *et al.*, 1994;

Krishnasastry *et al.*, 1994; Menzies and Kernodle, 1994; Walker and Bayley, 1995) and might therefore have a similar role in the assembly of the S- and F-proteins. Since the physicochemical properties of the S- and F-proteins are quite similar, a slight contamination of a given class by the other, occurring during the purification, might have led to the confusing observations. On the other hand, the synergistic behaviour of these proteins can be used to assess cross-contamination of the compounds. In fact, our purified single proteins do not show any activity onto human PMNs even at 1 μM, whereas their synergistic action as a couple starts below 1 nM.

Binding of staphylococcal bicomponent leucotoxins exhibits two, probably independent, features: a specific affinity for some receptors, and an intrinsic propensity to interact with membrane bilayers, which is critical at least for pore formation. The molecular binding of S proteins is very efficient. Recently, the association of LukS-PV was revisited by flow cytometry using a fluorescein-labelled LukS-PV mutant, Gly10Cys, which retains full functionality (Gauduchon *et al.*, 2001). The apparent dissociation constant, K_d, estimated was 0.07 nM for PMNs and 0.02 nM for monocytes. Further experiments showed that HlgC might compete with the binding of the fluorescent LukS-PV at equimolar concentrations, indicating a shared binding site. No competition was observed with two other S proteins, HlgA and LukE. Furthermore, the binding site of LukS-PV was specific enough to be regulated by the modulation of protein kinase C expression. In fact, cells treated with phorbol-myristyl acetate, that enhances protein kinase C, did not bind LukS-PV, whereas binding was recovered when protein kinase C was inhibited by staurosporine. These features favoured the existence of a high affinity receptor for LukS-PV and HlgC (see Figure 1.4), not excluding it might also exist for other S proteins, for example, LukE. A maximal binding of 197,000 and 80,000 LukS-PV molecules per PMN and monocyte, respectively, was recorded. Non-specific binding of LukS-PV for these cells was negligible. For HlgA, instead, both specific and non-specific binding were reported.

Relevant for the specific binding could be that the three β-strands arrangement of the Stem of LukF-PV resembles the three-finger fold of venom toxins binding to type II activin receptor serine kinase (Rees and Bilwes, 1993; Greenwald *et al.*, 1999). However, disulfide bonds maintain the fold of the latter, while the leucotoxins are naturally devoid of cysteines.

Oligomerisation

As described in the previous section, oligomerisation of bicomponent leucotoxins occurs only after the binding of the S protein onto the target cell. In fact, the formation of dimers, or oligomers, of LukS-PV + LukF-PV in solution is almost absent (Werner *et al.*, 2001).

The α-toxin oligomer has been crystallised as a heptamer (Gouaux *et al.*, 1994; Song *et al.*, 1996), but a hexamer may also be formed under certain conditions (Czajkowski *et al.*, 1998). In the case of bicomponent leucotoxins, a heptamer might not fit with the necessary symmetry of the S and F protein inside the oligomer. Electron microscopy observations evidenced a major association of HlgA + HlgB and HlgC + HlgB into hexamers on RBC and human leucocytes (Sugawara *et al.*, 1997, 1999). SDS–PAGE experiments led to the characterisation of oligomers of about 200 kDa after insertion into PC-Cho small unilamellar vesicles (SUV) (Ferreras *et al.*, 1998). When boiled in the presence of SDS, these oligomers dissociated into equal amounts of both 31 and 35 kDa proteins. More recently, similar oligomers were identified also from human PMNs by immunoblotting (Werner, 2001;

Werner *et al.*, 2001). Oligomers of 180–200 kDa were identified both by anti-S and anti-F rabbit affinity-purified antibodies. As the two proteins differ by only 4–5 kDa in molecular mass, the corresponding oligomers are assumed to be hexamers composed of S and F proteins, probably in an equimolar ratio.

One important question is to determine whether, through oligomerisation, more than one S component can bind to a single receptor engaged by the first of them, or whether each S component has to bind to one separate receptor. In the second case, oligomer formation would require either that the sites are already correctly oriented or that they can diffuse exploiting membrane fluidity, as it occurs with aerolysin (Fivaz *et al.*, 1999). It would also imply that the maximum number of pores is smaller than that of binding sites. In favour of this hypothesis is that PMNs, preliminarily incubated with saturating amounts of fluorescein-labelled LukS-PV and then washed, did not acquire more fluorescence if further incubated with LukF-PV and fluorescent LukS-PV (D. A. Colin, unpublished result). However, pore formation by some leucotoxins, HlgA + HlgB and HlgC + HlgB, occurs even in the absence of specific receptors, since these toxins can permeabilise synthetic membranes. Binding in such models could differ from that on target cells, and could exploit the interaction with specific membrane lipids.

Biological activity

At sub-lytic doses, leucotoxins induce cell swelling, loss of refringence, nucleus rounding and the reduction of membrane pseudopodes in target cells (Gladstone and Van Heyningen, 1957). Both Woodin (1972) and Noda *et al.* (1981, 1982) proposed that the biological activity of leucotoxins involves transmembrane fluxes of monovalent and divalent cations. New investigative techniques, like fluorescence and flow cytometry, allowed a more detailed analysis of these effects. Using Ca^{2+} sensitive fluorescent probes (Fura or Fluo), and ethidium (which becomes fluorescent after penetrating into cells and combining with nucleic acids), Finck-Barbançon *et al.* (1993) and Meunier *et al.* (1995) reported that PVL induced both Ca^{2+} and ethidium entry into human PMNs (Figure 1.4). The two pathways are distinct, since only ethidium entry was blocked with 1 mM $CaCl_2$ plus 0.2 mM $ZnCl_2$. Neutralising, affinity-purified, antibodies directed against LukS-PV or LukF-PV, prevented instead both actions. Interestingly, when ethidium entry is blocked, no cell lysis occurs. Staali *et al.* (1998, 2000) demonstrated the independence of the two phenomena by showing that Ca^{2+} entry started within 100 s from the application of PVL, whereas ethidium entry was significant only after 600 s (see Figure 1.4). Furthermore, calcium induction was inhibited by adenosine or verapamil analogues, that did not block ethidium or K^+ entries and cell lysis. The marked selectivity of leucotoxin pores for monovalent over divalent cations and anions is a further difference with staphylococcal α-toxin (Krasilnikov *et al.*, 1986; Menestrina, 1986; Korchev *et al.*, 1995a,b). α-toxin was reported to elicit Ca^{2+} entry, via a cell activation process, in PC12 cells (Fink *et al.*, 1989), but not in human fibroblasts (Valeva *et al.*, 2000). The minimal structure of the leucotoxin oligomer required for Ca^{2+} entry remains to be determined.

The peptide sequences of the leucotoxins that are certainly involved in pore formation are those forming the Stem, which are located in the middle of the protein (residues 110–150) and are among the most conserved within this family of toxins (Figure 1.3). The Stem domain is constituted by two β-strands formed by residues that will promote their association into a transmembrane β-barrel. Polar and apolar amino acids are alternated along these β-strands, so that all polar residues will be oriented towards the interior of the barrel (the lumen of the pore) whereas apolar residues will face the membrane. Figure 1.3 shows a schematic representation of these β-strands and their variability within the related proteins. It can be noticed that, despite forming a β-barrel, S and F β-hairpins share very few identities, whereas a higher similarity is present within each class of proteins. In the S proteins, the residues oriented towards the bilayer are more conserved than those oriented inside the lumen. The presence of a constant Lys facing the membrane might be noted, and a conserved proline residue is located at the loop of β-hairpins. Up to now, the exact location of this loop is not defined, although probably it interacts with the cytoplasmic side of the cell membrane. These proteins do not bear many electrostatic charges in the lumen of the pore, except for LukS-I, which has four charged amino acids (two Lys, one Asp and one Glu), three of which in the second β-strand, and HlgA that harbours one Asp and one Lys in the first β-strand. Also, in the Stem of F proteins, a few charges are present that should face the phospholipid bilayer, especially in LukF-I.

The fact that the cell membrane, that normally excludes ethidium, becomes permeable to this cation after the action of the leucotoxins, strongly suggests that the pore is transmembranous. By using Fourier-transformed infrared reflectance spectroscopy in the attenuated total reflectance configuration (FTIR-ATR), we confirmed that these toxins contain about 70% β-sheet structures. Examination of the aggregates formed onto PC-Cho small unilamellar vesicles suggested that only minor variations occur in the secondary structure during pore formation (Meunier *et al.*, 1997; Ferreras *et al.*, 1998). Data obtained with perpendicular or parallel polarisations were in accordance with the formation of a β-barrel perpendicular to the plane of the membrane.

The size of the pores formed by HlgA-HlgB in RRBC was assessed by electron microscopy (Sugawara *et al.*, 1997, 1999). A diameter of 19–21 Å was deduced. This was in agreement with flow cytometry experiments using polyethylene glycols, which indicated that molecules larger than 1,500 Da (hydrodynamic radius ⩾1.12 nm), were excluded from the pore, whereas those of 1,000 Da (radius 0.96 nm), were able to pass and induce osmotic imbalance and morphological changes in the cells (Baba Moussa *et al.*, 1999b). In the case of α-toxin, the crystallographic structure suggested a minimal diameter of 1.4–1.5 nm in its narrowest region (Song *et al.*, 1996). From conductance experiments, an effective diameter of 1.1–1.2 nm was estimated (Menestrina, 1986), whereas osmotic experiments suggested that sugars with a diameter of 2.7 nm could enter the pore (Krasilnikov *et al.*, 1992). Recently, pores formed by HlgC + HlgB were characterised in a similar system. In the presence of 1 M KCl they had a conductance around three times that of α-toxin, 2.5 nS vs 0.775 nS (Miles *et al.*, 2001), and a preference for cations instead of anions. These differences might result from real divergences between the two pores, regarding, for example, its effective diameter or the structural characteristics of the lumen. The β-hairpin of the α-toxin Stem contains more charged amino acids than those of the leucotoxins (Figure 1.3). Moreover, the transmembrane β-hairpin of the leucotoxins is shorter by two residues (F proteins) or four residues (S proteins). Compared to α-toxin, this might imply a reduced tilting, with respect to the symmetry axis, of the hairpins composing the leucotoxin β-barrel. However, it is noteworthy that estimated diameters varied a lot, from 1.2 to 2.0 nm, according to the experiment (Krasinilkov *et al.*, 1992; Bashford *et al.*, 1996), and to the state of protonation of the lumen (Kasianowicz and Bezrukov, 1995; Paula *et al.*, 1999). In all cases, these pores remain considerably larger than the electrostatic diameter of any ion like Cl^-, K^+ or Na^+. In fact, little is known about the mechanism of ion selectivity for these toxin pores, which should be clearly different from the steric selection occurring, for example, in the cell potassium channels (Doyle *et al.*, 1998).

Modifying the activity of leucotoxins

Available data on toxin modifications can be distinguished between those reporting only the overall activity and cell preference of modified toxins, and those which tried to evaluate most of the events involved in pore formation.

A natural truncate of the 17 C-terminal residues of HlgC was completely inactive (Nariya *et al.*, 1993). This deletion affects more than one element of the secondary structure. Chimeric proteins derived from HlgA (HγII) and HlgC (LukS) were constructed using conserved restriction sites, and were finally assayed, in association with HlgB, for leucotoxic and haemolytic activity. The exchange of the 123 C-terminal residues of HlgC by those of HlgA led to a leucotoxin with haemolytic activity (possibly on HRBC) (Nariya and Kamio, 1995). This exchange covered almost all the sequence downstream the β-hairpin forming the pore, which suggests that this portion of the sequence may delineate some binding sites for membrane ligands. In contrast, the substitution of the 24 C-terminal residues of HlgC by those of HlgA abolished leucocytolytic activity (Sugawara *et al.*, 1999). It may be suggested that the latter substitution was not conservative in structure. In another work, the authors added Asp12-Asp13, present in HlgC, at the corresponding position of LukS-PV and the new protein was able to lyse RRBC (Nariya *et al.*, 1997). Conversely, when the two amino acids were deleted from HlgC, the haemolytic activity was reduced to about 10%. Assuming similar structures for S and F components, positions 12 and 13 might be located in the β-sandwich core at the interface with LukS-PV. Obviously, the structure of the oligomer would be of great help in understanding the structural rationales sustaining these gains or losses of activity.

The same group recently suggested that a mutation of LukF-PV (Thr71Tyr) can be sufficient to acquire haemolytic activity on human erythrocytes when combined with HlgA (Yokota and Kamio, 2000), supposedly due to a better binding (which, however, was not demonstrated). The substitution corresponds to one of the sequence differences between LukF-PV and HlgB. The residue is at the bottom of the Rim and should be located in the vicinity of the membrane in the assembled oligomer. We have also observed that a small deletion in the loop of the hairpin of LukF-PV (ΔSer125-Gly127) was sufficient to make this protein able to lyse RRBC, and to a lesser degree HRBC, in combination with HlgA (Werner, 2001). This observation suggested that some features of the β-hairpins might influence differently the activity on different cells, for example, erythrocytes vs PMNs. Such features may be related with the interactions between regions of the proteins constituting the leucotoxin pore, or with the interaction with some specific membrane components, or with the transformation of the prefolded Stem into the functional β-barrel.

Mutagenesis of the Stem domain of LukF-PV indeed demonstrated the possibility to discouple the biological activities of the Panton–Valentine leucocidin onto human PMNs preserving only Ca^{2+} entry. Two LukF-PV mutants, ΔSer125-Ile128 and Gly130Asp, though keeping a normal binding onto previously bound LukS-PV, showed a substantial separation of the two activities (Baba Moussa *et al.*, 1999b). Ca^{2+} entry was rapidly promoted but Et^{+} entry was considerably reduced, although PEG molecules of a hydrodynamic diameter smaller than 1.12 nm crossed the pore in a way comparable to the wt toxin. The deletion in the β-hairpin presumably caused a rearrangement of the residues facing the lumen of the pore, thus affecting the traffic of ions, but not of neutral molecules. In the case of LukF-PV Gly130Asp, although the charged residue was introduced at a position that would face membrane phospholipids, it probably induced also a modification of the lumen of the pore disturbing the flow of cations. Both discoupled proteins were able to induce interleukin (IL)-8 secretion although less than similar doses of PVL. The difference could represent the amount of IL-8 normally released upon the lysis of the cells. When intradermally injected, the modified toxins failed to induce dermonecrosis in the rabbit skin, indicating that the lytic process may be essential for the full expression of the virulence normally associated with these toxins.

Human diseases associated with bicomponent leucotoxins

Staphylococcus aureus is a bacteria very well adapted to humans. From a frequent commensal on human skin or anterior nose, it may become a broad pathogen able to infect every organ (Kloos and Bannerman, 1999). Even if most infections occur in patients with other underlying diseases, some can also be primary, like staphylococcal scalded skin syndrome (SSSS) (Ladhani *et al.*, 1999; Gravet *et al.*, 2001), furuncles and other skin infections (Hedström, 1985), staphylococcal toxic shock syndrome (TSS) and staphylococcal enterotoxin food-poisoning (Dinges *et al.*, 2000). In fact, the bacteria is able to produce several toxic compounds: superantigens, proteases, pore-forming toxins and a bulk of adhesion factors which may be present in different combinations according to the strain (Kuroda *et al.*, 2001).

Among these families of molecules, some are constant in isolates, as the γ-haemolysin (Prévost *et al.*, 1995b), and therefore cannot be associated with any specific disease, whereas others are non-ubiquitous. In the case of bicomponent leucotoxins, Van der Velde (1894) was the first to report a leucocidal action consequent to a pleural infection in rabbit. Panton and Valentine (1932) confirmed that a secreted product, isolated from strains associated with furuncles, had a peculiar leucocidal action. Later, Ward and Turner (1980) showed that such culture supernatants were able to induce dermonecrosis of the rabbit skin. Eventually, it was shown that highly purified PVL could alone induce such a dermonecrosis, mimicking the lesions observed for furuncles, and that the presence of both LukS-PV and LukF-PV was required (Cribier *et al.*, 1992). PVL-immunised rabbits developed little inflammation after toxin injection, but no tissue necrosis was observed even at high toxin doses. Epidemiological studies in Strasbourg University Hospital (Couppié *et al.*, 1994; Prévost *et al.*, 1995b), in French Guyana (Gravet *et al.*, 2001) or other medical centres (Baba Moussa *et al.*, 1999a), evidenced that *S. aureus* isolated from furuncles produced PVL in about 95% of the cases. Conversely, among routine isolates, about 90% of *S. aureus* producing PVL were sampled from furuncles. In some cases, these strains may also harbour a reduced sensitivity, or even resistance, to antibiotics. Recently, it was shown that furuncle diseases might evolve to bacteremia with pulmonary abscesses (Couppié *et al.*, 1997). Indeed, some community-acquired pneumonia, were reported to be caused by strains producing PVL (Lina *et al.*, 1999). A critical point would be to understand the pathological process that is involved in the emergence of such diseases and whether it is related to the specific background of each patient.

Other implications with clinical syndromes, that is, post-antibiotic diarrhoea (Gravet *et al.*, 1999) and SSSS or *impetigo contagiosa* (Gravet *et al.*, 2001), have been observed for a less frequent leucotoxin, LukE + LukD, in association with other toxins. For the first of these diseases, a historical introduction may be important. *S. aureus* was originally associated with some pseudomembranous enterocolitis (Hinton *et al.*, 1960) and some antibiotic-associated diarrhoea (AAD) (Altemeier *et al.*, 1963). However, this association became less considered later because of the lack of a satisfying experimental model and for the characterisation and isolation of *Clostridium difficile* and its toxins as the major etiological agent of the disease (George, 1984; Barbut *et al.*, 1996). Recently, however, we considered those cases of diarrhoea where *S. aureus* was either predominant or in pure culture. During a two-year study within a 2,500-bed hospital, a series of 47 cases of elderly patients (mean: 65.8 y.o.) with AAD was analysed. All isolates were methicillin-resistant, and produced LukE + LukD and enterotoxin A in 44 of the cases (94%), whereas the natural occurrence of this association of toxins is 15–20% in routine isolates (30% of occurrences for LukE + LukD and 50% for enterotoxin A). All patients were previously treated with an antibacterial therapy involving fluoroquinolones. Fatal cases amounted to 20%, but when an appropriate anti-staphylococcal treatment, for example, oral vancomycin or stop of antibiotics, was administered, diarrhoea mostly disappeared. Different strains of *S. aureus* were responsible for these diseases, but epidemic isolates, previously known to be involved in nosocomial infections, were recognised. Complications such as bacteremia, developing after the beginning of diarrhoea, were shown to be due to similar strains. In the few cases in which histological investigations were performed, the destruction of intestinal villosities was revealed. *S. aureus* isolated from stools of patients without diarrhoea did not produce the two toxins. The amount of cases of post-antibiotic diarrhoea associated with *S. aureus* represented one-sixth of those associated with *C. difficile*. Recently, we could show that the injections of these two different toxins in rabbit ligated ileal loops cooperatively caused a time- and dose-dependent intestinal damage and fluid accumulation. LukE + LukD were the most acute and specific in producing these effects among the leucotoxins considered (Gravet *et al.*, 2001).

Another recent study reported that different isolates originating from impetigo or SSSS cases, in metropolitan France and French Guyana, produced LukE + LukD in association with at least one of the epidermolysins in 76% of the cases (Gravet *et al.*, 2001). The two toxins have no genetical link, as in the case of enterotoxin A. Impetigo is a primary skin infection, which essentially results in a cleavage between cell layers of the epidermis of new-borns and young infants (Gentilhomme *et al.*, 1990; Ladhani *et al.*, 1999). Exudative bullae are produced that expose patients to secondary infections. Among symptoms, exudates and erythema are not observed when purified epidermolysins are injected into the new-born mouse experimental model (Gemmell *et al.*, 1995). It is possible that in humans, these two symptoms might result from the association of the two toxins. The characterisation of specialised target cells for LukE + LukD in the epidermis or the intestine would help to understand the physiopathology of the relying diseases and to confirm the finding that the receptor for LukE is not the same as that for LukS-PV (Gauduchon *et al.*, 2001).

Concerning the other leucotoxins, LukM + LukF′-PV has never been characterised yet from human isolates in Eastern France, French Guyana or sub-equatorial Africa (Baba Moussa *et al.*, 1999a; Gravet *et al.*, 2001). Only two isolates from bovine mastitis were identified to produce this toxin. LukS-I + LukF-I from *S. intermedius* seemed to be constant in this species (Prévost *et al.*, 1995a). Considering that common antigens among leucotoxins are not many, the constant production of γ-haemolysin may induce immunity against few epitopes, thus favouring the action of other, non-ubiquitous and non-cross-reacting, leucotoxins.

In an experimental model of post-traumatic endophtalmitis, every S–F couple, injected into the vitreous humour of rabbit eye, induced a huge inflammatory reaction that could extend to the whole eyeball and the surrounding tissues, depending on the dose and the time after injection (Siqueira *et al.*, 1997). PVL and HlgA + LukF-PV were the most potent couples, suggesting that also heterologous pairs may indeed have an importance in the virulence. A study with a knock-out mutant, lacking γ-haemolysin, showed that the infection was less developed, and the count of bacteria in a given time-interval was lower (Supersac *et al.*, 1998). However, the Newman derivative strain, which was used in that study, also produced LukE + LukD that could have contributed to facilitate the infection since that toxin is pro-inflammatory too.

Several evidences indicate that the pore-forming function might be essential for the virulence of leucotoxins and bacteria. (i) LukF-PV Gly130Asp, which is unable to form functional pores, neither induced dermonecrosis in rabbit, nor was very potent in the rabbit vitreous humour. (ii) In the latter experimental model, leucotoxins are produced in vivo at no more than 10^5 CFU/ml of vitreous humour, or at late exponential growth phase in vitro. Interestingly, siderophores are present from the exponential growth phase (Morrissey *et al.*, 2000). Leucotoxins with their lytic functions may relay and amplify the accessibility of the ferric ion. (iii) Disruption of one gene encoding a leucotoxin can hamper bacterial development. The importance of the lytic function is possibly to amplify cell inflammation, but cell lysis should also liberate nutrients for bacteria, and locally produce a physical barrier of ghost cells that may protect the multiplication of the bacteria during the early phases of primary infection. Contrary to the leucotoxins, staphylococcal α-toxin is produced in vitro during the exponential growth phase, but its expression in vivo remains to be documented.

Inflammatory mediators generated by the action of staphylococcal leucotoxins

The impact of intravenous injection of leucotoxin on haemostasis (Szmigielski *et al.*, 1966, 1968) and the distribution of the toxin into organs (Grojec, 1979) were investigated in different experimental models. Despite the fact that the toxin used was uncertain and possibly contaminated, leucotoxins were recognised to elicit disorders on the PMN counts.

The cell response generated by leucotoxins is pleiotropic, as it could be suspected in view of the induced influx of Ca^{2+} and the pivotal role of this ion in inflammation (Figure 1.4). The intensity of cell response will depend on the leucotoxin considered. Secretion of granule contents was promoted, as evidenced by measuring lysozyme and β-glucuronidase activities (Colin *et al.*, 1994; König *et al.*, 1997). Around necrotic lesions caused by leucotoxin injections, chemotactic factors, as leucotriene B_4 (Hensler *et al.*, 1994b) and IL-8 (König *et al.*, 1994; Baba Moussa *et al.*, 1999b), are secreted from target cells and promote cell recruitment. Release of the vasodilator histamine (König *et al.*, 1995, 1997) might correspond to the vasodilatation phenomenon observed after intradermal injection in the rabbit skin (Cribier *et al.*, 1992). It should be noticed that leucotriene B_4 may induce PMNs to produce superoxide ions that would explain tissue necrosis. Besides secretion of IL-8, recent experiments also showed a slightly enhanced secretion of IL-6 and IL-12, whereby PVL was a better inducer than HlgA-HlgB (Figure 1.5). Interferon-α was instead not generated (Werner, 2001). In this cascade of inflammatory responses, phosphoinositides metabolism (Wang *et al.*, 1990), heterotrimeric GTP-binding proteins (Hensler *et al.*, 1994a) and heat-shock proteins (Köller *et al.*, 1993) are probably involved.

Conclusions

Staphylococcal bicomponent leucotoxins represent an important subfamily of pore-forming toxins, probably developed by *S. aureus* to refine its virulence capacities. At least two combinations of these toxins, but usually more, are produced by any *S. aureus* isolate. Some of the non-ubiquitous leucotoxins are strongly associated, in routine isolates, to some diseases and nosocomial infections, not only at the epidemiological level but also because they can generate, in experimental models, lesions comparable to the associated clinical syndrome.

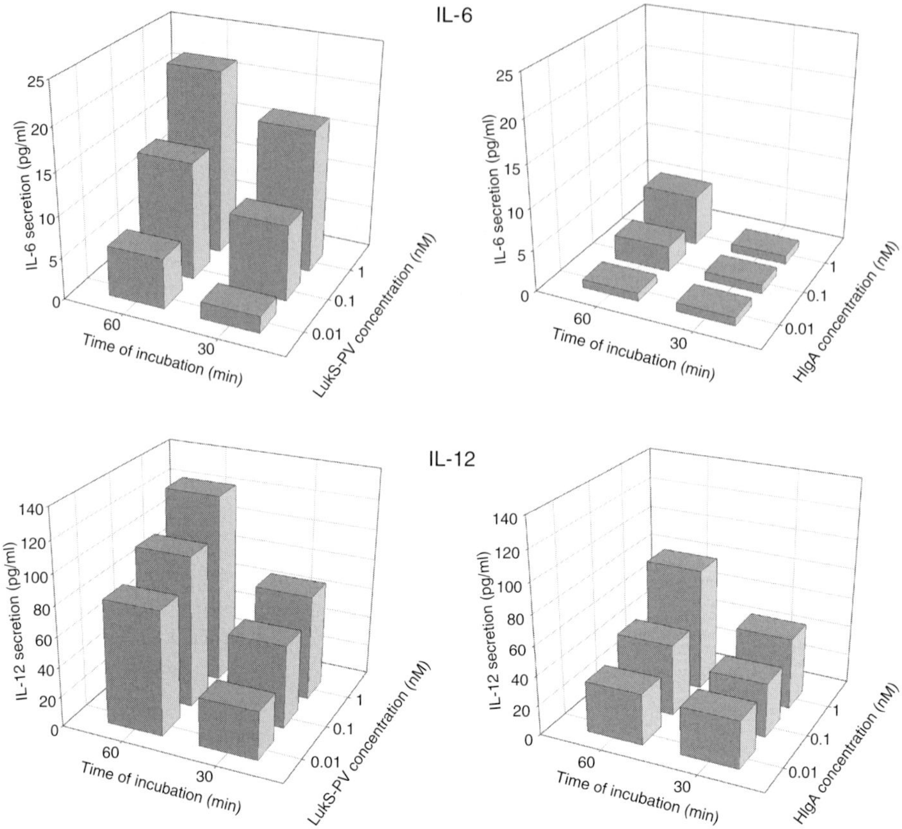

Figure 1.5 Bicomponent leucotoxins may elicit more IL-12 than IL-6 secretion. Here, 5×10^5 human PMNs were incubated for 30 and 60 min with various concentrations of LukS-PV and 5 nM of LukF-PV, or various concentrations of HlgA and 5 nM of HlgB ($n=3$). Cells were pelleted and ILs released in the supernatant were titrated using ELISA kits (Bender MedSystems). Non-incubated cells did not secrete significant amounts of either IL-6 or IL-12.

Despite being known for a long time as α-toxin, they have been unsteadily studied. Even if they are related to other toxins (Gouaux, 1998), their binary composition renders them an interesting, complex and unique model to explore some molecular aspects of the mechanism leading to the formation of the functional pore. They might also offer an opportunity to develop biotechnological applications such as: biosensors (Braha *et al.*, 1997; Gu *et al.*, 1999), permeabilising tools to eliminate malignant cells or to direct the uptake of specific intracellular drugs (Kasianowicz *et al.*, 1996; Panchal *et al.*, 1996; Eroglu *et al.*, 2000).

Acknowledgements

The authors appreciate the skilful technical assistances of R. Girardot and D. Keller. This work was supported by grant EA-3432 from the 'Direction pour la Recherche et les Etudes Doctorales (DRED)', and funds of the Institute of Bacteriology, Medicine Faculty of Strasbourg.

References

Ahnert-Hilger, G., Bhakdi, S. and Gratzl, M. (1985) Minimal requirements for exocytosis: a study using PC12 cells permeabilized with staphylococcal alpha-toxin. *Journal of Biological Chemistry*, **260**, 12730–12734.

Alouf, J. E. and Freer, J. H. (1999) *The Comprehensive Sourcebook of Bacterial Protein Toxins*, 2nd edition London: Academic Press.

Altemeier, W. A., Hummel, R. P. and Hill, E. O. (1963) Staphylococcal enterocolitis following antibiotic therapy. *Annals of Surgery*, **157**, 847–858.

Baba Moussa, L., Sanni, A., Dagnra, A. Y., Anagonou, S., Prince-David, M., Edoh, V., Befort, J. J., Prévost, G. and Monteil, H. (1999a) Approche épidémiologique de l'antibiorésistance et de la production de leucotoxines par les souches de *Staphylococcus aureus* isolées en Afrique de l'Ouest. *Médecine et Maladies Infectieuses*, **29**, 689–696.

Baba Moussa, L., Werner, S., Colin, D. A., Mourey, L., Pédelacq, J. D., Samama, J. P., Sanni, A., Monteil, H. and Prévost, G. (1999b) Discoupling the Ca^{2+}-activation from the pore-forming function of the bi-component Panton–Valentine leucocidin in human PMNs. *FEBS Letters*, **461**, 280–286.

Baida, G., Budarina, Z. I., Kuzmin, N. P. and Solonin, A. S. (1999) Complete nucleotide sequence and molecular characterization of hemolysin II gene from *Bacillus cereus*. *FEMS Microbiol. Letters*, **180**, 7–14.

Barbut, F., Corthier, G., Charpak, M., Cerf, M., Monteil, H., Fosse, T., Trévoux, A., de Barbeyrac, B., Boussougant, Y., Tigaud, S., Tytgat, F., Sédaillant, A., Duborgel, S., Collignon, A., Le Guern, M., Bernasconi, P. and Petit, J. C. (1996) Prevalence and pathogenicity of *Clostridium difficile* in hospitalized patients – a French multicenter study. *Archives of Internal Medicine*, **156**, 1449–1454.

Bashford, C. L., Alder, G. M., Fulford, L. G., Korchev, Y. E., Kovacs, E., MacKinnon, A., Pederzolli, C. and Pasternak, C. A. (1996) Pore formation by *S. aureus* alpha-toxin in liposomes and planar lipid bilayers: effects of nonelectrolytes. *Journal of Membrane Biology*, **150**, 37–45.

Bhakdi, S., Füssle, R., Utermann, G. and Tranum-Jensen, J. (1983) Binding and partial inactivation of *S. aureus* alpha-toxin by plasma low density lipoprotein. *Journal of Biological Chemistry*, **258**, 5899–5904.

Bhakdi, S., Muhly, M., Mannhardt, U., Hugo, F., Klappetek, K., Mueller-Eckardt, C. and Roka, C. (1988) Staphylococcal alpha-toxin promotes blood coagulation via attack on human platelets. *Journal of Experimental Medicine*, **168**, 527–542.

Bhakdi, S., Bayley, H., Valeva, A., Walev, I., Walker, B., Weller, U., Kehoe, M. and Palmer, M. (1996) Staphylococcal alpha-toxin, streptolysin-O, and *Escherichia coli* haemolysin: prototypes of pore-forming bacterial cytolysins. *Archives in Microbiology*, **165**, 73–79.

Braha, O., Walker, B., Cheley, S., Kasianowicz, J. J., Song, L., Gouaux, J. E. and Bayley, H. (1997) Designed protein pores as components for biosensors. *Chemistry and Biology*, **4**, 497–505.

Bronner, S., Stoessel, P., Gravet, A., Monteil, H. and Prévost, G. (2000) Variable expressions of *Staphylococcus aureus* bicomponent leucotoxins semiquantified by competitive reverse transcription-PCR. *Applied and Environmental Microbiology*, **66**, 3931–3938.

Colin, D. A., Mazurier, I., Sire, S. and Finck-Barbançon, V. (1994) Interaction of the two components of leukocidin from *Staphylococcus aureus* with human polymorphonuclear leukocyte membranes: sequential binding and subsequent activation. *Infection and Immunity*, **62**, 3184–3188.

Colin, D. A., Meunier, O., Staali, L., Prévost, G. and Monteil, H. (1997) Bi-component leukotoxins from *Staphylococcus aureus*. In *Microbial Pathogenesis and Host Response*: Proceedings of the Cold Spring Harbor Laboratory on Microbial Pathogenesis and Host Response, edited by Maloy, S. *et al.*, p. 150. New York: Cold Spring Harbor Laboratory Press.

Cooney, J., Kienle, Z., Foster, T. J. and O'Toole, P. W. (1993) The gamma-haemolysin locus of *Staphylococcus aureus* comprises three linked genes, two of which are identical to the genes for the F and S components of leukocidin. *Infection and Immunity*, **61**, 768–771.

Couppié, P., Cribier, B., Prévost, G., Grosshans, E. and Piémont, Y. (1994) Leucocidin from *Staphylococcus aureus* and cutaneous infections: an epidemiological study. *Archives of Dermatology*, **130**, 1208–1209.

Couppié, P., Hommel, D., Prévost, G., Godart, M. C., Moreau, B., Sainte-Marie, D., Peneau, C., Hulin, A., Monteil, H. and Pradinaud, R. (1997) Septicémie à *Staphylococcus aureus*, furoncle et leucocidine de Panton et Valentine: 3 observations. *Annales de Dermatologie et de Vénéréologie*, **124**, 684–686.

Cribier, B., Prévost, G., Couppié, P., Finck-Barbançon, V., Grosshans, E. and Piémont, Y. (1992) *Staphylococcus aureus* leukocidin: a new virulence factor in cutaneous infections. *Dermatology*, **185**, 175–180.

Czajkowsky, D. M., Sheng, S. T. and Shao, Z. F. (1998) Staphylococcal alpha-haemolysin can form hexamers in phospholipid bilayers. *Journal of Molecular Biology*, **276**, 325–330.

Dinges, M. M., Orwin, P. M. and Schlievert, P. M. (2000) Exotoxins of *Staphylococcus aureus*. *Clinical Microbiology Reviews*, **13**, 16–34.

Doyle, D. A., Morais-Cabral, J., Pfuetzner, R. A., Kuo, A., Gulbis, J. M., Cohen, S. L., Chait, B. T. and McKinnon, R. (1998) The structure of the potassium channel: molecular basis of K^+ conduction and selectivity. *Science*, **280**, 69–77.

Elbaze, P. and Ortonne, J. P. (1988) Chronic furunculosis. *Annales de Dermatologie et de Vénéréologie*, **115**, 859–864.

Eroglu, A., Russo, M. J., Bieganski, R., Fowler, A., Cheley, S., Bayley, H. and Toner, M. (2000) Intracellular trehalose improves the survival of cryopreserved mammalian cells. *Nature Biotechnology*, **18**, 163–167.

Ferreras, M., Menestrina, G., Foster, T., Colin, D. A., Prévost, G. and Piémont, Y. (1996) Permeabilisation of lipid bilayers by *Staphylococcus aureus* γ-toxins. In *Bacterial Protein Toxins*, edited by P. L. Frandsen *et al.*, Zentralblatt für Bakteriology, Supp. 28, pp. 105–106. Stuttgart, Jena, New York: Gustav Fischer Verlag.

Ferreras, M., Höper, F., Dalla Serra, M., Colin, D. A., Prévost, G. and Menestrina, G. (1998) The interaction of *Staphylococcus aureus* bi-component gamma haemolysins and leucocidins with cells and model membranes. *Biochimica and Biophysica Acta*, **1414**, 108–126.

Finck-Barbançon, V., Prévost, G. and Piémont, Y. (1991) Improved purification of leucocidin from *Staphylococcus aureus* and toxin distribution among hospital strains. *Research in Microbiology*, **142**, 75–85.

Finck-Barbançon, V., Duportail, G., Meunier, O. and Colin, D. A. (1993) Pore formation by a two-component leukocidin from *Staphylococcus aureus* within the membrane of human polymorphonuclear leukocytes. *Biochimica and Biophysica Acta*, **1182**, 275–282.

Fink, D., Contreras, M. L., Lelkes, P. I. and Lazarovici, P. (1989) *Staphylococcus* α-toxin activates phospholipases and induces a Ca^{2+} influx in PC12 cells. *Cellular Signaling*, **1**, 387–393.

Fivaz, M., Abrami, L. and van der Goot, F. G. (1999) Landing on lipid rafts. *Trends in Cell Biology*, **9**, 212–213.

Foster, T. J. and Höök, M. (1998) Surface proteins adhesins of *Staphylococcus aureus*. *Trends in Microbiology*, **6**, 484–488.

Freer, J. H., Arbuthnott, J. P. and Bernheimer, A. W. (1968) Interaction of staphylococcal alpha-toxin with artificial and natural membranes. *Journal of Bacteriology*, **95**, 1153–1168.

Gauduchon, V., Werner, S., Prévost, G., Monteil, H. and Colin, D. A. (2001) Flow cytometric determination of Panton–Valentine leucocidin S component binding. *Infection and Immunity*, **69**, 2390–2395.

Gemmell, C. G. (1995) Staphylococcal scalded skin syndrome. *Journal of Medical Microbiology*, **43**, 318–327.

Gentilhomme, E., Faure, M., Piémont, Y., Binder, P. and Thivolet, J. (1990) Action of staphylococcal exfoliative toxins on epidermal cell cultures and organotypic skin. *Journal of Dermatology*, **17**, 526–532.

George, W. L. (1984) Antimicrobial agent-associated colitis and diarrhea: historical background and clinical aspects. *Reviews in Infectious Diseases*, **(Suppl 1)**, S208–S213.

George, W., Sutter, V., Citron, D. and Finegold, S. (1979) Selective and differential medium for isolation of *Clostridium difficile*. *Journal of Clinical Microbiology*, **9**, 214–219.

Gladstone, G. P. and Fildes, P. (1940) A simple culture medium for general use without meat extract or peptone. *British Journal of Experimental Pathology*, **21**, 161–173.

Gladstone, G. P. and Van Heyningen, W. E. (1957) Staphylococcal leucocidin. *British Journal of Experimental Pathology*, **38**, 125–137.

Gladstone, G. P. and Glencross, E. J. G. (1960) Growth and toxin production of staphylococci in cellophane sacs in vivo. *British Journal of Experimental Pathology*, **41**, 313–333.

Gouaux, J. E., Braha, O., Hobaugh, M. R., Song, L., Cheley, S., Shustak, C. and Bayley, H. (1994) Subunit stoichiometry of staphylococcal α-haemolysin in crystals and on membranes: a heptameric transmembrane pore. *Proceedings of the National Academy of Sciences* (*USA*), **91**, 12828–12831.

Gouaux, E. (1997a) Channel-forming toxins: tales of transformation. *Current Opinion in Structural Biology*, **7**, 566–573.

Gouaux, J. E., Hobaugh, M. and Song, L. (1997b) α-haemolysin, γ-haemolysin and leukocidin from *Staphylococcal aureus*: distant in sequence but similar in structure. *Protein Science* **6**, 2631–2635.

Gouaux, E. (1998) α-haemolysin from *Staphylococcus aureus*: an archetype of β-barrel, channel-forming toxins. *Journal of Structural Biology*, **121**, 110–122.

Gravet, A. (2001) Characterisation of the LukE-LukD leucotoxin from Staphylococcus aureus and in vitro, in vivo analysis of its expression. PhD Thesis, Institute of Bacteriology, Louis Pasteur University, Strasbourg, France.

Gravet, A., Colin, D. A., Keller, D., Girardot, R., Monteil, H. and Prévost, G. (1998) Characterization of a novel structural member, LukE-LukD, of the bi-component leucotoxins family. *FEBS Letters*, **436**, 202–208.

Gravet, A., Rondeau, M., Harf-Monteil, C., Grunenberger, F., Monteil, H., Scheftel, J. M. and Prévost, G. (1999) Predominant *Staphylococcus aureus* isolated from antibiotic-associated diarrhea is clinically relevant and produces enterotoxin A and the bicomponent toxin LukE-LukD. *Journal of Clinical Microbiology*, **37**, 4012–4019.

Gravet, A., Couppié, P., Meunier, O., Clyti, E., Moreau, B., Pradinaud, R., Monteil, H. and Prévost, G. (2001) *Staphylococcus aureus* isolated from impetigo produces both epidermolysins A or B and LukE-LukD in 76% of the cases. *Journal of Clinical Microbiology*, **39**, 4349–4356.

Gray, G. S. and Kehoe, M. (1984) Primary sequence of the α-toxin gene from *Staphylococcus aureus* Wood 46. *Infection and Immunity*, **46**, 615–618.

Greenwald, J., Fischer, W. H., Vale, W. W. and Choe, S. (1999) Three-finger toxin fold for the extracellular ligand-binding domain of the type II activin receptor serine kinase. *Nature Structural Biology*, **6**, 18–22.

Grojec, P. (1979) Distribution of ^{131}I-labelled staphylococcal leukocidin in mouse organs. *Medycyna Do wiadczalna i Mikrobiologica*, **31**, 209–216.

Gu, L. Q., Braha, O., Conlan, S., Cheley, S. and Bayley, H. (1999) Stochastic sensing of organic analytes by a pore-forming protein containing a molecular adapter. *Nature*, **398**, 686–690.

Guyonnet, F. and Plommet, M. (1970) Hémolysine gamma de *Staphylococcus aureus*: purification et propriétés. *Annales de Institut Pasteur*, **118**, 19–33.

Hedström, S. A. (1985) Treatment and prevention of recurrent staphylococcal furunculosis: clinical and bacteriological follow-up. *Scandinavian Journal of Infectious Diseases*, **17**, 55–58.

Hensler, T., Köller, M., Prévost, G., Piémont, Y. and König, W. (1994a) GTP-binding proteins are involved in the modulated activity of human neutrophils treated by the Panton-Valentine leucocidin from *Staphylococcus aureus*. *Infection and Immunity*, **62**, 5281–5289.

Hensler, T., König, B., Prévost, G., Piémont, Y., Köller, M. and König, W. (1994b) LTB4- and DNA fragmentation induced by leukocidin from *Staphylococcus aureus*. The protective role of GM-CSF and G-CSF on human neutrophils. *Infection and Immunity*, **62**, 2529–2535.

Hinton, N. A., Taggart, J. G. and Orr, J. H. (1960) The significance of the isolation of coagulase-positive staphylococci from stool. With special reference to the diagnosis of staphylococcal enteritis. *Annual Journal of Clinical Pathology*, **33**, 505–510.

Humphrey, W., Dalke, A. and Schulten, K. (1996) VMD – Visual Molecular Dynamics, *Journal of Molecular Graphics*, **14**, 33–38.

Hunter, S. E. C., Brown, J. E., Oyston, P. C. F., Sakurai, J. and Titball, R. W. (1993) Molecular genetic analysis of beta-toxin of *Clostridium perfringens* reveals sequence homology with alpha-toxin, gamma toxin, and leukocidin of *Staphylococcus aureus*. *Infection and Immunity*, **61**, 3958–3965.

Jonas, D., Walev, I., Berger, T., Liebetrau, M., Palmer, M. and Bhakdi, S. (1994) Novel path to apoptosis: small transmembrane pores created by staphylococcal α-toxin in T lymphocytes evoke internucleosomal DNA degradation. *Infection and Immunity*, **62**, 1304–1312.

Jursch, R., Hildebrand, A., Hobom, G., Tranum-Jensen, J., Ward, R., Kehoe, M. and Bhakdi, S. (1994) Histidine residues near the N-terminus of *Staphylococcus* alpha-toxin as reporters of regions that are critical for oligomerization of pore-formation. *Infection and Immunity*, **62**, 2249–2256.

Kamio, Y., Rahman, A., Nariya, H., Ozawa, T. and Izaki, K. (1993) The two staphylococcal bi-component toxins, leukocidin and gamma-haemolysin, share one component in common. *FEBS Letters*, **321**, 15–18.

Kaneko, J., Kimura, T., Kawakami, Y., Tomita, T. and Kamio, Y. (1997a) Panton–Valentine leukocidin genes in a phage-like particle isolated from mytomycin C-treated *Staphylococcus aureus* V8 (ATCC 49775). *Biosciences Biotechnology and Biochemistry*, **61**, 1960–1962.

Kaneko, J., Muramoto, K. and Kamio, Y. (1997b) Gene of LukF-PV-like component of Panton–Valentine leukocidin in *Staphylococcus aureus* P83 is linked with *luk*M. *Biosciences Biotechnology and Biochemistry*, **61**, 541–544.

Kasianowicz, J. J. and Bezrukov, S. M. (1995) Protonation dynamics of the alpha-toxin ion channel from spectral analysis of pH-dependent current fluctuations. *Biophysical Journal*, **69**, 94–105.

Kasianowicz, J. J., Brandin, E., Branton, D. and Deamer, D. W. (1996) Characterization of individual polynucleotide molecules using a membrane channel. *Proceedings of the National Academy of Sciences* (*USA*), **93**, 13770–13773.

Katsumi, H., Tomita, T., Kaneko, J. and Kamio, Y. (1999) Vitronectin and its fragments purified as serum inhibitors of *Staphylococcus aureus* γ-haemolysin and leucocidin, and their specific binding to the Hlg2 and the LukS components of the toxins. *FEBS Letters*, **460**, 451–456.

Kloos, W. E. and Bannerman, T. L. (1999) *Staphylococcus* and *Micrococcus*. In *Manual of Clinical Microbiology*, edited by P. R. Murray, E. J. Baron, M. A. Pfaller, F. C. Tenover and R. H. Yolken (s), 7th edition, pp. 264–282. Washington, DC: ASM Press.

Köller, M., Hensler, T., König, B., Prévost, G., Alouf, J. and König, W. (1993) Induction of heat-shock proteins by bacterial toxins, lipid mediators and cytokines in human leucocytes. *Zentralblatt für Bakteriologie*, **278**, 365–376.

König, B., Köller, M., Prévost, G., Piémont, Y., Alouf, J. E., Schreiner, A. and König, W. (1994) Activation of human effector cells by different bacterial toxins (leukocidin, alveolysin, erythrogenic toxin A): generation of interleukin-8. *Infection and Immunity*, **62**, 4831–4837.

König, B., Prévost, G., Piémont, Y. and König, W. (1995) Effects of *Staphylococcus aureus* leucocidins inflammatory mediator release from human granulocytes. *Journal of Infectious Diseases*, **171**, 607–613.

König, B., Prévost, G. and König, W. (1997) Composition of staphylococcal bi-component toxins determines pathophysiological reactions. *Journal of Medical Microbiology*, **46**, 479–485.

Korchev, Y. E., Alder, G. M., Bakhramov, A., Bashford, C. L., Joomun, B. S., Sviderskaya, E. V., Usherwood, P. N. R. and Pasternak C. A. (1995a) *Staphylococcus aureus* alpha-toxin-induced pores: channel-like behavior in lipid bilayers and clamped cells. *Journal of Membrane Biology*, **143**, 143–151.

Korchev, Y. E., Bashford, C. L., Alder, G. M., Kasianowicz, J. J. and Pasternak, C. A. (1995b) Low conductance states of a single ion channel are not 'closed'. *Journal of Membrane Biology*, **147**, 233–239.

Krasilnikov, O. V., Ternovskii, V. I., Sabirov, R. Z., Zaripova, R. K. and Tashmukhamedov, B. A. (1986) Cationic–anionic selectivity of staphylotoxin channels in lipid bilayer. *Biophysics*, **31**, 658–663.

Krasilnikov, O. V., Sabirov, R. Z., Ternovsky, O. V., Merzlyak, P. G. and Muratkhodjaev, J. N. (1992) A simple method for the determination of the pore radius of channels in planar lipid bilayer membranes. *FEMS Microbiology and Immunology*, **105**, 93–100.

Krishnasastry, M., Walker, B., Braha, O. and Bayley, H. (1994) Surface labelling of key residues during assembly of the transmembrane pore formed by staphylococcal α-haemolysin. *FEBS Letters*, **356**, 66–71.

Kuroda, M., Ohta, T., Ushiyama, I., Baba, T., Yuzawa, H., Kobayashi, I. *et al.* (2001) Whole genome sequencing of methicillin-resistant *Staphylococcus aureus*. *Lancet*, **357**,1225–1239.

Ladhani, S., Joannou, C. L., Lochrie, D. P., Evans, R. W. and Poston, S. M. (1999) Clinical, microbial, and biochemical aspects of the exfoliative toxins causing staphylococcal scalded-skin syndrome. *Clinical Microbiology Reviews*, **12**, 224–242.

Lee, S. J., Kim, D. M., Bae, K. H., Byun, S. M. and Chung, J. H. (2000) Enhancement of secretion and extracellular stability of staphylokinase in *Bacillus subtilis* by wprA gene disruption. *Applied and Environmental Microbiology*, **66**, 476–480.

Lesieur, C., Vécsey-Semjén, B., Abrami, L., Fivaz, M. and van der Goot, F. G. (1997) Membrane insertion: the strategies of toxins. *Molecular Membrane Biology*, **14**, 45–64.

Lina, G., Piémont, Y., Godail-Gamot, F., Bès, M., Peter, M.-O., Vandenesch, F. and Jérôme, E. (1999) Involvement of Panton–Valentine leukocidin-producing *Staphylococcus aureus* in primary skin infections and pneumonia. *Clinical Infectious Diseases*, **29**, 1128–1132.

Lund, T., De Buyser, M. L. and Granum, P. E. (2000) A new cytotoxin from *Bacillus cereus* that may cause necrotic enteritis. *Molecular Microbiology*, **38**, 254–261.

Menestrina, G. (1986) Ionic channels formed by *Staphylococcus aureus* alpha-toxin: voltage dependent inhibition by di- and trivalent cations. *Journal of Membrane Biology*, **90**, 177–190.

Menestrina, G. and Vécsey-Semjén, B. (1999) Biophysical methods and model membranes for the study of bacterial pore-forming toxins. In *The Comprehensive Sourcebook of Bacterial Protein Toxins*, edited by J. E. Alouf and J. H. Freer, pp. 287–309. London, San Diego, Boston, New York, Sydney, Tokyo, Toronto: Academic Press.

Menestrina, G., Dalla Serra, M. and Prévost, G. (2001) Mode of action of β-barrel pore-forming toxins of the staphylococcal α-toxin family. *Toxicon*, **39**, 1661–1672.

Menzies, B. E. and Kernodle, D. S. (1994) Site-directed mutagenesis of the alpha-toxin gene of *Staphylococcus aureus*: role of histidines in toxin activity in vitro and in a murine model. *Infection and Immunity*, **62**, 1843–1847.

Meunier, O., Falkenrodt, A., Monteil, H. and Colin, D. A. (1995) Application of flow cytometry in toxinology: pathophysiology of human polymorphonuclear leucocytes damaged by a pore-forming toxin from *Staphylococcus aureus*. *Cytometry*, **21**, 241–247.

Meunier, O., Ferreras, M., Supersac, G., Hoeper, F., Baba Moussa, L., Monteil, H., Colin, D. A., Menestrina, G. and Prévost, G. (1997) A predicted β-sheet from class S components of staphylococcal γ-haemolysin is essential for the secondary interaction of the class F component. *Biochimica et Biophysica Acta*, **1326**, 275–289.

Meunier, O. (2000) Specificity and mode of action of staphylococcal leucotoxins. PhD Thesis, Institut de Bacteriologie, Université Louis Pasteur, Strasbourg, France.

Miles, G., Cheley, S., Braha, O. and Bayley, H. (2001) The staphylococcal leukocidin bicomponent toxin forms large ionic channels. *Biochemistry*, **40**, 8514–8522.

Morrissey, J. A., Cockayne, A., Hill, P. J. and Williams, P. (2000) Molecular cloning and analysis of a putative ABC transporter from *Staphylococcus aureus*. *Infection and Immunity*, **68**, 6281–6288.

Narita, S., Kaneko, J., Chiba, J. I., Piémont, Y., Jarraud, S., Etienne, J. and Kamio, Y. (2001) Phage conversion of Panton–Valentine leucocidin in *Staphylococcus aureus*: molecular analysis of a PVL-converting phage, ϕSLT. *Gene*, **268**, 195–206.

Nariya, H., Izaki, K. and Kamio, Y. (1993) The C-terminal region of the S component of staphylococcal leukocidin is essential for the biological activity of the toxin. *FEBS Letters*, **229**, 219–222.

Nariya, H. and Kamio, Y. (1995) Identification of the essential region for LukS- and HgII-specific functions of staphylococcal leukocidin and γ-hemolysin. *Bioscience Biotechnology and Biochemistry*, **59**, 1603–1604.

Nariya, H., Shimatani, A., Tomita, T. and Kamio, Y. (1997) Identification of the essential amino acids residues in LukS for the hemolytic activity of staphylococcal leukocidin towards rabbit erythrocytes. *Bioscience Biotechnology and Biochemistry*, **61**, 2095–2099.

Noda, M., Kato, I., Matsuda, F. and Hyrayama, T. (1980) Fixation and inactivation of staphylococcal leukocidin by phosphatidylcholine and ganglioside GM1 in rabbit polymorphonuclear leucocytes. *Infection and Immunity*, **29**, 678–684.

Noda, M., Kato, I., Matsuda, F. and Hirayama, T. (1981) Mode of action of staphylococcal leucocidin: relationship between binding of ^{125}I-labeled S and F components of leucocidin to rabbit polymorphonuclear leukocytes and leucocidin activity. *Infection and Immunity*, **34**, 362–367.

Noda, M., Kato, I., Hyrayama, T. and Matsuda, F. (1982) Mode of action of staphylococcal leucocidin: effects of the S and F components on the activities of membrane-associated enzymes of rabbit polymorphonuclear leukocytes. *Infection and Immunity*, **35**, 38–45.

Olson, R., Nariya, H., Yokota, K., Kamio, Y. and Gouaux, E. (1999) Crystal structure of staphylococcal LukF delineates conformational changes accompanying formation of a transmembrane channel. *Nature Structural Biology*, **6**, 134–140.

Ozawa, T., Kaneko, J., Nariya, H., Izaki, K. and Kamio, Y. (1994) Inactivation of γ-haemolysin HγII component by addition of monosialoganglioside GM1 to human erythrocyte. *Bioscience Biotechnology Biochemistry*, **58**, 602–605.

Ozawa, T., Kaneko, J. and Kamio, Y. (1995) Essential binding of LukF of staphylococcal γ-hemolysin followed by the binding of HγII for the hemolysis of human erythrocytes. *Bioscience Biotechnology Biochemistry*, **59**, 1181–1183.

Panchal, R. G., Cusak, E., Cheley, S. and Bayley, H. (1996) Tumor protease-activated, pore-forming toxins from a combinatorial library. *Nature Biotechnology*, **14**, 852–856.

Panton, P. N. and Valentine, F. C. O. (1932) Staphylococcal toxin. *Lancet*, **222**, 506–508.

Paula, S., Akeson, M. and Deamer, D. (1999) Water transport by the bacterial channel α-haemolysin. *Biochimica Biophysica Acta*, **1418**, 117–126.

Pédelacq, J. D., Maveyraud, L., Prévost, G., Baba-Moussa, L., González, A., Courcelle, E., Shepard, W., Monteil, H., Samama, J. P. and Mourey, L. (1999) The structure of a *Staphylococcus aureus* leucocidin component (LukF-PV) reveals the fold of the water-soluble species of a family of transmembrane pore-forming toxins. *Structure*, **7**, 277–287.

Prévost, G. (1999) The bi-component staphylococcal leucotoxins and γ-haemolysins (toxins). In *The Comprehensive Sourcebook of Bacterial Protein Toxins*, edited by J. E. Alouf, and J. H. Freer, pp. 402–418. London, San Diego, Boston, New York, Sydney, Tokyo, Toronto: Academic Press.

Prévost, G. (2000) Staphylococcal bi-component toxins: clinical significance, structure and function. *Medical Microbiology and Immunology*, **189**, 45.

Prévost, G., Supersac, G., Colin, D. A., Couppié, P., Sire, S., Hensler, T., Petiau, P., Meunier, O., Cribier, B., König, W. and Piémont, Y. (1994) The new family of leucotoxins from *S. aureus*: structural and biological properties. In *Bacterial Protein Toxins*, edited by J. Freer *et al.* Zentralblatt für Bakteriologie, Supp. 24, pp. 284–293. Stuttgart, Jena, New York: A Gustav Fischer Verlag.

Prévost, G., Bouakham, T., Piémont, Y. and Monteil, H. (1995a) Characterization of a synergohymenotropic toxin from *Staphylococcus intermedius*. *FEBS Letters*, **376**, 135–140.

Prévost, G., Couppié, P., Prévost, P., Gayet, S., Petiau, P., Cribier, B., Monteil, H. and Piémont, Y. (1995b) Epidemiological data on *Staphylococcus aureus* strains producing synergohymenotropic toxins. *Journal of Medical Microbiology*, **42**, 237–245.

Prévost, G., Cribier, B., Couppié, P., Petiau, P., Supersac, G., Finck-Barbançon, V., Monteil, H. and Piémont, Y. (1995c) Panton–Valentine leucocidin and gamma-haemolysin from *Staphylococcus aureus* ATCC 49775 are encoded by distinct genetic loci and have different biological activities. *Infection and Immunology*, **63**, 4121–4129.

Prévost, G., Colin, D. A., Mourey, L. and Menestrina, G. (2001) Staphylococcal pore-forming toxins. In *Pore-forming Toxins*, edited by F. G. van der Goot. *Current Topics in Microbiology and Immunology*, **257**, 53–83.

Rees, B. and Bilwes, A. (1993) Three-dimensional structures of neurotoxins and cardiotoxins. *Chemical Research in Toxicology*, **6**, 385–406.

Seeger, W., Birkenmeyer, R. G., Ermert, N., Suttorp, N., Bhakdi, S. and Dunker, H. R. (1990) Staphylococcal alpha-toxin-induced vascular leakage in isolated perfused rabbit lungs. *Laboratory Investigations*, **63**, 341–349.

Siqueira, J. A., Speeg-Schatz, C., Freitas, F. I. S., Sahel, J., Monteil, H. and Prévost, G. (1997) Channel-forming leucotoxins from *Staphylococcus aureus* cause severe inflammatory reactions in a rabbit eye model. *Journal of Medical Microbiology*, **46**, 486–494.

Smith, M. L. and Price, S. A. (1938) *Staphylococcus* γ-haemolysin. *Journal of Pathological Bacteriology*, **47**, 379–393.

Song, L., Hobaugh, M. R., Shustak, C., Cheley, S., Bayley, H. and Gouaux, J. E. (1996) Structure of staphylococcal alpha-haemolysin, a heptameric transmembrane pore. *Science*, **274**, 1859–1866.

Song, L. and Gouaux, E. (1998) Crystallization of the alpha-haemolysin heptamer solubilized in decyldimethyl- and decyldiethylphosphine oxide. *Acta Crystallographica D*, **54**, 276–278.

Staali, L., Monteil, H. and Colin, D. A. (1998) The pore-forming leukotoxins from *Staphylococcus aureus* open Ca^{2+} channels in human polymorphonuclear neutrophils. *Journal of Membrane Biology*, **162**, 209–216.

Staali, L., Monteil, H. and Colin, D. A. (2000) Staphylococcal bi-component leukotoxins induce the opening of Ca^{2+}-activated K^+ and Cl^- channels and form pores specific for monovalent cations (K^+, Na^+). In *Proceedings of the 9th European Workshop on Bacterial Protein Toxins*, edited by C. Locht *et al. International Journal of Medical Microbiology*, **290**(Supp. 30), A21. A Gustav Fischer Verlag. Stuttgart, Iena, New York.

Sugawara, N., Tomita, T. and Kamio, Y. (1997) Assembly of *Staphylococcus aureus* gamma-haemolysin into a pore-forming ring-shaped complex on the surface of human erythrocytes. *FEBS Letters*, **410**, 333–337.

Sugawara, N., Tomita, T., Sato, T. and Kamio, Y. (1999) Assembly of *Staphylococcus aureus* leukocidin into a pore-forming ring-shaped oligomer on human polymorphonuclear leukocytes and rabbit erythrocytes. *Bioscience Biotechnology and Biochemistry*, **63**, 884–891.

Supersac, G., Prévost, G. and Piémont, Y. (1993) Sequencing of leucocidin R from *Staphylococcus aureus* P83 suggests that staphylococcal leucocidins and gamma-haemolysin are members of a single, two-component family of toxins. *Infection and Immunity*, **61**, 580–587.

Supersac, G., Piémont, Y., Kubina, M., Prévost, G. and Foster, T. J. (1998) Assessment of the role of gamma-toxin in experimental endophthalmitis using a Δ*hlg* deficient mutant of *Staphylococcus aureus*. *Microbial Pathogenesis*, **24**, 241–251.

Szmigielski, S., Jeljaszewicz, J., Wiszinski, J. and Korbecki, M. (1966) Reaction of rabbit leucocytes to staphylococcal (Panton-Valentine) leukocidin in vivo. *Journal of Pathology and Bacteriology*, **84**, 599–604.

Szmigielski, S., Jeljaszewicz, J. and Zak, C. (1968) Leucocyte system of rabbits receiving repeated doses of staphylococcal leukocidin. *Pathology Microbiology*, **31**, 328–336.

Szmigielski, S., Sobiczewska, E., Prévost, G., Monteil, H., Colin, D. A. and Jeljaszewicz, J. (1998) Effects of purified staphylococcal leukocidal toxins on isolated blood polymorphonuclear leukocytes and peritoneal macrophages in vitro. *Zentralblatt für Bakteriologie*, **288**, 383–394.

Taylor, A. G. and Bernheimer, A. W. (1974) Further characterization of staphylococcal gamma-hemolysin. *Infection and Immunity*, **10**, 54–59.

Thibodeau, A., Yao, X. and Forte, J. G. (1994) Acid secretion in α-toxin-permeabilized gastric glands. *Biochemistry and Cell Biology*, **72**, 26–35.

Valeva, A., Palmer, M. and Bhakdi, S. (1997) Staphylococcal alpha-toxin: formation of the heptameric pore is partially cooperative and proceeds through multiple intermediate stages. *Biochemistry*, **36**, 13298–13304.

Valeva, A., Walev, I., Gerber, A., Klein, J., Palmer, M. and Bhakdi, S. (2000) Staphylococcal alpha-toxin: repair of a calcium-impermeable pore in the target cell membrane. *Molecular Microbiology*, **36**, 467–476.

Van der Velde, H. (1894) Etude sur le mécanisme de la virulence du staphylocoque pyogène. *La Cellule*, **10**, 401–460.

Van der Vijver, J. C. M., van Es-Boon, M. and Michel, M. F. (1972) Lysogenic conversion in *Staphylococcus aureus* to leucocidin production. *Journal of Virology*, **10**, 318–319.

Vécsey-Semjén, B., Lesieur, C., Möllby, R. and van der Goot, F. G. (1997) Conformational changes due to membrane binding and channel formation by staphylococcal α-toxin. *Journal of Biological Chemistry*, **272**, 5709–5717.

Walev, I., Martin, E., Jonas, D., Mohamadzadeh, M., Müller-Klieser, W., Kunz, L. and Bhakdi, S. (1993) Staphylococcal alpha-toxin kills human keratinocytes by permeabilizing the plasma membrane for monovalent ions. *Infection and Immunity*, **61**, 4972–4979.

Walev, I., Palmer, M., Martin, M., Jonas, D., Weller, U., Höhn-Bentz, H., Husmann, M. and Bhakdi S. (1994) Recovery of human fibroblasts from attack by the pore-forming α-toxin of *Staphylococcus aureus*. *Microbial Pathogenesis*, **17**, 187–201.

Walker, B. and Bayley, H. (1995) Key residues for membrane binding, oligomerization, and pore forming activity of staphylococcal alpha-haemolysin identified by cysteine scanning mutagenesis and targeted chemical modification. *Journal of Biological Chemistry*, **270**, 23065–23071.

Wang, X., Noda, M. and Kato, I. (1990) Stimulatory effect of staphylococcal leucocidin on phosphoinositide metabolism in rabbit polymorphonuclear leucocytes. *Infection and Immunity*, **58**, 2745–2749.

Ward, P. D. and Turner, W. H. (1980) Identification of staphylococcal Panton–Valentine leukocidin as a potent dermonecrotic toxin. *Infection and Immunity*, **28**, 393–397.

Werner, S. (2001) Structural and functional aspects of the interactions of leucotoxins from *Staphylococcus aureus* with human leucocytes. PhD Thesis, Institut of Bacteriology, Louis Pasteur University, Strasbourg, France.

Werner, S., Colin, D. A., Coraiola, M., Menestrina, G., Monteil, H. and Prévost, G. (2002) Retrieving biological activity from LukF-PV mutants combined with different S-components implies compatibility between stem domains of these staphylococcal bi-component leucotoxins. *Infection and Immunity*, **70**, 1310–1318.

Woodin, A. M. (1960) Purification of the two components of leukocidin from *Staphylococcus aureus*. *Biochemical Journal*, **75**, 158–165.

Woodin, A. M. (1972) The staphylococcal leukocidin. In *The Staphylococci*, edited by J. O. Cohen, pp. 281–289. New York, London, Sydney, Toronto: Wiley-Interscience.

Yokota, K. and Kamio, Y. (2000) Tyrosine72 residue at the bottom of rim domain in LukF crucial for the sequential binding of the staphylococcal gamma-hemolysin to human erythrocytes. *Bioscience Biotechnology and Biochemistry*, **64**, 2744–2747.

2 The formation of ion-permeable channels by RTX-toxins in lipid bilayer membranes

Basis for their biological activity

Roland Benz

Introduction

The RTX-toxins (Repeats in ToXin) represent a class of bacterial toxins that possess hemolytic and/or cytolytic activity on eukaryotic target cells and share common structural features. Alpha-hemolysin (HlyA) of *Escherichia coli* comprising 1,024 amino acids is the prototype of this family of cytolysins from Gram-negative bacteria (see Ludwig and Goebel (1999) for a recent review). Typical for them is a domain in the C-terminal half of the toxin protein that contains a variable number repeats of a glycine-rich nonapeptide of the form (L/I/F)–X–G–G–X–G–(N/D)–D–X, where X is an arbitrary amino acid (Ludwig and Goebel, 1999). The repeat domain binds calcium and is essential for biological activity via receptor-recognition on the mammalian target cell (Baumann *et al.*, 1993; Ostolaza *et al.*, 1995). The toxins are amphipathic proteins that contain different domains responsible for export and biological activity (see Figure 2.1). Essential for hemolytic and cytotoxic activity of the RTX-toxins is the amphipathic domain near the N-terminal end that may form α-helical structures (Ludwig *et al.*, 1987, 1991). For some of the RTX-toxins it is known that they are posttranslationally modified by cytoplasmic proteins (HlyC and CyaC) that act as acyltransferases together with an acyl carrier protein (Issartel *et al.*, 1991; Stanley *et al.*, 1994). The acylation of at least one (K690) out of two lysines near the repeat domain (K564 and K690) is essential for biological activity as cytolysin and/or hemolysin (Ludwig *et al.*, 1996).

The export signal out of the bacterial cell is localized within the last 60 amino acids from the C-terminus (Gray *et al.*, 1986). The export from the cytoplasm to the surface of the cells is mediated via a type 1 secretion system that contains different components (see Figure 2.2). This system is formed of an inner membrane ABC-transporter protein (HlyB in the case of *E. coli*), another protein that contains two of the inner membrane spanning domains and links the inner membrane to the outer membrane export device (HlyD) and the outer membrane protein TolC (Wandersman and Delepelaire, 1990). TolC represents a unique structure because its trimers contain two different parts: an outer membrane part that forms a 4 nm long β-stranded cylinder with 12 β-strands and an α-helical cage with 12 α-helices that has a length of about 10 nm and spans presumably the whole periplasmic space (Koronakis *et al.*, 2000). Most RTX-toxins of *Enterobacteriaceae* are exported through TolC homologous proteins that are not directly linked to their type 1 secretion systems. The operons encoding for other toxins such as cyclolysin (ACT, CyaA) of *Bordetella pertussis* and the metalloprotease PrtA of *Erwinia chrysanthemi* contain also the genes for TolC homologous outer membrane proteins (*cyaE* and *prtF*) essential for export of the toxins out of the cells (Glaser *et al.*, 1988; Delepelaire and Wandersman, 1991) (see Figure 2.2).

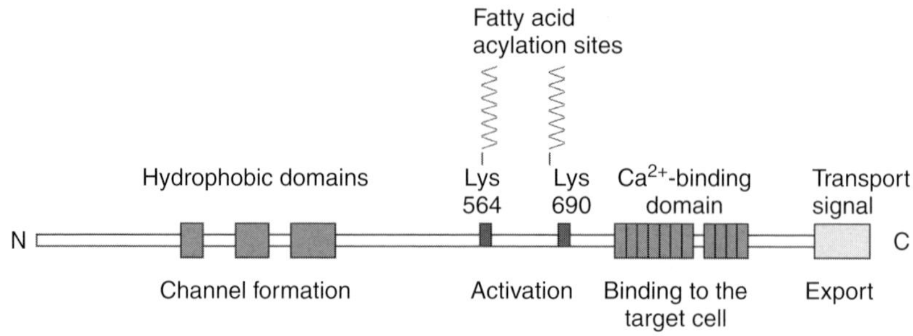

Figure 2.1 Structure of the mature form of HlyA of *E. coli*. The different domains essential for channel-forming activity, export and activation are indicated.

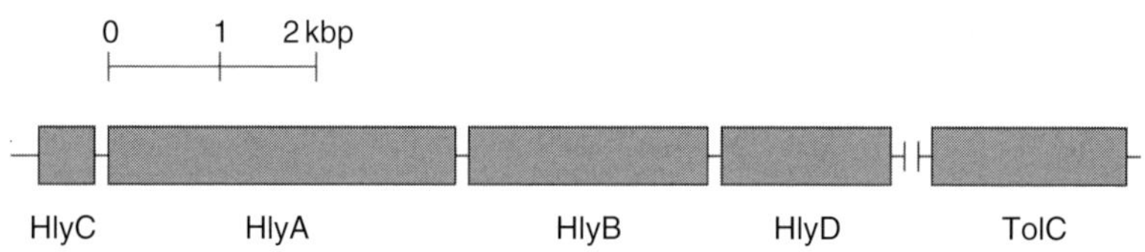

Figure 2.2 Map of the organization of the different genes needed for biosynthesis, activation and export of HlyA of *E. coli* (plasmid or chromosomal). Open boxes present the different genes. The scale is given in kilobasepairs (kbp). Note that the gene for TolC is not part of the HlyA operon but this protein is essential for HlyA export out of the cell. A variety of hemolysins and/or cytolysins from *E. coli* (EHEC), *Pasteurella hemolytica*, *Proteus mirabilis/vulgaris*, *Morganella morganii* and *Actinobacillus* spp. show the same genetic organization. The operon needed for biosynthesis, activation and export of adenylate cyclase toxin (CyaA) shows a similar organization, but *cyaE*, the gene encoding for the outer membrane channel (the TolC analog) is part of the operon.

Extensive similarity has been demonstrated between the four different *hly* genes of *E. coli* and those of the *Proteacae* (Koronakis *et al.*, 1987; Welch, 1987), and also to those of the hemolysins and leukotoxins of *Actinobacillus* (Chang *et al.*, 1989; Frey *et al.*, 1991; Lally *et al.*, 1989) and *Pasteurella hemolytica* (Lo *et al.*, 1987) as well as the cyclolysin of *B. pertussis* (Glaser *et al.*, 1988). HlyA-proteins from different organisms exhibit immunological cross-reaction (Devenish *et al.*, 1989). Several in vitro assays have shown that *E. coli* hemolysin is able to generate pores in mammalian target cell membranes of both normal tissue and the immune system, for example, human mast cells, and leukocytes such as granulocytes, causing cell lysis, or at sub-lytic levels the stimulation of leukotriene release and a disruption in cellular activity, which may be accompanied with oscillation of cellular calcium concentration (Bhakdi *et al.*, 1989, 1990; König *et al.*, 1994; Uhlen *et al.*, 2000). Recently, similar activities have been demonstrated with the hemolysin of *Morganella morganii*, *Proteus vulgaris*, *Actinobacillus pleuropneumoniae* and *Pasteurella hemolytica* (Koronakis *et al.*, 1987; Lally *et al.*, 1989; Lalonde *et al.*, 1989; Eberspächer *et al.*, 1990; Reimer *et al.*, 1995; Stevens and Czuprynski, 1995). Cyclolysin (CyaA, ACT) from *B. pertussis* is special because it contains besides a weak hemolytic activity, which is essential for its transport into the target cell, an enzymatic part also. Upon activation by the intracellular calmodulin (CaM), the invading enzyme catalyzes uncontrolled formation of cAMP and as a result, the microbicidal capacities of the intoxicated cells are debilitated (see the review by Hanski and Coote, 1991). Some of the toxins are a common cause of opportunistic, particularly nosocomial infections in humans (Stamm *et al.*, 1977; Senior, 1983). Other activities of RTX-toxins are combined with the lysis of red blood cells. The lysis of these cells by *E. coli* HlyA is characterized by three phases, an initial lag, an accelerated rate of hemolysis followed by a plateau, and then a decline (Rennie *et al.*, 1974). An initial adsorption step possibly due to the repeat domain (Welch *et al.*, 1992) may determine the lag step of erythrocyte lysis. Studies with erythrocytes have demonstrated that a pore of between 1 and 3 nm in diameter is formed either by hemolysin monomers or oligomers (Bhakdi *et al.*, 1986; Schmidt *et al.*, 1996). Interestingly, the channel formed by CyaA in red blood cells has a much smaller diameter as judged from similar experiments (Ehrmann *et al.*, 1991).

RTX-toxins form in lipid bilayer membranes and eukaryotic target cell membranes ion-permeable channels that are responsible for the biological activity of the toxins. In the following section, some basic features of the channel formation by RTX-toxins in lipid bilayer membranes are reviewed. These channels have a limited lifetime and are cation-selective caused by negative point charges in or near the channel. Lipid bilayer membranes composed of different lipids are rather inactive targets for channel formation but channels are frequently formed in lipid bilayer membranes formed of asolectin that represents a lipid mixture from soybeans. The analysis of the single-channel data suggests that the RTX-toxins form aqueous channels with different diameters. The channel-forming properties of the different RTX-toxins are discussed with respect to their biological activity.

Experimental procedures

Concentration of the RTX-toxins from the culture-supernatants

The channel-forming properties of the RTX-toxins can be studied in lipid bilayer membranes using either supernatants of bacterial cultures producing RTX-toxins or concentrated and purified toxin. The concentration was performed as follows (Döbereiner *et al.*, 1996): the sterile cell-free supernatant of bacteria producing RTX-toxins is precipitated with 22 g solid PEG 4,000 per 100 ml culture supernatant or by 40% (mass/vol.) ammonium sulfate under stirring for 45 min at 4°C. The toxins are harvested by centrifugation and resuspended in 1/100th of the original culture volume (25 mM Hepes-KOH pH 8, 5 mM EGTA, 1 mM dithiothreitol or 10 mM Tris–HCl, pH 7) and the protein solutions are frozen to −20°C. In frozen form the RTX-toxins remain active for at least eight weeks. Inactive samples can be activated again by treatment with 8 M urea. In other experiments, the supernatants of the cultures are used directly for the study of channel-forming activity of the RTX-toxins. The precipitation by PEG 4,000 represents another method to concentrate the RTX-toxins by a factor of at least 100. The loss of the channel-forming activity of the RTX-toxins is very small using these procedures (10–20%) as compared to that of the culture supernatants.

For the use in the lipid bilayer assay the adenylate cyclase toxin of *B. pertussis* (ACT or CyaA) is produced in the presence of the activating protein CyaC, using the *E. coli* strain XL1-Blue (Stratagene) transformed with the appropriate plasmid construct, derived from pCACT3. Exponential 500 ml cultures are induced with IPTG (1 mM) to produce CyaA. The proteins are extracted with 8 M urea in 50 mM Tris–HCl pH 8.0, 0.2 mM $CaCl_2$ (buffer A) from insoluble cell debris after sonication and purified by single-step affinity chromatography on calmodulin-agarose (Sigma) as described previously (Sebo *et al.*, 1991). The purified proteins are eluted in 8 M urea, 50 mM Tris–HCl pH 8.0, 2 mM EGTA (buffer B) and stored at −20°C. Using this procedure CyaA is essentially free of contaminant protein.

Formation of lipid bilayer membranes

The methods used for black lipid bilayer experiments in the presence of RTX-toxins have been described previously (Benz *et al.*, 1978, 1989). Membranes are formed from a 1% solution of different lipids in *n*-decane. The following lipids are used for membrane formation: diphytanoyl phosphatidylcholine, phosphatidylethanolamine, phosphatidylserine (Avanti Polar Lipids, Alabaster AL) and asolectin (soybean lecithin type IV-S from Sigma, St Louis MO). Membranes are formed from a 1% (mass/vol.) solution of these lipids in *n*-decane in a Teflon cell consisting of two aqueous compartments connected by a circular hole with a surface area of about 0.4 mm^2. The RTX-toxins are added in a concentration of about 10 ng to 1 μg/ml to one or both sides of the membrane. The aqueous salt solutions (analytical grade, Merck, Darmstadt, FRG) were in most cases used unbuffered and had a pH around 6.0 if not indicated otherwise. The temperature was kept at 20°C throughout. The membrane current was measured with a pair of calomel electrodes switched in series with a voltage source and an electrometer (Keithley 602). For single-channel recordings the electrometer was replaced by a current amplifier (Keithley 427). Zero-current membrane potential measurements were performed with a Keithley 617 electrometer by establishing a salt gradient across membranes containing 100–1,000 hemolysin channels (Benz *et al.*, 1979).

Results

Effect of RTX-toxins on the conductance of lipid bilayer membranes

The effect of RTX-toxins was studied on membranes formed from a variety of different pure lipids such as phosphatidylcholine, phosphatidylserine and asolectin. Membranes formed of the pure lipids were rather inactive targets for the study of channel formation by the RTX-toxins. Highest conductance increase was observed with asolectin membranes that contain approximately 40% phosphatidylcholine but also many other lipids from soybean. The difference in the conductance increase in asolectin membranes as compared with membranes made of pure lipids was at least 100-fold at an RTX-toxin concentration of 100 ng/ml. The difference between the effect of hemolysin on membranes made of pure lipids and asolectin may be explained by assuming that a component exists in the asolectin membranes that facilitates either toxin binding or channel formation.

The conductance increase observed with different lipids was a function of time after the addition of the hemolysin to membranes in the black state or of time after the blackening of the membranes if the toxin was present in the aqueous phase prior to membrane formation. The increase of the membrane current was steep for about 10–20 min. Only a small further increase (as compared with the initial one) occurred after that time. The increase of the membrane conductance was observed irrespective of whether the toxins were added to one or both sides of the membrane. Interestingly, purified toxin or supernatants of the cell cultures had approximately the same effect on the membrane conductance when the toxin concentration in the aqueous phase was the same. The effect on lipid bilayer membranes of

RTX-toxins produced by different *E. coli* strains including the uropathogenic EHEC-hemolysin of *E. coli* O157 : H7 and the *Proteacae* were very similar. Experiments with the different RTX-toxins of *A. pleuropneumoniae* demonstrated that ApxI had the highest effect on the membrane conductance followed by ApxIII, and that ApxII had the smallest effect. Cyclolysin from *B. pertussis* was special because very high membrane activity was combined with the formation of small channels (see next paragraph).

Small differences in the RTX-toxin concentration resulted in considerable variations of the specific membrane conductance. An increase of the hemolysin concentration by a factor of two resulted in an about four- to eightfold higher membrane conductance. This means that the relationship between hemolysin-induced conductance increase in membranes and toxin concentration was not linear and could be fitted to a straight line with a slope of approximately 2–3 on a double-logarithmic scale (see Figure 2.3 for the effect of HlyA of *P. vulgaris* on asolectin membranes). The steep conductance–concentration relationship may be explained by the assumption of an association–dissociation reaction between non-conducting hemolysin monomers and conducting oligomers. The increase of the HlyA concentration by a factor of two leads in such a case to a conductance increase by 2^2–2^3, that is, by a factor of 4–8. It is noteworthy that similar steep conductance–concentration curves have been found for ApxI of *A. pleuropneumoniae* (Maier *et al.*, 1996), for HlyA of *E. coli* (Benz *et al.*, 1989), for HlyA of *M. morganii* (Benz *et al.*, 1994a) and for CyaA of *B. pertussis* (Szabo *et al.*, 1994; Osickova *et al.*, 1999). This means probably that several RTX-toxin molecules could be involved in the formation of the conductive pathway in the membranes.

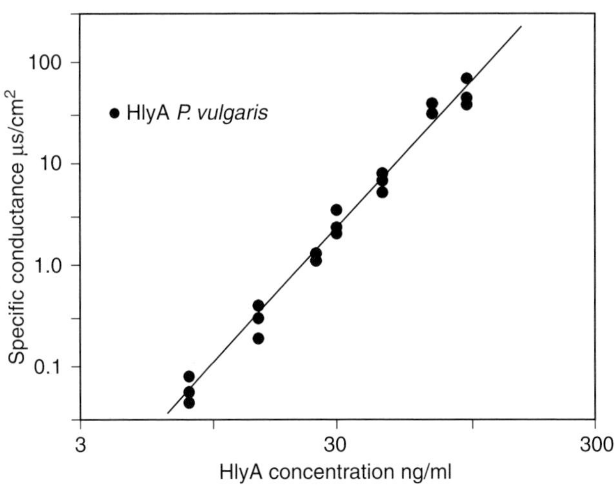

Figure 2.3 The steep conductance–concentration dependence of HlyA of *P. vulgaris* indicates that an association–dissociation reaction is responsible for the formation of conductive oligomers from non-conductive monomers. The aqueous phase contained 150 mM KCl (pH 6) and the protein at the given concentrations. The applied membrane potential was 20 mV; $T = 20$°C and the membranes were formed from asolectin dissolved in *n*-decane. The membrane conductance was taken 30 min after the addition of the protein, when further conductance increase in time was already minimal (Benz *et al.*, 1994a). The conductance–concentration relationship was plotted on a double-logarithmic scale. The straight line corresponds to a slope of about three on the double logarithmic scale.

RTX-toxins form ion-permeable channels in lipid bilayer membranes

Single-channel experiments revealed that the membrane activity described in the previous section was caused by the formation of ion-permeable channels by the toxins. Figure 2.4 shows a single-channel recording in the presence of HlyA of *E. coli.* The RTX-toxin was added to a black asolectin/*n*-decane membrane in a concentration of about 100 ng/ml. After a delay of some minutes, probably caused by slow aqueous diffusion and/or rearrangement of the toxin, we observed, first, small transient channels and then larger channels. The large conductance fluctuations always seemed to start from small steps. They also had a limited lifetime (mean lifetime about 2 s; see next section) and usually decayed back to the small state as Figure 2.4 clearly indicates. A histogram of the conductance fluctuations observed with HlyA of *E. coli* is shown in Figure 2.5. HlyA formed channels that had on average a single-channel conductance of about 520 pS and the initial small prestate one of 80 pS (in 150 mM KCl and at 20 mV membrane potential). The contribution of the prestates to the histogram are not included in Figure 2.5. Interestingly, the prestate had a much longer lifetime in salt solutions of higher ionic strength.

Lifetime of the RTX-toxin channels

The lifetime of the RTX-toxin channels is limited as Figure 2.4 clearly indicates for HlyA of *E. coli*. It was studied in experiments, in which only one or two channels formed by HlyA were present in lipid bilayer membranes that the opening and closing processes of the single channels could be detected at a membrane voltage of 20 mV and 150 mM KCl. The lifetime of the HlyA channels was obtained by averaging over all fluctuations observed in these experiments. For this a large number of current fluctuations were recorded from asolectin/*n*-decane membranes and the duration of the fluctuations was measured. The number of channels with a lifetime within a given time interval (e.g. 0–0.5, 0.51–1.0, 1.1–1.51 s, etc.) was plotted as a function of the mean lifetime within the interval, for example, 0.25, 0.75, 1.25 s, etc. The bar graph could be fitted to the equation $N = N_0 \exp(-t/\tau)$ with $\tau = 2.86$ s (Benz *et al.*, 1989). The mean lifetime was $2.86 \times \ln 2 = 2$ s. The mean lifetime of the HlyA channels of *M. morganii* and *P. vulgaris* were 27 and 32 s,

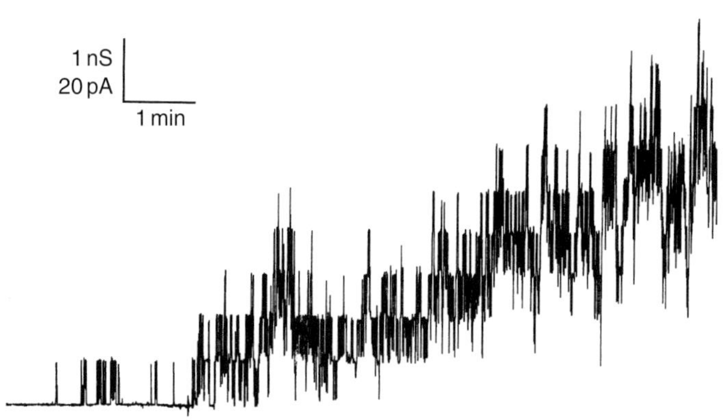

Figure 2.4 Single-channel recording of asolectin/*n*-decane membranes in the presence of HlyA of *E. coli*. The aqueous phase contained 150 mM KCl (pH 6) and 100 ng/ml HlyA. The applied membrane potential was 20 mV; $T = 20$°C.

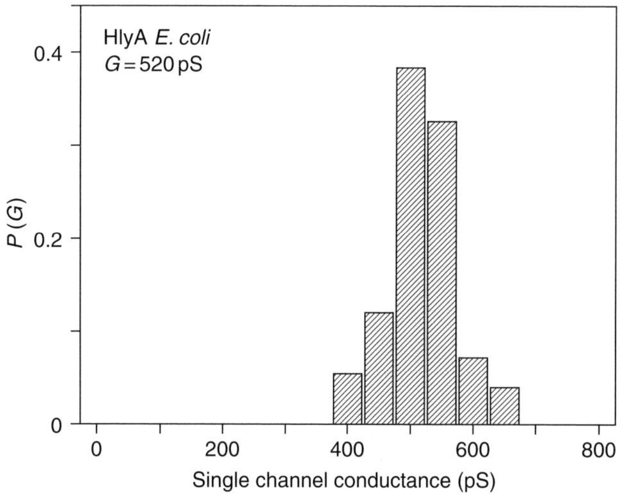

Figure 2.5 Histogram of the probability *P*(G) for the occurrence of a given conductivity unit observed with membranes formed of asolectin/*n*-decane in the presence of 100 ng/ml HlyA from *E. coli*. *P*(G) is the probability that a given conductance increment G is observed in the single-channel experiments. It was calculated by dividing the number of fluctuations with a given conductance increment by the total number of conductance fluctuations. The aqueous phase contained 150 mM KCl. The applied membrane potential was 20 mV; $T = 20°C$. The average single-channel conductance was 520 pS for 487 single-channel events.

respectively, under the same conditions, which means that they had a much longer lifetime than the HlyA channels of *E. coli* (Benz *et al.*, 1994a). The estimation of the mean lifetime also allowed the derivation of the dissociation rate constant for the dissociation of the channel-forming complex according to $k_D = 1/\tau = 0.35$ 1/s. It is noteworthy that the dissociation rate constants for the channel-forming HlyA complexes of *M. morganii* and *P. vulgaris* were with 0.026 and 0.022 1/s much lower (Benz *et al.*, 1994a). A similar low dissociation rate constant has also been found for EHEC-hemolysin that has in vitro a higher hemolytic activity than the plasmid-encoded HlyA showed here in a single-channel recording (Schmidt *et al.*, 1996). It is noteworthy that HlyA of *E. coli* also forms transient channels in the membranes of human macrophages (Menestrina *et al.*, 1996).

Single-channel conductances of the channels formed by RTX-toxins

RTX-channels are permeable for a variety of ions indicating that they are wide and water-filled. An example for the single-channel conductance of the channels formed by the different Apx channels of *A. pleuropneumoniae* in salts composed of different anions and cations is given in Table 2.1 (Maier *et al.*, 1996). All three Apx-toxins have molecular masses between 105 and 120 kDa. They show the typical features of RTX-toxins including the strongly hydrophobic domain and the repeat domain. ApxI is strongly hemolytic and cytotoxic and contains 13 glycine-rich nonapeptide repeats (Frey and Nicolet, 1990; Kamp *et al.*, 1991). It shows high amino acid similarity with the *E. coli* hemolysin HlyA (Frey *et al.*, 1991). ApxII is an RTX-toxin having typical RTX domains but only eight glycine-rich

Table 2.1 Average single-channel conductance, G, of the Apx-channels of A. *pleuropneumoniae* (Maier *et al.*, 1996) and the CyaA-channel (Benz *et al.*, 1994b) in different salt solutions[a]

Salt	*Concentration/M*	*G/nS*			
		ApxI	*ApxII*	*ApxIII*	*CyaA*
LiCl	0.15	0.18	0.29	0.037	0.015
NaCl	0.15	0.36	0.42	0.068	0.018
KCl	0.01	0.13	0.11	0.029	
	0.05	0.35	0.36	0.068	
	0.1				0.0048
	0.15	0.54	0.62	0.095	
	0.3				0.011
	0.5	0.85	1.3	0.17	
	1.0	1.5	1.5	0.30	0.027
	3.0	4.2	4.2	0.62	0.048
KCH_3COO(pH 7)	0.15	0.34	0.64	0.087	0.029
RbCl	0.15	0.64	0.42	0.105	0.025
CsCl	0.15	0.58	0.38	0.090	
NH_4Cl	0.15	0.49	0.58	0.089	
$N(CH_3)_4Cl$	0.15	0.24	0.21	0.020	
$N(C_2H_5)_4Cl$	0.15	0.17	0.19	0.014	
TrisCl	0.15	0.093	0.10	0.013	n.d.

Notes

a The membranes were formed of asolectin dissolved in *n*-decane. The aqueous solutions were unbuffered and had a pH of 6 unless otherwise indicated. The applied voltage was 50 mV, and the temperature was 20°C. The average single-channel conductance, G, was calculated from at least 80 single events. The table contains only the single-channel data of the open state of the four different toxins.

n.d. means not detectable.

nonapeptide repeats. It is weakly hemolytic and weakly cytotoxic and shows high similarity to the *P. haemolytica* leukotoxin LktA (Chang *et al.*, 1989; Kamp *et al.*, 1991; Frey *et al.*, 1992). ApxIII has an RTX structure with 13 glycine-rich repeats and shows significant amino acid similarity to ApxI and HlyA (Chang *et al.*, 1993; Jansen *et al.*, 1994). It is, however, non-hemolytic but strongly cytotoxic (Kamp *et al.*, 1991; Rycroft *et al.*, 1991; Cullen and Rycroft, 1994).

The influence of the cations on the single-channel conductances in different 150 mM salt solutions suggested that the Apx-channels are cation-selective. The mobility sequence for the different cation within the Apx-channels was $Cs^+Rb^+K^+ > Na^+ > Li^+ > Tris^+$ and $NH_4^+ > N(CH_3)_4^+ > N(C_2H_5)_4^+$, which means that their permeability through the channels followed approximately the mobility sequence of these ions in the aqueous phase. Despite the close homology of the primary sequence of all Apx-toxins to one another and to other RTX-toxins (Frey *et al.*, 1991; Kamp *et al.*, 1991; Frey *et al.*, 1992; Chang *et al.*, 1993; Jansen *et al.*, 1994), their single-channel conductance differed considerably. ApxI and ApxII had approximately the same single-channel conductance whereas ApxIII formed considerably smaller channels under the same conditions. ApxI and ApxII had single-channel conductances of 540 and 620 pS, respectively. The conductance of ApxIII was five times smaller and was only about 100 pS in 150 mM KCl. This means that the channels had a different permeability for ions, which may also be the basis of their different biological activity. ApxI is highly hemolytic and cytotoxic, whereas ApxII is only weakly hemolytic and ApxIII is only cytotoxic (Frey and Nicolet, 1990; Kamp *et al.*, 1991). Table 2.1 shows also the average single-channel conductance, G, of the Apx-channels as a function of the KCl concentration in the aqueous phase. Similarly, as in the case of other RTX-toxins of *E. coli* (Benz *et al.*, 1989; Ropele and Menestrina, 1989), *P. vulgaris* and *M. morganii* (Benz *et al.*, 1994a) the relationship between conductance and KCl-concentration was not linear. Instead, the slope of the conductance vs concentration curves on a double logarithmic scale was approximately 0.5, which indicated the influence of point net charges localized in or near the channels (see also section on "Discussion" and Figure 2.7).

Selectivity of the channels formed by the RTX-toxins

Zero-current membrane potential measurements allow the calculation of the permeability ratio P_{cation} divided by P_{anion} in multichannel experiments. This type of experiments was also performed for different RTX-toxins. One example is shown in Table 2.2 for the EHEC-hemolysin of the uropathogenic *E. coli* O157 : H7. For all salts (KCl, LiCl and KCH_3COO) tested in experiments with the toxin, the more diluted side of the membrane became positive, which indicated preferential movement of cations through the channels (Schmidt *et al.*, 1996). The zero-current membrane potentials for the salts mentioned above were between 40 mV (LiCl) and 45 mV (Kacetate). Analysis of the zero-current membrane potentials of Table 2.2 using the Goldman–Hodgkin–Katz equation suggested that anions could also have a small permeability through the HlyA-channels of *E. coli* O157 : H7 because the ratios of the permeabilities were between 10 and 15. On the other hand it is also possible that the channels are highly selective for cations but that the selectivity coefficient is influenced by the presence of point net charges in or near the channel (see section "Discussion"). It is noteworthy that the channels formed by other RTX-toxins of Gram-negative bacteria, such as the Apx of *A. pleuropneumoniae* (Maier *et al.*, 1996), HlyA of *E. coli* and the *Proteaceae* (Benz *et al.*, 1989, 1994a) and CyaA of *B. pertussis* (Benz *et al.*, 1994b) are also highly selective for cations over anions. This is consistent with the small single-channel conductance of the Apx-channels and other RTX-toxins in Tris–HCl (see Table 2.1).

Table 2.2 Zero-current membrane potentials, V_m, of asolectin/*n*-decane membranes in the presence of EHEC-hemolysin of *E. coli* O157 measured for a 10-fold gradient of different salts (Schmidt *et al.*, 1996)

	V_m (mV)	P_{cation}/P_{anion}
KCl	42	12
LiCl	40	10
Kacetate (pH 7)	45	15

Note
V_m is defined as the difference between the potential at the dilute side (10 mM) and the potential at the concentrated side (100 mM). The pH of the aqueous salt solutions was 6 unless otherwise indicated; $T = 20°C$. The permeability ratio P_{cation}/P_{anion} was calculated with the Goldman– Hodgkin–Katz equation (Benz *et al.*, 1979) on the basis of at least three individual experiments.

Discussion

RTX-toxins form channels in membranes

The results of the experiments with lipid bilayer membranes and biological membranes define the RTX-toxins from a variety of Gram-negative bacteria including *E. coli* (Bhakdi *et al.*, 1986; Menestrina *et al.*, 1987; Benz *et al.*, 1989; Ropele and Menestrina, 1989), *P. vulgaris* and *M. morganii* (Eberspächer *et al.*, 1990; Benz *et al.*, 1994a), *P. haemolytica* (Forestier and Welch, 1990; Menestrina *et al.*, 1994), *A. pleuropneumoniae* (Frey, 1994; Tascon *et al.*, 1994; Jansen *et al.*, 1995; Maier *et al.*, 1996), *Actinobacillus actinomycetemcomitans* (Lear *et al.*, 1995; Karakelian *et al.*, 1998) and adenylate cyclase toxin of *B. pertussis* (Benz *et al.*, 1994b; Szabo *et al.*, 1994) as channel-forming components. Most of these toxins act by their insertion into the cytoplasmic membrane and subsequently by causing permeability changes. The binding of the toxins to receptors in the plasma membrane is a prerequisite for the insertion of the RTX-toxins into the biological membranes. Some of the receptors are known, such as glycophorin that is essential for binding of HlyA of *E. coli* to erythrocytes or other cells (Cortajarana *et al.*, 2001), or $M^{\beta}2$ integrin for the binding of CyaA to target cells (Guermonprez *et al.*, 2001). Calcium bound to the repeat domain of the RTX-toxins seems to be essential for the receptor-mediated binding (Ludwig *et al.*, 1988; Ostolaza *et al.*, 1995). It is noteworthy, however, that the repeat domain of HlyA of *E. coli* is not important for channel formation in general because deletions of the repeat domain have demonstrated that this RTX-toxin loses its biological activity but not the ability to form channels in lipid bilayer membranes (Ludwig *et al.*, 1988; Döbereiner *et al.*, 1996). Important for channel-forming activity in biological and artificial membranes is also the HlyC-promoted modification of lysines (564 and 690) near the repeat domain of HlyA of *E. coli* by fatty acid acylation (Issartel *et al.*, 1991) and similarly lysine 983 by palmitoylation in the case of CyaA of *B. pertussis* (Hackett *et al.*, 1994). On the other hand, this modification, similarly to the repeats, has nothing to do with channel formation because channels may form without acylation, albeit with a much lower probability (Ludwig *et al.*, 1996). Similarly CyaA can also form channels when the acylation sites are deleted but needs a much higher toxin concentration (Benz *et al.*, 1994b).

Upon channel formation, the osmotic balance across the cytoplasmic membrane of the target cells is disturbed; they swell and lyse if pumps responsible for the active transport of ions cannot balance the permeability increase. The in vivo action of RTX-toxins ultimately cause physical damage to a variety of eukaryotic cells. Dependent on a calcium-dependent recognition of the target cells the toxins are either hemolytic or cytotoxic or both (Felmlee and Welch, 1988; Forestier and Welch, 1991; McWinney *et al.*, 1992). Besides the important cytotoxic activities, sublytic concentrations of RTX-toxins are believed to modulate the immune response after infection with a given pathogen (Bhakdi *et al.*, 1989, 1990; Welch, 1991; König *et al.*, 1994) or they cause calcium oscillation in target cells (Uhlen *et al.*, 2000). In some cases, RTX-toxins were shown to be the major pathogenicity factors of their producers (Tascon *et al.*, 1994). From the primary sequences of RTX-toxins no exact data were available about sequence motifs responsible for target cell specificity other than the repeat motifs, however, biophysical characteristics (i.e. lipid specificity, charge dependence) may contribute to this phenomenon. In addition, different cellular effects caused by the various RTX-toxins may be dependent on those characteristics such as size of the channel and pH dependence (Bhakdi *et al.*, 1994).

The RTX-channels are formed by toxin oligomers

The RTX-channels studied so far in reconstitution experiments with lipid bilayer membranes and in patch-clamp measurements have a limited lifetime as Figure 2.4 clearly demonstrates. This means that the channel-forming units are not rigid structures similar to α-toxin of *Staphylococcus aureus* (Gouaux, 1998), to aerolysin of different *Aeromonas* spp. (Chakraborty *et al.*, 1990; Wilmsen *et al.*, 1990) and to Gram-negative bacterial porins (Benz, 1994). Instead, they undergo molecular changes which result in rapid current fluctuations in artificial and biological membranes (Menestrina *et al.*, 1996). The mean lifetime of the HlyA channels of *E. coli* was about 2 s under the experimental conditions of Figure 2.4 (20 mV). Other RTX-channels have longer lifetimes (Benz *et al.*, 1994a; Schmidt *et al.*, 1996). Assuming that a complex of several HlyA molecules is responsible for the formation of the conductive unit, the dissociation rate constant of the oligomeric channel, k_D, is about 0.35 1/s. The assumption of a pore-forming oligomer responsible for the action of the RTX-toxins seems also to be justified by the steep conductance–concentration curve observed for a variety of toxins, which suggests that several HlyA molecules form the conductive unit and that an association–dissociation equilibrium exists between non-conductive monomers and conductive oligomers. Attempts to show pores in the electron microscopical analysis of erythrocytes lysed by high doses of HlyA of *E. coli* (Bhakdi *et al.*, 1986) or in lipid vesicles (Soloaga *et al.*, 1999) were not successful, which also argues in favor of unstable channels formed by hemolysin oligomers. It has to be noted, however, that it has been suggested in other investigations that hemolysin of *E. coli* acts on the basis of a "single hit mechanism" in biological and artificial membranes, that is, the channel is formed by a monomer (Menestrina, 1988; Ropele and Menestrina, 1989). Similarly, it has been suggested that CyaA (ACT) of *B. pertussis* is taken up as a monomer into the target cell to exhibit the enzymatic activity there (Osickova *et al.*, 1999). These possibilities cannot be ruled out completely on the basis of the lipid bilayer data. It is noteworthy, however, that hemolytically inactive HlyA-mutants become hemolytic when they are expressed together within one *E. coli* cell (Ludwig *et al.*, 1993). Such a complementation is only possible if several HlyA-molecules form one channel. So far it is not clear if the channels form in the aqueous phase or at the membrane water interface. The structure of the channels is also completely open. It is clear, however, that hydrophobic stretches of HlyA that are localized within amino acids 200–410 from the N-terminal end could form α-helices. These hydrophobic structures are essential for both membrane activity of RTX-toxins and translocation of ACT into the target cell.

Size of the channels formed by RTX-toxins

There exist several different methods to study the size of the channels formed by RTX-toxins in artificial and target cell membranes. The size of the hemolysin channels can be studied in osmotic protection experiments performed in the following way. Human erythrocyte suspensions (2%) were prepared in 90 mM NaCl – 45 mM KCl – 12.5 mM phosphate buffer (pH 7.2) supplemented with 30 mM osmolyte to counterbalance the colloid osmotic pressure inside the cells (Bhakdi *et al.*, 1986). When the osmolyte is permeable through the RTX-channels it is not able to protect the cells from osmotic lysis. Using this method the diameter of the EHEC-hemolysin channel ranges within the diameter of raffinose (1.2–1.4 nm) to that of dextran 4 (2–3 nm) (Schmidt *et al.*, 1996). Similar results have been derived from osmotic protection experiments with HlyA from *E. coli* (Bhakdi *et al.*, 1986). These result suggest that the channels are wide and water-filled. It is noteworthy that CyaA of *B. pertussis* appears to have a much smaller channel diameter in similar experiments (Ehrmann *et al.*, 1991).

The single-channel conductance of reconstituted channels cannot be used to evaluate its diameter because the field strength inside the channel may influence its conductance. However, single-channel data allow the estimation of the diameter when the channel is wide and water-filled and conducts only one sort of ions, either anions or cations. The estimation is based on the same assumptions, which have been used previously for the derivation of the Renkin correction factor (Renkin, 1954). The Renkin correction factor has successfully been applied on the diffusion of neutral molecules through general diffusion porins (Nikaido and Rosenberg, 1981) and of ions through channels formed by cell wall proteins of *Mycobacterium chelonae* and *M. smegmatis* in lipid bilayer membranes (Trias and Benz, 1993, 1994). The single-channel conductance can be used to calculate the Renkin correction factor when the entry of a hydrated ion into effective area, A, of the channel mouth (total cross section A_0) is the rate-limiting step of ion permeation (and not the diffusion of the hydrated ion through the channel itself). In such a case, the permeability of a cylindrical channel (radius r) for solutes is proportional to the aqueous diffusion coefficient, D, multiplied with the Renkin correction factor given by:

$$A/A_0 = [1-(a/r)]^2 \cdot [1-2.104 \cdot (a/r)+2.09 \cdot (a/r)^3-0.95 \cdot (a/r)^5] \tag{1}$$

where A is the effective area of the channel mouth, A_0 is the total cross-sectional area of the channel and a is the radius of the hydrated ions or substrates passing through the channel. For the use of the Renkin correction factor to estimate the channel size, the radii a of the hydrated ions and their diffusion coefficients, D, in the aqueous phase must be known (Table 2.3). The radii, a, of the hydrated ions can be calculated from the limiting molar conductivities, λ_i, by using the Stokes equation:

$$a = Fez_i^2/(6\pi\eta\lambda_i) \tag{2}$$

Table 2.3 Radii of the hydrated cations and relative permeability of cations (relative to Rb^+) through the ApxI and ApxIII channels of *A. pleuropneumoniae*

Cation	*Limiting molar conductivity* λ_i [mS/M]	*Hydrated ion radius of cation* a [nm]	*Permeability of cations relative to* Rb^+	
			ApxI	*ApxIII*
Li^+	38.68	0.216	0.281	0.352
Na^+	50.10	0.163	0.563	0.648
K^+	73.50	0.110	0.844	0.905
NH_4^+	73.55	0.110	0.766	0.848
Rb^+	77.81	0.105	1.00	1.00
Cs^+	77.26	0.106	0.906	0.857
$(CH_3)_4N^+$	44.92	0.182	0.375	0.190
$(C_2H_5)_4N^+$	32.66	0.250	0.266	0.133
$Tris^+$	25.50	0.321	0.145	0.124

Note
The data for the limiting conductivities of the different ions were taken from Castellan (1982). The limiting conductivity for $Tris^+$ was calculated on the basis of the specific conductivity of 10 mM Tris–HCl in the aqueous phase. The single-channel conductances of ApxI and ApxIII were taken from Table 2.1. The hydrated ion radii were calculated using the Stokes equation (Eqn (2)) and the limiting aqueous conductivity.

F ($F = 96{,}500$ As/mol) is the Faraday constant, e ($e = 1.602 \cdot 10^{-19}$ A s) is the elementary charge, z_i is the valency of the ions and η ($\eta = 1.002 \cdot 10^{-3}$ kg/(m s)) is the viscosity of the aqueous phase. The diffusion coefficient, D, of the hydrated ions is also given implicitly by the limiting molar conductivity, λ_i, of the different ions (Castellan, 1983). It is not necessary to calculate it (although this is possible) since both the Renkin correction factor times the diffusion coefficient (the relative rate of permeation) and the single-channel conductance may be given relative to the value of the hydrated Rb^+-ion because it has the smallest diameter. This corresponds to the dependence of the relative rate of permeation on solute radius in the liposome swelling assay with solutes of different size (Nikaido and Rosenberg, 1981).

The validity of the method was previously assessed by comparing the size of the cell wall channel of the *M. chelonae* as estimated from the method described above (i.e. from the single-channel conductance) and from the vesicle-swelling assay (Trias and Benz, 1994). It is noteworthy that the radii calculated from the single-channel data and the liposome-swelling assay shows satisfactory agreement. Figure 2.6A shows the best fit of the single-channel conductance of the ApxI channel from *A. pleuropneumoniae* with the Renkin correction factor times the aqueous diffusion coefficient of the corresponding cation. The data are given relative to the data for Rb^+ (relative permeability equal to unity) and the best fit of the single-channel conductances was obtained with $r = 1.2$ nm, which means that the diameter of the channel is 2.4 nm. The data lie within the range from $r = 1.0$ nm to $r = 1.4$ nm as shown in Figure 2.6A. A diameter of 2.4 nm is very similar to the diameter of the HlyA-channel of *E. coli* as has been derived from osmotic protection experiments (Bhakdi *et al.*, 1986) and to that of EHEC-hemolysin (Schmidt *et al.*, 1996). A similar fit of the data of ApxIII (see Figure 2.6B) suggested that the diameter of this channel is somewhat smaller (about 1.8 nm) than that of ApxI although the fit of the data was less satisfactory as compared to Figure 2.6A. The reason for this is not completely clear but it may indicate that the large organic quaternary ammonium ions tend to interact with the channel interior, an interaction that is not included in Renkin's correction factor (Renkin, 1954). These results are consistent with a defined size for the channels formed by the RTX-toxins. On the other hand, it has been suggested that HlyA of *E. coli* forms membrane lesions at higher toxin concentrations and/or longer times (Moayeri and Welch, 1994). Similarly, it has been intimated that HlyA has a more detergent like action on membranes because it does not disturb the structure of the hydrocarbon core of a membrane (Soloaga *et al.*, 1999) but interacts mostly with the surface of the membranes. This is easy to understand because only a small part of the 1,024 amino acids long HlyA monomer is involved in the membrane channel and spans the membrane while the rest is localized on the membrane surface.

Effects of point net charges on the channels formed by RTX-toxins

The single-channel conductance of the Apx-toxins and of other RTX-toxins did not saturate at high salt concentration, and the conductance was not a linear function of the bulk aqueous concentration (see Table 2.1). Instead a dependence of the single-channel conductance on the square root of the salt concentration in the aqueous phase was obtained. This means (i) that the cation specificity of the Apx-channels is not related to the presence of a binding site and (ii) that point charges are involved in ion selectivity as was demonstrated previously for a variety of RTX-toxins (Benz *et al.*, 1989, 1994a,b; Ropele and Menestrina, 1989) and for mycobacterial porins (Trias and Benz, 1993, 1994).

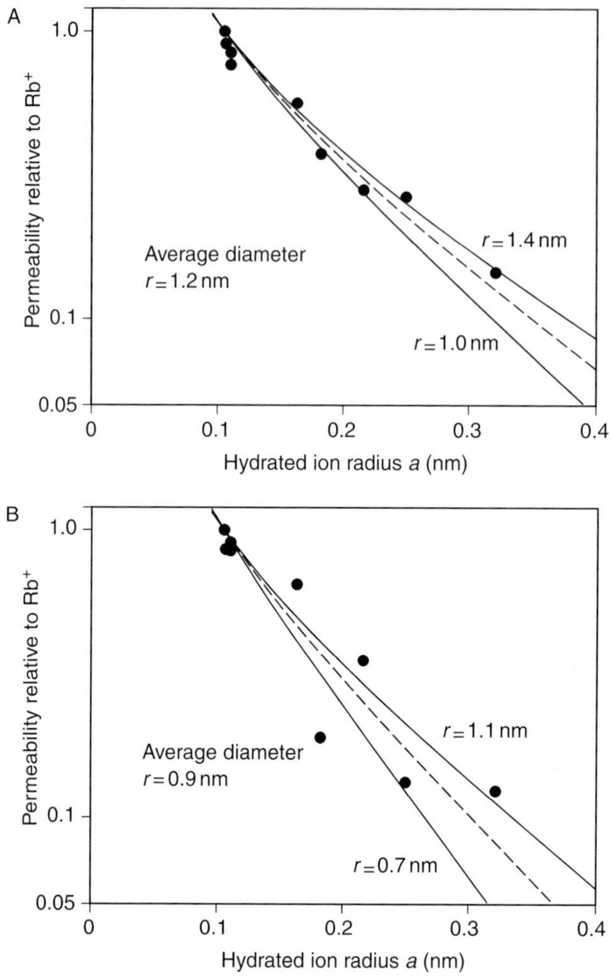

Figure 2.6 Fit of the single-channel conductance data of the ApxI- and ApxIII-channels by using the Renkin correction factor times the aqueous diffusion coefficients of the different cations. The values were normalized to 1 for $a = 0.105$ nm (radius of the hydrated Rb^+-ion). Single-channel conductances were normalized to the ones of Rb^+ and plotted vs the hydrated ion radii taken from Table 2.1. The single-channel conductances correspond to Li^+, Na^+, K^+, NH_4^+, Cs^+, $N(CH_3)_4^+$, $N(C_2H_5)_4^+$ and $Tris^+$, which were all used for the pore diameter estimation. (A) The fits (solid lines) are shown for ApxI-channels with $r = 1.4$ nm (upper line) and $r = 1.0$ nm (lower line). The best fit of all data was achieved with $r = 1.2$ nm (average diameter = 2.4 nm), which corresponds to the broken line. (B) The fits (solid lines) are shown for ApxIII-channels with $r = 1.1$ nm (upper line) and $r = 0.7$ nm (lower line). The best fit was achieved with $r = 0.9$ nm (average diameter = 1.8 nm), which corresponds to the broken line.

The influence of discrete charges on ion distribution on the membrane surface was considered in theory and practice (Nelson and McQuarrie, 1975; Menestrina and Antolini, 1981; Jordan, 1987). When they are localized in or near a channel they result in substantial ionic strength-dependent surface potentials at the channel opening, which attract cations and repel anions. Accordingly, they influence both single-channel conductance and zero-current membrane potential. Interestingly, the single-channel conductance is much larger than expected from the dimensions of the channel (Menestrina and Antolini, 1981). A quantitative description of the effect of the point charges on the single-channel conductance may be given on the basis of the the Debeye–Hückel Theorie or on the formalism proposed by Nelson and McQuarrie (1975). The former describes the effect of point charge in an aqueous environment and the latter on the surface of a membrane. A charge, q, in an aqueous environment creates a potential ϕ, which is dependent on the distance, r, from the point charge according to:

$$\phi = \frac{q \cdot e^{r/l_D}}{4\pi \cdot \varepsilon_0 \cdot \varepsilon \cdot r} \tag{3}$$

ε_0 ($=8.85 \times 10^{-12}$ F/m) and ε ($=80$) are the absolute dielectric constant of vacuum and the relative constant of water, respectively, and l_D is the so called Debeye length which controls the decay of the potential (and of the accumulated positively charged ions) in the aqueous phase:

$$l_D^2 = \frac{\varepsilon \cdot \varepsilon_0 \cdot R \cdot T}{2F^2 \cdot c} \tag{4}$$

c is the bulk aqueous salt concentration, and R, T and F ($RT/F = 25.2$ mV at 20°C) have the usual meaning. The potential ϕ created by a point charge on the surface of a membrane is larger caused by the generation of an image force on the opposite side of the low dielectric membrane (Nelson and McQuarrie, 1975). The potential, ϕ, at the mouth of a channel with a radius, r, and a total charge, q (in As), is given by :

$$\phi = \frac{2q \cdot e^{r/l_D}}{4\pi \cdot \varepsilon_0 \cdot \varepsilon \cdot r} \tag{5}$$

The concentration of the monovalent cation near the point charge increases because of the negative potential. Its concentration, c_0^+, at the channel mouth is in both cases given by:

$$c_0^+ = c \cdot e^{-\phi \cdot F/R.T} \tag{6}$$

The cation concentration c_0^+ at the mouth of the pore, can now be used for the calculation of the effective conductance concentration curve. A best fit of the data of Table 2.1 was obtained for the solid line of Figure 2.7 assuming that 2.3 negatively charged groups ($q = -3.7 \times 10^{-19}$ As) are located at the pore mouth and that its radius is approximately 1 nm. When the single-channel conductance is corrected for the presence of point charges an almost linear relationship is obtained between the calculated cation concentration at the channel mouth and the single-channel conductance (broken line Figure 2.7). The data also demonstrate that the influence of the surface charges is rather small at high ionic strength, c, that is, small l_D (see Eqn (4)). Similar fits could be performed of the single-channel conductances of the other Apx-channels and a variety of other RTX-toxin as functions of the salt concentration (data not shown). In these cases, the number of point charges was smaller

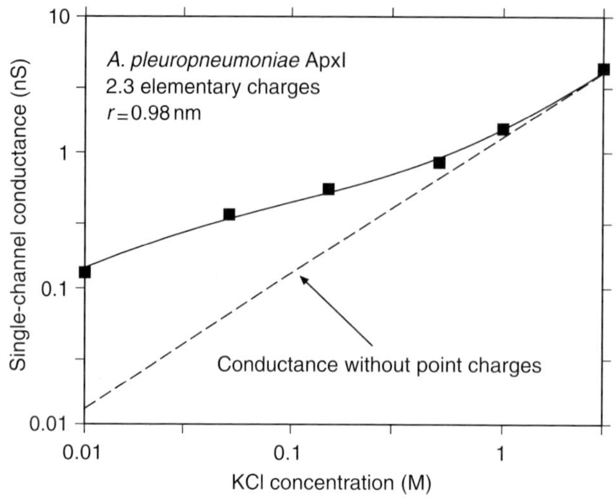

Figure 2.7 Single-channel conductance of ApxI of *A. pleuropneumoniae* as a function of the KCl-concentration in the aqueous phase (full squares). The solid line represents the fit of the single-channel conductance data of Table 2.1 with the Nelson and McQuarrie (1975) formalism assuming the presence of negative point charges (2.3 negative charges; $q = -3.7 \times 10^{-19}$ As) at the channel mouth on both sides of the membrane and assuming a channel diameter of about 2 nm. The broken (straight) line shows the single-channel conductance of the ApxI-channel in the absence of point charges and corresponds to a linear function between channel conductance and bulk aqueous concentration.

for ApxIII and was combined with a somewhat smaller diameter (1.7 nm) of the channel, whereas ApxI and ApxII had approximately the same charge density and similar diameters. This means that the fit of the single-channel data using charge distribution and Renkin equation showed reasonable agreement.

Comparison of channel properties formed by the RTX-toxins

Channel formation by RTX-toxins in membranes represents the basis of their biological activity. Table 2.4 shows the single-channel conductance of different RTX-toxins in artificial lipid bilayer membranes. Most of the channels have a conductance around 400–600 pS in 150 mM KCl. This means that many of them have approximately the same size. Exceptions are ApxIII of *A. pleuropneumoniae* and CyaA of *B. pertussis*, which have a smaller single-channel conductance and also a smaller size as judged from the use of the Renkin equation and also from from osmotic protection experiments. Common to all RTX-toxins is the cation selectivity based on the presence of negatively charged groups in or near the channel. Included in Table 2.4 is also the information about the membrane activity of these toxins (the rate of channel formation at a given toxin concentration).

Table 2.4 demonstrates also that the adenylate cyclase toxin of *B. pertussis* (CyaA) has the smallest conductance of all RTX-channels studied to date. Interestingly, the small single-channel conductance is combined with a high membrane activity, which results in contrast to ApxIII also in a small hemolytic activity. In general, it can be concluded that essential for

Table 2.4 Comparison of the channel properties of different RTX-toxins in biological and artificial membranes

Organism	*RTX-toxin*	*G (pS) in 150 mM KCl*	*Selectivity Pc/Pa in KCl*	*Channel diameter* (nm)	*Membrane activity*
E. coli[a]	HlyA	520	14	2.6	+++
E. coli O157[b]	EHEC	550	12	2.6	++++
M. morganii[c]	HlyA	520	10	2.0	+++
P. vulgaris[c]	HlyA	500	9.5	2.0	+++
A. pleuropneumoniae[d]	ApxI	540	5.7	2.4	++++
	ApxII	620	—	2.4	++
	ApxIII	95	9.6	1.8	+++++
B. pertussis[e]	CyaA	9.0	10	—	++++++

Notes

The data were taken from the following publications:

a Benz *et al.*, 1989.

b Schmidt *et al.*, 1996.

c Benz *et al.*, 1994a.

d Maier *et al.*, 1996.

e Benz *et al.*, 1994b.

It is noteworthy that the channel diameters estimated from the single-channel conductances in different salt solutions or in different KCl-solution show good agreement to those derived from the osmotic protection experiments with the same toxins.

hemolytic activity is not only the binding to erthrocytes but also the induced permeability change of the target cell membrane. For *E. coli* hemolysin it has been shown that sub-lytic amounts are strongly stimulating the neutrophil oxidative burst, degranulation and eicosanoid production, and interleukin-1 release from human monocytes, while larger amounts of this RTX toxin were deleterious to the cells (Bhakdi *et al.*, 1989, 1990; Jonas *et al.*, 1993; König *et al.*, 1994). It is possible to speculate that the pore-forming activities of RTX-toxins also play a role in the stimulation of monocytes, neutrophiles and other cells of the immune system. For this stimulation it is obviously sufficient to create a small permeability change of the target cell membrane, for instance the influx of calcium, which could result in calcium oscillations (Uhlen *et al.*, 2000). The activity and the biophysical character of the channels formed by the various RTX-toxins are hence speculated to have an influence on the specificity and potency of the toxin thus indirectly providing to the RTX-toxins a strong impact in bacterial virulence.

Acknowledgment

The financial support of the authors' own research by the Deutsche Forschungsgemeinschaft (Project A 5 of the Sonderforschungsbereich 487) and the Fonds der Chemischen Industrie is gratefully acknowledged.

References

Baumann, U., Wu, S., Flaherty, K. M. and McKay, D. B. (1993) Three-dimensional structure of the alkaline protease of *Pseudomonas aeruginosa*: a two domain protein with a calcium binding parallel beta roll motif. *EMBO J.*, **12**, 3357–3364.

Benz, R. (1994) Solute uptake through bacterial outer membranes. In *Bacterial Cell Wall*, edited by R. Hackenbek and J.-M. Ghuysen, pp. 397–423. Amsterdam: Elsevier.

Benz, R., Janko, K. and Läuger, P. (1979) Ionic selectivity of pores formed by the matrix protein (porin) of *Escherichia coli*. *Biochim Biophys Acta*, **551**, 238–247.

Benz, R., Janko, K., Boos, W. and Läuger, P. (1978) Formation of large, ion-permeable membrane channels by the matrix protein (porin) of *Escherichia coli*. *Biochim Biophys Acta*, **511**, 305–319.

Benz, R., Schmid, A., Wagner, W. and Goebel, W. (1989) Pore formation by the *Escherichia coli* hemolysin: evidence for an association–dissociation equilibrium of the pore-forming aggregates. Infect. *Infect. Immun.*, **57**, 887–895.

Benz, R., Hardie, K. R. and Hughes, C. (1994a) Pore-formation in artificial membranes by the secreted hemolysins of *Proteus vulgaris* and *Morganella morganii*. *Eur. J. Biochem.*, **220**, 339–347.

Benz, R., Maier, E., Ullmann, A. and Sebo, P. (1994b) Adenylate cyclase toxin (CyaA) of *Bordetella pertussis*: evidence for the formation of small ion-permeable channels and comparison with HlyA of *Escherichia coli*. *J. Biol. Chem.*, **269**, 27231–27239.

Bhakdi, S., Mackman, N., Nicaud, J.-M. and Holland, I. B. (1986) *Escherichia coli* hemolysin may damage target cell membranes by generating transmembrane pores. *Infect. Immun.*, **52**, 63–69.

Bhakdi, S., Greulich, S., Muhly, M., Eberspächer, B., Becker, H., Thiele, A. and Hugo, F. (1989) Potent leukocidal action of *Escherichia coli* hemolysin mediated by permeabilization of target cell membranes. *J. Exp. Med.*, **169**, 737–754.

Bhakdi, S., Muhly, M., Korom, S. and Schmidt, G. (1990) Effects of *Escherichia coli* haemolysin on human monocytes. Cytocydal action and stimulation of interleukin 1 release. *J. Clin. Invest.*, **85**, 1746–1753.

Bhakdi, S., Grimminger, F., Suttorp, N., Walmrath, D. and Seeger, W. (1994) Proteinaceous bacterial toxins and pathogenesis of sepsis syndrome and septic shock: the unknown connection. *Med. Microbiol. Immunol.*, **183**, 119–144.

Castellan, G. W. (1983) The ionic current in aqueous solutions. In *Physical Chemistry*, pp. 769–780. Reading MA: Addison Wesley.

Chakraborty, T., Schmid, A., Notermans, S. and Benz, R. (1990) Aerolysin of *Aeromonas sobria*: evidence for the formation of ion-permeable channels and comparison with alpha-toxin of *Staphylococcus aureus*. *Inf. Immun.*, **58**, 2127–2132.

Chang, Y. F., Young, R. and Struck, D. K. (1989) Cloning and characterization of a hemolysin gene from *Actinobacillus* (*Haemophilus*) *pleuropneumoniae*. *DNA*, **8**, 635–647.

Chang, Y. F., Shi, J., Ma, D., Shin, S. and Lein, D. H. (1993) Molecular analysis of the *Actinobacillus pleuropneumoniae* RTX toxin-III gene cluster. *DNA Cell. Biol.*, **12**, 351–362.

Cortajarena, A. L., Goni, F. M. and Ostolaza, H. (2001) Glycophorin as a receptor for *Escherichia coli* alpha-hemolysin in erythrocytes. *J. Biol. Chem.*, **276**, 12513–12519.

Cullen, J. M. and Rycroft, A. N. (1994) Phagocytosis by pig alveolar macrophages of *Actinobacillus pleuropneumoniae* serotype 2 mutant strains defective in haemolysin II (ApxII) and pleurotoxin (ApxIII). *Microbiol.*, **140**, 237–244.

Delepelaire, P. and Wandersman, C. (1991) Characterization, localization and transmembrane organization of the three proteins PrtD, PrtE and PrtF necessary for protease secretion by the Gram-negative bacterium *Erwinia chrysanthemi*. *Mol. Microbiol.*, **5**, 2427–2434.

Devenish, J., Rosendal, S., Johnson, R. and Hubler, S. (1989) Immunoserological comparison of 104-kilodalton proteins associated with hemolysis and cytolysis in *Actinobacillus pleuropneumoniae*, *Actinobacillus suis*, *Pasteurella haemolytica*, and *Escherichia coli*. *Infect. Immun.*, **57**, 3210–3213.

Döbereiner, A., Ludwig, A., Schmid, A., Goebel, W. and Benz, R. (1996) Effect of calcium and other di- and trivalent cations on channel formation by *Escherichia coli* alpha-hemolysin in red blood cells and lipid bilayer membranes. *Eur. J. Biochem.*, **240**, 454–460.

Eberspächer, L., Hugo, F., Pohl, M. and Bhakdi, S. (1990) Functional similarity between the haemolysins of *Escherichia coli* and *Morganella morganii*. *J. Med. Micro.*, **33**, 165–170.

Ehrmann, I. E., Gray, M. C., Gordon, V. M., Gray, L. S. and Hewlett, E. L. (1991) Hemolytic activity of adenylate cyclase toxin from *Bordetella pertussis*. *FEBS Lett.*, **278**, 79–83.

Felmlee, T. and Welch, R. A. (1988) Alterations of amino acid repeats in the *Escherichia coli* hemolysin affect cytolytic activity and secretion. *Proc. Natl. Acad. Sci. USA*, **85**, 5269–5273.

Forestier, C. and Welch, R. A. (1990) Nonreciprocal complementation of the hlyC and lktC genes of the *Escherichia coli* hemolysin and *Pasteurella haemolytica* leukotoxin determinants. *Infect. Immun.*, **58**, 828–832.

Forestier, C. and Welch, R. A. (1991) Identification of RTX toxin target cell specificity domains by use of hybrid genes. *Infect. Immun.*, **59**, 4212–4220.

Frey, J. (1994) RTX-toxins in *Actinobacillus pleuropneumoniae* and their potential role in virulence. In *Molecular Mechanisms of Bacterial Virulence*, edited by C. I. Kado and J. H. Crosa, Dordrecht, pp. 325–340. Boston, London: Kluwer Academic Publishers.

Frey, J. and Nicolet, J. (1990) Hemolysin patterns of *Actinobacillus pleuropneumoniae*. *J. Clin. Microbiol.*, **28**, 232–236.

Frey, J., Meier, R., Gygi, D. and Nicolet, J. (1991) Nucleotide sequence of the hemolysin I gene from *Actinobacillus pleuropneumoniae*. *Infect. Immun.*, **59**, 3026–3032.

Frey, J., van den Bosch, H., Segers, R. and Nicolet, J. (1992) Identification of a second hemolysin (HlyII) in *Actinobacillus pleuropneumoniae* serotype 1 and expression of the gene in *Escherichia coli*. *Infect. Immun.*, **60**, 1671–1676.

Glaser, P., Sakamoto, H., Bellalou, J., Ullmann, A. and Danchin, A. (1988) Secretion of cyclolysin, the calmodulin-sensitive adenylate cyclase-haemolysin bifunctional protein of *Bordetella pertussis*. *EMBO J.*, **7**, 3997–4004.

Gouaux, E. (1998) Alpha-hemolysin from *Staphylococcus aureus*: an archetype of beta-barrel, channel-forming toxins. *J. Struct. Biol.*, **121**, 110–122.

Gray, L., Mackman, N., Nicaud, J. M. and Holland, I. B. (1986) The carboxy-terminal region of haemolysin 2001 is required for secretion of the toxin from *Escherichia coli*. *Mol. Gen. Genet.*, **205**, 127–133.

Guermonprez, P., Khelef, N., Blouin, E., Rieu, P., Ricciardi-Castagnoli, P., Guiso, N., Ladant, D. and Leclerc, C. (2001) The adenylate cyclase toxin of *Bordetella pertussis* binds to target cells via the $\alpha_M\beta_2$ integrin (CD11b/CD18). *J. Exp. Med.*, **193**, 1035–1044.

Hackett, M., Guo, L., Shabanowitz, J., Hunt, D. F. and Hewlett, E. L. (1994) Internal lysine palmitoylation in adenylate cyclase toxin from *Bordetella pertussis*. *Science*, **266**, 433–435.

Hanski, E. and Coote, J. G. (1991) *Bordetella pertussis* adenylate cyclase toxin. In *Sourcebook of Bacterial Toxins*, edited by J. E. Alouf and J. H. Freer, first edition, pp. 349–366. London: Academic Press.

Issartel, J. P., Koronakis, V. and Hughes, C. (1991) Activation of *Escherichia coli* prohaemolysin to the mature toxin by acyl carrier protein-dependent fatty acylation. *Nature*, **351**, 759–761.

Jansen, R., Briaire, J., Vangeel, A. B. M., Kamp, E. M., Gielkens, A. L. J. and Smits, M. A. (1994) Genetic map of the *Actinobacillus pleuropneumoniae* RTX-toxin (Apx) operons: characterization of the ApxIII operons. *Infect. Immun.*, **62**, 4411–4418.

Jansen, R., Briaire, J., Smith, H. E., Dom, P., Haesebrouck, F., Kamp, E. M., Gielkens, A. L. J. and Smits, M. A. (1995) Knockout mutants of *Actinobacillus pleuropneumoniae* serotype 1 that are devoid of RTX toxins do not activate or kill porcine neutrophils. *Infect. Immun.*, **63**, 27–37.

Jonas, D., Schultheis, B., Klas, C., Krammer, P. H. and Bhakdi, S. (1993) Cytocidal effects of *Escherichia coli* hemolysin on human T lymphocytes. *Infect. Immun.*, **61**, 1715–1721.

Jordan, P. C. (1987) How pore mouth charge distributions alter the permeability of transmembrane ionic channels. *Biophys. J.*, **51**, 297–311.

Kamp, E. M., Popma, J. K., Anakotta, J. and Smits, M. A. (1991) Identification of hemolytic and cytotoxic proteins of *Actinobacillus pleuropneumoniae* by use of monoclonal antibodies. *Infect. Immun.*, **59**, 3079–3085.

Karakelian, D., Lear, J. D., Lally, E. T. and Tanaka, J. C. (1998) Characterization of *Actinobacillus actinomycetemcomitans* leukotoxin pore formation in HL60 cells. *Biochim. Biophys. Acta*, **1406**, 175–187.

König, B., Ludwig, A., Goebel, W. and König, W. (1994) Pore formation by the *Escherichia coli* alpha-hemolysin: role for mediator release from human inflammatory cells. *Infect. Immun.*, **62**, 4611–4617.

Koronakis, V., Cross, M., Senior, B., Koronakis, E. and Hughes, C. (1987) The secreted hemolysins of *Proteus mirabilis*, *Proteus vulgaris* and *Morganella morganii* are genetically related to each other and to the alpha-hemolysin of *Escherichia coli*. *J. Bacteriol.*, **169**, 1509–1515.

Koronakis, V., Sharff, A., Koronakis, E., Luisi, B. and Hughes, C. (2000) Crystal structure of the bacterial membrane protein TolC central to multidrug efflux and protein export. *Nature*, **405**, 914–919.

Lally, E. T., Golub, E. E., Kieba, I. R., Taichman, N. S., Rosenbloom, J., Rosenbloom, J. C., Gibson, C. W. and Demuth, D. R. (1989) Analysis of the *Actinobacillus actinomycetemcomitans* leukotoxin gene. Delineation of unique features and comparison to homologous toxins. *J. Biol. Chem.*, **264**, 15451–15456.

Lalonde, G., McDonald, T. V., Gardner, P. and O Hanley, P. D. (1989) Identification of a hemolysin from *Actinobacillus pleuropneumoniae* and characterization of its channel properties in planar phospholipid bilayers. *J. Biol. Chem.*, **264**, 13559–13564.

Lear, J. D., Furblur, U. G., Lally, E. T. and Tanaka, J. C. (1995) *Actinobacillus actinomycetemcomitans* leukotoxin forms large conductance,voltage-gated ion channels when incorporated into planar lipid bilayers. *Biochim. Biophys. Acta*, **1238**, 34–41.

Lo, R. Y., Strathdee, C. A. and Shewen, P. E. (1987) Nucleotide sequence of the leukotoxin of genes of *Pasteurella haemolytica* A1. *Infect. Immun.*, **55**, 1987–1996.

Ludwig, A. and Goebel, W. (1999) The family of the multigenic encoded RTX toxins. In *The Comprehensive Sourcebook of Bacterial Protein Toxins*, edited by J. E. Alouf and J. H. Freer, second edition, pp. 330–348. London: Academic Press.

Ludwig, A., Benz, R. and Goebel, W. (1993) Oligomerization of *Escherichia coli* haemolysin (HlyA) is involved in pore formation. *Mol. Gen. Genet.*, **241**, 89–96.

Ludwig, A., Vogel, M. and Goebel, W. (1987) Mutations affecting activity and transport of haemolysin in *Escherichia coli*. *Mol. Gen. Genet.*, **206**, 238–245.

Ludwig, A., Jarchau, T., Benz, R. and Goebel, W. (1988) The repeat domain of *E. coli* hemolysin is responsible for its Ca-dependent binding to erythrocytes. *Mol. Gen. Genet.*, **214**, 553–561.

Ludwig, A., Schmid, A., Benz, R. and Goebel, W. (1991) Mutations affecting pore formation by haemolysin from *Escherichia coli*. *Mol. Gen. Genet.*, **226**, 198–208.

Ludwig, A., Garcia, F., Bauer, S., Jarchau, T., Benz, R., Hoppe, J. and Goebel, W. (1996) Analysis of the in vivo activation of hemolysin (HlyA) from *Escherichia coli*. *J. Bacteriol.*, **178**, 5422–5430.

Maier, E., Reinhard, N., Benz, R. and Frey, J. (1996) Channel-forming activity and channel size of the RTX toxins ApxI, ApxII, and ApxIII of *Actinobacillus pleuropneumoniae*. *Infect. Immun.*, **64**, 4415–4423.

McWhinney, D. R., Chang, Y. F., Young, R. and Struck, D. K. (1992) Separable domains define target cell specificities of an RTX hemolysin from *Actinobacillus pleuropneumoniae*. *J. Bacteriol.*, **174**, 291–297.

Menestrina, G. (1988) *Escherichia coli* hemolysin permeabilizes small unilamellar vesicles loaded with calcein by a single-hit mechanism. *FEBS Letters*, **232**, 217–220.

Menestrina, G. and Antolini, R. (1981) Ion transport through hemocyanin channels in oxidized cholesterol membranes. *Biochim. Biophys. Acta*, **643**, 616–625.

Menestrina, G., Mackman, N., Holland, I. B. and Bhakdi, S. (1987) *Escherichia coli* haemolysin forms voltage-dependent ion channels in lipid membranes. *Biochim. Biophys. Acta*, **905**, 109–117.

Menestrina, G., Moser, C., Pellet, S. and Welch, R. (1994) Pore-formation by *Escherichia coli* hemolysin (HlyA) and other members of the RTX toxins family. *Toxicology*, **87**, 249–267.

Menestrina, G., Pederzolli, C., Dalla Sera, M., Bregante, M. and Gambale, F. (1996) Permeability increase induced by *Escherichia coli* haemolysin A in human macrophages is due to the formation of ionic pores: a patch-clamp characterization. *J. Membrane Biol.*, **149**, 113–121.

Moayeri, M. and Welch, R. A. (1994) Effects of temperature, time and toxin concentration on lesion formation by the *Escherichia coli* hemolysin. *Infect. Immun.*, **62**, 4124–4134.

Nelson, A. P. and McQuarrie, D. A. (1975) The effect of discrete charges on the electrical properties of a membrane. *J. Theor. Biol.*, **55**, 13–27.

Nikaido, H. and Rosenberg, E. Y. (1981) Effect of solute size on diffusion through the transmembrane pores of the outer membrane of *Escherichia coli*. *J. Gen. Physiol.*, **77**, 121–135.

Osickova, A., Osicka, R., Maier, E., Benz, R. and Sebo, P. (1999) An amphipathic alpha-helix including glutamates 509 and 516 is crucial for membrane translocation of adenylate cyclase toxin and modulates formation and cation selectivity of its membrane channels. *J. Biol. Chem.*, **274**, 37644–37650.

Ostolaza, H., Soloaga, A. and Goni, F. M. (1995) The binding of divalent cations to *Escherichia coli* alpha-haemolysin. *Eur. J. Biochem.*, **228**, 39–44.

Reimer, D., Frey, J., Jansen, R., Veit, H. P. and Inzana, T. (1995) Molecular investigation of the role of ApxI and ApxII in the virulence of *Actinobacillus pleuropneumoniae* serotype 5. *Microb. Pathog.*, **18**, 197–209.

Renkin, E. M. (1954) Filtration, diffusion, and molecular sieving through porous cellulose membranes. *J. Gen. Physiol.*, **38**, 225–243.

Rennie, R. P., Freer, J. H. and Arbuthnott, J. P. (1974) The kinetics of erythrocyte lysis by *Escherichia coli* haemolysin. *J. Med. Microbiol.*, **7**, 189–195.

Ropele, M. and Menestrina, G. (1989) Electrical properties and molecular architecture of the channel formed by *Escherichia coli* hemolysin in planar lipid membranes. *Biochim. Biophys. Acta*, **985**, 9–18.

Rycroft, A. N., Williams, D., Cullen, J. M. and Macdonald, J. (1991) The cytotoxin of *Actinobacillus pleuropneumoniae* (pleurotoxin) is distinct from the haemolysin and is associated with a 120 kDa polypeptide. *J. Gen. Microbiol.*, **137**, 561–568.

Schmidt, H., Maier, E., Karch, H. and Benz, R. (1996) Pore-forming properties of the plasmid-encoded hemolysin of enterohemorrhagic *Escherichia coli* O157 : H7. *Eur. J. Biochem.*, **241**, 594–601.

Sebo, P., Glaser, P., Sakamoto, H. and Ullmann, A. (1991) High-level synthesis of active adenylate cyclase toxin of *Bordetella pertussis* in a reconstructed *Escherichia coli* system. *Gene*, **104**, 19–24.

Senior, B. W. (1983) *Proteus morganii* is less frequently associated with urinary tract infections than *Proteus mirabilis* – an explanation. *J. Med. Microbiol.*, **16**, 317–322.

Soloaga, A., Veiga, M. P., Garcia-Segura, L. M., Ostolaza, H., Brasseur, R. and Goni, F. M. (1999) Insertion of *Escherichia coli* alpha-haemolysin in lipid bilayers as a non-transmembrane integral protein: prediction and experiment. *Mol. Microbiol.*, **31**, 1013–1024.

Stamm, W. E., Martin, S. and Bennett J. V. (1977) Epidemiology of nosocomial infections due to Gram-negative bacilli: aspects relevant to development and use of vaccine. *J. Infect. Dis.*, **136**, S151–S160.

Stanley, P., Packman, L. C., Koronakis, V. and Hughes, C. (1994) Fatty acylation of two internal lysine residues required for the toxic activity of *Escherichia coli* hemolysin. *Science*, **266**, 1992–1996.

Stevens, P. and Czuprynski, C. (1995) Dissociation of cytolysis and monokine release by bovine mononuclear phagocytes incubated with *Pasteurella haemolytica* partially purified leukotoxin and lipopolysaccharide. *Can. J. Vet. Res.*, **59**, 110–117.

Szabo, G., Gray, M. C. and Hewlett, E. L. (1994) Adenylate cyclase toxin from *Bordetella pertussis* produces ion conductance across artificial lipid bilayers in a calcium- and polarity-dependent manner. *J. Biol. Chem.*, **269**, 22496–22499.

Tascon, R. I., Vazquezboland, J. A., Gutierrezmartin, C. B., Rodriguezbarbosa, I. and Rodriguezferri, E. F. (1994) The RTX haemolysins ApxI and ApxII are major virulence factors of the swine pathogen *Actinobacillus pleuropneumoniae*: evidence from mutational analysis. *Mol. Microbiol.*, **14**, 207–216.

Trias, J. and Benz, R. (1993) Characterization of the channel formed by the mycobacterial porin of *Mycobacterium chelonae* in lipid bilayer membranes: demonstration of voltage-dependent regulation and the presence of negative point charges at the channel mouth. *J. Biol. Chem.*, **268**, 6234–6240.

Trias, J. and Benz, R. (1994) Permeability of the cell wall of *Mycobacterium smegmatis*. *Mol. Microbiol.*, **14**, 283–290.

Uhlen, P., Laestadius, A., Jahnukainen, T., Soderblom, T., Backhed, F., Celsi, G., Brismar, H., Normark, S., Aperia, A. and Richter-Dahlfors, A. (2000) Alpha-haemolysin of uropathogenic *E. coli* induces Ca^{2+} oscillations in renalepithelial cells. *Nature*, **405**, 694–697.

Wandersman, C. and Delepelaire, P. (1990) TolC, an *Escherichia coli* outer membrane protein required for hemolysin secretion. *Proc. Natl. Acad. Sci.*, **87**, 4776–4780.

Welch, R. A. (1987) Identification of two different hemolysin determinants in uropathogenic Proteus isolates. *Infect. Immun.*, **55**, 2183–2190.

Welch, R. A. (1991) Pore-forming cytolysins of gram-negative bacteria. *Mol. Microbiol.*, **5**, 521–528.

Welch, R. A., Forestier, C., Lobo, A., Pellett, S., Thomas, W. Jr and Rowe, G. (1992) The synthesis and function of the *Escherichia coli* hemolysin and related RTX exotoxins. *FEMS Microbiol. Immunol.*, **5**, 29–36.

Wilmsen, H. U., Pattus, F. and Buckley, J. T. (1990) Aerolysin, a hemolysin from *Aeromonas hydrophila* forms voltage-gated channels in planar lipid bilayers. *J. Membrane Biol.*, **115**, 71–81.

3 Properties of the monomeric and the pentameric *Pseudomonas* cytotoxin

Annette Sliwinski-Korell, Takako Nakada, Dietmar Linder, Monica Linder, Sabine Geis, Marita Langewische and Frieder Lutz

Cytotoxin of *Pseudomonas aeruginosa*, a 29 kDa protein rich in β-structure, oligomerizes into a pentamer of 145 kDa to form hydrophilic membrane pores on eucaryotic cells. The activated monomer and the pentamer are both resistant to further proteolysis with trypsin. Proteinase K and protease V8 cleave the monomer between E122–I123 or T124–S125 into an N-terminal 13.5 and a C-terminal 15.5 kDa fragment, respectively. The membrane associated pentamer was digested by several proteases to a fragment of about 130 kDa. Further digestion with papain and proteinase K led to a product, representing the core region, which gave a single band of about 65–70 kDa in SDS–PAGE, but was shown by amino acid sequence analysis to be composed of two distinct fragments: One starts with A237 and represents the C-terminal 30 amino acids, the other starts in the region of S56–Q61 and has by calculation a mass of 10 kDa. These results imply that different regions of the monomer and the pentamer are exposed to the protein surface and are therefore accessible to proteolysis. Also, the two fragments of the core region must still be connected in the pentamer by intra- and intermolecular interactions. Single site mutations of H156 to P or H254 to P caused extensive inhibition of oligomerization, indicating that those two histidines promote oligomerization.

The region E108–S132 of the long core fragment is proposed as the membrane-spanning part of the cytotoxin pentamer, because: (a) It is the only complete, predicted amphipathic β-strand-loop-β-strand structure of the detected fragment. (b) It shows structural homologies to the membrane-spanning part of aerolysin as well as α-hemolysin. (c) The single mutation of V111 to E reduced membrane binding and pore formation to 50%, while V107 to E had no effect. (d) A mutant with a deletion from amino acid V99–V111 containing additionally a single mutation of P116 to S resulted in a loss of structure, so that activation by trypsin resulted in a total digestion of the monomer.

Introduction

Several strains of *P. aeruginosa* produce the phage-encoded 31.7 kDa pro-cytotoxin, which is activated by bacterial autolysis of about 20 C-terminal amino acids. The mature, water-soluble, acidic 29 kDa cytotoxin is classified as a member of the pore-forming toxins that are rich in β-sheet structure (50%) and must oligomerize in order to insert into target cell membranes. The resulting oligomer of 145 kDa was identified as a pentamer. During pentamerization a reorganization of β-structural associations was detected by ATR–FTIR spectroscopy (Sliwinski-Korell *et al.*, 1999).

In this chapter we report on structural changes involved in the oligomerization process which were examined by limited digestion of the monomeric and the pentameric cytotoxin with several proteases and analyzed by SDS–PAGE, N-terminal amino acid sequencing and matrix assisted laser desorption/ionization–time of flight (MALDI–TOF) mass spectrometry (MS). Additionally, a deletion mutant and various single site mutations of valine and histidine were tested for membrane binding, oligomerization and pore-forming activity. Comparisons of cytotoxin with other pore-forming toxins support the model of the focussed regions.

Experimental procedures

Cytotoxin. Cytotoxin was purified from *P. aeruginosa* strain 158 according to the method described by Lutz (1979). The toxin was stored in PBS (137 mM NaCl, 2.7 mM KCl, 0.5 mM $MgCl_2$, 9.3 mM phosphate, pH 7.4) at −20°C. To form pentamers of cytotoxin, rat erythrocyte membranes (70 μg protein) in 15 mM NaCl/phosphate, pH 6.0, were incubated with 15 μg cytotoxin for 1 h at 37°C.

Protease treatment. Cytotoxin monomer (3 μg) was partially digested at 37°C by chymotrypsin (EC 3.4.21.1; Serva, 0.3 U), papain (EC 3.4.22.2; Serva, 0.018 U), pronase E (EC 3.4.24.31; Sigma, 0.008 U), protease V8 (EC 3.4.21.19; ICN, 0.4 U), proteinase K (EC 3.4.21.64; Boehringer–Ingelheim, 0.018 U), subtilisin (EC 3.4.21.62; Roche, 0.005 U) and trypsin (EC 3.4.21.4; Serva, 0.6 U). For digestion of the membrane-associated pentamer (1–2 μg) and the membrane proteins (60 μg) the 20-fold protease activities were used. Digestion was stopped after 1 h by a protease inhibitor cocktail, resulting in final concentrations of 1 mM NaN_3, 5 mM EDTA, 1 mM iodacetamide, 10 mM 4-(2-aminoethyl)-benzenesulfonyl fluoride–HCl (Pefa-Bloc SC), 200 μM phenylmethylsulfonyl fluoride, 260 μM TLCK, and 4 μM leupeptin. The monomer samples were immediately heated in SDS–PAGE buffer at 95°C for 5 min. The pentamer samples were first centrifuged for 10 min at 22,000 *g* and 4°C to collect the membranes with the bound pentameric toxin. The pellets were treated with 5 μl protease inhibitor cocktail for 15 min and incubated in SDS–PAGE sample buffer at 60°C for 20 min. To effectively stop the digestion of the pentamer, pellets were treated with 50 μl 15 mM Pefa-Bloc SC (proteinase K) or with 20 μl 25 mM glutathion (papain) after centrifugation.

SDS–PAGE. Electrophoresis was performed according to Laemmli (1970) using a gradient of 12–18% for digests of the monomer and of 6–12% for pentamer digests. Fluka marker (charge 335516/1294), LMW electrophoresis kit from Pharmacia (5050446011) and the Sigma ultralow marker (66H9407) were used for calibration. Proteins were detected by Coomassie blue staining.

Peptide analysis. Peptides of about 13.5 and 15.5 kDa obtained by digestion of the monomer with papain, protease V8 and proteinase K as well as the 65–70 kDa fragment obtained after papain and proteinase K digestion were separated by SDS–PAGE and blotted onto a PVDF-membrane. Edman degradation was performed of about 50 pmol peptide on an Applied Biosystems pulsed-liquid-phase sequencer model 477A (Linder *et al.*, 1994).

Mass spectrometry of the digested pentamer. The pentamer was digested with proteinase K and stopped with 4-(2-aminoethyl)-benzenesulfonyl fluoride as described in the preceding paragraph. After centrifugation at 22,000 *g* and 4°C for 10 min, the pellet (containing about 2–5 pmol protein as well as the lipid) was dissolved in 5 μl 0.1% trifluoroacetic acid/30% acetonitril and centrifuged again. The supernatant (0.5 or 1.0 μl) was mixed with 1 μl of matrix solution (9 mg 2,5-dihydroxybenzoic acid and 1 mg 2-hydroxy-5-methoxybenzoic acid per ml 0.1% (v/v) trifluoroacetic acid, 30% (v/v) acetonitrile) and allowed to air-dry. Molecular masses were determined by MALDI–TOF-MS on a Vision 2000 mass spectrometer (Finnegan MAT). Ions were generated by irradiation with a pulsed nitrogen laser (emission wavelength: 337 nm; laser power density about 10^6 W/cm^2) and positive ions were accelerated and detected in the reflector mode. Spectra were calibrated using subtilisin (Sigma) as external standard.

Site-directed mutagenesis. Point and deletion mutations of *ctx* in the vector pSN3 (Orlik-Eisel *et al.*, 1990) were generated by site-directed mutagenesis using the QuikChange Site-directed Mutagenesis Kit (Stratagene). Primers used are described in Table 3.1. DNA was prepared from single colonies after growth in LB-ampicillin (100 μg/ml) medium for 16 h at 37°C with the Qiagen Midi Kit. The generated mutant clones were confirmed by sequence analysis (primers see Table 3.1), using the automated dye terminator cycle sequencing method (Applied Biosystems DNA Sequencer 373A).

Cytotoxin expression, partial purification and activation. Expression of *ctx* and its mutants was performed in *Escherichia coli* XL1-Blue cells (Stratagene) for 48 h at 37°C. P^{trc}-promotor of the plasmid pSN3 was induced by adding IPTG to the medium after 3 h of cultivation. The bacterial suspensions were centrifuged for 1 h at 4°C and 12,100 g, the sediments were resuspended in PBS and finally homogenized. Lysis of bacterial cells was performed by sonication.

Pro-cytotoxin was partly purified from sonicated cell lysates by precipitation with ammonium sulfate according to the method of Lutz (1979) and activated with 0.055 U trypsin per 200 μl lysate for 2 h at 37°C (Orlik-Eisel *et al.*, 1990). The reaction was stopped by adding 0.38 U trypsin inhibitor from soybean (Sigma) and EDTA to a final concentration of 10 mM.

Table 3.1 Primers used for generation of mutated cytotoxin and for sequencing

			Nucleotide (5′-3′)
Mutant generation	H156 > P	F	GTGCTTATTTCCCCA.C**C**T. AGCCACGGCTGG
		R	CCAGCCGTGGCT.A**G**G.TGGGGAAATAAGCAC
	H156 > L	F	GTGCTTATTTCCCCA.C**T**T.AGCCACGGCTGG
		R	CCAGCCGTGGCT.A**A**G.TGGGGAAATAAGCAC
	H158 > L	F	CTTATTTCCCCACATAGC.C**T**C.GGCTGGGGAGAG
		R	CTCTCCCCAGCC.G**A**G.GCTATGTGGGGAAATAAG
	H242 > L	F	GCCACAGGCACCTTC.C**T**T.AGCAGCGTGGGC
		R	GCCCACGCTGCT.A**A**G.GAAGGTGCCTGTGGC
	H254 > R	F	CTTGAAGACCATCGCG.C**G**C.GAGCGGCCCTATCC
		R	GGATAGGGCCGCTC.G**C**G.CGCGATGGTCTTCAAG
	H254 > P	F	CTTGAAGACCATCGCG.C**C**C.GAGCGGCCCTATCC
		R	GGATAGGGCCGCTC.G**G**G.CGCGATGGTCTTCAAG
	DelV99–V111	F	CACCTTCACCTGGAGC.AAGGCGAACATTCCCC
	+ P116 > S*	R	GGGGAATGTTCGCCTT.GCTCCAGGTGAAGGTG
Sequencing		F1	TCCGGCTCGTATAATGTGTG
		R1	TCTCTCATCCGCCAAAACAG
		F2	CAACGAAGGCAGCCAAGTG
		R2	CACTCCCAAAATACCTGCTC
		F3	GGTTGCCTTGAGCTTTACTG
		R3	TGGTGATCTCAGCGCCGC
		F4	AGGCAATGGAGTGTTGGC
		R4	TATAGCTGCGAGTGTCCTTG

Notes
Point mutation is given in bold letter.
* Mutation P116 > S occurred spontaneously.

Biological activities of the cytotoxin mutants. Binding to rat erythrocyte membranes was tested as a competition experiment and the toxicity on human granulocytes as stated by Xiong *et al.* (1991). Functional oligomerization of the mutants and the relative amounts of the pentamer were determined according to Sliwinski-Korell *et al.* (1999).

Results and conclusions

Digestion of the monomeric cytotoxin by various proteases. Limited digestion of the cytotoxin monomer (29 kDa) with papain, pronase E, proteinase K, subtilisin and protease V8 (Figure 3.1, lanes 1–3, 5 and 6, respectively) always resulted in the formation of two fragments of about 13.5 and 15.5 kDa (Figure 3.1, see symbol ←). N-terminal amino acid sequencing after digestion with papain, protease V8 or proteinase K showed that the 13.5 kDa fragment represents exactly the N-terminal part of cytotoxin. The other part of the monomer of 15.5 kDa starts with I123 after protease V8 (Figure 3.6, see symbol ❘) or with S125 after proteinase K treatment (Figure 3.6, see symbol ⇓). It represents therefore the C-terminal part of the monomeric toxin. The cleavage site by papain was not exactly determined. It is most likely between L117–V118, G119–G120 or G120–A121 (Figure 3.6, see symbol ↓), because the C-terminal part has a slightly higher molecular mass as is shown by SDS–PAGE (Figure 3.1, lane 1). Therefore, the region from L117 to S125 must be exposed at the surface of the cytotoxin monomer. Interestingly, all these cleavage sites are located in the center of the proposed membrane-spanning region of the cytotoxin pentamer (Figure 3.6). It should be mentioned that the monomer of *Staphylococcus aureus* α-hemolysin also possesses several cleavage sites for proteinase K in this region (Song *et al.*, 1996), as is indicated in Figure 3.5, by ⇓.

Digestion of the membrane-associated pentamer by proteases. Incubation of *Pseudomonas* cytotoxin with rat erythrocyte membranes for 1 h at 37°C resulted in the formation of pentamers of 145 kDa (Figure 3.3, lane 1). All membrane proteins were completely digested by each protease as is shown for papain and proteinase K in Figure 3.3. Digestion of the

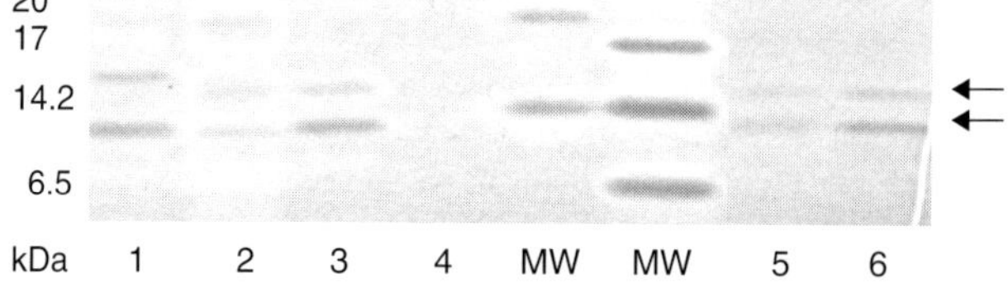

Figure 3.1 Digestion of the *Pseudomonas* cytotoxin monomer by proteases. Cytotoxin without protease treatment (lane 4) and limited digestion by papain (lane 1), pronase E (lane 2), proteinase K (lane 3), subtilisin (lane 5) and protease V8 (lane 6) resulted in two fragments of about 13.5 and 15.5 kDa (←). SDS–PAGE, 12–18%.

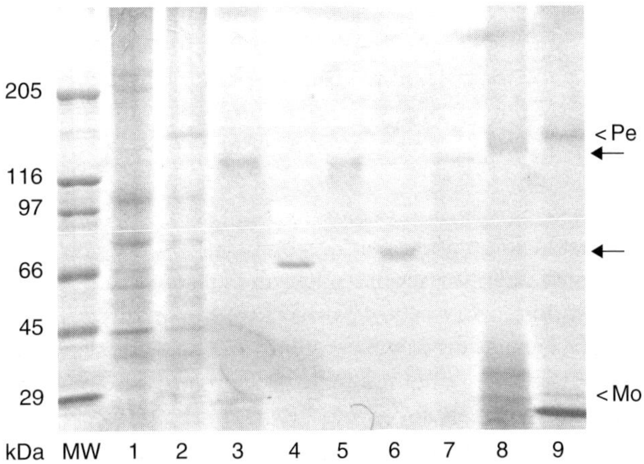

Figure 3.2 Digestion of the *Pseudomonas* cytotoxin pentamer by various proteases. Lane 1 demonstrates pure rat erythrocyte membranes and lane 2 pentamer formation (Pe). Digestion of the membrane-associated pentamer by chymotrypsin (lane 3), pronase E (lane 5), subtilisin (lane 7) and protease V8 (lane 8) resulted in a band of about 125–135 kDa (upper ←). Papain (lane 4) and proteinase K treatment (lane 6) led to a band of 65–70 kDa (lower ←). Trypsin treatment (lane 9) demonstrates that the monomer (Mo) and the pentamer (Pe) were not digested. The membrane proteins were totally digested by every protease, as is demonstrated for papain and proteinase K in Figure 3.3. SDS–PAGE, 6–12%.

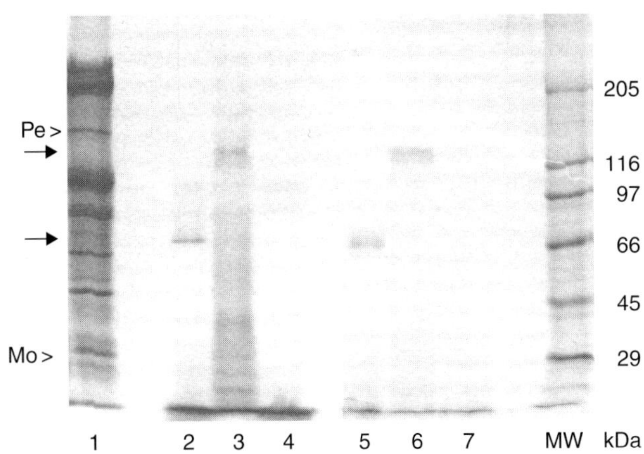

Figure 3.3 Digestion of the membrane-associated pentamer with papain or proteinase K without and with effective protease stop before SDS–PAGE. Lane 1 demonstrates the cytotoxin monomer (Mo) and pentamer (Pe) formation, lanes 2–4 the digestion with papain and lanes 5–7 with proteinase K. If the digest was not stopped at all (lanes 2 and 5) the pentamer was further digested to a band of 65–70 kDa (lower → see Figure 3.2). Treatment with high concentrations of protease inhibitor before SDS-PAGE resulted in a band of about 130 kDa (upper → lanes 3 and 6). Lanes 4 and 7 demonstrate the total digestion of rat erythrocyte membranes not exposed to cytotoxin. SDS–PAGE, 6–12%.

membrane-associated pentamer by chymotrypsin, pronase E, subtilisin and protease V8 displays a fragment of 125–135 kDa in SDS–PAGE (Figure 3.2, lanes 3, 5, 7 and 8, respectively). Thus, only a small part, in total about 15 kDa, representing approximately 26 amino acids per monomer, was cut off. Papain and proteinase K which are both active in the presence of SDS, led to further digestion during SDS–PAGE, resulting in a 65–70 kDa band (Figure 3.2, lanes 4 and 6, respectively), as the protease inhibitor cocktail was not efficient enough. If the papain digestion, however, was stopped effectively by glutathion and by Pefa-Bloc SC in case of proteinase K digestion before the addition of SDS–PAGE sample buffer, the pentamer was digested to a band of about 130 kDa only (Figure 3.3, lanes 3 and 6), thus resembling the behaviour of other proteases shown in Figure 3.2. It ensues that the 65–70 kDa band resulted from digestion of the 125–135 kDa fragment continuing in SDS which destroys the membrane association of the pentamer and partially denatures the toxin. Hence, the membrane-associated pentamer is largely stable to protease digestion.

Two distinct fragments of the 65–70 kDa band could be distinguished by N-terminal amino acid sequencing. One starts with S56 or N57 after papain digestion (Figure 3.6, see symbol ↓) and with F58 or Q61 after proteinase K digestion (Figure 3.6, see ⇓), respectively. The other fragment begins with A237 for both proteases (Figure 3.6, see ↓ and ⇓). Therefore, the core region of the pentamer consists of two fragments that must be connected by strong intra- and intermolecular interactions, which are resistant to SDS. The above-determined surface exposed region L117–S125 of the cytotoxin monomer is hidden in the pentamer and thus resistant to protease digestion. This is in analogy to the staphylococcal α-hemolysin, whose glycine-rich region is not attacked by proteinase K in its heptameric form (Song *et al.*, 1996).

Activated cytotoxin monomer and pentamer are not digested by trypsin. After processing the 20 C-terminal amino acids (Figure 3.6, crossed out sequence) the cytotoxin monomer as well as the pentamer (Figure 3.2, lane 9) are both resistant to further trypsin digestion. This indicates that possible cleavage sites for trypsin (Figure 3.6, see symbol ∨) must be inaccessibly buried inside the cytotoxin monomer and that pentamerization is not connected with the exposure of these cleavage sites to the protein surface.

Mass determination of the pentamer core region. The trypsin or proteinase K digested pentamer was examined by MALDI–TOF-MS. As neither the 145 kDa pentamer nor the 65–70 kDa core region, respectively, could be detected, it has to be concluded that the pentamer disintegrates into its monomeric components during MS. Therefore, the pentamer and the SDS-stable core region of cytotoxin are not resistant to the conditions of MALDI–TOF-MS.

One detected peak after proteinase K digestion represents with 3,286 Da (Figure 3.4) the 30 C-terminal amino acids of cytotoxin, starting with the amino acid A237 (Figure 3.6, see symbol ⇓ and underlined sequence), which was detected by N-terminal amino acid sequencing. The second monomeric component starting with F58 or Q61 (Figure 3.6, see symbol ⇓) should have a calculated remaining mass of about 10 kDa and consist of 94–103 amino acids (Figure 3.6, underlined sequence), therefore including the three cleavage sites of the monomer in the region L117–T124. But MS of the proteinase K digest (Figure 3.4) shows three different fragments with higher masses than 10 kDa: 13.6, 14.5 and 15.2 kDa, demonstrating partial digestion by proteinase K without any SDS contact and thus a momentary picture of limited digestion. The N-terminal ends of these fragments have not yet been determined.

The membrane-spanning region is confirmed by protease digestion and site-directed mutagenesis. Rossjohn *et al.* (1998) published the structure-based sequence alignment of the long loop in

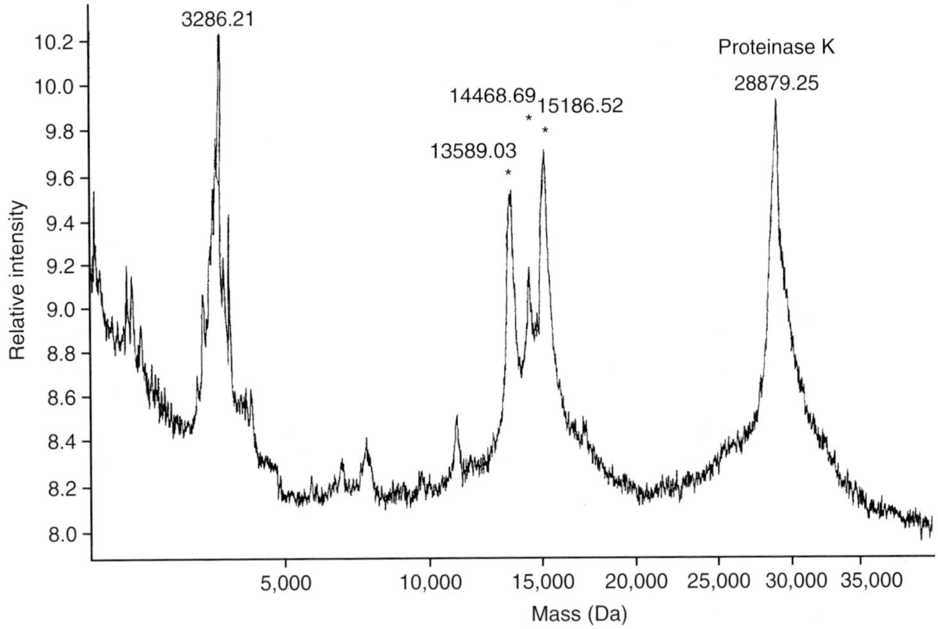

Figure 3.4 MALDI–TOF mass determination of the pentameric core region after proteinase K digestion. The mass of 3,286 Da corresponds to the C-terminal 30 amino acids of the cytotoxin monomer, that were detected by N-terminal amino acid sequencing. The peaks at 13.6, 14.5 and 15.2 kDa (*) show a momentary picture of limited digestion without protease inhibitor of the long fragment of the core region.

domain 3 of *Aeromonas hydrophila* aerolysin with the loops observed in the crystal structures of *S. aureus* α-hemolysin and *Bacillus anthracis* anthrax toxin protective antigen; the aligned sequence of *Pseudomonas* cytotoxin bases on sequence similarities to aerolysin (Figure 3.5). This alignment is supported by the result that the cleavage sites of the monomer (L117–S125) are located in the proposed membrane-spanning region and are not accessible to protease digestion in the pentamer. Additionally, the region from V105 to S132 represents the only β-strand-loop-β-strand motif with alternating hydrophilic and hydrophobic amino acids in the 100 amino acid long, protease-resistant core of the pentamer.

A single point mutation of V111 to E reduced membrane binding and pore formation by 50%, while V107 to E had not such an effect (not shown). This indicates that V111 might be involved in the membrane-spanning part of the β-sheet, while V107 is not (Figure 3.5). The deletion of amino acids V99–V111, also containing a single mutation of P116 to S, changed the structure of the monomer so extensively, that pro-cytotoxin activation by trypsin resulted in total digestion (not shown).

Histidines are involved in oligomerization. As shown earlier, modification of cytotoxin with diethylpyrocarbonate reduced oligomerization extensively. This was reversible by hydroxylamine, so that histidine residues must be involved in the oligomerization process (Sliwinski-Korell *et al.*, 1999). The two fragments after proteinase K digestion of the pentamer, detected by N-terminal amino acid sequencing and MALDI–TOF-MS (Figure 3.6, underlined sequences), possess two histidines each (Figure 3.6, H at positions 156, 158, 242

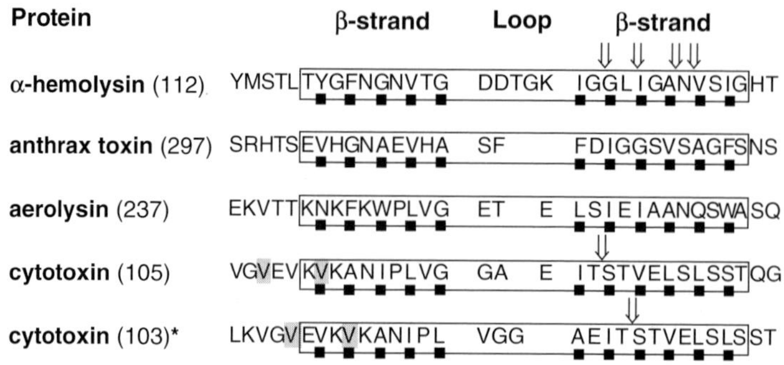

Figure 3.5 Structure-based sequence alignment of *S. aureus* α-hemolysin, *B. anthracis* anthrax toxin, *A. hydrophila* aerolysin and *P. aeruginosa* cytotoxin according to Rossjohn *et al.* (1998). The number of the first residue in each sequence is given in parenthesis. Residues that point into the hydrophobic core of the bilayer are marked with ■. Cleavage sites of proteinase K in the monomer of α-hemolysin (refer to Song *et al.*, 1996, their figure 3C) and cytotoxin are indicated by ⇓. Point mutations of cytotoxin from V to E are pointed out by V. The membrane-spanning region is boxed. * This alignment is additionally proposed by the authors.

and 254). They might be responsible for the intra- and intermolecular interactions of the two fragments.

The effect of single-site mutations of these histidines and their relative effect on membrane binding, oligomerization and pore formation is shown in Table 3.2. Mutations to L, P or R obtained, either increased or moderately diminished cytotoxin binding to rat erythrocyte membranes. Mutations of H156 or H254 to P extensively disturbed oligomerization and therefore pore formation. H156, H158 and H242 mutated to L reduced oligomerization to 70–80%, but resulted in a highly decreased pore-forming activity, perhaps due to faulty oligomerization. Mutation of H254 to R diminished oligomerization and pore formation to only 70% and 60%, respectively. Thus, the strong SDS-resistant interactions between the fragments in the pentamer might be due to hydrogen bonds and salt links of particularly H156 and H254, but also of H158 or H242 to other polar residues, similar to other pore-forming toxins.

Discussion

Limited digestion of the monomer with different proteases indicated that *Pseudomonas* cytotoxin is split at the surface-exposed region of L117 to S125 into an N-terminal and a C-terminal fragment (Figure 3.6). Yet, this region is not accessible to the same proteases after pentamer formation on rat erythrocyte membranes. As the cleavage sites of the monomer are located in the postulated membrane-spanning region, the reorganization of β-structural elements, detected by ATR–FTIR spectroscopy (Sliwinski-Korell *et al.*, 1999), might be relevant to a rearrangement of this region during pentamerization. This feature is very similar to the staphylococcal α-hemolysin, where the monomer is split by proteinase K in the glycine-rich region (Figure 3.5), while the heptameric toxin is not attacked (Song *et al.*, 1996).

Although the cytotoxin pentamer is very resistant to proteolysis, the proteases papain and proteinase K demonstrated further digestion of the pentamer after membrane solubilization

```
                 v v                                v  v   v
  1 MNDIDTITNA WGRWKTAQYG TTCWFTESTQ YGRNKDTRSY MQHQTNVSAP
    LLL                    /\/\/\      L   LL          LLL

     v    ↓↓⇓   ⇓
 51 KDLVYSNFVQ QDGGSTLLGQ YDMINEGSQV IELAVNLQQG LVDTFTWSVT
    L /\/\/\/\  LLL /\/\/\/\                        /\/\/

        v      v  v     ↓ ↓↓   | ⇓                v          v
101 EQLKVGVEVK VKANIPLVGG AEITSTVELS LSSTQGASTS KSSNYGASTK
       /\/\/\/\/\/ L    LL  /\/\/\/\/\/ LLLLL   LLLLLLL /\

                                v                   v
151 VLISPHSHGW GEVALSFTEL RTQWVGNVGL QGYVAIWFNN KVALNNDGDY
    /\/\ LLLLL /\/\/\/\/\/\/\/\/\    /\/\/\/\    /\/ LLLL

                                               ↓
                                               ⇓               v
201 HYLWFIPVEQ VFWECVQHNI VNTSGYVVQG NGVLAQATGT FHSSVGLNLK
    /\/\/\/\/\/\/\/\/\/\/\/    /\/\/  LL /\/\   /\/\/\

         v         v
251 TIAHERPYPE TSEAVRTFYN YASLVPDLET RVRSAE
       LLLLL
```

Figure 3.6 *Pseudomonas* cytotoxin amino acid sequence, specifying structural and functional important regions. The secondary structure prediction concerning β-sheet (/\/\/\) and loop (LL) of the EMBL, Heidelberg (Rost and Sander, 1993, 1994), is given underneath the amino acid sequence. The 20 C-terminal amino acids cut off by bacterial autolysis are crossed out (~~XXX~~). Above the amino acid sequence the cleavage sites of the different proteases are indicated: possible trypsin cleavage sites (v) or determined after digestion with papain (↓), proteinase K (⇓) and protease V8 (|) in the monomer (C-terminal of L117, G119, G120, E122 or T124) and the pentamer (C-terminal of Y55, S56, N57, Q60 or Q236). The two fragments of the pentamer core region are underlined. The proposed membrane-spanning region aligned to aerolysin and α-hemolysin (see Figure 3.5) is boxed. The deletion of V99–V111 leads to structural changes, so that monomeric cytotoxin was accessible to trypsin digestion. Single mutations of valine V107 and V111 (V) and histidine H156, H158, H242 and H254 (H) influenced oligomerization and pore formation. A cytotoxin mutant, where the C-terminal part was deleted (   ), showed only very reduced pore formation (Xiong *et al.*, 1991).

and protein denaturation by SDS. Two fragments of this core region could be identified: One includes the 30 C-terminal amino acids and the other fragment starts between S56 and Q61 and is about 100 amino acids long. The C-terminal end of cytotoxin is known to be functionally important for membrane binding and pore formation: Xiong *et al.* (1991) showed that a deletion mutant of the last 43 amino acids S224–R266 (Figure 3.6, grey background) resulted in only 35% membrane binding and 6% pore formation as compared to the complete pSN3-expressed cytotoxin. As the 30 C-terminal amino acids are very resistant to proteolysis and an amphipathic β-strand is predicted from primary structure (Figure 3.6), this region might play a special role in membrane association. Gouaux (1997) argued that an ideal β-barrel of pore-forming toxins should possess a shear number of strands and therefore suggested 21 or 28 strands for the heptameric channel of aerolysin

Table 3.2 Activities of histidine mutants, performed in the pentameric core region

Histidine mutation		*Relative activity (%)*		
	Codon	*Membrane binding*	*Oligomerization*	*Pore formation*
156>P	CAT>CCT	160	10	0
156>L	CAT>CTT	110	80	10
158>L	CAC>CTC	50	70	15
242>L	CAT>CTT	50*	70	30
254>R	CAC>CGC	170	70	60
254>P	CAC>CCC	50*	0	0

Notes

Relative activity to pSN3-expressed cytotoxin on binding to and oligomerization on rat erythrocyte membranes as well as pore formation on human granulocytes are displayed.

* Binding to membranes was estimated from SDS–PAGE instead of competition experiments with ^{125}I-cytotoxin.

with an estimated radius of 1.9–2.3 nm. The cytotoxin pentamer therefore might need a shear number of 15 or 20 strands to create a pore diameter of about 1.5–2.0 nm (Lutz *et al.*, 1987). To what extent the C-terminal end might be involved in this process is still open.

The other long fragment of the core region includes the proposed membrane-spanning part (E108–S132), which is characterized by a β-strand-loop-β-strand structure with alternating hydrophobic and hydrophilic amino acids (Figures 3.5 and 3.6). Additionally, the protease cleavage sites of the monomer (L117–S125), which are situated in this region, are not accessible to protease digestion in the pentamer. Even in the presence of SDS, which destroys the membrane association and denatures the toxin, this region was not accessible to papain and proteinase K digestion and thus this region must be involved in the reorganization of the β-structural elements. Single mutations of V107 and V111 to the negatively charged E indicate that V111 might be involved in the membrane-inserted part of the β-sheet, while V107 is not (Figure 3.5). The total digestion of the deletion mutant (V99–V111 and P116 to S) by activation with trypsin suggests that the elimination of nearly one β-strand of the proposed membrane-spanning region causes extensive structural changes. These might be due to the abolition of the antiparallel β-sheet formation.

The strong intra- and intermolecular contacts between the digested fragments of the pentamer should be due to a network of nonpolar and polar interactions. Mutagenesis of histidine residues demonstrated an important role of particularly H156 and also H254. But on the other hand both mutations to the rigid proline (Table 3.2) might induce structural changes and therefore induce sterical hindrance between the protomers.

Taken together the properties of the monomeric and the pentameric cytotoxin, the structural details presented by Sliwinski-Korell *et al.* (1999) as well as the obvious similarities to other pore-forming toxins lead to the proposal of the following model of cytotoxin oligomerization: Membrane contact of the activated cytotoxin monomers induce a reorganization of β-structural elements so that 2, 3 or 4 β-strands of one monomer form antiparallel β-sheets in the pentamer, which becomes very resistant to protease treatment. In the presence of SDS the pentamer was digested to the core region of 65–70 kDa, composed of two fragments, still connected by strong intra- and intermolecular interactions. The region

E108–S132 was identified as one membrane-spanning part; the C-terminal end might include a second membrane interactive region of the pentamer.

Acknowledgments

This work was supported by the "Deutsche Forschungsgemeinschaft" (Lu 252/4-1 and the "Graduiertenkolleg" Molecular Biology and Pharmacology) and the "Wilhelm-Sander-Stiftung" (97.091.1). We thank M. Kröger for the suggestions concerning the site-directed mutagenesis and D. Grizan for bringing forth the images and figures.

References

Gouaux E. (1997) Channel-forming toxins: tales of transformation. *Curr. Opin. Struct. Biol.* **7**, 566–573.

Laemmli U. K. (1970) Cleavage of structural proteins during the assembly of the head of bacteriophage T4. *Nature* **227**, 680–685.

Linder M., Wenzel V., Linder D. and Stirm S. (1994) Structural elements in glycoprotein 70 from polytropic friend mink cell focus-inducing virus and glycoprotein 71 from ecotropic friend murine leukemia virus, as defined by disulfide-binding pattern and limited proteolysis. *J. Virol.* **68**, 5133–5141.

Lutz F. (1979) Purification of a cytotoxic protein from *Pseudomonas aeruginosa*. *Toxicon* **17**, 467–475.

Lutz F., Maurer M. and Failing K. (1987) Cytotoxic protein from *Pseudomonas aeruginosa*: formation of hydrophilic pores in Ehrlich ascites tumor cells and effect on cell viability. *Toxicon* **25**, 293–305.

Orlik-Eisel G., Lutz F., Henschen A., Eisel U., Struckmeier M., Kräuter J. and Niemann H. (1990) The cytotoxin of *Pseudomonas aeruginosa*: cytotoxicity requires proteolytic activation. *Arch. Microbiol.* **153**, 561–568.

Rossjohn J., Feil S. C., McKinstry W. J., Tsernoglou D., van der Goot G., Buckley J. T. and Parker M. W. (1998) Aerolysin – a paradigm for membrane insertion of beta-sheet protein toxins? *J. Struct. Biol.* **121**, 92–100.

Rost B. and Sander C. (1993) Prediction of protein secondary structure at better than 70% accuracy. *J. Mol. Biol.* **232**, 584–599.

Rost B. and Sander C. (1994) Combining evolutionary information and neural networks to predict protein secondary structure. *Proteins* **19**, 55–77.

Sliwinski-Korell A., Engelhardt H., Kampka M. and Lutz F. (1999) Oligomerization and structural changes of the pore-forming *Pseudomonas aeruginosa* cytotoxin. *Eur. J. Biochem.* **265**, 221–230.

Song L., Hobaugh M. R., Shustak C., Cheley S., Bayley H. and Gouaux E. (1996) Structure of staphylococcal α-hemolysin, a heptameric transmembrane pore. *Science* **274**, 1859–1866.

Xiong G., Struckmeier M., Alberti U. and Lutz F. (1991) Relationship of primary structure to functional properties of the cytotoxin protein from *Pseudomonas aeruginosa*. *Naunyn-Schmiedeberg's Arch. Pharmacol.* **343**, 538–541.

4 *Helicobacter pylori* vacuolating toxin VacA

Channel-forming properties and cell intoxication

Mario Zoratti, Francesco Tombola, Cesare Montecucco, Lucantonio Debellis and Emanuele Papini

The cytotoxin VacA of *Helicobacter pylori* is a channel-forming protein. Following a pH-dependent activation step, this protein inserts into lipids bilayers and forms a voltage-dependent, low-conductance, anion-selective pore, which is inhibited by some blockers. In this chapter we describe the physico-chemical properties of the VacA channel, the information on the pore structure derived from the analysis of the interaction with various anion-channel blockers and the major differences between m1 and m2 VacA isotypes. Experimental evidence for the essential role of the VacA pore in the induction of known major cytotoxic effects on cells, in vitro epithelial models and gastric tissue from frogs is also described and discussed. A model is summarized of a mechanism of cell vacuolation, based on endocytosis of VacA channels, followed by an osmotic unbalance of late endosomes. It is proposed that permeabilization of the apical domain of the epithelial cell plasma membrane to critical small molecules, such as bicarbonate and urea, represents the main in vivo role of VacA, aimed at ensuring an efficient gastric colonization by *H. pylori*.

Introduction

Infection of the stomach mucosa by the Gram-negative bacterium *H. pylori* concerns a large portion of the human population (Covacci *et al.*, 1999). Mostly asymptomatic, the presence of this parasite leads, in a minority of cases, to a strong mucosal inflammation often accompanied by tissue damage, gastric ulcer and adenocarcinoma (Dunn *et al.*, 1997). Many virulence and pathogenic factors produced by *H. pylori* cooperate in the task of stomach colonization and contribute to tissue inflammation and alterations (Montecucco *et al.*, 1999; Montecucco and Rappuoli, 2001). The protein VacA is one such agent (Atherton *et al.*, 2001). Epidemiological evidence and experiments conducted in animal models show that VacA increases the severity of *H. pylori*-induced tissue damage (Telford *et al.*, 1994; Gerhardt *et al.*, 1999; Ogura *et al.*, 2000). According to a recent study in mouse (Salama *et al.*, 2001), VacA-producing *H. pylori* strains are much more effective than their VacA$^-$ isogenic mutants in the colonization of the stomach. VacA is also referred to as *vacuolating toxin* because its first recognized biological effect, characterized in vitro, was the induction of many large, round vacuoles in the cell cytoplasm, derived from late endosomes (Leunk *et al.*, 1988; Cover and Blaser, 1992; Papini *et al.*, 1994).

VacA is a secreted protein of 95 kDa, formed by an NH_2-terminal fragment (p37) and a COOH-terminal fragment (p58) linked by a protease-sensitive loop, often cleaved upon storage of the purified protein (Telford *et al.*, 1994). The encoding gene (*vaca*) forms a large family of allelic variants, expressed in different strains with diverse geographic distributions (Atherton *et al.*, 1995; Covacci *et al.*, 1999). The comparison of their sequences reveals many differences and deletions scattered along the whole protein length. However, the highest variability is found in a region belonging to p58, called m region. Two major classes of alleles have been identified, called m1 (mostly present in the western population) and m2 (more frequent in far-eastern Asia).

Purified VacA appears to be assembled into oligomeric structures of up to 12 polypeptides, having negligible activity (Lupetti *et al.*, 1996; Cover *et al.*, 1997). Freezing and thawing in phosphate buffer or treatment at acidic (<5.5) or alkaline pHs (>9.5) are necessary to activate the protein (de Bernard *et al.*, 1995; Yahiro *et al.*, 1999). This process entails oligomer disassembly and exposure of hydrophobic regions on the monomers, favoring their insertion into membranes (Cover *et al.*, 1997; Molinari *et al.*, 1998). Besides vacuolation, VacA has been shown to induce a variegated pattern of intoxicated phenotypes in vitro, in many cases depending on the cell line employed (Papini *et al.*, 2001). Patch-clamp analysis has demonstrated that VacA forms ion channels in the cell plasma membrane of HeLa cells, allowing the inclusion of this protein in the category of pore-forming toxins (Szabò *et al.*, 1999). In the present chapter, we focus on the channel-forming property of VacA, and on its relationship to the main effects of the toxin on cells and tissues, in various models. Finally, we propose a model of the functional role of the VacA pore in *H. pylori* infection and pathology.

The VacA ion channel

As anticipated in the previous section, VacA induces a variety of phenomena in cells and epithelia. Its only known molecular activity, however, is the formation of pores. Indeed, acid-activated VacA dramatically increased the ionic permeability of both planar bilayers (Tombola *et al.*, 1999a,b) and HeLa cells (Szabò *et al.*, 1999), while unactivated VacA had little or no effect. The properties of these pores, established relying mainly on planar lipid membrane experiments, are described below.

Incorporation. Upon addition of activated VacA a transmembrane current develops after a variable lag period (minutes) and increases in sigmoidal fashion up to a plateau (Figure 4.1A), which is roughly proportional to the amount of activated toxin added. It can be easily calculated from single-channel conductance estimates (see following paragraphs) that the current intensity at plateau corresponds to the incorporation of only a minor fraction of the activated toxin added to the bilayer chamber. The most likely explanation for this may be the competing formation of unproductive species, presumably including dodecamers and tetradecamers, which have been observed to be the predominant complexes in solutions of m1 VacA 60190 (Cover *et al.*, 1997).

The characteristics of the VacA incorporation process vary depending on the toxin isoform. m1 VacA 17874 readily inserts into membranes containing only zwitterionic phospholipids, such as PC, PE or DPhPC, as well as into mixed-composition bilayers such as those made of azolectin (Tombola *et al.*, 1999a). On the other hand, incorporation of m1 VacA 60190 (ATCC 49503) requires negatively charged lipids (Czajkowsky *et al.*, 1999; Iwamoto *et al.*, 1999), and incorporation of m2 VacA 95-54 is much more efficient with azolectin than with various defined mixtures of phospholipids. The preference of 95-54 for azolectin is not due to the presence of negative charges in the lipid, but rather presumably reflects an interaction with an unidentified minor component of the complex lipid mixture (Tombola *et al.*, 2001a). Incorporation under the conditions of these experiments

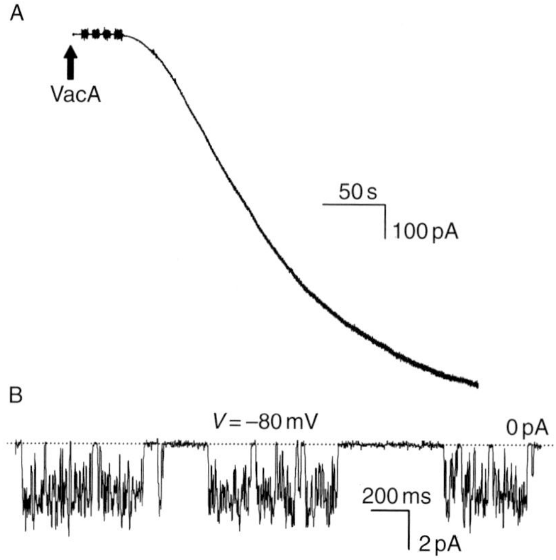

Figure 4.1 (A) Kinetics of VacA-elicited transmembrane current development. In this illustrative planar bilayer experiment 21 nM acid-activated VacA was added on the *cis* side of an azolectin planar bilayer when indicated. After stirring, the conductance of the membrane increased in sigmoidal fashion. Medium: 500 mM KCl, 0.5 mM $MgCl_2$, 0.5 mM $CaCl_2$, 10 mM Hepes/K^+, pH 7.2. $V(cis)$: −40 mV. (B) VacA single channel activity recorded at $V(cis)=-80$ mV with a DPhPC planar membrane and 2 M KCl. Sampling: 2 kHz. Filtering: 100 Hz.

(no proteinaceous receptors) reflects an association of VacA with phospholipids that may be akin to the unsaturable, "non-specific" binding to cells recently described by McClain *et al.* (2000) and Ricci *et al.* (2000).

The sigmoidal kinetics of VacA-induced current formation are consistent with a process in which VacA monomers, formed upon dissociation of complexes, assemble into channel-forming oligomers, with a stoichiometry of about six (Iwamoto *et al.*, 1999; Tombola, PhD thesis). This model has received support from data suggesting the formation of mixed oligomers upon intoxication of cells with two different VacA types (Vinion-Dubiel *et al.*, 1999). Whether the genesis of channels involves the assembly of monomers already inserted into the bilayer, or the formation of membrane-bound pre-pores which then incorporate, remains unclear, but some observations suggest that the latter type of mechanism may apply. Besides lipid composition (see above), the rate of conductance increase (i.e. pore formation) in bilayer experiments is influenced by the transmembrane voltage: voltages negative on the side of toxin addition favor the process (Tombola *et al.*, 1999a), which is 2–3 times faster at $V(cis)=-40$ mV than at $V(cis)=+40$ mV. This implies that the (at least partly) rate-determining step for pore formation involves the movement of negative charge along the field lines. Such a displacement of charge is more likely to be associated with penetration of the protein into the bilayer, than with sidewise diffusion and association to form oligomeric structures either on the surface or within the membrane. Binding and insertion have at any rate been reported to be two separate processes with different pH optima (Pagliaccia *et al.*, 2000).

Insertion takes place with a 100%-specific orientation. This is most clearly shown by the fact that SITS blocks VacA 17874 channels when present on the toxin addition side of the planar bilayer (*cis*), but not at all when added on the opposite (*trans*) side (Tombola *et al.*, 2000) (see the following paragraphs for more details on inhibition by stilbene derivatives).

Conductance. The properties of VacA pores have been examined in multi- and single-channel planar lipid bilayer experiments. VacA is, by any reckoning, a small channel, with a conductance in the 5–30 pS range (2 M KCl), depending on the experimental conditions and VacA isoform tested. The single-channel conductance may vary from one experiment to the next, and a given channel may adopt more than one conductance level in the course of an experiment (Tombola *et al.*, 2001a). These observations suggest that VacA pores are best considered as a family of related structures, some of which may interconvert.

We have carried out a statistical comparison of the channels formed by m1 VacA 17874, m2 VacA 95-54 and by the construct derived from 17874 by deletion of the loop connecting the p37 and p58 domains (m1del46). The latter reportedly forms only hexamers in solution (Burroni *et al.*, 1998), while the parent toxin, 17874, is present as approximately 70% heptamers and 30% hexamers (Lupetti *et al.*, 1996; Lanzavecchia *et al.*, 1998). The ranges of single-channel conductance values measured using these two forms overlapped to some extent due to a greater scatter of the m1 17874 values. However, the average, voltage-adjusted conductance of m1del46 channels was lower than that of m1 channels by 23% (Tombola *et al.*, 2001a). Interestingly, if the percentage of hexamers is taken as 100% for m1del46 and 70% for m1 17874, the channel cross-section is modeled as a regular hexagon or heptagon with a constant side length, and conductance is assumed to be proportional to its area, one calculates an expected decrease of the average conductance by 22%. This suggests that heptamers (as well as hexamers) might form in membranes as well as in solution, producing channels with a somewhat higher mean conductance and more scattered individual values. On the other hand, only hexamers were detected by Atomic Force Microscopy observation of bilayer-inserted m1 VacA 60190, which in solution formed oligomers with seven-fold symmetry in about 20% of cases (Czajkowsky *et al.*, 1999).

Only hexamers have been reported for mica-adsorbed 95–54 (Cover *et al.*, 1997). The average V-adjusted conductance of the channels produced by this isoform was only about 44% of that of m1 17874 (Tombola *et al.*, 2001a), indicating that factors other than stoichiometry intervene to determine the rate of ion permeation through the pore. These are likely to be related to the sequence of the m region, where the vast majority of the sequence variations between the two forms are located.

Kinetics. The channels display bursting kinetics (Figure 4.1B), with the superimposition of very fast (kHz) and relatively slow (Hz) gating modes. The parameters characterizing the latter kinetic mode vary from experiment to experiment, and show little or no voltage dependence. A kinetic analysis of the fast mode would be difficult and it has not been attempted. In rare instances involving the m1del46 construct we have observed an approximately 20 pS (2 M KCl) well-behaved channel lacking the fast mode. This suggests that the latter may be associated with the rapid movement of a domain which might be blocked in some uncommon oligomeric structures, or perhaps lost sometimes by limited proteolysis, known to take place before or during toxin harvesting (Nguyen *et al.*, 2001).

Selectivity. An important characteristic of VacA pores is that they are anion-selective. A permeability sequence for VacA 17874 is: Cl^-, HCO_3^- > pyruvate > D-gluconate > K^+, Li^+, Ba^{2+} > NH_4^+, with Cl^- permeating some four times more readily than pyruvate and roughly 24 times better than K^+ (Tombola *et al.*, 1999a). Thus, VacA allows the passage of anions (HCO_3^-, pyruvate, Cl^-) that are potentially relevant for the survival of *H. pylori* on the gastric mucosa. The loop-deleted form, m1del46, has the same Cl^-/K^+ permeability ratio as the parent toxin, while m2 VacA 95-54 is somewhat less selective, with a ratio of about 10 (Tombola *et al.*, 2001a). This again points to a role of the m region in determining channel properties, and suggests that it might actually contribute to the formation of the lumenal wall. For m1 VacA 60190, the following sequence has been determined: $SCN^- > I^- > Br^- > Cl^- > F^- > Na^+$, with Cl^- permeating some four times better than Na^+ (Iwamoto *et al.*, 1999). This sequence is of the "weak field strength" type, which is generally considered to reflect mostly the energy required by the various ions for dehydration during transport (Hille, 1992). This implies the absence of a high-affinity binding site for anions in the channel lumen, and suggests that selectivity may originate from "weak" interactions between the ion and dipoles and partial charges of the protein, in analogy to many other anion-selective channels.

Permeability to urea. Tracer studies have shown that VacA pores allow the passive diffusion of urea across model epithelia and cell membranes (Tombola *et al.*, 2001b). Transport selectivity (glycerol and mannitol were excluded) as well as inhibitor and electrophysiological experiments supported the conclusion that the pore itself provides the permeation pathway. The phenomenon is thus distinct from the aspecific "loosening" of intercellular junctions also caused by the toxin (see below).

Voltage dependence. VacA displays a weak voltage dependence: the conductance of a toxin-doped planar bilayer or of an intoxicated cell membrane increases as the voltage is made more negative on the side of toxin addition (*cis*) (Tombola *et al.*, 1999a, 2000). The extent of the effect depends on the charge-carrying specie (Tombola *et al.*, 1999a). Single-channel *i/V* curves also exhibit a slight rectification. The voltage dependence observed in multi-channel experiments appears therefore to reflect variations in channel conductance, rather than in open probability. While the behavior of m1del46 is indistinguishable from that of m1 17874, the voltage dependence of m2 VacA 95-54 is somewhat more marked (Tombola *et al.*, 2001a), again suggesting a role of the m region in determining the channel properties.

Functional linkage between the VacA channel and the toxin's cellular and tissue effects

The observations summarized in the previous paragraph show that VacA can permeabilize the biological membrane by forming a pore. As already mentioned in the introduction, VacA can induce a striking phenotype in cells: the appearance of large vacuolar compartments (Figure 4.2A,B). In addition it has also been documented that VacA can increase the epithelial permeability to ions and small non-permeant molecules like mannitol and sucrose (Figure 4.2C) (Papini *et al.*, 1998). This second effect occurs in high resistance in vitro epithelia (such as MDCK I and T84), is vacuolation-independent and is not due to general cytopathic alterations leading to monolayer disruption. In fact it is size-selective and presumably reflects a modulation of the resistance of cell–cell junctions along the so-called paracellular pathway.

VacA forms channels in the plasma membrane of cells, and the permeabilization process precedes the appearance of vacuoles (Szabò *et al.*, 1999) (Figure 4.2D). It is therefore important to establish whether there is a relation between pore formation and the two different cellular phenotypes described above. A priori, the formation of channels might be supposed to be a collateral result of a diphtheria toxin-style self-translocation process, or, at any rate, a secondary effect. By analogy with A/B type toxins, pore formation may be

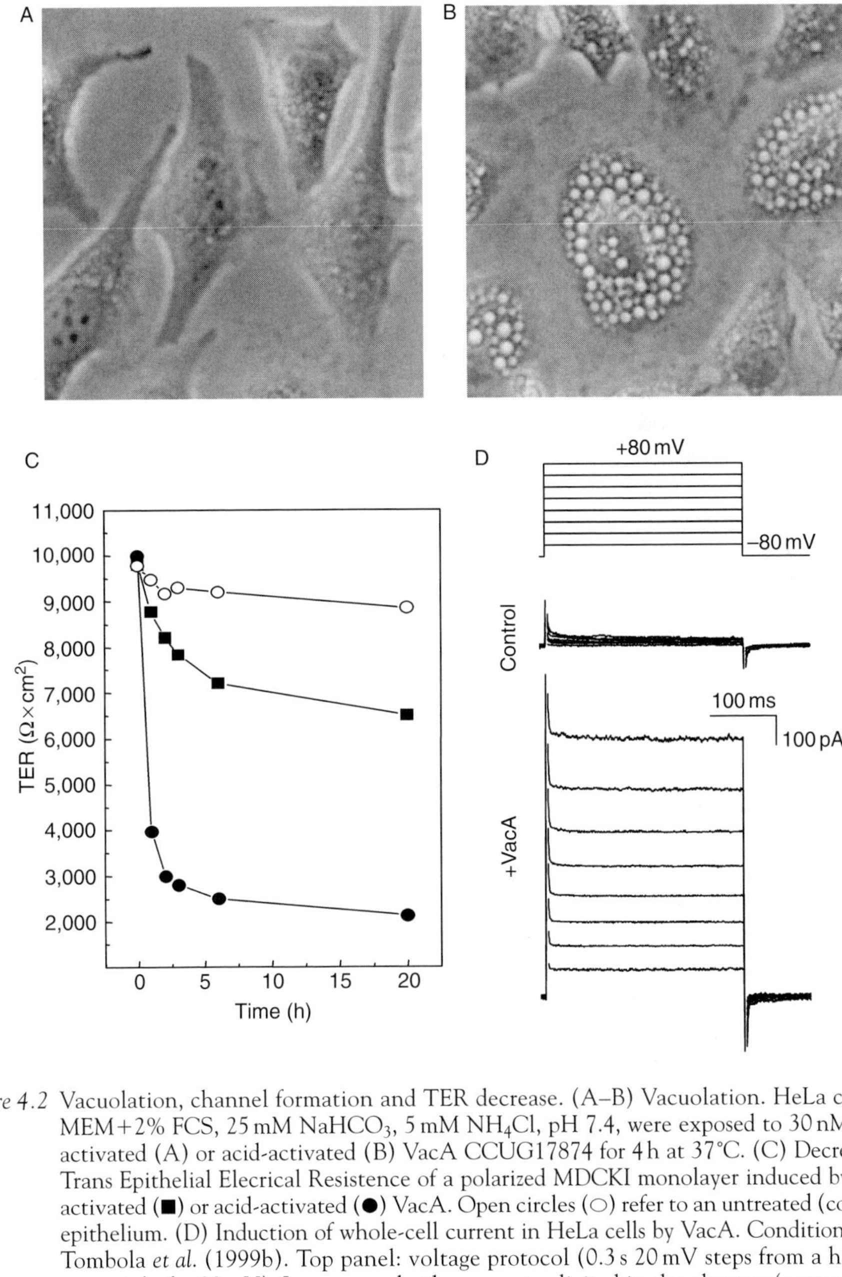

Figure 4.2 Vacuolation, channel formation and TER decrease. (A–B) Vacuolation. HeLa cells in MEM+2% FCS, 25 mM $NaHCO_3$, 5 mM NH_4Cl, pH 7.4, were exposed to 30 nM non-activated (A) or acid-activated (B) VacA CCUG17874 for 4 h at 37°C. (C) Decrease of Trans Epithelial Elecrical Resistence of a polarized MDCKI monolayer induced by non-activated (■) or acid-activated (●) VacA. Open circles (○) refer to an untreated (control) epithelium. (D) Induction of whole-cell current in HeLa cells by VacA. Conditions as in Tombola *et al.* (1999b). Top panel: voltage protocol (0.3 s 20 mV steps from a holding potential of −80 mV). Lower panels: the currents elicited in the absence (control) and presence of toxin.

expected to be a property of the C-terminal p58 domain, which mediates cell binding (Garner and Cover, 1996; Pagliaccia *et al.*, 1998; Reyrat *et al.*, 1999; Ji *et al.*, 2000; Wang and Wang, 2000) and might be thought to mediate translocation of the putative A (p37) domain. Purified p58 however forms only dimers, binds to cells but is not internalized

(Reyrat *et al.*, 1999), and does not form channels (Tombola *et al.*, 1999a) or induce cell vacuolation. Furthermore, McClain *et al.* (2000) have provided evidence that VacA is present inside cells as the full-size mature form, and have concluded that internalization of the whole protein is required for cytotoxicity. Thus, it seems now unlikely that VacA might be an A/B type toxin. Vinion-Dubiel *et al.* (1999) have shown that the deletion of a few N-terminal amino acids results in a much lowered propensity to form channels and abrogates toxin-induced vacuolation, even though the protein binds to, and is taken up normally by, the cells. In a model system, the amino-terminal hydrophobic region of VacA was able to promote insertion and dimerization of TOX–CAT fusion proteins in the inner membrane of *Escherichia coli* (McClain *et al.*, 2001), suggesting that this segment of VacA may play a similar role in the process of membrane-channel formation. de Bernard *et al.* (1998a) and Ye *et al.* (1999), studying the effects of cytosolically expressed VacA-derived constructs, had also observed that deletion of the extreme N-terminal portion results in loss of vacuolating activity. These studies have indicated that the p37 domain and a portion of the p58 region are both required for cytoxicity. Structural studies had previously led to the conclusion that both regions are necessary for the formation of VacA's oligomeric ring structure (Lupetti *et al.*, 1996; Lanzavecchia *et al.*, 1998). Thus, there is a close connection between the ability to form channels and that to cause vacuolation of HeLa cells. Since the A/B paradigm does not fit very well the available data, it is likely that VacA channels cause cell vacuolation *per se*.

Previous evidence suggested that VacA intoxication determines alterations of ion homeostasis during vacuolation. In fact, the activity of the V-ATPase was found to be strictly required to determine and to maintain the vacuolar structures (Papini *et al.*, 1993) indicating that active proton pumping into the lumen of the forming vacuole is essential. In addition, careful elimination of weak bases in the incubation media revealed that ammonia or equivalent membrane-permeant amines are necessary for this effect (Ricci *et al.*, 1997). This suggests that VacA intoxication somehow results in an increased uptake of osmotically active acidotropic molecules, such as NH_4Cl, into the lumen of late endosomes, followed by water influx and compartment swelling.

The possibility that the channel formed by VacA is responsible for cell vacuolation is supported by the fact that other recognized channel-forming toxins, such as aerolysin from *Pseudomonas aeruginosa* or *Vibrio Cholerae* hemolysin, induce abnormal vacuole structures originating from intracellular organelles (Abrami *et al.*, 1998; Coelho *et al.*, 2000). Moreover, endocytosis of VacA and its accumulation in endo-lysosomes and in the vacuoles, where it persists for many hours (Ricci *et al.*, 1997), is necessary for the development of vacuolar degeneration of cells (McLain *et al.*, 2000). Tombola *et al.* (1999a) put together all this previous information and the data on the properties of the VacA channel in a model of vacuole genesis described in the following paragraphs (Figure 4.3).

According to this model, VacA channels form in the cell plasma membrane. Following endocytosis, they would then appear in the endosomal membrane, where, by selectively increasing its permeability to chloride and HCO_3^-, they would be expected to stimulate the V-ATPase whose electrogenic proton pumping is limited by the potential created by its own activity. A flux of anions towards the endosomal lumen would tend to dissipate the electrical component of the gradient, thus releasing the constrain on V-ATPase turnover. Hence, the presence of VacA pores would lead to an increased rate of HCl transport from the cytosol to the endo-lysosomal lumen. In the presence of ammonia, or other permeant bases, these excess pumped protons would combine with NH_3 to give NH_4^+. The latter would accumulate inside the endosome because, while NH_3 is freely permeable to membrane, NH_4^+ is less permeant. NH_4Cl accumulation is predicted to determine an influx of water

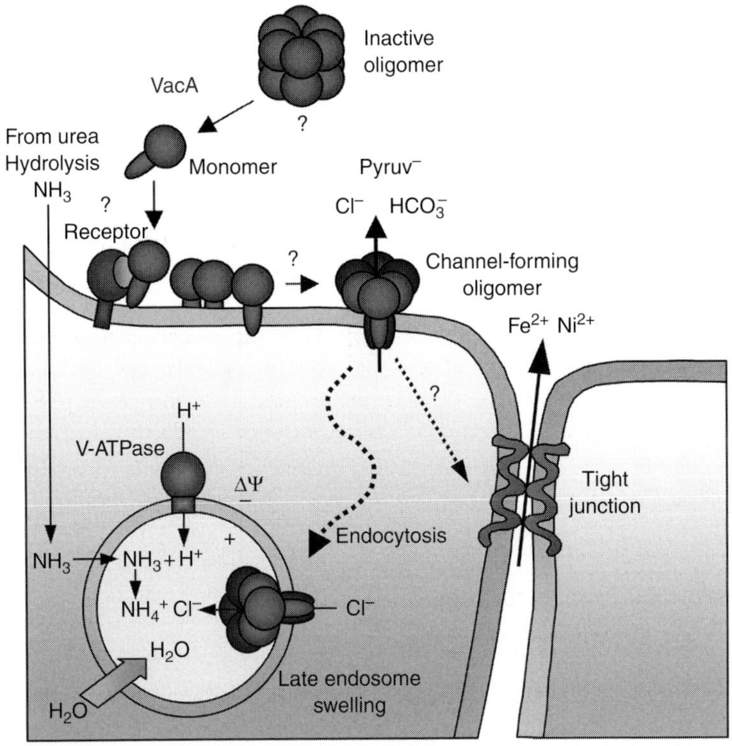

Figure 4.3 The current mechanistic model for VacA-induced cell vacuolation. See text for details. (See Colour Plate III.)

from the cell cytosol, driven by the osmotic pressure created in the endolysosomal compartment. This model is consistent with the observation that the conduction of NH_4^+ ions by the VacA channel is very low, when compared to that of Cl^- (Tombola *et al.*, 1999a). This scheme implies that vacuole development is energy-dependent and driven by active proton pumping, and therefore also explains the inhibitory effect of V-ATPase inhibitors, such as bafilomycins (Papini *et al.*, 1993).

The model is backed by additional experimental evidence. As mentioned, patch–clamp experiments have shown that low doses of activated toxin lead to the rapid appearance of an anionic current across the cell plasma membrane, causing its depolarization (Szabò *et al.*, 1999; Tombola *et al.*, 1999b). The voltage dependence and anion selectivity of this current is very similar to that induced by VacA in the planar lipid bilayer. The indirect activation of endogenous anionic pores by VacA, for example, Ca^{2+} and stretch-activated Cl^- channels, was ruled out. These data showed that one of the earliest effects of VacA intoxication is the interaction with the cell plasma membrane and assembling of channels, as in artificial lipid bilayers.

VacA channels were inhibited by a series of typical chloride channel blockers. In the case of VacA 17874, Tombola *et al.* (1999b) determined the following order of inhibitor efficacy (for inhibition from the *cis* side and in the lower inhibitor concentration range): 5-Nitro-2-(3-phenylamino)benzoate (NPPB) > DIDS > flufenamic acid > niflumic acid > N-phenylanthranilic acid > SITS > IAA-94. These compounds blocked VacA channels

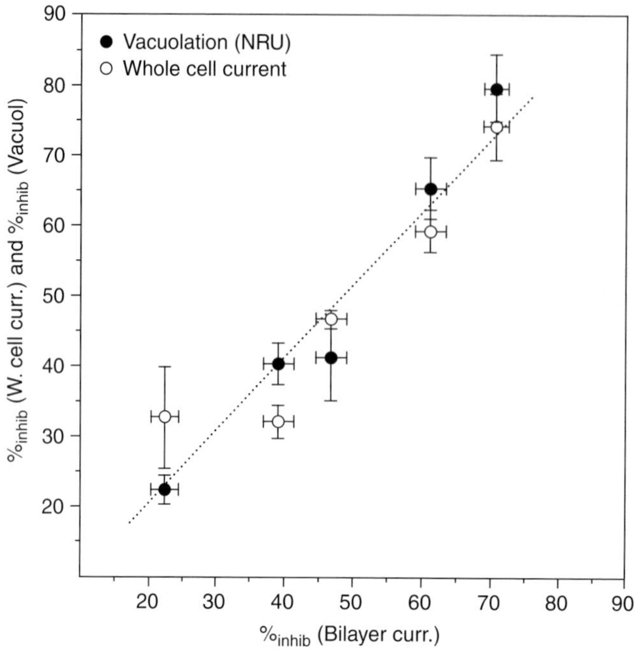

Figure 4.4 Correlation between the inhibition of VacA-elicited transmembrane current in planar bilayer experiments (abscissa) and that of whole cell current (○ patch-clamp) and Neutral Red Uptake (NRU) (●) by HeLa cells. Error bars represent the standard error of the means. Graded inhibition was caused by (in order of increasing inhibitory power): IAA-94, NPA, Niflumic acid, Flufenamic acid and NPPB added at a concentration of 100 μM.

(both in planar lipid bilayers and in the plasma membrane of HeLa cells) and also inhibited vacuole formation, with the same relative effectiveness (Figure 4.4), while they did not interfere with intracellular acidification or VacA internalization. Moreover, cell vacuolation induced by cytosolically expressed VacA was also impaired by NPPB. With regard to this latter experimental system, it seems quite possible that VacA, produced as mature monomers in the cytoplasm, may insert into the membrane of endosomes (and possibly other organelles). This hypothesis is consistent with the need for V-ATPase activity (de Bernard *et al.*, 1997) and for weak bases also for vacuolation by intracellularly expressed VacA.

The formation of VacA channels in the apical membrane of polarized cells in high-resistance epithelia appears also essential for the toxin-induced permeabilization of the monolayers to ions and neutral molecules via the paracellular pathway. In fact, this effect of VacA is counteracted and partially reverted by NPPB, and, with lower efficacy, by IAA-94 (Szabò *et al.*, 1999). In this case no hypothesis on the mechanism linking the channel to the modulation of cell–cell junction is at present formulated.

A considerable amount of supporting evidence for the model above has thus been gathered using model and cellular systems. Nevertheless, very little is known about the acute functional alterations induced by the toxin in the actual target of the bacterium: the gastric epithelium. The isolated gastric mucosa of *Rana esculenta* has recently been used to study the effects of acute topical addition of VacA by electrophysiological techniques applied to oxyntic (OCs) and superficial epithelial cells (SECs) (Debellis *et al.*, 2001). Using the same preparation, it was also possible to monitor pH changes in the lumen of single gastric glands and variations of intracellular [Cl^-] and pH in SECs. Treatment of the gastric tissue with activated VacA induced a significant (25%) apparent decrease of histamine-stimulated acid secretion. Measurement of net alkaline secretion and microelectrode monitoring of HCl secretion by gastric glands demonstrated that this decrease was due to an enhancement of HCO_3^- secretion by SECs. This effect may be attributed to the formation of anion-selective VacA channels in the apical membranes of SECs. To test this hypothesis, single superficial cells were impaled with selective microelectrodes, and the variation of cytoplasmic [HCO_3^-] or [Cl^-] was monitored upon application of a relevant transmembrane ion gradient (Figure 4.5). Exposure to VacA led to a significant intracellular acidification or alkalization upon luminal [HCO_3^-] reduction or increase, respectively (Figure 4.5A). These data indicate that, following VacA addition, the apical membrane of SECs became permeable to HCO_3^-. Moreover, following sudden luminal [Cl^-] reduction, cytosolic [Cl^-] decreased in cells treated with activated VacA (Figure 4.5B). In analogy to observations with artificial planar lipid bilayers (Tombola *et al.*, 1999a) and HeLa cells (Szabò *et al.*, 1999; Tombola *et al.*, 1999b), NPPB significantly decreased the pH_i response to luminal [HCO_3^-] reduction, while IAA-94 had little effect. This further supports the proposal that VacA can permeabilize gastric cells much as it does HeLa cells.

Increased urea and HCO_3^- efflux from the gastric superficial cells, to which bacterial cells are intimately bound, may represent a strategy employed by *H. pylori* to colonize the stomach. Urea is critically required for the survival of infecting *H. pylori*: the bacterium urease converts it to ammonia and bicarbonate, thus preventing lethal acidification of the cytosol

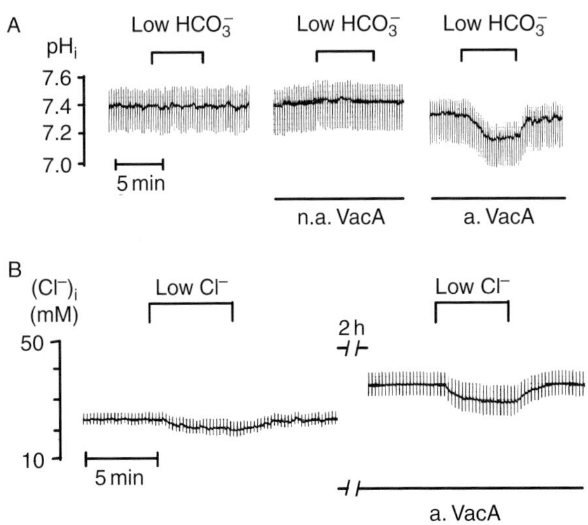

Figure 4.5 VacA permeabilizes frog gastric cells to Cl^- and HCO_3^-. SECs of isolated stomach from *Rana esculenta* were impaled with double-barreled pH (A) or Cl^- (B) -sensitive microelectrodes and treated with nothing, non-activated VacA (n.a. VacA) or activated VacA (a. VacA) (40 nM). After 2–3 h, for the period indicated, the lumenal concentration of HCO_3^- was reduced from 17.8 to 2 mM (A), or that of Cl^- from 91.4 to 2 mM (B).

and of the immediate environment of the cell. Low pH-activated VacA presumably acts to increase urea supply when needed. An outward HCO_3^- alkaline flux is expected to modify the pH gradient across the mucus layer covering the gastric epithelium, with the creation of a microenvironment more suitable for *H. pylori* life. Moreover, since it is known that HCO_3^- is also an important metabolic substrate for *H. pylori* (Burns *et al.*, 1995), necessary for its growth, it can be speculated that a VacA-dependent efflux of this compound from the gastric mucosa may increase the efficacy of *H. pylori* metabolism.

Biophysical studies on the VacA channel

The mechanism of block by NPPB. NPPB, an amphiphilic molecule, is a popular inhibitor of anion-selective channels whose action often displays unexpected features. In the case of VacA, single-channel recordings indicated that NPPB was acting as a fast blocker, in agreement with the general characterization of this and related compounds as blockers. The extent of inhibition in multi-channel planar bilayer experiments turned out to depend on the applied voltage, but not on the side of addition of the inhibitor. Inhibition was lower at $V(cis) > 0$, no matter whether NPPB was present on the *cis* or *trans* side. This result was counter-intuitive and at variance with that of the well-known "classical" blocker DIDS, which acts by entering the channel lumen and binding to a blocking site. In the case of this latter compound, an anion at neutral pHs, the effect of a change of the side of addition was analogous to that of an inversion of the field polarity.

An investigation sparked by these observations (Tombola *et al.*, 2000), led to their rationalization within a model which can also explain the variable features of the inhibition of other channels. Due to its amphipatic properties, NPPB appears to partition into the membrane, where it reaches an average concentration determined by the average concentrations in the two aqueous phases bathing the bilayer, regardless of the side of addition. From the membrane, the compound then migrates to the blocking site within the channel by a voltage-independent pathway, presumably by diffusing in the plane of the bilayer and infiltrating between toxin monomers. This suggests a certain degree of "looseness" in the barrel-like structure of the oligomer, which would allow molecules in the 300 Da range to squeeze through between the "staves" (Figure 4.6). Efflux from the channel takes place, at least in part, by voltage-dependent diffusion along the channel lumen, thus accounting for the existence of a voltage effect. The blocking site for NPPB in VacA was determined to be situated in the distal (*trans*-side facing) half of the complex, on the basis of an analysis of the concentration- and voltage-dependence of block. The model ought to be of general applicability: the exact characteristics of the inhibition of other channels by NPPB may be determined by the relative importance of the various ways available to reach and leave the blocking site. Thus, a completely voltage-independent block may be due to an exclusively "sidewise" permeability of the oligomeric channel, while a classical-type voltage dependence would indicate that NPPB only diffuses as a charged species along the channel lumen.

Structural hints from blockade studies. Blockers have often been used as gauging tools to gain structural information about channels. SITS and in particular DIDS, two well-known stilbene derivatives and blockers, have been used in this way on VacA.

As mentioned, SITS inhibited m1 VacA 17874 from the *cis*, but not from the *trans* side (Tombola *et al.*, 2000). DIDS was reported to behave analogously with m1 VacA 60190 (Iwamoto *et al.*, 1999), but studies with 17874 (Tombola *et al.*, 2001a) revealed complexities.

When this toxin isoform was incorporated in planar bilayers made of zwitterionic lipids, DIDS inhibited both from the *cis* and from the *trans* side, although it was considerably less

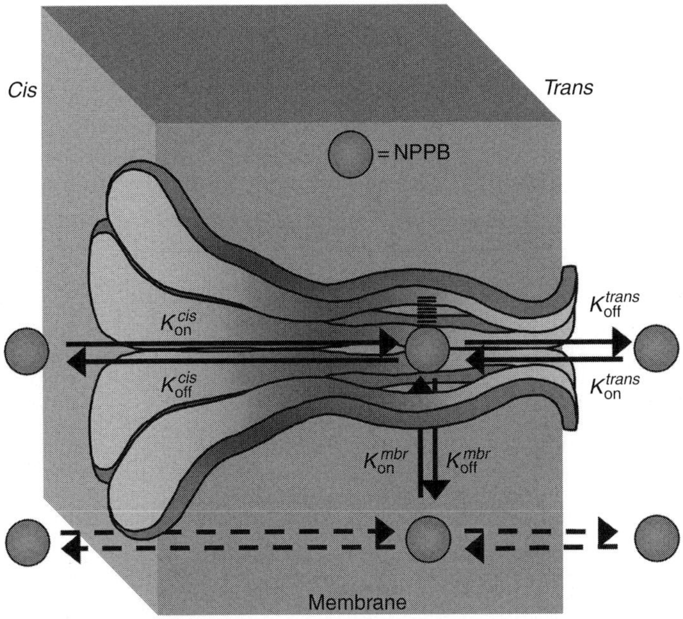

Figure 4.6 A model of VacA inhibition by NPPB. A cartoon of a cross-section of a VacA hexamer, illustrating the pathways for block by NPPB. Inside the channel, the blocker is positioned at the approximate location of the blocking site. The rate constants for the various binding equilibria are indicated. The length of the corresponding arrows is arbitrary and it is not meant to reflect the relative magnitudes of the constants. In particular, K_{on}^{mbr} is thought to be larger than either K_{on}^{cis} or K_{on}^{trans}. "*Cis*" indicates the side from which the toxin inserted into the membrane. (See Colour Plate IV.)

effective when in *trans*. This suggests that for medium-sized molecules such as DIDS access to the lumen of the VacA channel through the *trans*-facing opening is more difficult than entry from the opposite extremity. That is, assuming that the effect is due to structural characteristics rather than to charge interactions, the channel may be imagined to have a roughly funnel-like shape, with the wider opening on the side from which the toxin incorporated. An analysis of the voltage dependence of inhibition indicated that the blocking site for this compound is located in the *cis* side facing half of the VacA channel, that is, it is distinct from the one for NPPB.

In azolectin, DIDS still inhibited to a comparable extent from the *cis* side, but was nearly ineffective when in *trans*. This difference can be at least partly attributed to the presence of negatively charged lipids, since their addition to zwitterionic membranes led to a partial loss of inhibition by *trans*-DIDS. Conversely, masking azolectin charges with polylysine partially restored it in that system. The effect cannot, however, be due simply to a repulsive interaction between membrane charges and the negatively charged blocker, because lipid composition had only a minor effect when block took place from the *cis* side. A possible explanation is that negative charges in the bilayer induce a conformational change of the *trans*-facing portion of VacA, such as to restrict access to DIDS and thus prevent block. One can further speculate that a repulsive interaction may take place between membrane charges and deprotonated acidic residues of the protein. The latter might indeed be responsible for the lower effectiveness of *trans*-side SITS and DIDS even in zwitterionic membranes.

Concluding remarks

Since the identification of a vacuolating activity in the extracellular medium of *H. pylori* cultures (Leunk *et al.*, 1988), the crucial relevance for *H. pylori* disease of the protein responsible for that phenotype (VacA) has clearly emerged (Atherton *et al.*, 2001). The genetic and biochemical properties of VacA were defined thanks to the work of many laboratories while other intoxicated phenotypes, distinct from vacuolation, were described and characterized in detail.

About seven years after its purification (Cover and Blaser, 1992) VacA was recognized as a pore-forming toxin (Tombola *et al.*, 1999a). In this chapter, we have summarized the present knowledge on the channels properties of VacA, from the physical–chemical, structural and biological points of view. Targeting endo-lysosomes of membrane-inserted VacA pores, together with their anion selectivity, provided the basis for the formulation of an hypothesis of vacuole genesis. In short, this model hypothesizes that VacA channels in the endo-lysosomal membranes favor the accumulation of Cl^- and NH_4^+ in the lumen of the organelles, and the influx of water.

However, although channel action explains much of the mechanism of vacuole development, it is reasonable to ask whether VacA is made by *H. pylori* for the sole purpose of triggering vacuole formation.

Although vacuolation has been important in revealing the presence of VacA, this cellular effect is clearly dependent on a combination of conditions: the presence of weak bases, low serum concentrations, low density of cell culture (de Bernard *et al.*, 1998b).

Moreover, in models closer to the in vivo target of VacA, the polarized epithelium, vacuolation does not take place (Papini *et al.*, 1998; Pelicic *et al.*, 1999). There is not at present a clear explanation of this fact, but, since vacuolation requires VacA endocytosis, it is reasonable to assume that the rate of VacA endocytosis at the apical membrane and the transport of the toxin to late endosomes is low.

Based on these observations and also on studies in gastric tissue models in frogs (Debellis *et al.*, unpublished), we propose that the target and the functional meaning of VacA is not vacuolation *per se*, but rather the permeabilization of the apical plasma membrane of gastric epithelial cells. As represented schematically in Figure 4.3, the anion-selective channel of VacA may be functional to the adaptation of *H. pylori* to the environment of the mucus layer covering the gastric epithelium. It may increase the efflux of metabolic substrates for *H. pylori* like pyruvate or HCO_3^-. The conduction of urea, along with the increased efflux of HCO_3^-, confirmed in the frog gastric model, is supposed to allow the creation of a micro-environment closer to pH neutrality around the parasite. The anion-selective VacA pore inserted in the apical cell membrane of gastric cells is, in summary, proposed to help *H. pylori* colonize the stomach and its persistence.

In agreement with a recent study in animal models (Salama *et al.*, 2001), our data suggest that the link between VacA and pathology is mostly attributable to the positive effect of the channel on gastric infection and colonization by *H. pylori*.

Acknowledgments

This work was supported in part by Cofin2000 program of MURST of the University of Bari and by CNR Progetto Finalizzato Biotecnologie (97.01168.PF 49).

References

Abrami, L., Fivaz, M., Glauser, P. E., Parton, R. G. and van der Goot, F. G. (1998) A pore-forming toxin interacts with a GPI-anchored protein and causes vacuolation of the endoplasmic reticulum. *J. Cell Biol.*, **140**, 525–540.

Atherton, J. C., Cao, P., Peek, R. M., Jr., Tummuru, M. K., Blaser, M. J. and Cover, T. L. (1995) Mosaicism in vacuolating cytotoxin alleles of *Helicobacter pylori*. Association of specific vacA types with cytotoxin production and peptic ulceration. *J. Biol. Chem.*, **270**, 17771–17777.

Atherton, J. C., Cover, T. L., Papini, E. and Telford, J. L. (2001) Vacuolating cytotoxin. In *Helicobacter pylori: Physiology and Genetics*, edited by H. L. T. Mobley, G. L. Mendz and S. L. Hazell, Chapter 9, pp. 97–110. Pub ASM Press, Washington, D. C.

Burns, B. P., Hazell, S. L. and Mendz, G. L. (1995) Acetyl-CoA carboxylase activity in *Helicobacter pylori* and the requirement of increased CO_2 for growth. *Microbiology*, **141**, 3113–3118.

Burroni, D., Lupetti, P., Pagliaccia, C., Reyrat, J.-M., Dallai, R., Rappuoli, R. *et al.* (1998) Deletion of the major proteolytic site of the *Helicobacter pylori* cytotoxin does not influence toxin activity but favours assembly of the toxin into hexameric structures. *Infect. Immun.*, **66**, 5547–5550.

Coelho, A., Andrade, J. R. C., Vincente, C. A. and DiRita, V. J. (2000) Cytotoxic cell vacuolating activity from *Vibrio cholerae* Hemolysin. *Infect. Immun.*, **68**, 1700–1705.

Cover, T. L. and Blaser, M. J. (1992) Purification and characterization of the vacuolating toxin from *Helicobacter pylori*. *J. Biol. Chem.*, **267**, 10570–10575.

Cover, T. L., Hanson, P. I. and Heuser, J. E. (1997) Acid-induced dissociation of VacA, the *Helicobacter pylori* vacuolating cytotoxin, reveals its pattern of assembly. *J. Cell. Biol.*, **138**, 759–769.

Covacci, A., Telford, J. L., Del Giudice, G., Parsonnet, J. and Rappuoli, R. (1999) *Helicobacter pylori* virulence and genetic geography. *Science*, **288**, 1329–1333.

Czajkowsky, D. M., Iwamoto, H., Cover, T. L. and Shao, Z. (1999) The vacuolating toxin from *Helicobacter pylori* forms hexameric pores in lipid bilayers at low pH. *Proc. Natl. Acad. Sci. USA*, **96**, 2001–2006.

Debellis, L., Papini, E., Caroppo, R., Montecucco, C. and Curci, S. (2001) *Helicobacter pylori* cytotoxin VacA increases alkaline secretion in gastric epithelial cells. *Am. J. Physiol. Gastrointest. Liver Physiol.*, **281**, G1440–1448.

de Bernard, M., Papini, E., de Filippis, V., Gottardi, E., Telford, J. Manetti, R. *et al.* (1995) Low pH activates the vacuolating toxin of *Helicobacter pylori*, which becomes acid and pepsin resistant. *J. Biol. Chem.*, **270**, 23937–23940.

de Bernard, M., Aricò, B., Papini, E., Rizzuto, R., Grandi, G., Rappuoli, R. *et al.* (1997) *Helicobacter pylori* toxin VacA induces vacuole formation by acting in the cell cytosol. *Mol. Microbiol.*, **26**, 665–674.

de Bernard, M., Burroni, D., Papini, E., Rappuoli, R., Telford, J. and Montecucco, C. (1998a) Identification of the *Helicobacter pylori* VacA toxin domain active in the cell cytosol. *Infect. Immun.*, **66**, 6014–6016.

de Bernard, M., Moschioni, M., Papini, E., Telford, J. L., Rappuoli, R. and Montecucco, C. (1998b) Cell vacuolization induced by *Helicobacter pylori* VacA toxin: cell line sensitivity and quantitative estimation. *Toxicol. Lett.* , **99**, 109–115.

Dunn, B. E., Cohen, H. and Blaser, M. J. (1997) *Helicobacter pylori*. *Clin. Microbiol. Rev.*, **10**, 720–741.

Garner, J. A. and Cover, T. L. (1996) Binding and internalization of the *Helicobacter pylori* vacuolating cytotoxin by epithelial cells. *Infect. Immun.*, **64**, 4197–4203.

Gerhardt, M., Lehn, N., Neumayer, N., Boren, T., Rad, R., Schepp, W. *et al.* (1999) Clinical relevance of the *Helicobacter pylori* gene for blood-group antigen-binding adhesin. *Proc. Natl. Acad. Sci. USA*, **96**, 12778–12783.

Hille, B. (1992) *Ionic Channels of Excitable Membranes*, 2nd edition. Sinauer Assoc., Sunderland, MA.

Iwamoto, H., Czajkowsky, D. M., Cover, T. L., Szabo, G. and Shao, Z. (1999) VacA from *Helicobacter pylori*: a hexameric chloride channel. *FEBS Lett.*, **450**, 101–114.

Ji, X., Fernandez, T., Burroni, D., Pagliaccia, C., Atherton, J. C., Reyrat, J. M. *et al.* (2000) Cell specificity of *Helicobacter pylori* cytotoxin is determined by a short region in the polymorphic midregion. *Infect. Immun.*, **68**, 3754–3757.

Lanzavecchia, S., Bellon, P. L., Lupetti, P., Dallai, R., Rappuoli, R. and Telford, J. L. (1998) Three-dimensional reconstruction of metal replicas of the *Helicobacter pylori* vacuolating cytotoxin. *J. Struct. Bio.*, **121**, 9–18.

Leunk, R. D., Johnson, P. T., David, B. C., Kraft, W. G. and Morgan, D. R. (1988) Cytotoxic activity in broth-culture filtrates of *Campylobacter pylori*. *J. Med. Microbiol.*, **26**, 93–99.

Lupetti, P., Heuser, J. E., Manetti, R., Massari, P., Lanzavecchia, S., Bellon, P. L. *et al.* (1996) Oligomeric and subunit structure of the *Helicobacter pylori* vacuolating cytotoxin. *J. Cell. Biol.*, **133**, 801–807.

McClain, M. S., Schraw, W., Ricci, V., Boquet, P. and Cover, T. L. (2000) Acid activation of *Helicobacter pylori* vacuolating cytotoxin (VacA) results in toxin internalization by eukaryotic cells. *Mol. Microbiol.*, **37**, 433–442.

McClain, M. S., Cao, P. and Cover, T. L. (2001) The amino-terminal hydrophobic region of *Helicobacter pylori* vacuolating cytotoxin (VacA) mediates transmembrane protein dimerization. *Infect. Immun.*, **69**, 1181–1184.

Molinari, M., Galli, C., de Bernard, M., Norais, N., Ruysschaert, J. M., Rappuoli, R. and Montecucco, C. (1998) The acid activation of *Helicobacter pylori* toxin VacA: structural and membrane binding studies. *Biochem. Biophys. Res. Commun.*, **248**, 334–340.

Montecucco, C., Papini, E., de Bernard, M., Telford, J. L. and Rappuoli, R. (1999) *Helicobacter pylori* vacuolating cytotoxin and associated pathogenic factors. In *The Comprehensive Sourcebook of Bacterial Protein Toxins*, edited by J. E. Alouf and J. H. Freer, pp. 264–283. Academic Press, San Diego, USA.

Montecucco, C. and Rappuoli, R. (2001) Living dangerously: how *Helicobacter pylori* survives in the human stomach. *Nat. Rev. Mol. Cell Biol.*, **2**, 457–466.

Nguyen, V. Q., Caprioli, R. M. and Cover, T. L. (2001) Carboxy-terminal proteolytic processing of *Helicobacter pylori* vacuolating toxin. *Infect. Immun.*, **69**, 543–546.

Ogura, K., Maeda, S., Nakao, M., Watanabe, T., Tada, M., Kyutoku, T. *et al.* (2000) Virulence factors of *Helicobacter pylori* responsible for gastric diseases in Mongolian gerbil. *J. Exp. Med.*, **192**, 1601–1610.

Pagliaccia, C., de Bernard, M., Lupetti, P., Ji, X., Burroni, D., Cover, T. L. *et al.* (1998) The m2 form of the *Helicobacter pylori* cytotoxin has cell type-specific vacuolating activity. *Proc. Natl. Acad. Sci. USA*, **95**, 10212–10217.

Pagliaccia, C., Wang, X. M., Tardy, F., Telford, J. L., Ruysschaert, J. M. and Cabiaux, V. (2000) Structure and interaction of VacA of *Helicobacter pylori* with a lipid membrane. *Eur. J. Biochem.*, **267**, 104–109.

Papini, E., Bugnoli, M., De Bernard, M., Figura, N., Rappuoli, R. and Montecucco, C. (1993) Bafilomycin A1 inhibits *Helicobacter pylori*-induced vacuolization of HeLa cells. *Mol. Microbiol.*, **7**, 323–327.

Papini, E., de Bernard, M., Milia, E., Bugnoli, M., Zerial, M., Rappuoli, R. *et al.* (1994) Cellular vacuoles induced by *Helicobacter pylori* originate from late endosomal compartments. *Proc. Natl. Acad. Sci. USA*, **91**, 9720–9724.

Papini, E., Satin, B., Norais, N., de Bernard, M., Telford, J. L., Rappuoli, R. *et al.* (1998) Selective increase of the permeability of polarized epithelial cell monolayers by *Helicobacter pylori* vacuolating toxin. *J. Clin. Invest.*, **102**, 813–820.

Papini, E., Zoratti, M. and Cover, T. L. (2001) In search of the *Helicobacter pylori* VacA mechanism of action. *Toxicon*, **39**, 1757–1767.

Pelicic, V., Reyrat, J. M., Sartori, L., Pagliaccia, C., Rappuoli, R., Telford, J. L. *et al.* (1999) *Helicobacter pylori* VacA cytotoxin associated with the bacteria increases epithelial permeability independently of its vacuolating activity. *Microbiology*, **145**, 2043–2050.

Reyrat, J. M., Lanzavecchia, S., Lupetti, P., de Bernard, M., Pagliaccia, C., Pelicic, V. *et al.* (1999) 3D imaging of the 58 kDa cell binding subunit of the *Helicobacter pylori* cytotoxin. *J. Mol. Biol.*, **290**, 459–470.

Ricci, V., Sommi, P., Fiocca, R., Romano, M., Solcia, E. and Ventura, U. (1997) *Helicobacter pylori* vacuolating toxin accumulates within the endosomal-vacuolar compartment of cultured gastric cells and potentiates the vacuolating activity of ammonia. *J. Pathol.*, **183**, 453–459.

Ricci, V., Galmiche, A., Doye, A., Necchi, V., Solcia, E. and Boquet, P. (2000) High cell sensitivity to *Helicobacter pylori* VacA toxin depends on a GPI-anchored protein and is not blocked by inhibition of the clathrin-mediated pathway of endocytosis. *Mol. Cell. Biol.*, **11**, 3897–3909.

Salama, N. R., Otto, G., Tompkins, L. and Falkow, S. (2001) Vacuolating cytotoxin of *Helicobacter pylori* plays a role during colonization in a mouse model of infection. *Infect. Immun.*, **69**, 730–736.

Szabò, I., Brutsche, S., Tombola, F., Moschioni, M., Satin, B., Telford, J. L., *et al.* (1999) Formation of anion-selective channels in the cell plasma membrane by the toxin VacA of *Helicobacter pylori* is required for its biological activity. *EMBO J.*, **18**, 5517–5527.

Telford, J. L., Ghiara, P., Dell'Orco, M., Comanducci, M., Burroni, D., Bugnoli, M. *et al.* (1994). Gene structure of the *Helicobacter pylori* cytotoxin and evidence of its key role in gastric disease. *J. Exp. Med.*, **179**, 1653–1658.

Tombola, F., Carlesso, C., Szabò, I., de Bernard, M., Reyrat, J. M., Telford, J. L. *et al.* (1999a) *Helicobacter pylori* vacuolating toxin forms anion-selective channels in planar lipid bilayers: possible implications for the mechanism of cellular vacuolation. *Biophys. J.*, **76**, 1401–1419.

Tombola, F., Oregna, F., Brutsche, S., Szabo, I., Del Giudice, G., Rappuoli, R. *et al.* (1999b) Inhibition of the vacuolating and anion channel activities of the VacA toxin of *Helicobacter pylori*. *FEBS Lett.*, **460**, 221–225.

Tombola, F., Del Giudice, G., Papini, E. and Zoratti, M. (2000) Blockers of VacA provide insights into the structure of the pore. *Biophys. J.*, **79**, 863–873.

Tombola, F., Pagliaccia, C., Campello, S., Telford, J. L., Montecucco, C., Papini, E. and Zoratti, M. (2001a) How the loop and middle regions influence the properties of *Helicobacter pylori* VacA channels. *Biophys. J.*, **81**, 3204–3215.

Tombola, F., Morbiato, L., Del Giudice, G., Rappuoli, R., Zoratti, M. and Papini, E. (2001b) The *Helicobacter pylori* VacA toxin is a urea permease that promotes urea diffusion across epithelia. *J. Clin. Invest.*, **108**, 929–937.

Yahiro, K., Niidome, T., Kimura, M., Hatakeyama, T., Aoyagi, H., Kurazono, H. *et al.* (1999) Activation of *Helicobacter pylori* VacA toxin by alkaline or acid conditions increases its binding to a 250-kDa receptor protein–tyrosine phosphatase beta. *J. Biol. Chem.*, **274**, 36693–36699.

Ye, D., Willhite, D. C. and Blanke, S. R. (1999) Identification of the minimal intracellular vacuolating domain of the *Helicobacter pylori* vacuolating toxin. *J. Biol. Chem.*, **274**, 9277–9282.

Vinion-Dubiel, A. D., McClain, M. S., Czajkowsky, D. M., Iwamoto, H., Ye, D., Cao, P. *et al.* (1999) A dominant negative mutant of *Helicobacter pylori* vacuolating toxin (VacA) inhibits VacA-induced cell vacuolation. *J. Biol. Chem.*, **274**, 37736–37742.

Wang, H. J. and Wang, W. C. (2000) Expression and binding analysis of GST–VacA fusions reveals that the C-terminal approximately 100-residue segment of exotoxin is crucial for binding in HeLa cells. *Biochem. Biophys. Res. Commun.*, **278**, 449–454.

5 Delta-endotoxin mode of action

Doron Gerber and Yechiel Shai

The δ-endotoxins are highly potent insecticidal toxins produced by *Bacillus thuringiensis* bacteria as parasporal crystals. The tertiary structure of these toxins consists of three domains. The N-terminal domain (Domain I) is believed to be responsible for pore formation, and the other two domains are responsible for specific binding to the receptor. The mode of action of the pore-forming domain fits the Umbrella model. After receptor binding, domain I swings away from the protein and binds to the target membrane. This is followed by a drastic structural rearrangement, during which a hairpin formed by α4-loop-α5 penetrates into the core of the membrane. The rest of domain I helices spread on the surface of the membrane, much like the ribs of an umbrella, hence the name of the model. Upon oligomerization an active pore is formed. The evidence to support this model was derived mainly through using synthetic segments in model systems and combining the results with functional information on the whole protein. The latter were achieved through conventional methods, such as site directed mutagenesis. The superb correlation between peptide and whole-protein strategies greatly enhanced our view of the δ-endotoxin structure and function within the membrane. As long as we cannot obtain high-resolution structural information on membrane protein, strategies employing short segments remain a powerful tool to probe the structural and functional character of membrane proteins.

General introduction

Membrane interacting toxins are the major defence and attack strategies of bacteria (Cramer *et al.*, 1990; Bayley, 1997; Lesieur *et al.*, 1997). The interaction of the toxin with the target cell membrane frequently results in permeation of the membrane or translocation of proteins into the cytoplasm. Many of these toxins adopt an alpha helical bundle structure in order for the protein to remain soluble (Lesieur *et al.*, 1997). This kind of structure strategy allows the toxin to hide the hydrophobic membrane binding surfaces, within the core of the protein, while the hydrophilic surfaces of the protein face the aqueous medium surrounding the target cell. These proteins have a very high similarity in the 3D structure of the pore-forming domain, although sequence homology is usually poor (Parker and Pattus, 1993) (Figure 5.1). Upon binding to membrane these proteins undergo a metamorphosis into membrane proteins. This demands a drastic rearrangement of the structure, which results in the insertion of part of the protein into the membrane, oligomerization and finally pore formation.

The myriad of high-resolution protein structures available to date is misleading when it comes to membrane proteins. Out of approximately 16,000 solved structures less than a 100 belong to membrane proteins and much less were solved in a natural membrane environment.

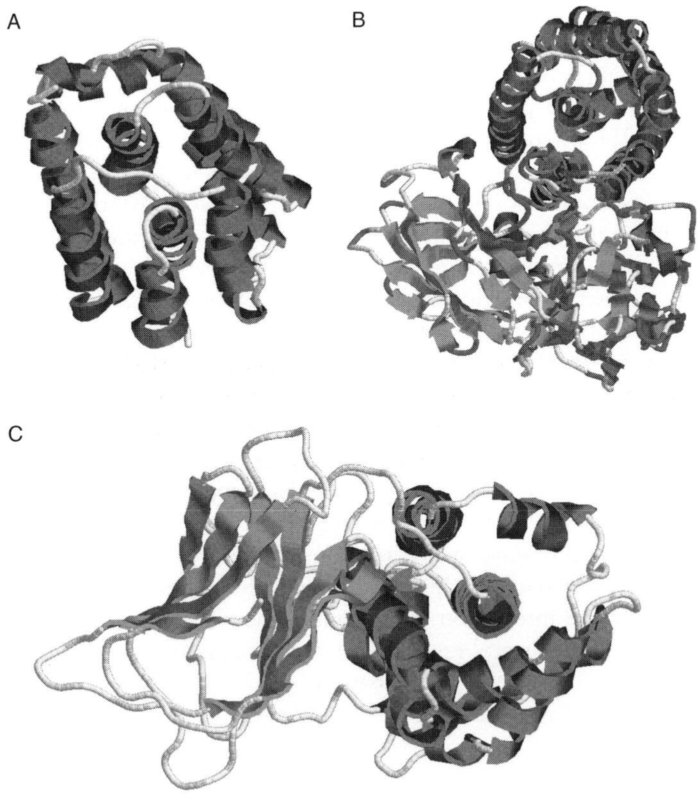

Figure 5.1 A cartoon representation of the helical bundle structures of the pore-forming domains from different toxin families. The representative molecules are: (A) *Citrobacter freundii* colicin A pore-forming domain (Tucker *et al.*, 1986); (B) *B. thuringiensis* Cry3A δ-endotoxin (Li *et al.*, 1991); (C) *Clostridium perfringens* α toxin (Basak *et al.*, 1994, 1998). (See Colour Plate V.)

Considering that the first 3D structure of a membrane bound protein was solved 20 years ago it is safe to say that high-resolution structure of membrane proteins is very difficult to obtain. The main questions facing us when investigating a membrane toxin are: What is the active form of the toxin (monomer or oligomer)? Which parts insert into the membrane? What is the oligomeric state of the toxin in the pore and how is it assembled? What is the substrate of this pore (is it at all specific) and where do they bind? In order to answer these questions we need to resort to molecular biology and to biochemical methods such as site directed mutagenesis, channel and pore-formation measurements and toxicity screening. For example, fluorescent labeling of the protein can help determine the membrane binding site by using membrane sensitive probes. The oligomeric state can also be determined using such labeling strategies. But unfortunately, specific labeling of whole proteins with fluorescent probes is inefficient making the investigation of the protein orientation and assembly within the membrane limited. To solve this problem one must incorporate in vitro model systems together with the traditional methods, more specifically, the use of synthetic protein segments and phospholipid vesicles. This combination can greatly facilitate our understanding of the overall picture.

The use of peptide segments in the investigation of membrane proteins has been well established as a unique and powerful tool. Perhaps the most striking examples are the studies done with bacteriorhodopsin (Popot *et al.*, 1986), lactose permease (Bibi and Kaback, 1990) and potassium channel (Ben-Efraim and Shai, 1996, 1997; MacKinnon *et al.*, 1998), although many other membrane protein examples are available (Kliger *et al.*, 2000; Peisajovich and Shai, 2001). Bacteriorhodopsin was the first protein to have its high-resolution structure within the membrane solved (Michel *et al.*, 1980; Ceska and Henderson, 1990). Reconstitution of several polypeptide segments of bacteriorhodopsin into the bacterial membrane resulted in a functional protein with a structure similar to the wild type (Popot *et al.*, 1986). Similarly, segments of the lactose permease were shown to reassemble and form an active protein with wild type activity (Bibi and Kaback, 1990). In the case of the KcsA channel, the high-resolution structure was solved in a membrane environment (MacKinnon *et al.*, 1998). The orientation and assembly of the channel within the membrane was in excellent agreement with a previous model obtained for a close homolog channel, ROMK1, based on studies done with synthetic peptides and their fluorescent labeled analogs (Ben-Efraim and Shai, 1996, 1997).

The synthetic peptide approach allows us to expand our basic tools. We can now investigate membrane-binding affinity of different segments and determine the membrane localization of the protein. The secondary structure in membrane environments can be determined by methods such as circular dichroism (CD) or infrared spectroscopy (FTIR). We can investigate assembly and interaction of different segments in the protein by labeling them with different fluorescent probes and measuring energy transfer between the different segments. Furthermore, it is possible to "zoom in" on the active site of the protein, that is, the pore-forming segment in the case of membrane-bound toxins.

The insecticidal δ-endotoxin protein family

The δ-endotoxins are highly potent insecticidal toxins produced by *B. thuringiensis* bacteria. These toxins belong to the crystal insecticidal proteins (Cry) gene family, which is very diverse in proteins (Schnepf *et al.*, 1998). The Cry family members are cataloged into subfamilies according to their sequence homology (Dean *et al.*, 1996; Crickmore *et al.*, 1998; Rajamohan *et al.*, 1998). While the overall homology of the Cry toxins is quite poor, the mode of toxicity seems to be very similar and so does the tertiary structure (Parker and Pattus, 1993) (Figure 5.2). They are produced as protoxins in parasporal crystals to be digested by the target insect. Digestion by "trypsin-like" proteases solubilizes the inclusions and activates the toxin close to the target cells. In turn, this promotes recognition followed by binding to a specific receptor in a reversible fashion. Following receptor binding the toxin undergoes a conformational change, inserts into the membrane irreversibly and forms a pore (Aronson and Shai, 2001).

The advantage of using δ-endotoxins as pesticides rather than the conventional chemical pesticides is obvious. Current available pesticides are non-specific chemicals and are thus not very efficient. Furthermore, the damage to the environment caused by these chemicals is becoming alarmingly obvious. They enter the food chain and poison the local ecosystem, seeping into underground water reservoirs. The long-term damage done by the broad use of chemical pesticides is very difficult to assess, as some of the toxic effects, like cancer, develop only years after the initial damage was done. A replacement to these toxic chemical compounds is badly needed and it has to be target-specific and efficient. Ironically, nature came to our rescue. The diversity of the δ-endotoxin family allows for specific and

A

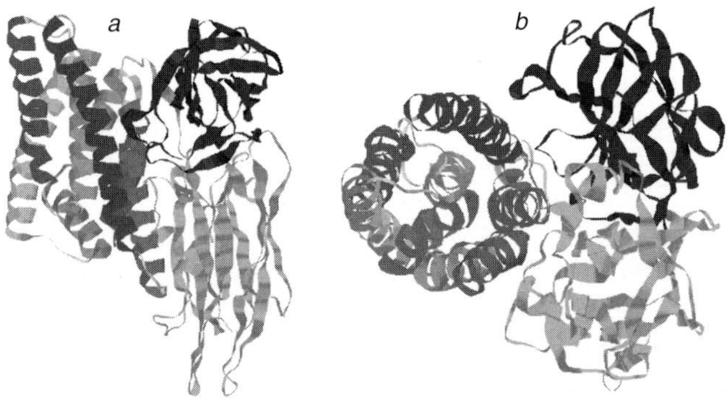

B α1 X–NH–KDVIQKGISVVGDLLGV–COOH

α2 X–NH–GGALVSFYTNFLNT IWPSEDPWKAMENVEALM–COOH

α3 X–NH–DYAKNKALAELQGLQNNVEDYVSALSSWQK–COOH

α4 X–NH–HASQGRIRELFSQAESHFRNSMPSFA–COOH

α5 X–NH–FLTTYAQAANTHLFLLKDAQIYG–COOH
(|Z attached to the D of KDAQ)

α6 X–NH–KEDIAEFYKRQLKLTQEYTDHCVKWYNVGLDKL–COOH

α7 X–NH–YESWVNFNRYRREMTLTVLDLIALF–COOH

Figure 5.2 Structure of *B. thuringiensis* δ-endotoxin. (A) Cartoon representations of the *B. thuringiensis* Cry3A δ-endotoxin showing domain organization as determined by the crystal structure (Li *et al.*, 1991) from two perspectives (a and b): domain I (colored), the pore-forming domain; domain II (gray), the receptor-binding domain, and domain III (black). The helices of the pore-forming domain are colored in a rainbow direction: α1, red; α2, orange; α3, yellow; α4, green; α5, cyan; α6, blue; and α7, purple. (B) Sequences of the helices and their corresponding synthetic peptides. For unlabeled peptides, X = H and Z = H. For labeled peptides, X = NBD or Rho. Only for NBD-C-α5, X = acetyl and Z = NBD. (See Colour Plate VI.)

efficient targeting of a broad spectrum of insect pests while keeping environmental damage to a minimum. Moreover, their specificity makes them harmless to animals and humans (Taylor and Hefle, 2001). Another major advantage is that plants can be engineered to produce the δ-endotoxins endogenously. For these reasons, the δ-endotoxin is the perfect candidate to serve in new pesticide strategies (Schnepf *et al.*, 1998). Insight into the structure of the toxin in its various life stages can significantly advance our understanding of the mode of action the toxin uses to perform its task. In turn, this can advance the design of new and more potent toxins. Furthermore, the striking structural similarity of the pore-forming domain to that of other known toxin structures suggests that the δ-endotoxins can serve as a general paradigm for toxins mechanism of action (Figure 5.1).

Mode of action of the δ-endotoxin

The tertiary structures of δ-endotoxins are composed of three domains. Currently, there are X-ray structures available for Cry proteins from at least two different subfamilies cry1, cry3 (Li *et al.*, 1991; Grochulski *et al.*, 1995). All have very similar overall 3D structures that represent the soluble form of these toxins (Figure 5.2). Since these toxins act on the membranes of insects' midgut cells, the soluble conformation is only a small part of the big picture. Membrane binding of the toxin was found to have two steps (Schnepf *et al.*, 1998; Aronson and Shai, 2001). Through an extensive site directed and deletion mutagenesis, domains II and III were determined to affect the reversible binding step. This step corresponds to initial receptor binding, thus believed to control mainly receptor recognition. The latter determines the toxin's specificity toward different insect families. The major receptor target of the δ-endotoxins belongs to the aminopeptidase N receptors (Masson *et al.*, 1995; Jenkins *et al.*, 2000). Domain I mutants were shown to have an effect on the irreversible binding step, believed to correspond to membrane insertion and channel formation (Liang *et al.*, 1995; Schnepf *et al.*, 1998). While maintaining the initial binding step, these mutants could not form an active channel (Ahmad and Ellar, 1990; Wu and Aronson, 1992; Schwartz *et al.*, 1997; Kumar and Aronson, 1999; Masson *et al.*, 1999). On the other hand, hybrid toxins containing different homologs of domain I were shown to have wild-type activity while their specificity was altered according to the specificity of domains II and III (Bosch *et al.*, 1994; de Maagd *et al.*, 1996a,b, 1999a,b, 2000; Tabashnik *et al.*, 1996; Rang *et al.*, 1999; Naimov *et al.*, 2001).

This chapter will concentrate on the structure and function of domain I, and specifically in its membrane-bound active conformation. Studies with hybrid toxins and site directed mutagenesis all suggest that domain I forms a pore upon membrane binding (Ge *et al.*, 1989; Ahmad and Ellar, 1990; Wu and Aronson, 1992). The current available toxin structures suggest that, in solution, domain I is a bundle made of seven alpha helices (Li *et al.*, 1991). The most hydrophobic helix, α5, is in the middle and the rest of the helices surround it. These helices position their hydrophobic surfaces to face the core of the bundle while the hydrophilic surfaces face the aqueous environment; thus the membrane-inserting toxin can remain soluble until the time comes to insert into the target membrane.

The Umbrella model of the δ-endotoxin pore

Initially, there were two main models suggested for the δ-endotoxin mode of action. One is the "Penknife model" and the second is the "Umbrella model," both were previously suggested for colicin A (Tucker *et al.*, 1986; Pattus *et al.*, 1990; Lakey *et al.*, 1991). The Penknife model claims the insertion of a hairpin into the membrane from the side of domain I with little structural rearrangement for the rest of domain I. The Umbrella model is best supported by the experimental data accumulated on the δ-endotoxin mechanism of toxicity. David Ellar first suggested the model in 1991 after the crystal structure of the Cry3A δ-endotoxin was solved (Li *et al.*, 1991). Initial binding to the receptor is followed by "swinging out of domain I" (Schwartz *et al.*, 1997), irreversible binding, oligomerization, and insertion to the membrane (Schnepf *et al.*, 1998). These processes are concurrent with a "molten globule" like transition in domain I. The current data suggests that this transition is initiated by a signal started upon receptor binding and conveyed to domain I through helix number seven (Gazit and Shai, 1995). The transition to a molten globule exposes the hydrophobic faces of domain I helices to the membrane thus promoting the binding to the surface of the membrane. This is followed by fanning out of the helices in the membrane surface and membrane insertion of the α4-loop-α5 hairpin (Li *et al.*, 1991; Gazit *et al.*, 1998; Gerber and Shai, 2000). The structure formed closely resembles an umbrella. The umbrella handle is formed by α4-loop-α5, perpendicular to the plane of the membrane, and the rest of the helices form the ribs of the umbrella on the surface (Figure 5.3). The oligomerization of several hairpins is believed to be crucial for correct assembly of an active pore. The number of oligomers and the exact step at which this process occurs is yet to

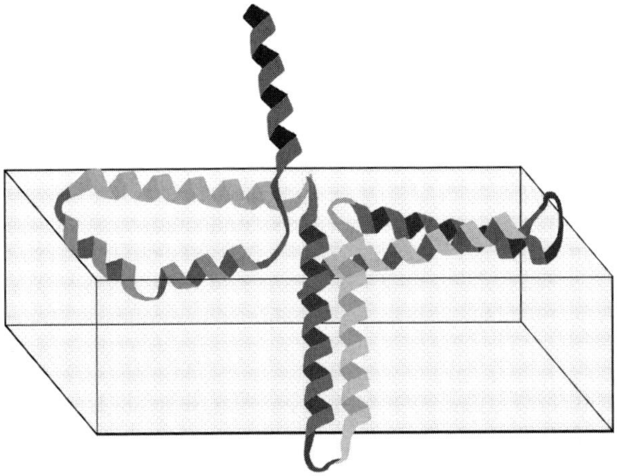

Figure 5.3 The Umbrella model – a schematic presentation of the δ-endotoxin interaction with the phospholipid membranes. The helices of the pore-forming domain are colored in a rainbow direction: α1, red; α2, orange; α3, yellow; α4, green; α5, cyan; α6, blue; and α7, purple. The loop connecting α4 and α5 may be either in an intracellular localization or may interact with the inner leaflet of the membrane because of its hydrophobicity. (See Colour Plate VII.)

be established. However, it has been shown that α5 plays an active role in the oligomerization process, as will be discussed in detail further on (Gazit and Shai, 1993a, 1995; Cummings *et al.*, 1994; Gerber and Shai, 2000).

Characterization of domain I membrane binding

What parts of domain I bind the membrane? This is the first question we ask ourselves when trying to distinguish which model suits the δ-endotoxin better. If the Penknife model is correct then there is a specific membrane docking site which involves only parts of domain I. If the Umbrella model is more accurate then a drastic change occurs in domain I, and most of the secondary structural elements will interact with the membrane. Part of the answer to this question lies in the intrinsic membrane binding capability of the individual helices of domain I.

The systematic characterization of domain I binding to membranes shed light on membrane docking and the localization of the channel within domain I. Seven peptides corresponding to the seven helices of Cry3A domain I were synthesized and labeled with a special fluorescent probe. When this probe, 4-chloro-7-nitrobenz-2-oxa-1,3-diazole (NBD), is localized in a non-polar environment, such as a phospholipid bilayer, its fluorescence increases (Rajarathnam *et al.*, 1989). Thus, partitioning into the membrane can be determined (Rapaport and Shai, 1991; Pouny *et al.*, 1992), based on a partition model (Schwarz *et al.*, 1986, 1987; Rizzo *et al.*, 1987). The membrane binding of the different peptides was characterized and the results are summarized in Table 5.1 (Gazit *et al.*, 1998). It seems that helix 1 cannot bind the membrane, which agrees with deletion mutants lacking this helix and maintain wild-type activity. All other peptides bind the membrane with a free energy suitable for membrane proteins. One striking observation was that the

Table 5.1 Derived partition coefficient (K_p) and free energy ($\Delta G_{binding}$) upon membrane binding

Peptide	K_p 3×10^4 M^{-1}	$\Delta G_{binding}$ kcal/mol
α1	No binding	0
α2	4.8 ± 0.9	-8.8 ± 0.3
α3	0.2 ± 0.04	-6.2 ± 0.2
α4	1.5 ± 0.3	-7.9 ± 0.2
α5	1.0 ± 0.2	-7.5 ± 0.2
α6	1.2 ± 0.2	-7.7 ± 0.2
α7	5.2 ± 0.4	-8.9 ± 0.1

α5 peptide, corresponding to the most conserved sequence, demonstrated cooperative binding (Gazit and Shai, 1993a). This suggests that the α5 helix has the intrinsic ability to oligomerize in a membrane environment. Furthermore, in the presence of α7 the binding was faster, suggesting a possible interaction between them (Gazit and Shai, 1995). Combined with the binding of all the other helices, but the exception of helix 1, the results point to a major rearrangement of domain I upon membrane binding.

A network of interactions between different domain I helices

In solution, most of domain I helices are in contact with α5 which is in the middle of the bundle, and with at least two more helices on their sides. This bundle-like pattern has a specific network of intra molecular helix–helix interactions. According to the Umbrella model, a drastic rearrangement of the helices occurs upon membrane binding. Thus, we expect to get a large change in the "network" of bundle-like helix–helix interactions. On the contrary, according to the Penknife model, we expect the pattern of bundle-like interactions to be maintained with the exception of changes involving two helices that flip over into the membrane from the side of domain I.

Interactions between the different helices of domain I were investigated using fluorescence energy transfer (FRET) (Fung and Stryer, 1978; Gazit and Shai, 1993b). The peptides were labeled with fluorescent probes and the energy transfer between a donor NBD labeled peptide and an acceptor Rhodamine labeled peptide was measured (Rajarathnam *et al.*, 1989). A schematic illustration of the system is demonstrated in Figure 5.4. Scanning the pattern of helix–helix interactions suggests a strong anti-parallel association between α4 and α5, which is compatible with the formation of a hairpin. Another strong interaction measured is between α5 and itself. The latter interaction suggests a role for α5 in channel oligomerization. Separate interactions between α7 and α4, five and six were also measured. Considering that α7 can promote faster binding of α4 and α5, it is certain that α7 plays the role of a lever. It conveys a signal from the reversible receptor binding to promote the irreversible membrane binding and insertion of domain I. Thus, the pattern of bundle-like interaction observed in solution is mostly missing in the membrane (Gazit and Shai, 1995; Gazit *et al.*, 1998; Gerber and Shai, 2000) (Figure 5.5).

Secondary structure of the membrane-bound protein

The high-resolution structure would give us a more direct answer to the last questions. But to date, the conventional methods for high-resolution structure determination are difficult

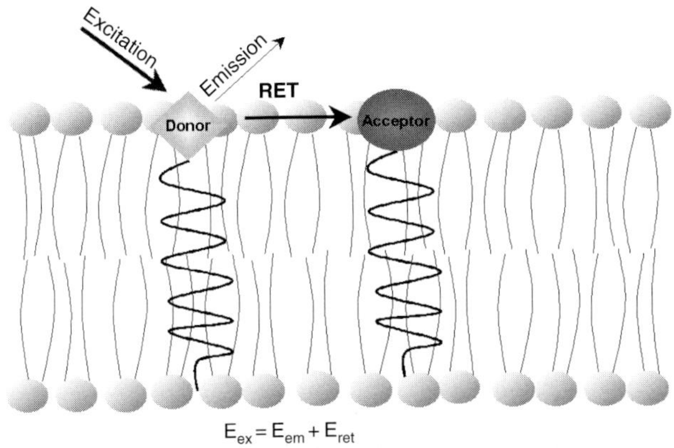

Figure 5.4 Illustration of FRET. When a donor molecule is excited in the appropriate wavelength we get fluorescence emission. If an acceptor molecule is close enough to the donor some of the energy will be transferred to the acceptor molecule via a resonance process. Thus, the intensity of donor emission will decrease. Using this phenomenon we can detect whether two molecules associate with each other specifically, or there is only random distribution.

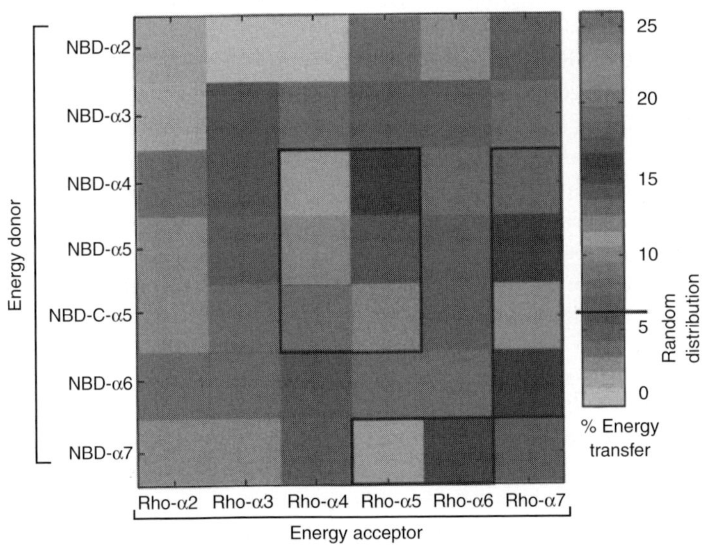

Figure 5.5 Fluorescence energy transfer between NBD-labeled peptides (donors) and Rho-labeled peptides (acceptors). The degree of energy transfer between acceptor and donor peptides of the 42 possible combinatorial acceptor–donor pairs is represented as a pseudocolor map. The thick lines mark the areas where the energy transfer was significantly higher than expected for random distribution of monomers. (See Colour Plate VIII.)

Table 5.2 ATR Fourier-transform analysis of α2–α7 helices within phospholipid multibilayers

Peptide	f[a]	*α-helicity (%)*[b]
α2	−0.28 ± 0.05	59.5 ± 3.9
α3	−0.26 ± 0.02	78.9 ± 6.7
α4	0.08 ± 0.02	62.3 ± 3.4
α5	0.12 ± 0.04	67.8 ± 4.2
α6	−0.24 ± 0.05	58.6 ± 6.3
α7	−0.20 ± 0.07	63.2 ± 4.8

Notes

a The order parameter f is a function of molecular orientations in the sample. It can be interpreted in terms of models with specific distribution functions. Theoretically, f ranges from −0.5 for $\theta = 90°$ and +1.0 for $\theta = 0°$. Practically, positive values for f relate to a perpendicular orientation and negative numbers below −0.2 are parallel with respect to the plane of the membrane (Tamm and Tatulian, 1997).

b The ± values mean the upper and lower limits of α-helicity calculated by either Voigt or Lorentzian line-shape analysis.

to apply to the membrane-bound toxin. Therefore, the secondary structure of cry3A domain I within the membrane is the best option we have to understand whether the secondary structure is maintained and to find the orientation of the bound toxin. These questions were answered by harnessing the power of "Attenuated Total Reflection Fourier Transform Infrared" (ATR-FTIR) spectroscopy (Jackson and Mantsch, 1995; Tamm and Tatulian, 1997; Goormaghtigh *et al.*, 1999). Domain I peptide segments were incorporated into phospholipid multibilayers and the ATR-FTIR spectra were determined (Gazit *et al.*, 1998). The frequencies of the amide I absorption peaks of the peptides indicated that all peptides are predominantly alpha helical (60–80%). The relative amount of various secondary structure elements was determined by deconvolution of the peptides' infrared spectra (Table 5.2). Furthermore, ATR dichroic ratio of polarized amide I at 0° and 90° spectra revealed that helices four and five adopt a transmembrane orientation while all the rest of the helices seem to be surface bound.

Domain I pore-formation analysis

The structural studies suggest that the δ-endotoxin adopts an umbrella-like structure in the membrane. Furthermore, α4 and α5 were implicated in forming a hairpin in a transmembrane orientation. To further validate these results a connection between the structural results and the functionality of the toxin was required. To this end, we scanned different domain I peptides and tested them for pore-forming ability. The results were integrated with data obtained from site directed mutagenesis and deletion mutants of the whole toxin. Using the synthetic approach, domain I was systematically scanned for pore-forming segments (Gazit *et al.*, 1998; Gerber and Shai, 2000). Pore formation was detected by the release of the fluorescent dye calcein (Pouny *et al.*, 1992) or by the release of potassium ions (Gazit and Shai, 1993a) trapped inside small unilamellar vesicles. The release of the calcein dye causes an increase in fluorescence as a consequence of a dequenching process. While in the case of the potassium ions, the release of the ions causes a dequenching process of fluorescent dye $DiSC_35$. Out of the seven helices only α4 and α5 showed weak activity, whereas the rest were inactive. When the two poorly active segments were combined into one longer segment the activity increased a hundred fold. This tremendous pore-forming cooperativity points to the formation of an active channel by α4-loop-α5 (Gerber and Shai, 2000) (Figure 5.6).

Site directed mutagenesis of domain I revealed that most inactivating mutations are localized in the region of α4-loop-α5 (Ahmad and Ellar, 1990; Wu and Aronson, 1992; Hussain *et al.*, 1996; Rajamohan *et al.*, 1996; Schwartz *et al.*, 1997; Kumar and Aronson, 1999). Furthermore, mutations within the α5 region were shown to affect toxin oligomerization within the membrane (Kumar and Aronson, 1999). This is supported by the ability of the wild-type α5 peptide to strongly self-associate, as seen in FRET experiments mentioned earlier. A peptide corresponding to an α5 proline mutant was found not to oligomerize (Gazit *et al.*, 1994). Accordingly, site directed mutagenesis showed that the full-length protein was inactive. On the other hand, inactive mutations in α4 usually maintained the wild-type oligomeric state (Kumar and Aronson, 1999). This leads to the conclusion that α5 is responsible for the oligomerization of the pore while α4 is responsible for channel activity. Further, mutagenesis of charged amino acids within helix 4 suggest that the helix faces the lumen of the channel. For example, the restoration of a negative charge on a D to C mutant restored channel activity to normal levels (Masson *et al.*, 1999).

The accessibility of helix 4 to the lumen is further supported by the ability of an α4 peptide with a hyper active mutation to displace the wild-type helix, in a manner very much like immunity proteins, and abolish pore formation (Gerber and Shai, 2000). The α5 peptide cannot displace its counterpart in the channel, probably because it is on the outside

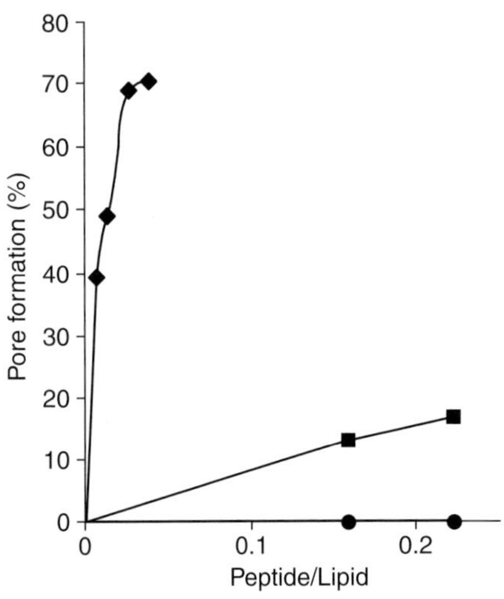

Figure 5.6 Pore formation induced by peptides. The peptides were added to calcein containing small unilamellar vesicles. Increase in the fluorescence vs molar ratio of peptide/lipid was recorded. ◆ α4-loop-α5; ■ α5; ● α4. Maximal calcein release (100%) was determined by adding 10 μl of 1% Triton-X. Membrane content: phosphatidyl choline : phosphatidyl serine 10 : 1.

of the channel. The displacement of an active α4-loop-α5 pore by a shorter segment of α4 can be a very powerful tool in the further characterization of the cry toxin pore. Indeed, a similar strategy was found to rescue mice injected with the anthrax toxins by pre-treating the mice with a dominant negative mutant of the pore-forming toxin (Leppla, 2001; Sellman *et al.*, 2001).

References

Ahmad, W. and Ellar, D. J. (1990) Directed mutagenesis of selected regions of a *Bacillus thuringiensis* entomocidal protein. *FEMS Microbiol. Lett.*, **56**, 97–104.

Aronson, A. I. and Shai, Y. (2001) Why *Bacillus thuringiensis* insecticidal toxins are so effective: unique features of their mode of action. *FEMS Microbiol. Lett.*, **195**, 1–8.

Basak, A. K., Howells, A., Eaton, J. T., Moss, D. S., Naylor, C. E., Miller, J. *et al.* (1998) Crystallization and preliminary X-ray diffraction studies of alpha-toxin from two different strains (NCTC8237 and CER89L43) of *Clostridium perfringens*. *Acta Crystallogr. D. Biol. Crystallogr.*, **54**, 1425–1428.

Basak, A. K., Stuart, D. I., Nikura, T., Bishop, D. H., Kelly, D. C., Fearn, A. *et al.* (1994) Purification, crystallization and preliminary X-ray diffraction studies of alpha-toxin of *Clostridium perfringens*. *J. Mol. Biol.*, **244**, 648–650.

Bayley, H. (1997) Toxin structure: part of a hole? *Curr. Biol.*, **7**, 763–767.

Ben-Efraim, I. and Shai, Y. (1996) Secondary structure, membrane localization, and coassembly within phospholipid membranes of synthetic segments derived from the N- and C-termini regions of the ROMK1 K+ channel. *Protein Sci.*, **5**, 2287–2297.

Ben-Efraim, I. and Shai, Y. (1997) The structure and organization of synthetic putative membranous segments of ROMK1 channel in phospholipid membranes. *Biophys. J.*, **72**, 85–96.

Bibi, E. and Kaback, H. R. (1990) In vivo expression of the lacY gene in two segments leads to functional lac permease. *Proc. Natl. Acad. Sci. USA*, **87**, 4325–4329.

Bosch, D., Schipper, B., van der Kleij, H., de Maagd, R. A. and Stiekema, W. J. (1994) Recombinant *Bacillus thuringiensis* crystal proteins with new properties: possibilities for resistance management. *Biotechnology (NY)*, **12**, 915–918.

Ceska, T. A. and Henderson, R. (1990) Analysis of high-resolution electron diffraction patterns from purple membrane labelled with heavy-atoms. *J. Mol. Biol.*, **213**, 539–560.

Cramer, W. A., Cohen, F. S., Merrill, A. R. and Song, H. Y. (1990) Structure and dynamics of the colicin E1 channel. *Mol. Microbiol.*, **4**, 519–526.

Crickmore, N., Zeigler, D. R., Feitelson, J., Schnepf, E., Van Rie, J., Lereclus, D. *et al.* (1998) Revision of the nomenclature for the *Bacillus thuringiensis* pesticidal crystal proteins. *Microbiol. Mol. Biol. Rev.*, **62**, 807–813.

Cummings, C. E., Armstrong, G., Hodgman, T. C. and Ellar, D. J. (1994) Structural and functional studies of a synthetic peptide mimicking a proposed membrane inserting region of a *Bacillus thuringiensis* delta-endotoxin. *Mol. Membr. Biol.*, **11**, 87–92.

de Maagd, R. A., Kwa, M. S., van der Klei, H., Yamamoto, T., Schipper, B., Vlak, J. M. *et al.* (1996a) Domain III substitution in *Bacillus thuringiensis* delta-endotoxin CryIA(b) results in superior toxicity for Spodoptera exigua and altered membrane protein recognition. *Appl. Environ. Microbiol.*, **62**, 1537–1543.

de Maagd, R. A., van der Klei, H., Bakker, P. L., Stiekema, W. J. and Bosch, D. (1996b) Different domains of *Bacillus thuringiensis* delta-endotoxins can bind to insect midgut membrane proteins on ligand blots. *Appl. Environ. Microbiol.*, **62**, 2753–2757.

de Maagd, R. A., Bakker, P., Staykov, N., Dukiandjiev, S., Stiekema, W. and Bosch, D. (1999a) Identification of *Bacillus thuringiensis* delta-endotoxin Cry1C domain III amino acid residues involved in insect specificity. *Appl. Environ. Microbiol.*, **65**, 4369–4374.

de Maagd, R. A., Bakker, P. L., Masson, L., Adang, M. J., Sangadala, S., Stiekema, W. *et al.* (1999b) Domain III of the *Bacillus thuringiensis* delta-endotoxin Cry1Ac is involved in binding to Manduca sexta brush border membranes and to its purified aminopeptidase N. *Mol. Microbiol.*, **31**, 463–471.

de Maagd, R. A., Weemen-Hendriks, M., Stiekema, W. and Bosch, D. (2000) *Bacillus thuringiensis* delta-endotoxin Cry1C domain III can function as a specificity determinant for Spodoptera exigua in different, but not all, Cry1–Cry1C hybrids. *Appl. Environ. Microbiol.*, **66**, 1559–1563.

Dean, D. H., Rajamohan, F., Lee, M. K., Wu, S. J., Chen, X. J., Alcantara, E. *et al.* (1996) Probing the mechanism of action of *Bacillus thuringiensis* insecticidal proteins by site-directed mutagenesis – a minireview. *Gene*, **179**, 111–117.

Fung, B. K. and Stryer, L. (1978) Surface density determination in membranes by fluorescence energy transfer. *Biochemistry*, **17**, 5241–5248.

Gazit, E. and Shai, Y. (1993a) Structural and functional characterization of the alpha 5 segment of *Bacillus thuringiensis* delta-endotoxin. *Biochemistry*, **32**, 3429–3436.

Gazit, E. and Shai, Y. (1993b) Structural characterization, membrane interaction, and specific assembly within phospholipid membranes of hydrophobic segments from *Bacillus thuringiensis* var. israelensis cytolytic toxin. *Biochemistry*, **32**, 12363–12371.

Gazit, E. and Shai, Y. (1995) The assembly and organization of the alpha 5 and alpha 7 helices from the pore-forming domain of *Bacillus thuringiensis* delta-endotoxin. Relevance to a functional model. *J. Biol. Chem.*, **270**, 2571–2578.

Gazit, E., Bach, D., Kerr, I. D., Sansom, M. S., Chejanovsky, N. and Shai, Y. (1994) The alpha-5 segment of *Bacillus thuringiensis* delta-endotoxin: in vitro activity, ion channel formation and molecular modelling. *Biochem. J.*, **304**, 895–902.

Gazit, E., La Rocca, P., Sansom, M. S. and Shai, Y. (1998) The structure and organization within the membrane of the helices composing the pore-forming domain of *Bacillus thuringiensis* delta-endotoxin are consistent with an "umbrella-like" structure of the pore. *Proc. Natl. Acad. Sci. USA*, **95**, 12289–12294.

Ge, A. Z., Shivarova, N. I. and Dean, D. H. (1989) Location of the *Bombyx mori* specificity domain on a *Bacillus thuringiensis* delta-endotoxin protein. *Proc. Natl. Acad. Sci. USA*, **86**, 4037–4041.

Gerber, D. and Shai, Y. (2000) Insertion and organization within membranes of the delta-endotoxin pore-forming domain, helix 4-loop-helix 5, and inhibition of its activity by a mutant helix 4 peptide. *J. Biol. Chem.*, **275**, 23602–23607.

Goormaghtigh, E., Raussens, V. and Ruysschaert, J. M. (1999) Attenuated total reflection infrared spectroscopy of proteins and lipids in biological membranes. *Biochim. Biophys. Acta*, **1422**, 105–185.

Grochulski, P., Masson, L., Borisova, S., Pusztai-Carey, M., Schwartz, J. L., Brousseau, R. *et al.* (1995) *Bacillus thuringiensis* CryIA(a) insecticidal toxin: crystal structure and channel formation. *J. Mol. Biol.*, **254**, 447–464.

Hussain, S. R., Aronson, A. I. and Dean, D. H. (1996) Substitution of residues on the proximal side of Cry1A *Bacillus thuringiensis* delta-endotoxins affects irreversible binding to Manduca sexta midgut membrane. *Biochem. Biophys. Res. Commun.*, **226**, 8–14.

Jackson, M. and Mantsch, H. H. (1995) The use and misuse of FTIR spectroscopy in the determination of protein structure. *Crit. Rev. Biochem. Mol. Biol.*, **30**, 95–120.

Jenkins, J. L., Lee, M. K., Valaitis, A. P., Curtiss, A. and Dean, D. H. (2000) Bivalent sequential binding model of a *Bacillus thuringiensis* toxin to gypsy moth aminopeptidase N receptor. *J. Biol. Chem.*, **275**, 14423–14431.

Kliger, Y., Peisajovich, S. G., Blumenthal, R. and Shai, Y. (2000) Membrane-induced conformational change during the activation of HIV-1 gp41. *J. Mol. Biol.*, **301**, 905–914.

Kumar, A. S. and Aronson, A. I. (1999) Analysis of mutations in the pore-forming region essential for insecticidal activity of a *Bacillus thuringiensis* delta-endotoxin. *J. Bacteriol.*, **181**, 6103–6107.

Lakey, J. H., Massotte, D., Heitz, F., Dasseux, J. L., Faucon, J. F., Parker, M. W. *et al.* (1991) Membrane insertion of the pore-forming domain of colicin A. A spectroscopic study. *Eur. J. Biochem.*, **196**, 599–607.

Leppla, S. H. (2001) A dominant-negative therapy for anthrax. *Nat. Med.*, **7**, 659–660.

Lesieur, C., Vecsey-Semjen, B., Abrami, L., Fivaz, M. and van der Goot, F. G. (1997) Membrane insertion: the strategies of toxins (review). *Mol. Membr. Biol.*, **14**, 45–64.

Li, J. D., Carroll, J. and Ellar, D. J. (1991) Crystal structure of insecticidal delta-endotoxin from *Bacillus thuringiensis* at 2.5 Å resolution. *Nature*, **353**, 815–821.

Liang, Y., Patel, S. S. and Dean, D. H. (1995) Irreversible binding kinetics of *Bacillus thuringiensis* CryIA delta-endotoxins to gypsy moth brush border membrane vesicles is directly correlated to toxicity. *J. Biol. Chem.*, **270**, 24719–24724.

MacKinnon, R., Cohen, S. L., Kuo, A., Lee, A. and Chait, B. T. (1998) Structural conservation in prokaryotic and eukaryotic potassium channels. *Science*, **280**, 106–109.

Masson, L., Lu, Y. J., Mazza, A., Brousseau, R. and Adang, M. J. (1995) The CryIA(c) receptor purified from Manduca sexta displays multiple specificities. *J. Biol. Chem.*, **270**, 20309–20315.

Masson, L., Tabashnik, B. E., Liu, Y. B., Brousseau, R. and Schwartz, J. L. (1999) Helix 4 of the *Bacillus thuringiensis* Cry1Aa toxin lines the lumen of the ion channel. *J. Biol. Chem.*, **274**, 31996–32000.

Michel, H., Oesterhelt, D. and Henderson, R. (1980) Orthorhombic two-dimensional crystal form of purple membrane. *Proc. Natl. Acad. Sci. USA*, **77**, 338–342.

Naimov, S., Weemen-Hendriks, M., Dukiandjiev, S. and de Maagd, R. A. (2001) *Bacillus thuringiensis* delta-endotoxin Cry1 hybrid proteins with increased activity against the Colorado potato beetle. *Appl. Environ. Microbiol.*, **67**, 5328–5330.

Parker, M. W. and Pattus, F. (1993) Rendering a membrane protein soluble in water: a common packing motif in bacterial protein toxins. *Trends Biochem. Sci.*, **18**, 391–395.

Pattus, F., Massotte, D., Wilmsen, H. U., Lakey, J., Tsernoglou, D., Tucker, A. *et al.* (1990) Colicins: prokaryotic killer-pores. *Experientia*, **46**, 180–192.

Peisajovich, S. G. and Shai, Y. (2001) SIV gp41 binds to membranes both in the monomeric and trimeric states: consequences for the neuropathology and inhibition of HIV infection. *J. Mol. Biol.*, **311**, 249–254.

Popot, J. L., Trewhella, J. and Engelman, D. M. (1986) Reformation of crystalline purple membrane from purified bacteriorhodopsin fragments. *Embo. J.*, **5**, 3039–3044.

Pouny, Y., Rapaport, D., Mor, A., Nicolas, P. and Shai, Y. (1992) Interaction of antimicrobial dermaseptin and its fluorescently labeled analogues with phospholipid membranes. *Biochemistry*, **31**, 12416–12423.

Rajamohan, F., Alzate, O., Cotrill, J. A., Curtiss, A. and Dean, D. H. (1996) Protein engineering of *Bacillus thuringiensis* delta-endotoxin: mutations at domain II of CryIAb enhance receptor affinity and toxicity toward gypsy moth larvae. *Proc. Natl. Acad. Sci. USA*, **93**, 14338–14343.

Rajamohan, F., Lee, M. K. and Dean, D. H. (1998) *Bacillus thuringiensis* insecticidal proteins: molecular mode of action. *Prog. Nucleic Acid Res. Mol. Biol.*, **60**, 1–27.

Rajarathnam, K., Hochman, J., Schindler, M. and Ferguson-Miller, S. (1989) Synthesis, location, and lateral mobility of fluorescently labeled ubiquinone 10 in mitochondrial and artificial membranes. *Biochemistry*, **28**, 3168–3176.

Rang, C., Vachon, V., de Maagd, R. A., Villalon, M., Schwartz, J. L., Bosch, D. *et al.* (1999) Interaction between functional domains of *Bacillus thuringiensis* insecticidal crystal proteins. *Appl. Environ. Microbiol.*, **65**, 2918–2925.

Rapaport, D. and Shai, Y. (1991) Interaction of fluorescently labeled pardaxin and its analogues with lipid bilayers. *J. Biol. Chem.*, **266**, 23769–23775.

Rizzo, V., Stankowski, S. and Schwarz, G. (1987) Alamethicin incorporation in lipid bilayers: a thermodynamic study. *Biochemistry*, **26**, 2751–2759.

Schnepf, E., Crickmore, N., Van Rie, J., Lereclus, D., Baum, J., Feitelson, J. *et al.* (1998) *Bacillus thuringiensis* and its pesticidal crystal proteins. *Microbiol. Mol. Biol. Rev.*, **62**, 775–806.

Schwartz, J. L., Juteau, M., Grochulski, P., Cygler, M., Prefontaine, G., Brousseau, R. *et al.* (1997) Restriction of intramolecular movements within the Cry1Aa toxin molecule of *Bacillus thuringiensis* through disulfide bond engineering. *FEBS Lett.*, **410**, 397–402.

Schwarz, G., Gerke, H., Rizzo, V. and Stankowski, S. (1987) Incorporation kinetics in a membrane, studied with the pore-forming peptide alamethicin. *Biophys. J.*, **52**, 685–692.

Schwarz, G., Stankowski, S. and Rizzo, V. (1986) Thermodynamic analysis of incorporation and aggregation in a membrane: application to the pore-forming peptide alamethicin. *Biochim. Biophys. Acta*, **861**, 141–151.

Sellman, B. R., Mourez, M. and Collier, R. J. (2001) Dominant-negative mutants of a toxin subunit: an approach to therapy of anthrax. *Science*, **292**, 695–697.

Tabashnik, B. E., Malvar, T., Liu, Y. B., Finson, N., Borthakur, D., Shin, B. S. *et al.* (1996) Cross-resistance of the diamondback moth indicates altered interactions with domain II of *Bacillus thuringiensis* toxins. *Appl. Environ. Microbiol.*, **62**, 2839–2844.

Tamm, L. K. and Tatulian, S. A. (1997) Infrared spectroscopy of proteins and peptides in lipid bilayers. *Quarterly Reviews of Biophysics*, **30**, 365–429.

Taylor, S. L. and Hefle, S. L. (2001) Will genetically modified foods be allergenic? *J. Allergy Clin. Immunol.*, **107**, 765–771.

Tucker, A. D., Pattus, F. and Tsernoglou, D. (1986) Crystallization of the C-terminal domain of colicin A carrying the voltage-dependent pore activity of the protein. *J. Mol. Biol.*, **190**, 133–134.

Wu, D. and Aronson, A. I. (1992) Localized mutagenesis defines regions of the *Bacillus thuringiensis* delta-endotoxin involved in toxicity and specificity. *J. Biol. Chem.*, **267**, 2311–2317.

6 Functional studies of helix α-5 region from *Bacillus thuringiensis* Cry1Ab δ-endotoxin

Mario Soberón, Rigoberto V. Pérez, María E. Nuñez-Valdez, Isabel Gómez, Jorge Sánchez, Leopoldo Güereca and Alejandra Bravo

The crystal insecticidal proteins (Cry) from *Bacillus thuringiensis* (Bt) are modular proteins comprised of three domains connected by single linkers. Domain I, is a seven α-helix bundle, with helix α5 in the center, surrounded by the other helices. This domain is involved in membrane insertion and pore formation. We hypothesized that highly conserved residues of α5 could play an important role in toxin insertion, oligomerization and/or pore formation. A total of 15 Cry1Ab mutants located in six conserved residues of Cry1Ab, Y153, Y161, H168, R173, W182 and G183, were isolated. Eleven mutants were located within helix α5, one mutant was located in the loop α4–α5 and three mutants W182P, W182I, G183C were located in the loop α5–α6. Their effect on binding, K^+ permeability, oligomerization and toxicity against *Manduca sexta* larvae was analyzed and compared.

To determine oligomerization we analyze the functional complementation of two Cry1Ab mutant proteins affected in different steps of their action (F371A in receptor binding and H168F in pore formation). These mutants were affected in their toxicity against *M. sexta* larvae. Binding analysis showed that F371A protein recovered binding to *M. sexta* brush border membrane vesicles (BBMV) when mixed with H168F in a one to one ratio and that the interaction between monomers occurs after receptor binding. These mutant proteins also recovered pore-formation activity and toxicity against *M. sexta* larvae when mixed, showing that monomers affected in different steps of their mode of action can form functional hetero-oligomers.

The results provide direct evidences that some residues located within α5 have an important role in stability of the toxin, but not in pore formation, while H168 has also an important role in pore formation but is not affected in oligomerization.

Introduction

The Cry from Bt are toxic to different insect larvae. Due to their high specificity and their safety for the environment, Cry toxins are valuable alternatives to chemical pesticides for control of insect pests in agriculture and forestry. It has been proposed that the rational use of Bt toxins will provide a variety of alternatives for insect control and for coping with the problem of insect resistance to pesticides (Van Rie *et al.*, 1990a,b).

After ingestion by susceptible insects, the crystal proteins are solubilized in the gut where midgut proteases process the protoxin into a 60 kDa toxic fragment. Cry toxins bind to specific high-affinity binding proteins on the surface of midgut epithelial cells (Hofmann *et al.*, 1988; Bravo *et al.*, 1992). After binding, the toxin inserts into the membrane and forms ionic channels (Lorence *et al.*, 1995; Martin and Wolfersberger, 1995) that leads to cell lysis by colloid osmotic lysis and eventually the death of the insect (Knowles and Ellar, 1987; Pietrantonio and Gill, 1996). Based on the observation of large conductance states formed by Cry1Ac, Cry3A, Cry3B and Cry1C toxins in synthetic planar lipid bilayers (Slatin *et al.*, 1990; Schwartz *et al.*, 1993; Von Tersch *et al.*, 1994) and the estimation of a pore size between 10–20 Å (Knowles, 1994), it has been proposed that the pore could be formed by an oligomer of Cry toxins containing four to six toxin monomers. The oligomeric state of some Cry toxins in solution has been analyzed. A recent report showed that Cry proteins in solution do not form oligomers of a defined subunit number suggesting that oligomers form after the toxin interacts with its receptor or after it is inserted into the membrane (Güereca and Bravo, 1999). Also, Cry1Ac oligomeric state was analyzed after binding to BBMV where multimers were identified (Aronson *et al.*, 1999).

The three-dimensional structure of the Cry3A and Cry1Aa toxins showed that both proteins are comprised of three domains connected by single linkers (Li *et al.*, 1991; Grochulski *et al.*, 1995). Domain I, extending from the N-terminus, is a seven α-helix bundle, where the highly conserved helix α5 is located in the center, surrounded by the other helices. It has been considered as the pore-forming domain. Domain II consists of three anti-parallel β-sheets and Domain III is a β-sandwich of two anti-parallel β-sheets. Domain II has been identified as the specificity-determining domain (Lu *et al.*, 1994; Smith *et al.*, 1994; Rajamohan *et al.*, 1996; Wu and Dean, 1996). Domain III is also involved in specificity (Burton *et al.*, 1999; de Maagd *et al.*, 1999; Lee *et al.*, 1999) and protection to further proteolysis within the gut (Li *et al.*, 1991).

Although the crystal structure has been known for more than nine years, the conformational changes of Cry toxin involved in insertion into the membrane and pore formation are still unclear. Site-directed mutagenesis in domain I has contributed to identify regions that might play an important role in the structure of the pore-forming domain within the membrane. There are several evidences supporting that the hairpin α4–α5 inserts into the membrane in an antiparallel manner, while other helices lie on the membrane surface. Gazit *et al.* (1998) compared the membrane interaction and the structural organization in the membrane plane of the α1–α7 synthetic peptides. These authors found that α4 and α5 have a transmembrane orientation and they co-assamble within the phospholipid membrane while helices α2, α3, α5, α6 and α7 display a membrane surface orientation. Schwartz *et al.* (1997) showed that restriction of mobility of α4 or α5 by creation of disulfide-bridges affected the ion-channel formation in synthetic planar lipid bilayers. On the other hand, mutations in the central residues of helix α5 decrease toxicity without affecting its binding capacities (Wu and Aronson, 1992). Also, mutations in α4 affected toxicity and reduced pore-formation activity (Uawithya, 1998; Manoj Kumar and Aronson, 1999; Masson *et al.*, 1999). Furthermore, restoration of the negative charge of the α4-Cry1Aa mutant D136C, with a charged thiol-specific reagent, restores wild type conductance in planar lipid bilayers (Masson *et al.*, 1999). In contrast, mutations in α2, α3 and α6 showed little or no effect on toxicity (Aronson *et al.*, 1995; Manoj Kumar and Aronson, 1999; Masson *et al.*, 1999). Overall, these results suggest that α4 and α5 are the key elements in pore formation by Cry toxins.

We hypothesized that highly conserved residues of α5 could play an important role in toxin insertion, oligomerization and/or pore formation. Functional analysis of site-directed mutants located in highly conserved residues within helix α5 as well as residues at the loops linking this helix with α4 and α6 in the Cry1Ab toxin were performed. Their effect on K^+ permeability, binding and toxicity against *M. sexta* larvae was compared. The results provide direct evidences that some residues located within α5 have an important role in stability of the toxin within the insect gut, while some others have also an important role in K^+ permeability. We show that mutant H168F, affected in pore formation, is not affected in oligomer formation, since it can form functional heteroligomers with a Cry mutant affected in receptor binding. Moreover our results suggest that oligomer formation occurs after receptor binding.

Isolation of Cry1Ab mutant toxins

Residues Y153, Y161, H168, R173, W182 and G183 which are highly conserved among Cry toxins were mutagenized by site-directed mutagenesis (Figure 6.1) (Nuñez-Valdez *et al.*, 2001).

```
Cry2A   YQLLLLPLFAQAANMHLSFIRDVILNADEWG
Cry11A  YEGVSIALFTQMCTLHLTLLKDGILAGSAWG
Cry10A  YRIPTLPAYAQIATWHLNLLKHAATYYNIWL
Cry4A   YNILVLSSYAQAANLHLTVLNQAVKFEAYLK
Cry4B   YELLLLPIYAQVANFNLLLIRDGLINAQEWS
Cry1Aa  YQVPLLSVYVQAANLHLSVLRDVSVFGQRWG
Cry1Ab  YQVPLLSVYVQAANLHLSVLRDVSVFGQRWG
Cry1Ac  YQVPLLSVYVQAANLHLSVLRDVSVFGQRWG
Cry1D   YEVALLSVYVQAANLHLSILRDVSVFGERWG
Cry1F   FEIPLLSVYVQAANLHLSLLRDAVSFGQGWG
Cry1I   EEVPLLPIYAQAANLHLLLLRDASIFGKEWG
Cry1B   QEVPLLMVYAQAANLHLLLLRDASLFGSEFG
Cry1G   LEVVNLSVYTQAANLHLSLLRDAVYFGERWG
Cry1H   FEIPLLSVYVQAANLHLSLLRDAVYFGQRWG
Cry1C   FEVPLLSVYAQAANLHLAILRDSVIFGERWG
Cry1J   FEVPLLSVYVQAANLHLALLRDSVIFGERWG
Cry1E   YQVPFLSVYVQAANLHLSVLRDVSVFGQAWG
Cry3A   YEVLFLTTYAQAANTHLFLLKDAQIYGEEWG
Cry3B   FEVLFLPTYAQAANTHLLLLKDAQVFGEEWG
Cry3C   YEVLFLPTYAQAANTHLLLLKDAQIYGTDWG
Cry7A   YEIPLLTVYAQAANLHLALLRDSTLYGDKWG
Cry8A   HEVLLLAVYAQAVNLHLLLLRDASIFGEEWG
Cry8B   FEVPFLTVYTMAANLHLLLLRDASIFGEEWG
Cry8C   YETPLLPTYAQAASLHLLVMRDVQIYGKEWG
Cry9C   QQVPLLSVYAQAVNLHLLLLKDASLFGEGWG
Cry9B   FQAQLLVVYAQAANLHLLLLADAEKYGARWG

        Y153C  Y161I  H168F  R173P  W182Y  G183C
               Y161R         R173N  W182P
               Y161G         R173T
               Y161L         R173C
                             R173L
                             R173D
```

Figure 6.1 Multiple alignment of the helix α-5 region from different Cry toxins.

Table 6.1 Effect of Cry1Ab mutations on toxicity to M. *sexta* larvae and in K^+ permeability across the brush border membrane vesicles (BBMV)

Toxin[a]	*Mortality (%)*[b]	$m_{tox} - m_{control}$[c]
Cry1Ab	100	0.145
Y153C	90	0.110
Y161R	45	0.135
Y161L	50	0.134
H168F	4	0.006
R173C	45	0.236
R173T	100	0.109
R173N	20	0.050
R173L	10	0.070
W182I	90	0.152
G183C	100	0.132
W182P	0	n.d.[e]
F371A	0	0.003
F371A + H168F[d]	25	0.139

Notes

a All proteins were tested as pure trypsin-activated toxins.

b Mortality as % of 24 larvae treated with 20 ng/cm^2 after five days.

c $m_{tox} - m_{control}$ = The slope (m) of the curves of changes in fluorescence (ΔF) (%) vs K^+ equilibrium potential (E_K+) (mV) or vs external K^+ concentration was determined in BBMV treated with 50 mM Cry1Ab or mutant toxins. The difference of the slope of the curve of each mutant toxin minus the slope of the control trace in which the same amount of buffer was added, are presented. Values are averages of four different measurements. Standard deviations were less than 5%.

d 1 : 1 protein mixture.

e n.d. = not determined.

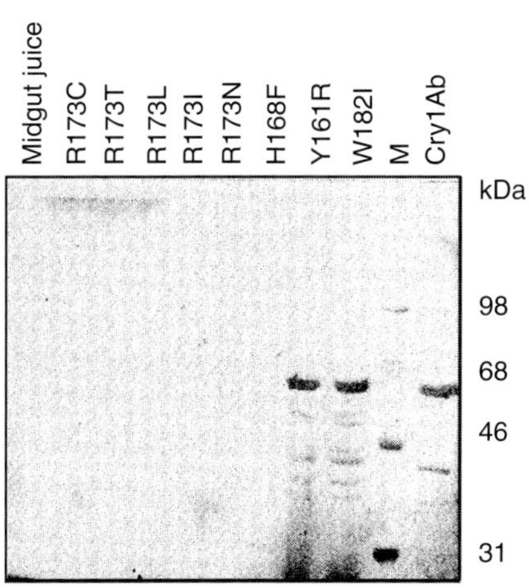

Figure 6.2 Stability of Cry1Ab mutants to midgut juice proteases. SDS–PAGE electrophoresis and staining with Coomassie brilliant blue of Cry1Ab mutant protoxins after 4 h treatment with 1 : 1 M. *sexta* midgut juice. From left to right lanes are as follows: lane 1, midgut juice; lane 2, R173C; lane 3, R173T; lane 4, R173L; lane 5, R173N; lane 6, H168F; lane 7, Y161R; lane 8, W182I; M, MW markers; lane 9, Cry1Ab.

A total of 15 Cry1Ab mutants were isolated. Eleven mutants were located within helix α5: Y161I, Y161R, Y161G, Y161L, H168F, R173P, R173N, R173T, R173C, R173L and R173D. One mutant, Y153C was located in the loop α4–α5 and three mutants, W182P, W182I and G183C were located in the loop α5–α6 (Figure 6.1).

Proteins about 130 kDa were observed in very low yields in mutants affected in H168 and R173. In contrast, mutants Y153C, Y161R, Y161L, W182I and G183C produced high yields of the 130 kDa protein (data not shown). Mutant W182P was processed by Bt intrinsic proteases to an insoluble 75 kDa fragment. This protein was non-toxic to M. *sexta* larvae (Table 6.1) and was not further analyzed. Mutants Y161G, Y161I, R173D and R173P were highly unstable to trypsin digestion and neither these were further analyzed. The rest of the mutants including those of H168 and R173, produced a stable toxin fragment when digested with trypsin. However, mutants affected in H168 and R173 residues were highly unstable when treated 4 h with M. *sexta* midgut juice (Figure 6.2).

Functional analysis of Cry1Ab mutant toxins

Mutants W182P, H168F, R173L and R173N were severely altered in toxicity showing 0–20% mortality against M. *sexta* larvae (Table 6.1) that corresponds to a 25–1000-fold higher LC_{50} than the Cry1Ab wild type toxin. In contrast, mutants Y153C, Y161R, Y161L, R173C and W182I had reduced insecticidal activities from 45% to 90% mortality that corresponds to 3–12 times higher LC_{50} than the Cry1Ab toxin. R173T and G183C had comparable toxicity to the wild type toxin (Table 6.1).

To determine if the mutations in the conserved amino acids affected receptor binding, heterologous binding competition experiments were performed. The binding of biotinylated mutant proteins to BBMV isolated from M. *sexta* larvae were analyzed in the presence and absence of 100-fold higher concentration of unlabeled toxin. All mutant toxins bound to M. *sexta* BBMV. The interaction of all mutants is specific since 100-fold excess of unlabeled Cry1Ab protein competed for binding (data not shown). The fact that all the mutant toxins were fully displaced as the wild type Cry1Ab with 100-fold excess of unlabeled toxin indicates that the affinity of the different mutants was not drastically changed.

Finally, in order to determine the role of the mutated residues in K^+ permeability, we analyzed the effect of the mutant toxins in the permeability of BBMV isolated from M. *sexta* larvae by determining the distribution of a fluorescent dye sensitive to changes in membrane potential, Dis-C_3-(5). Addition of 50 nM concentration of the activated Cry1Ab to BBMV loaded with 150 mM KCl produced a fast hyperpolarization (Figure 6.3) and also increased the response to KCl additions ($m = 0.225$) when compared to the control, in which the same amount of buffer was added ($m = 0.08$) (Figure 6.3). The observed hyperpolarization can be explained by the exit of potassium ions from inside the BBMV through the Cry toxin open permeability. Figure 6.3 shows that additions of 50 nM of the activated Cry1Ab mutant toxins to the BBMV produced different effects in permeability. The mutation H168F affected severely the interaction of the toxin with the membrane and the K^+ permeability, since it presented a 24-fold lower effect on K^+ permeability than Cry1Ab (Figure 6.3, Table 6.1). The mutants R173L and R173N showed reduced activity since their effect was 2–3-fold lower than the wild type toxin. In contrast, the mutants Y153C, Y161L, Y161R, R173T, R173C, W182I and G183C presented similar K^+ permeability as the wild type toxin (Figure 6.3, Table 6.1).

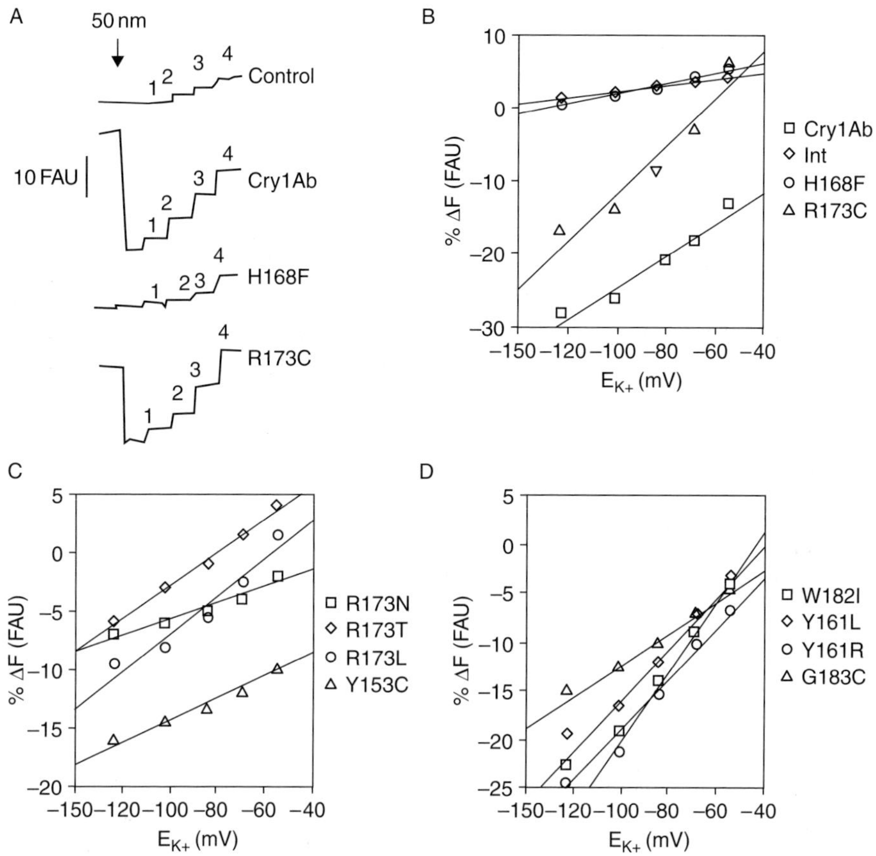

Figure 6.3 Analysis of the K^+ permeability across M. *sexta* BBMV induced by Cry1Ab activated toxin and different Cry1Ab mutants. Changes in distribution of a fluorescent dye sensitive to changes in membrane potential, Dis-C_3-(5) were recorded as described (Lorence *et al.*, 1995). Activated wild type or mutant toxins (50 nM) were added to BBMV (20 μg) loaded with 150 mM KCl, 2 mM EGTA, 0.5 mM EDTA, 10 mM HEPES–HCl pH 7.5 and suspended in 140 mM CsCl, 10 mM HEPES–HCl pH 9 buffer. Pre-equilibration with 1.5 mM Dis-C_3-(5) (5 min) is not shown. An upward deflection indicates a membrane potential depolarization, the opposite effect indicates an hyperpolarization. FAU = fluorescence arbitrary units. Final K^+ concentrations were (mM): 1 = 3; 2 = 6; 3 = 12 and 4 = 24. In the control, buffer A was added instead of toxin. (A) Fluorescence recording of K^+ permeability changes produced by selected different Cry1Ab mutants (B, C, D). The slope (m) of the curve ΔF (%) vs K^+ equilibrium potential (E_K+) (mV) or vs external K^+ concentration was determined for the indicated mutant or control vesicles (int). $E_K{}^+$ was calculated with the Nerst equation. Changes in fluorescence were determined four times.

From all mutated residues analyzed, only the mutant H168F showed a severe loss of the ability to induce K^+ permeability across the BBMV (Figure 6.3) and this data correlates with a severe loss in toxicity (1000-fold lower) (Table 6.1). Mutant H168F is produced in low yields and is quite unstable to the treatment with gut proteases (Figure 6.2). However, mutants in R173 were also highly unstable to gut proteases treatment but they showed similar K^+ permeability across the BBMV to the wild type toxin. Mutants R173L and R173N showed 2–3-fold reduced K^+ permeability activity across the *M. sexta* BBMV (Table 6.1). The data presented here suggest that conserved residues (R173, Y161 and W182) of $\alpha5$ are not critical for oligomer formation since several mutations in these residues are able to make pores in BBMV. Rather, our results indicate that these conserved residues are important for toxin stability within the insect gut, probably by contacts with neighbor helices, and structural integrity may be the main reason for low toxicity of these mutants.

In the closely related Cry1Ac toxin, several mutations in the conserved residue H168 showed altered toxicity to *M. sexta* larvae (Wu and Aronson, 1992). The positive charge seemed to be important for toxicity since a mutant with a change to negative charge (H168D) or a change with no charge (H168N) showed reduced toxicity in contrast with a conservative change (H168R) that showed threefold increased toxicity against *M. sexta* (Wu and Aronson, 1992). H168 is localized in the central part of helix $\alpha5$ and is the only polar amino acid located in the highly hydrophobic face of the helix $\alpha5$. It is possible that this residue plays an important structural role during the pore formation of Cry toxins. However, with the evidence of low K^+ permeability presented in Figure 6.3, we cannot rule out the possibility that this residue could be involved in oligomer formation. For this reason, we decide to analyze if mutant H168F could interact with other toxins to make oligomers.

Analysis of oligomer formation by Cry1Ab toxins

In order to gain evidences for the interaction of different Cry1Ab monomers during pore-formation activity and toxicity, we decided to explore if two independent mutants affected in different regions, could recover toxicity and pore formation when assayed as protein mixtures (Soberón *et al.*, 2000). Mutant F371A of Cry1Ab protein was chosen since previous reports showed that this mutant was affected on receptor binding and toxicity (Rajamohan *et al.*, 1995). The mutation F371A map in loop 2 of Domain II and it is affected primarily in the irreversible binding step of the toxin. Also, no effects in stability to trypsin treatment were observed in this mutant which presented a 400-fold reduction in toxicity against *M. sexta* larvae (Rajamohan *et al.*, 1995). Mutant H168F, affected in pore formation (Figure 6.3 and Table 6.1) was used as the second mutant to test if oligomer formation was affected in this mutant.

As stated above, the analysis of binding of biotinylated H168F protein to BBMV prepared from *M. sexta* showed that this protein was able to bind to BBMV (Figure 6.4A). We analyzed the binding of biotinylated F371A to *M. sexta* BBMV, we found that in our binding conditions, F371A did not bind to BBMV (Figure 6.4A). Nevertheless, binding of F371A was evident when a 100-fold excess of unlabeled Cry1Ab protein was incubated along with biotinylated F371A (Figure 6.4A). We also analyzed if F371A mutant could bind to BBMV when mixed with H168F. Figure 6.4B shows that biotinylated F371A bound efficiently to BBMV when mixed with H168F in a one to one ratio. F371A also bound to BBMV when mixed with 100-fold excess of H168F although less efficiently (Figure 6.4B). These results showed that F371A mutant bind efficiently to BBMV when mixed with toxin proteins that are not affected in receptor binding. In order to determine if the inter-molecular interaction between both mutant toxins occurs after receptor binding, a binding protocol was performed in which the interaction of labeled F371A with wild type Cry1Ab was analyzed

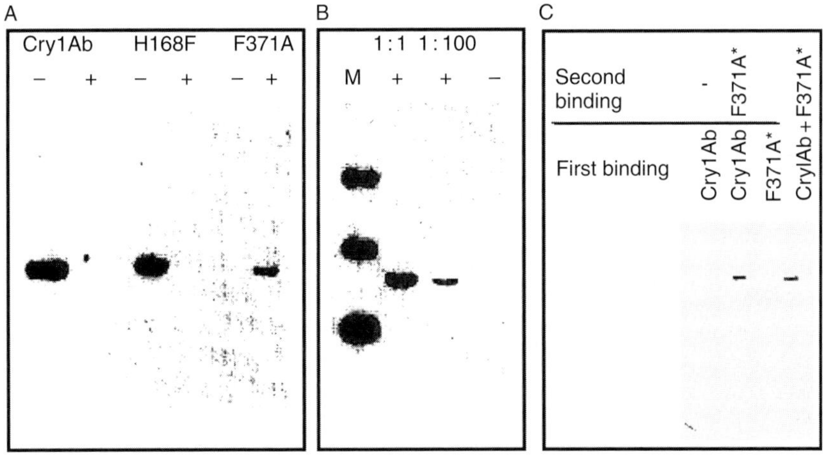

Figure 6.4 Homologous competition binding assays on BBMV isolated from M. *sexta* larvae. Panel A: Biotinylated Cry1Ab (lanes 1, 2), H168F (lanes 3, 4) and F371A (lanes 5, 6) toxins were incubated with the BBMV in the absence (−) or in the presence of 100-fold excess of unlabeled Cry1Ab toxin (+). After 1 h of incubation, unbound toxins were removed and vesicles containing bound toxins were loaded onto a SDS–PAGE and blotted to a nitrocellulose membrane. Labeled proteins were visualized by means of streptavidin-peroxidase conjugate. Panel B: Biotinylated F371A was incubated with the BBMV in the absence (−) or in the presence of 1 fold (1 : 1) or 100-fold (1 : 100) unlabeled H168F toxin. Molecular size markers were 98, 68 and 46 kDa respectively. Panel C: Sequential binding of wild type Cry1Ab and biotin labeled mutant toxin F371A. The BBMV were incubated with Cry1Ab and then unbound toxin was washed, these vesicles were used to bind biotinylated F371A in a second binding assay.

after binding of Cry1Ab toxin to BBMV (Figure 6.4C). Figure 6.4C shows that labeled F371A is able to bind to BBMV when the Cry1Ab toxin was previously bound to the receptor, suggesting that inter-molecular interaction between monomers occurs after receptor binding.

To determine if proteins F371A and H168F could form functional hetero-oligomers, the pore formation activity of wild type Cry1Ab and F371A–H168F mixtures was analyzed in BBMV from M. *sexta* larvae. In contrast with Cry1Ab toxin, proteins F371A ($m = 0.02$) or H168F ($m = 0.035$) did not induce an increased K^+ permeability when compared to control done with Cry1Ab toxin ($m = 0.170$) (Figure 6.5), showing that both mutants could not make functional pores on the BBMV. In contrast, addition of 100 nM of a 1 : 1 mixture of F371A–H168F to BBMV showed an increased K^+ permeability ($m = 0.174$) very similar to the one induced by wild type Cry1Ab toxin, and also produced a fast hyperpolarization of -80.4 ± 3 mV, $n = 3$. These data showed that the protein mixture of F371A and H168F interact with each other and recovered their capacity to form ionic channels *in vitro*.

Finally, the toxicity against M. *sexta* larvae was analyzed. M. *sexta* larvae were fed with 20 ng/cm^2 of a 1 : 1 mixture of F371A–H168F proteins. Table 6.1 shows that the mixture of both proteins recover toxicity to some extent since mortality was 25% in contrast with larvae fed with F371A or H168F mutants that showed no mortality or with Cry1Ab protein that showed a 100% mortality (Table 6.1).

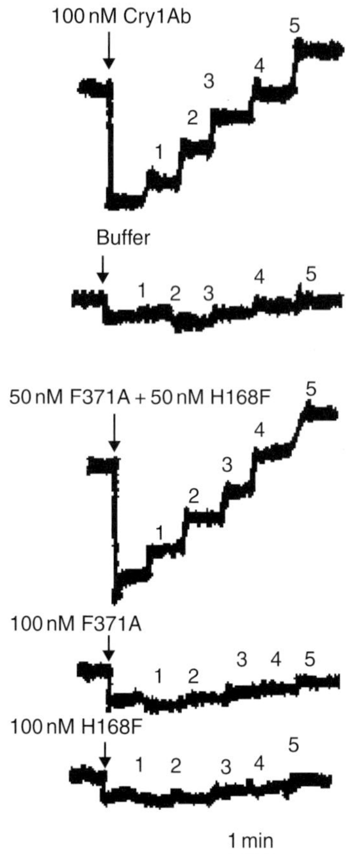

Figure 6.5 Effect of the Cry1Ab, F371A, H168F and F371A + H168F toxins on the membrane potential in M. *sexta* midgut BBMV. Membrane potential of BBMV (20 μg) loaded with (mM) 150 KCl, 2 EGTA, 0.5 EDTA, 10 HEPES–HCl pH 7.6 were recorded as described in Section 2. Pre-equilibration with 1.5 mM Dis-C_3-(5) (9 min) is not shown. The arrow on top of the traces corresponds to the time of toxin or buffer addition. Final K^+ concentrations were (mM): 1 = 6; 2 = 12; 3 = 24; 4 = 48 and 5 = 96. FAU = fluorescence arbitrary units. In the control, buffer A was added instead of toxin.

The pore-formation activity observed with the F371A–H168F protein mixture was very similar to the pore-formation activity of Cry1Ab. However, toxicity was only partially recovered by the F371A–H168F protein mixture (Figure 6.5, Table 6.1). This result could be explained by the reduced stability of H168F mutant toxin in the midgut, due to high sensitivity to midgut proteases (Figure 6.2). Nevertheless, the fact that the F371A–H168F protein mixture presented increased mortality in comparison with the toxicity displayed by individual mutant proteins, clearly shows that functional hetero-oligomers could be formed also *in vivo*. These data suggests that residue H168 is not involved in oligomer formation since H168F mutant is capable to interact with F371A protein and make functional pores.

Finally, these data are the first to show that the interaction of more than one monomer is fundamental for pore formation and toxicity of Cry1 proteins.

Concluding remarks

Cry toxins have five conserved blocks common to a large majority of the proteins. Among them α-helix 5 of domain I represents the first conserved block. As mentioned previously, several evidences indicate that this helix is an important structural element that has a role during the pore-formation activity induced by Cry toxins within the insect midgut cells. In this work, we described the analysis of some Cry mutant toxins, affected in several of the conserved residues of α-helix 5. The results presented here suggest that most conserved residues have an important function in preserving the structure of the protein within the midgut of susceptible insects. Only residue H168 has also an important role in the pore function. However, it is not implicated in oligomer formation. Still there remains the identification of the residues of Cry toxins involved in oligomer formation. The identification of such residues could give important clues on the structure of the toxin within the membrane and on the function of the pore induced by Cry toxins.

Acknowledgments

We thank Laura Lina for technical assistance. This work was partially supported by CONACyT No. G36505-N, DGAPA-UNAM IN206300 and IN216300, UC MEXUS and United States Department of Agriculture, No. 2002–35302–12539.

References

Aronson, A. I., Wu, D. and Zhang, C. (1995) Mutagenesis of specificity and toxicity regions of a *Bacillus thuringiensis* protoxin gene. *J. Bacteriol.*, **177**, 4059–4065.

Aronson, A. I., Geng, Ch. and Wu, L. (1999) Aggregation of *Bacillus thuringiensis* Cry1A toxins upon binding to target insect larval midgut vesicles. *Appl. Environ. Microbiol.*, **65**, 2503–2507.

Bravo, A., Jansens, S. and Peferoen, M. (1992) Immunocytochemical localization of *Bacillus thuringiensis* insecticidal crystal proteins in intoxicated insects. *J. Invertebr. Pathol.*, **60**, 237–246.

Burton, S. L., Ellar, D. J., Li, J. and Derbyshire, D. J. (1999) N-acetylgalactosamine on the putative insect receptor aminopeptidase N is recognized by a site on the domain III lectin-like fold of a *Bacillus thuringiensis* insecticidal toxin. *J. Mol. Biol.*, **287**, 1011–1022.

de Maagd, R. A., Bakker, P. L., Masson, L., Adang, M. J., Sangadala, S., Stiekema, W. *et al.* (1999) Domain III of the *Bacillus thuringiensis* delta-endotoxin Cry1Ac is involved in binding to *Manduca sexta* brush border membranes and to its purified aminopeptidase N. *Mol. Microbiol.*, **31**, 463–471.

Gazit, E., La Rocca, P., Sansom, M. S. P. and Shai, Y. (1998) The structure and organization within the membrane of the helices composing the pore-forming domain of *Bacillus thuringiensis* δ-endotoxin are consistent with an umbrella-like structure of the pore. *Proc. Natl. Acad. Sci. USA*, **95**, 12289–12294.

Grochulski, P., Masson, L., Borisova, S., Pusztai-Carey, M., Schwartz, J. L., Brousseau, R. *et al.* (1995) *Bacillus thuringiensis* CryIA(a) insecticidal toxin: crystal structure and channel formation. *J. Mol. Biol.*, **254**, 447–464.

Güereca, L. and Bravo, A. (1999) The oligomeric state of *Bacillus thuringiensis* Cry toxins in solution. *Biochim. Biophys. Acta.*, **1429**, 342–350.

Hofmann, C., Vanderbruggen, H., Höfte, H., Van Rie, J., Jansens, S. and Van Mellaert, H. (1988) Specificity of *Bacillus thuringiensis* δ-endotoxins is correlated with the presence of high affinity binding sites in the brush border membrane of target insect midgut. *Proc. Natl. Acad. Sci. USA*, **85**, 7844–7848.

Knowles, B. H. and Ellar, D. J. (1987) Colloid-osmotic lysis is a general feature of the mechanism of action of *Bacillus thuringiensis* δ-endotoxins with different insect specificity. *Biochim. Biophys. Acta*, **924**, 509–518.

Knowles, B. H. (1994) Mechanism of action of *Bacillus thuringiensis* insecticidal δ-endotoxins. *Adv. Insect. Physiol.*, **24**, 275–308.

Lee, M. K., You, T. H., Gould, F. L. and Dean, D. H. (1999) Identification of residues in domain III of *Bacillus thuringiensis* Cry1Ac toxin that affect binding and toxicity. *Appl. Environ. Microbiol.*, **65**, 4513–4520.

Li, J., Carroll, J. and Ellar, D. J. (1991) Crystal structure of insecticidal δ-endotoxin from *Bacillus thuringiensis* at 2.5 Å resolution. *Nature*, **353**, 815–821.

Lorence, A., Darszon, A., Díaz, C., Liébano, A., Quintero, R. and Bravo, A. (1995) δ-endotoxins induce cation channels in *Spodoptera frugiperda* brush border membranes in suspension and in planar lipid bilayers. *FEBS Lett.*, **360**, 217–222.

Lu, H. L., Rajamohan, F. and Dean, D. H. (1994) Identification of amino acid residues of *Bacillus thuringiensis* δ-endotoxin CryIAa associated with membrane binding and toxicity to *Bombyx mori*. *J. Bacteriol.*, **176**, 5554–5559.

Manoj Kumar, A. S. and Aronson, A. I. (1999) Analysis of mutations in the pore-forming region essential for insecticidal activity of a *Bacillus thuringiensis* δ-endotoxin. *J. Bacteriol.*, **181**, 6103–6107.

Martin, F. G. and Wolfersberger, M. G. (1995) *Bacillus thuringiensis* δ-endotoxin and larval *Manduca sexta* midgut brush border membrane vesicles act synergistically to cause very large increases in the conductance of planar lipid bilayers. *J. Exp. Biol.*, **198**, 91–96.

Masson, L., Tabashnik, B., Liu, Y.-B., Brusseau, R. and Schwartz, J. L. (1999) Helix 4 of the *Bacillus thuringiensis* Cry1Aa toxin lines the lumen of the ion channel. *J. Biol. Chem.*, **274**, 31996–32000.

Nuñez-Valdez, M.-E., Sánchez, Lina, L., Lorence, A., Güereca, L. and Bravo, A. (2001) Structural and functional studies of α-helix 5 region from *Bacillus thuringiensis* Cry1Ab δ-endotoxin. *Biochem. Biophys. Acta.*, **36370**, 1–10.

Pietrantonio, P. V. and Gill, S. S. (1996) *Bacillus thuringiensis* δ-endotoxins: action on the insect midgut. In *Biology of the Insect Midgut*, edited by M. J. Lehane and P. F. Billingsley, pp. 345–372. London: Chapman and Hall.

Rajamohan, F., Alcantara, E., Lee, M. K., Chen, X. J. and Dean, D. H. (1995) Single amino acid changes in domain II of *Bacillus thuringiensis* CryIAb δ-endotoxin affect irreversible binding to *Manduca sexta* midgut membrane vesicles. *J. Bacteriol.*, **177**, 2276–2282.

Rajamohan, F., Cotrill, J. A., Gould, F. and Dean, D. H. (1996) Role of Domain II, loop 2 residues of *Bacillus thuringiensis* CryIAb δ-endotoxin in reversible and irreversible binding to *Manduca sexta* and *Heliothis virescens*. *J. Biol. Chem.*, **271**, 2390–2396.

Schwartz, J. L., Garneau, L., Savaria, D., Masson, L., Brousseau, R. and Rousseau, E. (1993) Lepidopteran-specific δ-endotoxins from *Bacillus thuringiensis* form cation- and anion-selective channels in planar lipid bilayers. *J. Membrane Biol.*, **132**, 53–62.

Schwartz, J. L., Juteau, M., Grochulski, P., Cygler, M., Préfontaine, G., Brousseau, R. *et al.* (1997) Restriction of intramolecular movements within the Cry1Aa toxin molecule of *Bacillus thuringiensis* through disulfide bond engineering. *FEBS Lett.*, **410**, 397–402.

Slatin, S. L., Abrams, C. K. and English, L. H. (1990) δ-Endotoxins form cation-selective channels in planar lipid bilayers. *Biophys. Res. Commun.*, **169**, 765–772.

Smith, G. P. and Ellar, D. J. (1994) Mutagenesis of two surface-exposed loops of the *Bacillus thuringiensis* CryIC δ-endotoxin affects insecticidal specificity. *Biochem. J.*, **302**, 611–616.

Soberón, M., Pérez, R. V., Nuñez-Valdéz, M. E., Lorence, A., Gómez, I., Sánchez, J. *et al.* (2000). Evidences for inter-molecular interaction as a necessary step for pore-formation activity and toxicity of *Bacillus thuringiensis* Cry1Ab toxin. *FEMS Microbiol. Lett.*, **191**, 221–225.

Uawithya, P., Tuntitippawan, T., Katzenmeier, G., Panyim, S. and Angsuthanasombat, C. (1998) Effects on larvicidal activity of single proline substitutions in α-3 or α-4 of the *Bacillus thuringiensis* Cry4B toxin. *Biochem. Mol. Biol. Int.*, **44**, 825–832.

Van Rie, J., Jansens, S., Höfte, H., Degheele, D.and Van Mellaert, H. (1990a) Receptors on the brush border membrane of the insect midgut as determinants of the specificity of *Bacillus thuringiensis* delta-endotoxins. *Appl. Environ. Microbiol.*, **56**, 1378–1385.

Van Rie, J., McGaughey, W. H., Johnson, D. E., Barnett, B. D. and Van Mellaert, H. (1990b) Mechanism of insect resistance to the microbial insecticide *Bacillus thuringiensis*. *Science*, **247**, 72–74.

Von-Tersch, M. A., Slatin, S. L., Kulesza, C. A. and English, L. H. (1994) Membrane-permeabilizing activities of *Bacillus thuringiensis* coleopteran-active toxin CryIIIB2 and CryIIIB2 domain I peptide. *Appl. Environ. Microbiol.*, **60**, 3711–3717.

Wu, D. and Aronson, A. I. (1992) Localized mutagenesis defines regions of the *Bacillus thuringiensis* δ-endotoxin involved in toxicity and specificity. *J. Biol. Chem.*, **267**, 2311–2317.

Wu, S.-J. and Dean, D. H. (1996) Functional significance of loops in the receptor binding domain of *Bacillus thuringiensis* CryIIIA δ-endotoxin. *J. Mol. Biol.*, **255**, 628–640.

7 Colicin channels and protein translocation

Parallels with diphtheria toxin

Stephen L. Slatin and Paul K. Kienker

Introduction

Many protein toxins need to transfer a toxic factor into the cytoplasm of target cells, a process which is, in some cases, associated with pore formation (e.g. botulinum, pertussis, anthrax and diphtheria toxins) (Kagan *et al.*, 1981; Blaustein *et al.*, 1987; Blaustein *et al.*, 1989; Shannon and Fernandez, 1999). Such pores are thought to be part of the translocation machinery, and are not themselves lethal. However, other protein and peptide toxins kill by forming pores in the target cell membrane, disrupting the cell's permeability barrier, without further cytoplasmic involvement. In this chapter, we discuss recent work that shows that a class of proteins that kill strictly by pore formation, the pore-forming colicins, are themselves capable of extraordinary feats of protein translocation, by an unknown mechanism that may well not be unique to them. At least one toxin from the former group, diphtheria toxin (DT[1]), which kills by injecting a toxic enzyme into the cytoplasm and also forms pores, shares many properties with colicin. We conclude by considering the possibility that colicin and DT may catalyze translocation by a similar mechanism.

Colicin channels

Events preceding channel formation

Background

The colicins are a family of proteins toxic to *Escherichia coli* that are found, usually, on plasmids carried by various strains of *E. coli*. Each colicin is produced along with a matching immunity protein, which protects the producing cell from the particular colicin being produced. This system evidently confers a sufficient advantage on strains that possess it so that about 30% of *E. coli* are colicinogenic (Pugsley, 1984; Riley and Gordon, 1999).

Many colicin and immunity protein genes have been sequenced; in fact, colicin E1 was the first protein channel to be sequenced (Yamada *et al.*, 1982). We shall refer extensively to colicins A, E1, E2 and Ia, and the E2 immunity protein (Morlon *et al.*, 1983; Cole *et al.*, 1985; Mankovich *et al.*, 1986). The colicin proteins that have been examined to date are composed of three functional domains: the R domain binds to a receptor in the outer membrane of the target cell; the N-terminal N domain[2] interacts with certain periplasmic proteins; and the C-terminal C domain carries the actual toxicity (Ohno *et al.*, 1977; Yamamoto *et al.*, 1978; Dankert *et al.*, 1982). For a review, see Lazdunski *et al.* (1988). This last domain exhibits considerable variability among the different colicins. Some are enzymes that act in the target cell cytoplasm (and therefore must cross the inner membrane, by as yet unexplained mechanisms). Another group, the one we will focus on here, kills its target cells by forming a voltage-gated channel in the inner membrane. Cell death is a consequence of the unregulated loss of ions and small metabolites through this pore (Wendt, 1970). The sign of the voltage dependence is such that the open state of the channel is favored when the colicin is bound to the exterior surface of the inner membrane (the cytoplasm is electrically negative with respect to the cell exterior), while the closed state is favored when the colicin is bound to the interior surface of the inner membrane, thus providing a measure of protection for the producing cell. The immunity protein of each of the pore-forming colicins is a small inner membrane protein with three or four transmembrane helices that is constitutively expressed from the colicin plasmid and is present at about 500 copies per cell. It acts by interacting with the closed form of the colicin pore and presumably prevents its opening (Geli *et al.*, 1989; Song and Cramer, 1991; Geli and Lazdunski, 1992; Espesset *et al.*, 1996). The immunity proteins of enzymatic colicins, on the other hand, are small proteins that bind directly to their cognate colicins in the cytoplasm of the producing cell and block their enzymatic activity (James *et al.*, 1996).

A salient feature of the pore-forming colicins is that, as released by the cell, they are water-soluble, monomeric proteins (Cavard *et al.*, 1988), and yet they go on to become, effectively, membrane proteins of the target cell. They are also able to form channels in artificial lipid membranes, unaided by any host cell proteins, or any other proteins at all. That is, purified colicins, dissolved in aqueous buffer, can bind to pure lipid membranes and form voltage-dependent channels, a feat which has endeared them to membrane physiologists interested in studying channel-forming proteins free of the potentially muddling effects of any number of factors regularly found *in vivo* (Schein *et al.*, 1978).

Structures of the water-soluble forms of the pore-forming colicins

Colicin, then, has the ability to take on both soluble and membrane-associated conformations. The structure of the soluble form of several colicins has been solved, but to date these structures have been of limited use in deducing the structure of the membrane pore. The most complete solved structure is that of colicin Ia, one of the largest colicins, with 626 amino acids (Wiener *et al.*, 1997) (shown in Figure 7.4A). The structures of the C domains of three other colicins (A, E1 and N) have also been solved (Parker *et al.*, 1989; Elkins *et al.*, 1997; Vetter *et al.*, 1998). Each of the three, as well as the C domain of colicin Ia, was shown to consist of a tightly packed bundle of 10 alpha helices (Figure 7.1). The sole sustained hydrophobic segment of 35–49 amino acids found in all sequenced pore-forming colicins forms a helical hairpin (helices 8 and 9) sequestered from the aqueous phase by the other 8, mostly amphipathic, alpha helices, thus rendering the domain soluble. The isolated 10-helix bundle (prepared by either protease cleavage from the whole protein or by splicing out the relevant DNA) can itself undergo a soluble-to-membrane protein transition, and can form channels similar to those of the whole protein (but see below). Based on the colicin Ia structure, the rest of the protein is strikingly elongated (as had been predicted by hydrodynamic studies of colicin E1) (Schwartz and Helinski, 1971). The C domain is at one end of an extraordinarily long alpha helix (in fact, "helix 1" of the C domain is actually the final 20 amino acids of a 109-residue-long helix) that leads to the small receptor-binding domain, which is in turn separated by another long alpha helix from the N-terminal domain, which is important for delivering the toxic end of the molecule from the outer to the inner membrane. The two long alpha helices form a coiled-coil, bringing the C and N domains into contact. While these regions upstream from the 10-helix bundle do not

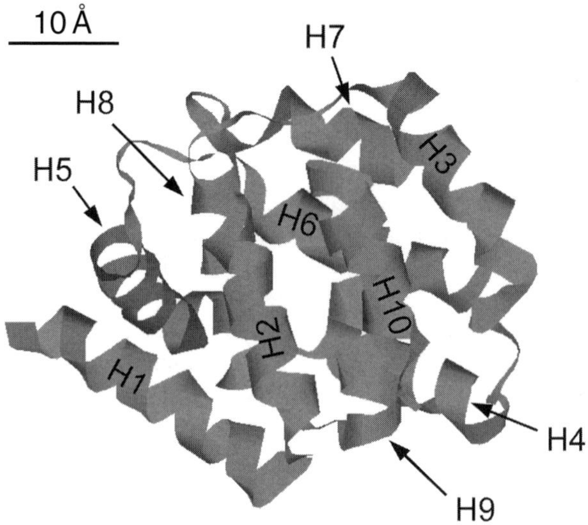

Figure 7.1 The crystal structure of colicin Ia, truncated to show only the C domain portion (residues 450–626). Helices H1 through H10 are labeled. The hydrophobic hairpin H8–H9 is colored orange and the rest of the C domain is blue. In the whole colicin Ia molecule, H1 would extend to the left for another 91 residues. The colicin Ia structure (Wiener *et al.*, 1997) was obtained from the Protein Data Bank (Berman *et al.*, 2000) (access code 1CII). The figure was produced with RasMol (v2.7.1). (See Colour Plate IX.)

participate in channel formation, we shall argue below that they can exert an important influence under certain conditions.

Transition to the membrane-bound state

The water-soluble protein or its C domain binds tightly to lipid membranes. Before the structure of the first colicin was solved it was proposed that the unique hydrophobic segment (helices 8 and 9) was important in this transformation (Cleveland *et al.*, 1983). Somehow, the other helices must reconfigure themselves so as to permit the hydrophobic segment to interact with the lipid and anchor the bound state. If these other 8 helices, many of which are amphipathic, splay out on the surface while retaining their secondary structure, the resulting bound state would resemble an umbrella, a model originally proposed for colicin A (Parker *et al.*, 1989, 1990). This model was supported by fluorescence resonance energy transfer (FRET) experiments, which showed that the distance between the hairpin formed by helices 1 and 2 (H1/2) and the rest of the molecule increased by 10–15 Å in the bound state compared to the solution state (Lakey *et al.*, 1991). However, subsequent work failed to support some details of the umbrella model and led to the "penknife" model, in which it was postulated that only H1/2 moves substantially away from the rest of the channel-forming domain, which sinks into the membrane to a depth sufficient to bury the (now exposed) hydrophobic hairpin (Lakey *et al.*, 1993). The penknife model was supported by disulfide-bond engineering experiments, in which a disulfide bond between two introduced cysteine residues was used to covalently link adjacent helices in the crystal structure (Duché *et al.*, 1994). The only disulfide links that blocked the insertion of colicin A into DOPG vesicles (determined by a fluorescence quenching assay) were those that prevented H1 or H2 from moving away from the rest of the domain. Unfortunately, different workers have used different colicins and lipids to study the bound state, making comparisons precarious. In any case, steady-state optical methods cannot reveal the pathway of unfolding, which undoubtedly involves a series of steps. Nor can it deconstruct a dynamic system, where the protein visits an ensemble of conformations under steady-state conditions, which may well be the case with colicin (see below). Time-resolved FRET and quenching experiments with colicin E1 (Zakharov *et al.*, 1998; Zakharov *et al.*, 1999; Lindeberg *et al.*, 2000) showed that initial unfolding at the membrane surface begins with a subtle rearrangement of H9/10, followed by the movement of the H1/2 hairpin away from the hydrophobic hairpin (in about 70 ms). Half a second later helices 3–7 have moved out away from the hydrophobic hairpin. The final structure is reminiscent of the umbrella model, in that the hydrophilic helices are all on the surface, but they are longer than those in the crystal, and they form a spiral, rather than spokes. The data allow some flexibility in arranging the helices, and in any case such a structure would be expected to be rather mobile, which may be related to the multiple closed states inferred for the channel from electrophysiological measurements.

While it is generally agreed that the hydrophobic segment is in contact with the lipid core in the membrane-bound state, its orientation with respect to the plane of the membrane remains unclear. For colicin Ia, at least, the hydrophobic segment forms a transmembrane element, presumably still a helical hairpin, in the open state of the channel and in at least some closed states. We base this on the observation that the loop between H8 and H9 can be detected on the *trans* side of the membrane in both open and closed states (see below).

Paradoxes of colicin channel formation

So little colicin protein, so much for it to do

At this point we must introduce the fundamental mystery of the colicin channel: apparently the pore has a large diameter, but there is not enough protein in the membrane to form such a large pore. A presumably related problem is even more troublesome: the same, small amount of colicin protein (or even less) forms a protein translocation pathway, which is demonstrably larger than the ion-conducting pore. (This will be discussed in a later section.) In this section, we endeavor to lay out the relevant issues concerning the structure of the channel.

General properties

The channel-forming colicins, or their isolated C domains, form well-defined channels in artificial lipid bilayers, in the absence of any other proteins (Schein *et al.*, 1978; Dankert *et al.*, 1982; Pattus *et al.*, 1982; Cleveland *et al.*, 1983; Martinez *et al.*, 1983; Nogueira and Varanda, 1988). While there are important differences among them, they all share certain features that must be explained by any viable model. The conductance of the channel is small, and generally pH-dependent. Typical values in 1 M KCl at pH 7 range from 15 pS for colicin A to about 50 pS for colicin Ia. In more physiological salts the conductance is of course smaller – smaller than that of many highly selective eukaryotic channels, such as sodium and potassium channels (Hille, 1992). But colicin channels are not very selective (see the following section), conducting all monovalent cations and anions that have been tested (Raymond *et al.*, 1985; Bullock *et al.*, 1992; Bullock and Kolen, 1995). All colicin channels are voltage-dependent, but good measurements of the relevant parameters have proved difficult to obtain due to the complexity of gating. Positive voltage opens them and negative voltage closes them. (By convention, the voltage is the potential of the *cis* side, the side to which colicin is added, with respect to that of the opposite, *trans* side.) Both the rate constants and the channel conductances reveal a complex, multistate system (Collarini *et al.*, 1987; Slatin, 1988; Cramer *et al.*, 1995; Cassia-Moura *et al.*, 2000).

Molecularity

Perhaps the most fundamental parameter required to build a model of the pore is the number of colicin molecules required. Ample evidence has accumulated to demonstrate that the channel is formed by a colicin monomer, beginning with the original work of Jacob *et al.* (1952) which showed that the biological activity exhibits "single hit killing", that is, toxicity is the first order in colicin concentration. When release of K^+, rather than cell death, was used as the marker for colicin action, the result was the same (Wendt, 1970). The possibility of a preformed multimer is belied by several studies that showed that colicin is a monomer in solution. Ultracentrifugation studies have reported that soluble colicin E1 (Schwartz and Helinski, 1971), and colicin A (Cavard *et al.*, 1988) are monomers at neutral pH, and chromatographic studies have reported that the C domain of colicin E1 is a monomer unless it is denatured (Steer and Merrill, 1997; Steer *et al.*, 1999). Likewise, a variety of studies in non-biological systems all point toward a monomeric pore. In the planar lipid bilayer system, colicin K (Schein *et al.*, 1978) and E1 (Slatin, 1988) formed channels in direct proportion to their concentration, implying a monomeric structure for the channels actually observed electrophysiologically. Colicin pores can permeabilize lipid vesicles, which thus provide another system to test the molecularity. Assays that measured initial efflux of Cl^- (Peterson and Cramer, 1987), or initial influx of Tl^+ (Bruggemann and Kayalar, 1986) as a function of colicin concentration were, again, linear. Levinthal *et al.* (1991) used an EPR assay to measure the discharge of the electrical potential across lipid vesicle membranes by low concentrations of the C domain of colicin E1. The peptide behaved as a monomer down to concentrations where there was less than one colicin molecule per vesicle. Recently, Tory and Merrill (1999) used a FRET assay to look for multimer formation in heterogeneously tagged colicin bound to lipid vesicles – they found none. While none of the above experiments, by itself, proves that a single molecule of colicin forms the channel, taken together, this extensive and concordant list of results would seem to settle the issue. Nevertheless, one might well remain somewhat skeptical, given the difficulty of envisioning a monomeric channel with the requisite properties.

The size of the lumen

One such property requiring explanation is the diameter of the lumen, which appears to be substantial, in spite of the relatively low conductance. Raymond *et al.* (1985) estimated the pore size of colicin E1 by measuring its permeability to a series of large ions, as determined by its reversal potential in a salt gradient. If a large ion is excluded from the pore on the basis of size, then, in a salt of that ion and a small counterion, the pore will be ideally selective for the counterion (assuming there are no other charge carriers in the system), and the current will reverse at the Nernst potential of the small ion, as it would for any channel that excludes the large ion on whatever basis. Thus, the gramicidin channel, say, which is ideally cation-selective, can serve as a standard of comparison for the exclusion of anions. By this technique, Raymond *et al.* found that large ions, such as tetraethyl ammonium (TEA), bis–tris propane and NAD could pass through the channel. Allowing for the most extended possible configuration of their asymmetric probes, such as NAD, these authors estimated that the colicin E1 pore was at least 8 Å in diameter, and that the pores formed by colicins A and Ib were that size or larger. More recently, and more compulsively, we have confirmed that colicins A and Ia are permeable to TEA (unpublished results). Bullock *et al.* (1992, 1995) extended these studies and concluded that the colicin E1 channel is at least 9 Å in diameter. Neither group was able to find a monovalent ion that blocked the channel. Using a different technique, Krasilnikov and his colleagues (1998) estimated both the size and shape of the colicin Ia pore. They looked at the effect of non-electrolytes added to either side of the membrane on the conductance of the channel to small ions. The technique essentially measures the loss in effective concentration of the charge carriers in the pore. If the non-electrolyte were freely permeant, it would have the same mole fraction effect in the pore as in bulk solution. To the extent that it is excluded from the pore, its effect is diminished. Using a series of non-electrolytes, these authors concluded that the colicin Ia channel was narrowest, 7 Å, at a point near the *trans* entrance, which was itself about 10 Å in diameter, while the *cis* entrance was 18 Å in diameter. Bearing in mind that they were studying colicin Ia and not E1, this result is not seriously at odds with the selectivity studies, which measure the narrowest part of the pore. This is smaller than several estimates for known multimeric toxins, but it is still large compared to the conductance (comparable to, say, the acetylcholine receptor channel, which is, however, formed by a pentamer of a total MW of approximately 300 kD). If the lumen is, indeed, funnel shaped, as the non-electrolyte experiments suggest, it would seem to require even more protein to construct than a cylinder.

The minimal pore-forming domain

The channel is formed by the crystallographically well-defined C domain, but several studies have shown that the entire domain is not required, hindering us further in our model-building attempts. Liu *et al.* (1986) made a series of colicin E1 C-terminal peptides by placing unique methionine residues in various positions downstream of the last native methionine (M370), and treating the mutant protein with cyanogen bromide, which cuts at methionine. The fragments were purified on SDS gels and tested in the planar bilayer system. They found that a fragment consisting of the final 94 amino acid residues formed voltage-dependent channels similar to channels formed by the whole colicin. This suggested that the channel formed by the whole colicin was, in fact, made by only the final 5 1/2 helices or so, rather than the entire C domain. However, the short peptide channels were not identical to native channels, and may not have the identical (presumably monomeric) structure. Cramer *et al.* (1990) reported that a slightly shorter cyanogen bromide peptide also makes voltage-dependent channels in bilayers. Noting that some of the gating properties, and the mix of conductance states of this channel, were not identical to native colicin E1 channels, these authors suggested that the short-peptide channels were formed by aggregates. While this interpretation is certainly tenable, subsequent findings show that it may not be necessary to invoke aggregation to explain the ability of such a short peptide to make a quasi-normal colicin channel (see below).

Baty *et al.* (1990) found that a peptide consisting of the final 136 residues of colicin A forms channels with a conductance identical to whole colicin A channels, but, again, with altered gating properties. Nardi *et al.* (2001) found that a shorter peptide encompassing the final 6 helices of colicin A makes channels with nearly normal conductance, but altered gating; and that a peptide encompassing the final 5 helices of colicin A makes channels with both altered conductance and altered gating properties. Considering, also, the translocation experiments with colicin Ia (see below), which suggest that helices 2–5 are not part of the channel, it seems reasonable to us to hypothesize that the normal colicin channel is formed by, at most, 6 helices from the original 10 found in the crystal structure.

The role of the hydrophobic segment in the channel

Apart from its role as a membrane anchor, the hydrophobic segment presumably plays a structural role in the channel, although there is actually very little evidence for this, other than the need to use any and all protein that may be available for the purpose (see above). Nevertheless, there is good evidence that the hydrophobic hairpin must be inserted into and across the membrane to permit the formation of the channel. *In vivo* experiments on colicin E1 showed that substituting single charged residues into the hydrophobic segment severely inhibited activity, unless they were in the loop between the two helices, suggesting that, at some stage that was crucial for intoxication, the segment was arranged in the membrane as a helical hairpin with the loop exposed to the *trans* aqueous phase (Song *et al.*, 1991). Interestingly, among the mutations that had sufficient activity to test in bilayers, none showed significantly altered ion selectivity, which might have been expected if the hydrophobic segment were part of the channel. Kienker *et al.* (1997) showed that the tip of the hydrophobic hairpin of colicin Ia is exposed to the *trans* side of the membrane by biotinylating a residue at the tip and showing that *trans* streptavidin altered the channel's gating properties. Closed channels so biotinylated could be grabbed by *trans* streptavidin even if they had never opened before, but some closed channels could be bound by *cis* streptavidin and prevented from opening. Thus, the hydrophobic segment has a transmembrane orientation in the open channel. When the channel is closed, but bound to the membrane, the hydrophobic segment can be in a transmembrane orientation, or an orientation that exposes the interhelical loop to the *cis* solution; and no doubt in several other orientations. This dynamic flexibility may account for some of the uncertainty in the steady-state spectroscopy data concerning the hydrophobic segment in closed channels: it is not in a fixed position, and the dwell time in any particular position probably depends on the conditions of the experiment.

Thus, we are left with the unenviable task of imagining how to build a stable pore with a diameter greater than 8 Å from a single peptide of no more than about 100 residues. (But, as we shall see, it gets worse.)

Protein translocation by colicin channels

Self-translocation

Early observations

Early on, it was observed that gating of the channel was associated with the movement of part of the protein into the membrane as the channel opens, and back out as it closes.

Working at the low pH requisite for colicin E1, Slatin *et al.* (1986b) showed that pepsin on the *cis* side of the membrane destroyed channels only if they were closed – open channels were protected – and concluded that gating involved the insertion of at least part of the channel structure. Furthermore, Raymond *et al.* (1986) showed that a site which, upon exposure to trypsin, converted the behavior of the channel from that of a long C-terminal fragment to that of a shorter one, crossed the membrane to the *trans* side in association with gating. Since protease action on this site did not destroy the channel, but merely altered its properties, they surmised that a part of the peptide that was not structurally part of the channel was translocated, in addition to a part that was.

Several techniques have now been applied to study this translocation phenomenon, which is clearly a crucial part of the gating mechanism. However, this does not preclude the possibility that there may be other gating mechanisms, as suggested by the observation that several of the channel's (poorly understood) kinetic states are fast – faster than most of the techniques which have been used to study translocation. Vesicle systems, which have proved essential for describing the binding steps, do not lend themselves as easily to the study of voltage-driven insertion, due to the difficulty of controlling the transmembrane voltage. Nevertheless, Merrill and Cramer (1990) were able to use lipophilic photoaffinity probes and a hydrophilic iodination catalyst to follow the movements of vesicle-bound colicin E1 in response to an imposed diffusion potential. They found small but significant changes in labeling that could be explained by the insertion of the protein into the membrane in response to positive voltage. They found that the change in labeling was limited to a 36-residue segment – roughly from the middle of helix 5 through helix 6.

Mapping the topology of colicin in the membrane

In order to be able to probe the movement of specific residues of the colicin channel in the planar bilayer system, where the channels under scrutiny can be effectively isolated and directly manipulated with voltage, Slatin *et al.* (1994) tagged colicin Ia mutants with biotin at a series of unique cysteine residues. Most of the biotinylated colicin mutants made normal channels, which could then be probed with streptavidin, a large, water-soluble biotin-binding protein. An effect of streptavidin added to one side of the membrane or the other would reveal the location of the tagged residue with respect to the membrane – *cis* or *trans*. It was found that, at many positions, the effect of streptavidin was to "freeze" the gating process: open channels could be locked in the open state by streptavidin added to the *trans* side, and closed channels could be prevented from opening by streptavidin added to the *cis* side. Figure 7.2 shows a representative experiment with *trans* streptavidin at one such position. Unexpectedly, these and subsequent experiments revealed that a large segment of the colicin, corresponding to helices 2–5 in the crystal structure, was exposed to the *trans* solution when the channel was open, and to the *cis* solution when it was closed, as shown in Figure 7.3 (Qiu *et al.*, 1996).

Two unforeseen inferences can be drawn from these experiments: (1) helices 2–5 are not in the membrane when the channel is open and thus are not part of the transmembrane channel structure, leaving only helices 1 and 6/7 as possible membrane-spanning elements upstream from the hydrophobic hairpin; and (2) in addition to its channel-forming properties, colicin has a protein translocation property. Somehow, probably in the course of forming the channel, a long stretch of hydrophilic peptide moved completely across the lipid bilayer (the translocated segment includes 15 positive and 8 negative charges at neutral pH). There is as yet no accepted explanation of either how or, teleologically, why the small

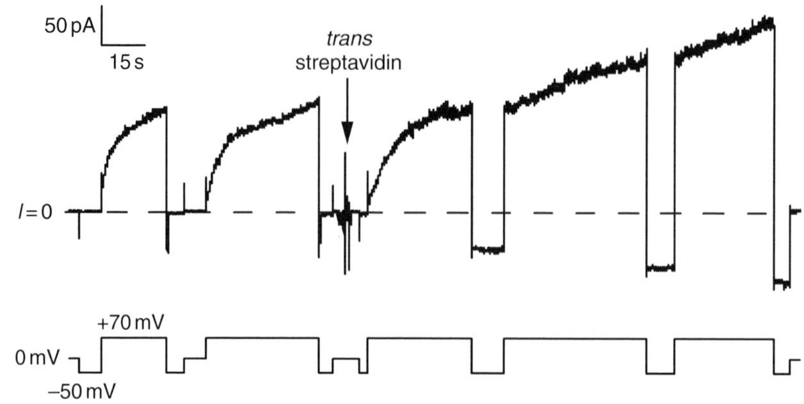

Figure 7.2 The effect of *trans* streptavidin on a colicin Ia mutant biotinylated in the translocated segment (at residue 511C), as in Qiu *et al.* (1996). The upper trace shows the current (with the zero-current level indicated by $I = 0$) and the lower trace shows the voltage across the bilayer. Time goes from left to right. This is a record with many channels in the membrane. The first two pulses show the normal voltage-dependent gating of colicin Ia, with channels opening at positive voltage and closing rapidly at negative voltage. After streptavidin was added to the *trans* solution, the channels no longer closed.

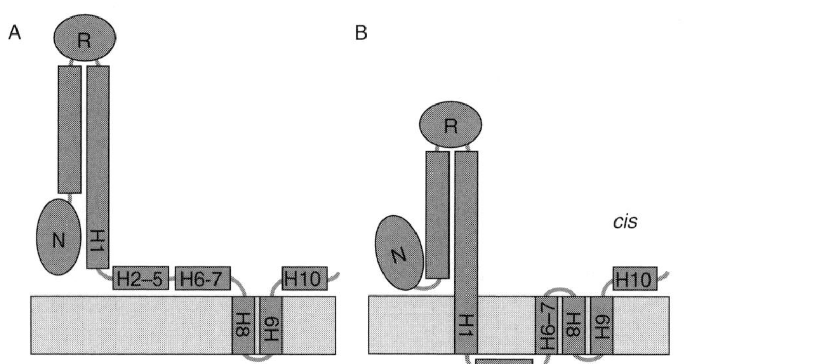

Figure 7.3 Topological model of colicin Ia, with respect to a planar bilayer, in the closed and open channel states, as in Qiu *et al.* (1996). (A) There is a closed state in which the hydrophobic hairpin (H8–H9) is inserted in the membrane and the rest of the molecule is on the *cis* side. (B) The voltage-dependent open state has two additional membrane-inserted segments, parts of H1 and H6–7. In addition, H2–5 are translocated across the membrane to the *trans* side.

colicin C domain transports 40% of itself across the membrane. (As will be seen, the upstream domains are not required for translocation.) The mechanism does not appear to mimic more familiar examples of protein translocation, which typically occurs through large, specialized protein pores in organelles such as the endoplasmic reticulum (Johnson and van Waes, 1999) and mitochondria (Pfanner and Geissler, 2001), or as carried out by protein toxins such as anthrax (Blaustein and Finkelstein, 1990) and pertussis (Shannon and Fernandez, 1999), which form multimeric structures.

In vivo *experiments*

Some insight into the possible mechanism of translocation comes from work of Duché, Baty and their colleagues, who studied deletion mutants of the C domain of colicin A fused to a signal sequence at the N terminus and to alkaline phosphatase at the C-terminus (Duché *et al.*, 1999). These constructs, expressed in *E. coli*, were directed across the bacterial inner membrane and into the periplasm by the signal sequence, where they could be detected by the enzymatic activity of the alkaline phosphatase. Intact C domain as well as constructs lacking H10, or H9–10, were protected from externally added protease, but if H8 was also missing, they were not. This protection was eliminated by a double point mutation in the loop between H5 and H6 that blocks channel formation in the intact, purified protein. If protease protection were the result of translocation, the implication of these results would be that helices 1–8 contain a functional translocation mechanism. Since, in colicin Ia anyway, helices 2–5 end up being translocated, H6–8 are implicated in the catalysis of the translocation. However, the H1–8 (or, for that matter, H1–9) constructs are not active toxins, whereas H1–10 is. This suggests that channel formation may not be necessary for translocation. If so, perhaps translocation is catalyzed by only helices 6–8, which somehow form a pathway through the membrane sufficiently amphipathic to allow co-insertion of alpha-helical protein (in the manner of the helical hairpin hypothesis (Engelman and Steitz, 1981)), resulting in net translocation in the presence of an electric field. A crucial unresolved question here is the temporal order of channel opening and translocation.

Translocation of regions upstream from the channel-forming domain

The extent of the translocation raises the question of what prevents the translocation of H1. That is, if translocation is catalyzed by a part of the protein downstream from the translocated segment, one might suppose that, after the translocation of H5, 4, 3 and 2, H1 would follow. There are apparently no stringent constraints on what sequences in the H3–4 region can be translocated (see below). Likewise, the part of helix 1 that spans the membrane in the open state of the channel ("H1") has no conspicuous feature that would trap it in the membrane. The question of whether translocation is arrested by a local feature of the H1 sequence or by a remote part of the protein was addressed by examining C-terminal fragments of colicin Ia (Figure 7.4). Two types of experiment demonstrated that removing remote parts of the colicin sequence allows H1, as well as some upstream regions, to be translocated (Kienker *et al.*, 2000).

In one type of experiment, with a colicin Ia fragment biotinylated near its N-terminal end, *trans* streptavidin was used to probe whether the attached biotin moved across the membrane to the *trans* side. For three different biotinylated C-terminal fragments, designated CT-S, CT-M and CT-L (174, 189, and 300 residues long, respectively), *trans* streptavidin had dramatic effects on channel gating. For the shorter biotinylated fragments CT-S and CT-M (roughly equal to the C domain), *trans* streptavidin induced a new conductance that turned on at *cis* negative voltages, the reverse of the normal voltage dependence of gating. Although the precise mechanism of this effect has not been determined, it provides clear evidence that, in these fragments, the N-terminal end of H1 is translocated. Even

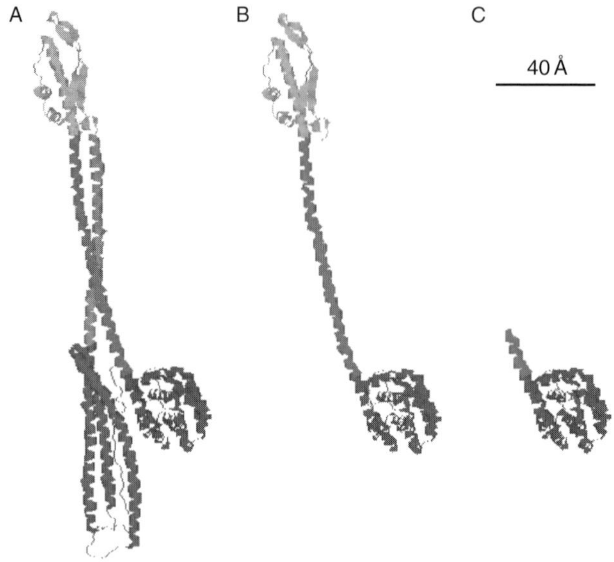

Figure 7.4 Whole colicin Ia and two of the C-terminal fragments used by Kienker *et al.* (2000). (A) The crystal structure of whole colicin Ia (Wiener *et al.*, 1997), obtained as in Figure 7.1. The N-terminal domain is colored red, the R domain is yellow, the C domain is blue, and the long inter-domain helices are orange and green. (B) The same structure, truncated to show the 345-residue fragment CT-XL. (C) The same structure, truncated to show the 189-residue fragment CT-M. The figure was produced with RasMol (v2.6-ucb). (See Color Plate X.)

more remarkably, *trans* streptavidin also had an effect on the much longer biotinylated fragment CT-L; in this case the effect was to inhibit channel closing at negative voltages. Thus, more than 120 residues upstream from the C domain can be translocated.

A second type of experiment, in which single channels were observed, appears to provide an alternative assay for translocation of the N terminus of colicin Ia C-terminal fragments. When a whole colicin Ia channel opens at a sufficiently large positive voltage ($\geq$50 mV), its conductance remains constant more or less indefinitely, aside from brief flickers to other conductance levels. In contrast, a fragment channel initially opens to the same conductance as whole colicin, but, after some delay (ranging from a fraction of a second to several minutes), the conductance drops to about 1/6 of the initial value (Figure 7.5). This "mini-channel" state is stable for as long as the positive voltage is maintained. The rate of transition from normal channels to mini channels appears to increase with larger voltage and lower pH. The mini channel turns off much more slowly than the normal channel at negative voltages. The transitions were seen for CT-S, CT-M, CT-L and a 345-residue fragment, CT-XL, which includes all of the long helix 1 and the receptor-binding domain, which contains both alpha-helical and beta-sheet structure. When CT-M or CT-L fragments biotinylated near their N-terminal end were bound to streptavidin, to hold the end on the *cis* side of the membrane, the transitions were abolished; this supports the inference that the transitions correspond to translocation of the end. Thus, it appears that at least 260 residues of colicin Ia (282–541) can be translocated, when the upstream residues are deleted.

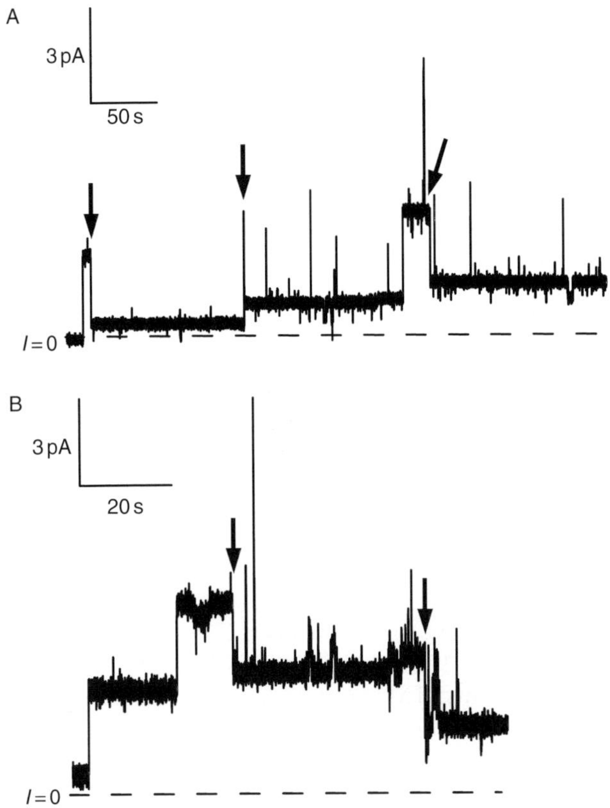

Figure 7.5 Colicin Ia C-terminal fragment channels making transitions to the mini-channel state, as in Kienker *et al.* (2000). (A) Single channels formed by the fragment CT-M. The record shows current vs time, while the voltage was held at +70 mV. Each of the three channels shown opened initially to the normal conductance level seen for whole colicin Ia channels (38–39 pS), but, after a few seconds, each dropped to the lower-conductance mini-channel state (7 pS), as highlighted by the arrows. For the second event, the transition occurred very quickly, so the opening to the normal conductance appears as a spike. (B) Single channels formed by the much longer fragment CT-XL. This record is similar to that in part A, except that the dwell time at the normal conductance tended to be longer. There was a mini channel already open at the start of the record. Two channels opened to the normal conductance level (42–43 pS). Then one of them dropped to about 8–9 pS (first arrow). Another mini channel opened, with no transition or spike visible, and, finally, the other normal channel dropped to the 8–9 pS mini-channel conductance (second arrow). The salt solution used for this experiment (and for the single-channel experiments in subsequent figures) was 1 M KCl, pH 6.2. (The CT-XL fragment had biotin chemically attached near the N terminus, but that is not relevant for this experiment.)

A brief digression on the gating of colicins A and E1

The translocation of the N terminus of colicin Ia fragments may help to explain several unusual gating phenomena observed in colicins E1 and A. At large positive voltages, open channels formed by whole colicin E1 close into an "inactivated" state, a state that is kinetically different from the "normal" closed states seen at negative voltage (Slatin *et al.*, 1986a); this may well be related to the mini-channel state seen in colicin Ia fragments. More dramatically, after inactivation of colicin E1 channels, some channels reappear with reversed voltage dependence, suggesting that they have crossed to the other side of the membrane; this may be analogous to the "flipping" of colicin Ia C-domain channels induced by the binding of *trans* streptavidin. Both these observations suggest that much of the protein, presumably including the hydrophobic segment, can be induced to reorient in the membrane.

Unlike colicins Ia and E1, whole colicin A channels have two distinct gating modes: in the "high-pH" mode, the channels turn off at moderate negative voltages, whereas in the "low-pH" mode, large negative voltages are required for turn-off (Collarini *et al.*, 1987). Colicin A C-domain channels are reported to exhibit only the low-pH mode. It is tempting to postulate that the high- and low-pH modes correspond to the normal and mini-channel states, respectively, of colicin Ia fragments. Presumably, at low pH, whole colicin A, but not whole colicin Ia, can translocate enough upstream sequence to enter a state that is more difficult to close, as is the mini channel. Perhaps differences in the propensity to translocate their upstream domains may account for some of the differences in gating among colicins.

Implications for channel structure

Figures 7.6A and 7.6B show the inferred topology of the open C domain channel in the "normal" and mini-channel configurations. It should be noted that the mini channel consists of only three transmembrane segments (TMS) per colicin molecule, compared with four for whole colicin Ia. It is a challenge to imagine how to make a channel with only three TMS (but see below). While H1 of the C domain cannot completely stop translocation, it does slow it down. The normal, 4 TMS channel, which has H1 in the membrane, can endure for many minutes before making the transition to the 3 TMS channel. Thus, the 4 TMS structure is somewhat stable even in the absence of any upstream elements. It is not known if the particular sequence of H1 accounts for this, or if it is due to a more global property of the channel.

So, what does stop translocation in whole colicin Ia? The experiments just described show that the determining factor cannot be a feature of the C domain alone. Rather, it must involve the upstream parts of the colicin, either the N-terminal N domain or the long helix that connects the N and R domains. Most likely, it is the interaction between the two long helices in the coiled coil that brings translocation to a halt; for instance, the coiled coil structure may be too big to be translocated and too stable to be unwound.

Translocation of exogenous substrates

Insertions in the translocated domain

Apart from whatever role they may play in channel formation, the ability of C-domain helices 6–10 to translocate protein is remarkable in its own right. To test whether the translocated segment (H2–5) had unique (and obscure) properties disposing it to be so easily translocated, Jakes *et al.* inserted epitopes of 8–12 amino acids into the loop between H3 and H4 of colicin Ia (Jakes *et al.*, 1998). These insertions had little effect on channel properties, and could be bound on the *trans* side by their respective antibodies when the channel was open, thereby inhibiting channel closing. Thus, sequences that differ

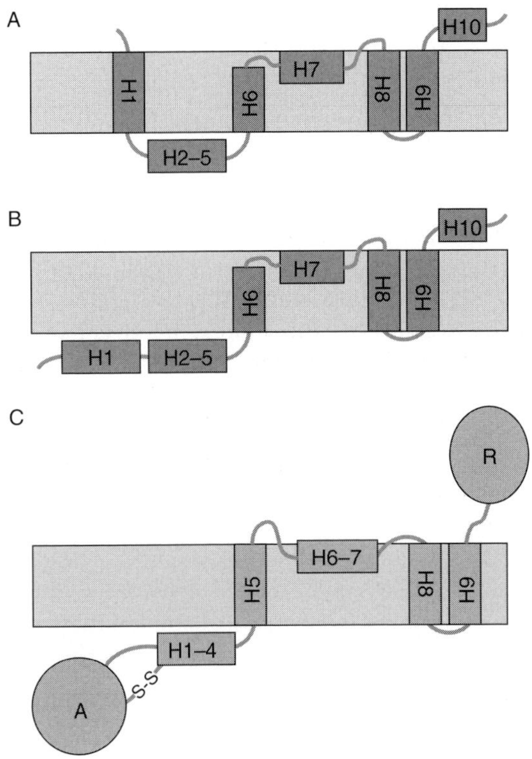

Figure 7.6 Models showing the parallels in the topology of colicin Ia C-terminal fragment channels and diphtheria toxin channels. (A) The transient "normal"-conductance state of a colicin Ia fragment, with four TMS, as in Kienker *et al.* (2000). (Helices H6–7 form one membrane-spanning segment, but it is not known if they are oriented precisely as drawn.) (B) The low-conductance mini-channel state of a colicin Ia fragment, with three TMS, and the N-terminal end translocated to the *trans* side. (C) The open channel state of diphtheria toxin (Senzel *et al.*, 2000), with three TMS, and the N-terminal region translocated to the *trans* side.

significantly from the native sequence of the translocated segment are still susceptible to translocation. One of these exogenous sequences, the FLAG epitope, contains 7 charged residues (out of 8), ruling out amphipathicity as a requirement for translocation. Furthermore, the duplication of 1 or 2 helices in the translocated segment does not block channel formation (S. Slatin and V. Geli, unpublished observations), suggesting that larger peptides may also be translocatable.

In order to test the potential of the system to translocate whole proteins, we made use of the non-pore-forming colicin, colicin E2 (S. Slatin, D. Duché, A. Nardi, D. Baty, unpublished observations). Colicin E2 kills its target cells via an enzymatic C-terminal domain (a DNAse, in this case), which takes the place of the channel-forming domain. In order to protect the producing cell from this DNAse, an immunity protein (ImmE2) is synthesized along with the colicin. The immunity protein binds tenaciously to the E2 and inhibits its enzymatic activity. Thus, E2 is only found in nature as part of a 1 : 1, E2 : ImmE2 complex. The binding affinity is extraordinarily tight – about 10^{15} M^{-1} – comparable to the biotin–streptavidin complex, rather than to typical antigen–antibody complexes, which bind with affinities roughly five orders of magnitude weaker. ImmE2 is composed of 86 amino acid residues, including, conveniently, a single cysteine. The crystal structure of the homologous colicin E9 immunity protein, bound to its cognate colicin E9 C-terminal, enzymatic domain, has been solved (Kuhlmann *et al.*, 2000).

We inserted the sequence of the entire 86 residue E2 immunity protein between H3 and H4 of colicin A. The resulting fusion protein (colA–ImmE2) formed channels of normal conductance, but somewhat abnormal gating, in planar bilayers. When CTE2, the isolated C-terminal domain of colicin E2, (prepared as a single protein, free from its endogenously bound immunity protein) was added to the *trans* side of the bilayer, it prevented the closing of open colA–ImmE2 channels (Figure 7.7). This block could not be reversed over tens of minutes by subsequent addition of excess free ImmE2, demonstrating that the off rate of the bound CTE2 is appropriately slow. If CTE2 was instead added to the *cis* chamber in the presence of closed colA–ImmE2 channels, it did not prevent channel opening, but, rather, rendered the channels that opened insensitive to *trans* CTE2. Thus, *cis* CTE2 binds to the ImmE2 insert and keeps it on the *cis* side. The fact that normal channels still form under these conditions shows that all of the necessary channel-forming apparatus is downstream from the insert. Comparable results were obtained by biotinylating the lone cysteine residue in colA–ImmE2 and probing with streptavidin instead of CTE2. These results demonstrate that the inserted E2 immunity protein is translocated across the bilayer along with the translocated segment, and that it is present in a functional form on the *trans* side, able to bind to its cognate CTE2 with high affinity. Figure 7.8 summarizes these results schematically.

Apparently, there is nothing special about the acidic, mostly alpha-helical ImmE2 that makes it a good substrate for translocation, since the roles of ImmE2 and CTE2 in this

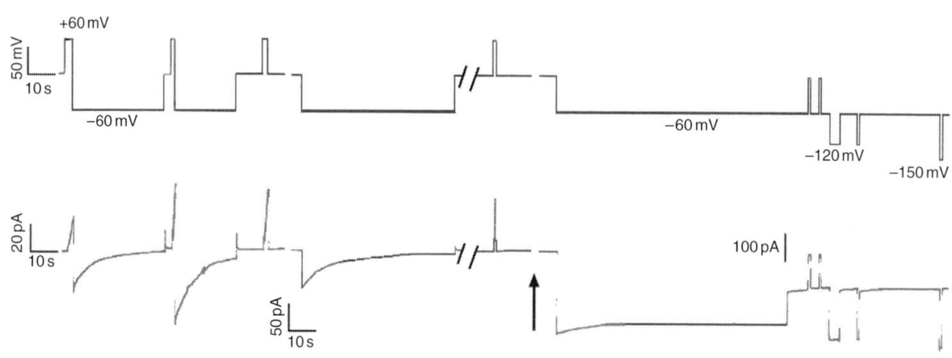

Figure 7.7 ColA–ImmE2 channels in planar lipid bilayers – translocation of the ImmE2 insert. Current (lower)/voltage (upper) traces begin at 0 pA and 0 mV, respectively. At the beginning of the trace, a series of pulses to +60 mV and −60 mV opened and closed a population of channels. (Individual channels are too small to resolve at this gain.) A few seconds after the first break in the record about 700 pS of conductance was turned on with a brief +60 mV pulse. During the second break (arrow) 4 μg of CTE2 was added to the *trans* side. A large fraction of the conductance failed to turn off at −60 mV after exposure to *trans* CTE2. Short pulses to more negative potentials (−120 and −150 mV) were also unable to close the channels.

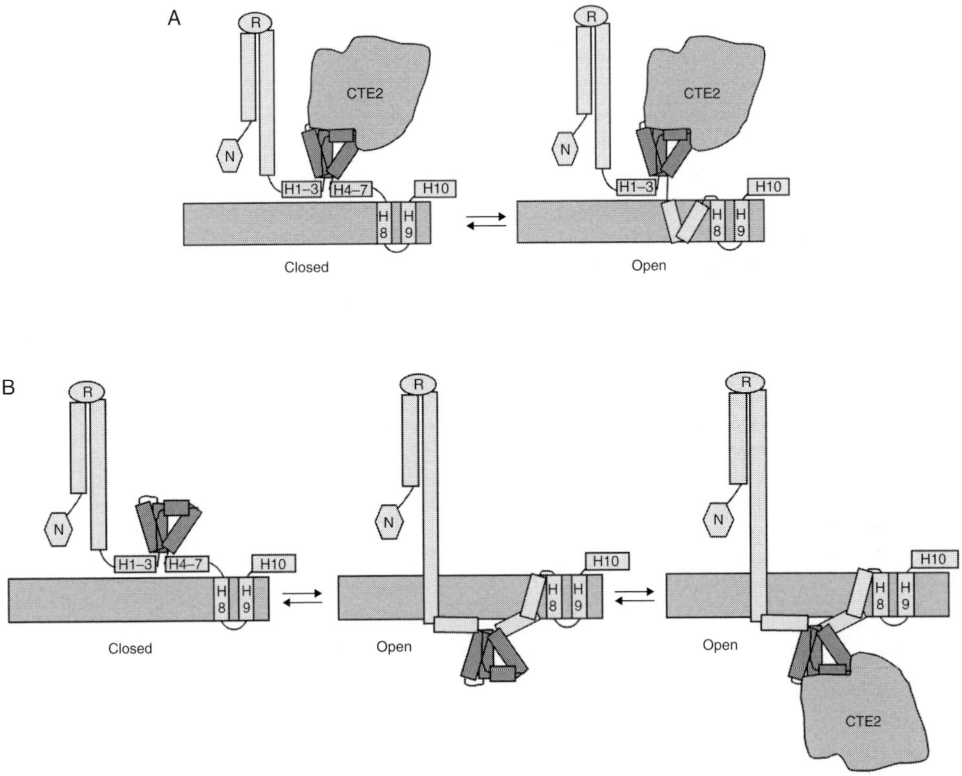

Figure 7.8 Topological model of colA–ImmE2 as it interacts with free CTE2. Colicin A, light grey; ImmE2 insert, dark grey; CTE2, medium grey. The crystal structure of ImmE9 (Kuhlmann *et al.*, 2000) was used as a guide to draw ImmE2. Colicin A is drawn schematically to conform to the structure of colicin Ia (Wiener *et al.*, 1997), except for the C-terminal domain. The transmembrane segments between ImmE2 and H8 in the open conformation are unlabeled in order to indicate the unknown nature of their structure. (A) *Cis* CTE2. The channel can open and close with *cis* CTE2 bound to the ImmE2 insert. Conductance of the CTE2-bound channel is normal, even though the topology of the helices in the membrane is not. (B) *Trans* CTE2. When CTE2 binds on the *trans* side to the translocated ImmE2 insert, the channel cannot close.

experiment can be switched. If the 134 residue CTE2 is inserted as the cargo protein and ImmE2 used as the *trans* probe, the same result is obtained – CTE2 is efficiently translocated and is functional (i.e. able to bind ImmE2 with high affinity) on the *trans* side. As in the first experiment, free ImmE2 added to the *cis* side does not prevent channel opening, but it does eliminate the *trans* ImmE2 effect, and free *trans* CTE2 added before, but not after, the addition of *trans* ImmE2 blocks its effect. Thus the colicin can translocate the larger, basic CTE2 (which includes both beta and alpha structure) seemingly as well as ImmE2. It is not unreasonable to suppose that it might be capable of moving a large variety of inserted proteins.

Sizing the translocation pathway with N-terminal probes

The colicin E2 inserts, as well as the upstream regions of colicin Ia discussed earlier, may cross the membrane in a wholly, or partially, unfolded state. Thus, if the translocation pathway has a size limitation, its value cannot be directly inferred from the size of such probes in their folded state. In order to better define the size limit of the translocation pathway taken by the upstream segments, Jakes *et al.* (2001) attached a series of "molecular stoppers" to the N terminus of the CT-M fragment. Each stopper was a peptide or protein that was held in its folded conformation by one or more disulfide bonds. The ability (or failure) of the colicin to translocate the attached stopper was assessed by the presence (or absence) of transitions from normal-conductance channels to mini channels (Figure 7.9). Stoppers of known structure were used in most cases; the diameter of the smallest cross-section was taken as the relevant size. It was found that a 9-residue peptide with a single disulfide bond (diameter ≈ 12 Å) and α-conotoxin (16 Å) (Jakes *et al.*, unpublished observations) were readily translocated; these stoppers were too small to stop translocation. On the other hand, charybdotoxin (28 Å) was too big to be translocated. Finally, apamin (21 Å) seemed to be just the right size; it could be translocated, but at a substantially slower rate than the smaller stoppers. As a control, each of the stoppers could be unfolded by reducing their disulfide bonds, in which case they were all rapidly translocated (shown in Figure 7.9C for charybdotoxin). This shows that it is the folded, globular structure of the stopper that prevents translocation, rather than any aspect of the primary sequence and is consistent with the notion that the physical size of the stopper is the relevant parameter for suitability as a transported cargo. It was estimated that helix 1 is translocated through a pathway (which may include the ion-conducting pore) approximately 21 Å in diameter.

The three pathways: ions, translocated segment, stoppers

The various studies on colicin channels and protein translocation reviewed here address three pathways across the bilayer, which, in principle, could be either distinct or overlapping. There are: (1) the pathway for ions, that is, the ion-conducting channel; (2) the pathway that allows the translocated segment (H2–5 in colicin Ia) to cross the membrane; and (3) the pathway that allows "stoppers" attached at the N terminus of the C domain to cross. The simplest view, in our opinion, is that pathways 1 and 2 are distinct and pathway 3 is 1 + 2. This is illustrated more concretely in the rather fanciful Figure 7.10A, which shows a view, from the *cis* side, of the four membrane-spanning helices of colicin Ia, H1, H6[3], H8 and H9, arranged around a pore containing a TEA molecule. This is not intended as a realistic model of the channel, but it allows us to think about the sizes of the helices and the pore. (In fact, there is reason to think that at least some of these segments are not alpha helices, as discussed below. In Figure 7.10A, we nevertheless use the helices from the crystal structure of colicin Ia, for lack of a practical alternative.) Pathway 1, the channel, is assumed to be just big enough to allow TEA to fit through (7–8 Å);[4] that is, no bigger than it has been proven to be. The size of pathway 2, the H2–5 translocation route, is given by the diameters of H2–5 (17–19 Å), as well as that of H1 (18 Å), since H1 is postulated to follow the translocated helices into this pathway. The outlines of pathways 1 and 2, in light blue and pink, respectively, are superimposed on apamin (Figure 7.10B), which, with a 21 Å diameter, was the largest stopper to traverse pathway 3 and on charybdotoxin (Figure 7.10C) which, at 28 Å, was too big to cross. Although the apamin cross-section does not precisely fit within the combined outlines, it is close enough in size to suggest that 1 + 2 = 3 may be an essentially correct description of the three pathways. Charybdotoxin, in

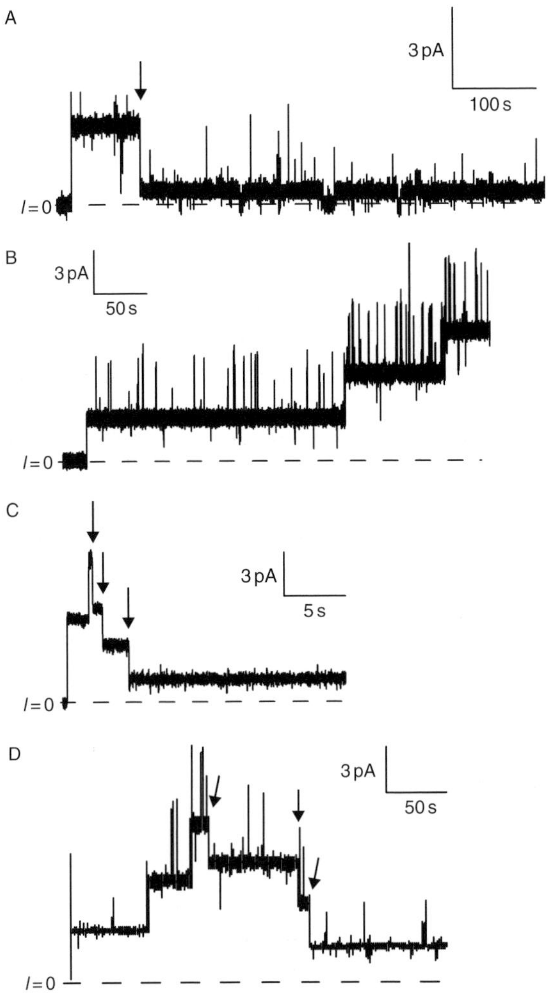

Figure 7.9 Translocation of N-terminal "stoppers" by the colicin Ia fragment, CT-M. (A) Translocation of a small stopper. A 9-residue, disulfide-bonded peptide (CGSGDGSGC) was attached at the N terminus of CT-M (Jakes *et al.*, 2001). The record shows a single channel opening to the normal conductance level of 41 pS, and then dropping (at the arrow) to the 7 pS mini-channel state. The voltage across the membrane was held at +70 mV for the entire record. (B) Folded charybdotoxin resists translocation by CT-M. Charybdotoxin was chemically attached near the N terminus of CT-M. The record shows three channels opening to the normal conductance level (each 42 pS) and staying there, except for brief flickers. (Note the lack of arrows.) (C) The same charybdotoxin-tagged CT-M, after unfolding of the charybdotoxin by treatment with a reducing agent at an elevated temperature. The record shows two channels opening together to the normal conductance level (average 41–42 pS), followed by the opening of a third, slightly larger channel (59 pS). At the first arrow, the 59 pS channel (apparently) made a transition to the mini channel (10 pS). The other two channels then dropped to mini channels (approximately 5–8 pS), at the second and third arrows. The voltage was held at +70 mV throughout both records. (D) Slow translocation of apamin by CT-M. Apamin was attached at the N-terminus of CT-M. The record shows three channels opening in succession to slightly larger than normal conductance levels (46, 46, and 53 pS, in order). After spending a couple of minutes in this state, first the two 46 pS channels each made transitions to the mini channel (10 and 11 pS), and finally the 53 pS channel dropped to a 12 pS mini channel, as indicated by the arrows. The voltage was held at +70 mV throughout the record.

contrast, is much bigger than the 1 + 2 outlines, as would be expected for a stopper that really stops translocation.

So, what forms the channel?

Of course, the problem of not having enough protein to make a big channel has not magically vanished; the four helices in Figure 7.10A were placed in proximity to one another,

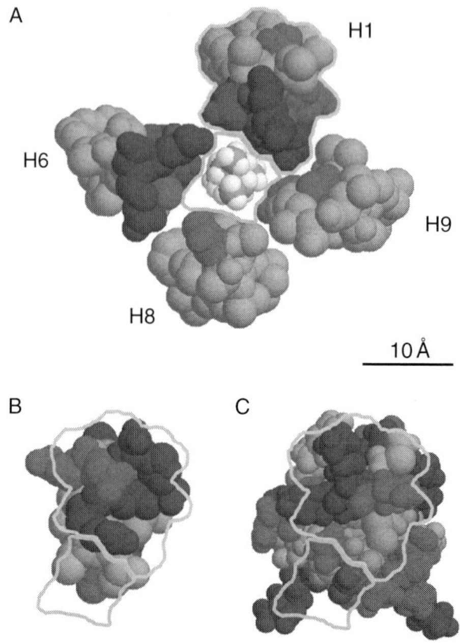

Figure 7.10 Comparison of pathways for ion conduction, "normal" protein translocation and translocation of N-terminal "stoppers" by colicin Ia and its C domain. (A) Fanciful model of the normal colicin Ia channel. The four membrane-spanning helices, H1, H6, H8 and H9, are viewed end-on from the *cis* side; they surround a pore which contains a TEA molecule. The helices were excised from the crystal structure, arranged in proximity to one another around a TEA molecule, and oriented with their polar faces lining the channel. The ion-conducting channel (pathway 1) is outlined in light blue, and the normal protein translocation pathway (pathway 2), represented by H1, is outlined in pink. No account of the interhelical forces required to hold this assemblage together was taken. (B) The structure of apamin, rotated to show the smallest cross section. Apamin is just small enough to be translocated through pathway 3 as a "stopper" attached to the colicin Ia C domain. The good match in size between apamin and pathways 1 + 2 (the superimposed outlines) suggests that the pathway for stopper translocation is simply the combination of the normal protein translocation pathway and the ion-conducting channel. (C) The structure of charybdotoxin, rotated as in B. Charybdotoxin is too big to be translocated as a stopper, and its cross-section is much bigger than the superimposed outlines (pathways 1 + 2). Residues are color-coded as follows: blue, positively charged; red, negatively charged; green, uncharged polar; grey, hydrophobic; yellow, cysteine. The colicin Ia (Wiener *et al.*, 1997) and charybdotoxin (Bontems *et al.*, 1992) structures were obtained from the Protein Data Bank (Berman *et al.*, 2000) (access codes 1CII and 2CRD, respectively), and the apamin coordinates were kindly provided by D. E. Wemmer (Pease and Wemmer, 1988). The TEA structure model was generated using the program "B" (N. White, http://www.scripps.edu/~nwhite/B). Hydrogen atoms are shown for charybdotoxin and TEA, but not for colicin Ia or apamin. The figures were produced with RasMol (v2.6-ucb). (See Colour Plate XI.)

but they are not packed together in an energetically plausible manner. More realistic models of packed alpha-helices around a pore suggest that a four-helix channel would be too small even to let alkali metals such as Li^+ through; a TEA-permeable channel would need something more like 6 helices (Lear *et al.*, 1988). A proposal that may lead out of this predicament would be that the membrane lipids form part of the structure of the channel, as suggested by, for example, Qiu *et al.* (1996). Although there is no direct evidence for this in colicin channels, recent results from magainin and other pore-forming peptides indicate that they work by recruiting lipids into a mixed protein/lipid structure resembling an inverted micelle that penetrates the bilayer (Matsuzaki, 1998). In a similar vein, a molecular dynamics study of melittin has suggested that four membrane-inserted peptide helices can induce the formation of a large pore, with much of the pore lined by lipids (Lin and Baumgaertner, 2000).

The inadequacy of the model in Figure 7.10A becomes even more pronounced when the mini channel with three TMS, formed by colicin Ia C-terminal fragments, is considered. Although it has not been established that this channel is monomeric, nor that it has a large pore diameter, its conductance is comparable to that of colicin A, and it clearly forms a stable pore. Furthermore, it has been shown that the normal upstream TMS of colicin Ia (H1) can be replaced by part of the segment normally translocated without major effect (Qiu *et al.*, 1996). If a biotinylated residue in the normally translocated segment (in H2, for example) is grabbed by *cis* streptavidin and thus forced to remain on the *cis* side, a channel can still be formed, but with reduced conductance and altered voltage dependence. Thus, the channel's properties are sensitive to the nature of H1 – changing it, or eliminating it, alters the channel – but it clearly is not crucial for channel formation. Are the 4 TMS and 3 TMS channels two fundamentally different structures, or is H1 a peripheral element that merely influences the channel's properties? Current research does not resolve this question. In any case, even 4 TMS, let alone 3, seem inadequate to the task of forming a pore with the necessary properties, implying that the channel is something other than a bundle of transmembrane helices.

The affiliation of colicin and diphtheria toxin channels

Background

Comparison to colicin

The colicin C domain has been reported to be homologous to several proteins, including phycocyanin and myoglobin (Holm and Sander, 1993), the Bcl-2 family of apoptosis proteins (Muchmore *et al.*, 1996), the large group of delta-endotoxins (Parker and Pattus, 1993), and, notably, diphtheria toxin (Choe *et al.*, 1992). Most of these systems are less well understood than is colicin, but significant progress has been made in the case of diphtheria toxin (DT), such that it may well be useful to consider it here. DT bears many obvious similarities to colicin, but it has not often been thought of as a kindred protein because its mechanism of action is fundamentally different. While DT does form low conductance channels, they are not the lethal agent of the toxin. The toxicity of DT depends on its "A fragment," which is a water-soluble enzyme that ADP-ribosylates elongation factor II, blocking protein synthesis and leading directly to cell death. Furthermore, DT channels have traditionally been assumed to be formed from multimers of the toxin, further distancing it from colicin, which, as we have seen, appears to form monomeric channels. Recent work, however, has suggested that DT may be more than superficially related to colicin, an insight that may be helpful in understanding both systems.

The soluble form of DT

Diphtheria toxin is a 535 residue protein that, like colicin, has three major domains in its crystal structure (Choe *et al.*, 1992; Bennett and Eisenberg, 1994) – the N-terminal catalytic domain, or A fragment; a C-terminal receptor-binding domain and a central, channel-forming domain. The catalytic domain is easily separated from the adjacent domain by the mild proteolysis of an arginine-rich segment between them, along with reduction of the disulfide bond between Cys 186 and Cys 201. The 176-residue, channel-forming, or "T" domain, is, like the comparably sized colicin C domain, a bundle of 10 alpha helices, with H8 and H9 forming a hydrophobic hairpin in the core. Biologically, DT acts by insinuating its lethal catalytic domain into the cytoplasm of target cells via a pathway that involves endocytosis into a low pH compartment (for reviews, see Madshus and Stenmark, 1992; Falnes and Sandvig, 2000). Once in the cytoplasm, the A fragment is liberated to do its sinister work. The job of the T domain is to transfer the A fragment across the endosomal membrane into the cytoplasm.

DT Channels

Description

The T domain forms channels *in vivo* and also in planar bilayers. In bilayers, DT conductance properties are quite similar to those of colicin channels, although those of colicin channels are somewhat less reproducible. Whole DT or isolated T-domain channels are a few tens of pS in 1 M KCl (the exact value depending on the pH), well within the range seen for colicin channels. Both channels are also somewhat selective for monovalent cations over anions, more so at high than at low pH. The gating of T-domain channels is, however, rather different from that of colicin channels, although both are highly sensitive to pH. The complex nature of the voltage dependence of both systems has discouraged quantitative analysis, but the hyperlinear dose response of DT (in contrast to the linear curve for colicin) supported the notion of a multimeric pore, a notion that was easy to accept, conforming, as it does, to the postulated structure of many other protein toxins, as well as to common sense. Nevertheless, the connection between the dose-response curve, which goes as less than the square of the DT concentration, and a multimeric pore is actually rather weak. Such a hyperlinear curve can arise from cooperative steps prior to actual pore formation. Below, we discuss evidence supporting a monomeric structure for the DT channel, along with results that speak to protein translocation and channel formation by T domain as compared to colicin.

Mapping the topology of DT in the membrane

Since the function of the T domain is to translocate the A fragment, and since the T-domain channel is implicated in the mechanism of translocation, it is worthwhile to review the process as it occurs in lipid bilayers, free of any other proteins. Due to a serendipitous observation, it was possible to map out the topology of the open DT channel with respect to the membrane (Senzel *et al.*, 1998), much as the topology of the colicin Ia channel was mapped using site-directed biotinylation. The now nearly universal protein purification technique of adding a 6-histidine (6-His) tag to one end of a protein in order to subsequently use a Ni^{++} affinity column for purification led to the discovery that 6-His is a potent voltage-dependent blocker of DT channels. Channels formed from N-terminal His-tagged T domain closed rapidly at negative voltages, and re-opened rapidly at positive voltage, whereas untagged T-domain channels did not. This closure turned out to be due to the presence of the 6-His tag on the *trans* side of the membrane, most likely because the electric potential drives it into the pore, blocking it (Senzel *et al.*, 1998). The assignment of the 6-His tag to the *trans* side was verified by showing that *trans* Ni^{++} blocked the effect, as did manipulations on the *trans* side that physically removed the N-terminal region, such as trypsin cleavage. If the protein was biotinylated near the N-terminal 6-His segment, *trans* streptavidin blocked the 6-His-induced rapid closure. Thus, the N-terminus of the T domain is on the *trans* side of the bilayer when the channel is open. Even more dramatically, the same phenomenon was observed in whole DT, with the 6-His tag directly attached to the N terminus of the catalytic domain, 200 residues removed from the T domain (Oh *et al.*, 1999a). If the DT was protease-nicked between the A fragment and T domains (leaving them attached by their inter-domain disulfide bond), reduction by the membrane-impermeant reagent tris-2-carboxyethyl phosphine on the *trans* side eliminated the rapid closure, showing that the C terminus of the A fragment was also on the *trans* side when the channel was open. In fact, this and other experiments have shown that the entire A fragment is translocated by the T domain in planar bilayers, without the benefit of any other proteins it may encounter *in vivo* (Senzel *et al.*, 1998).

The 6-His tag block has been exploited further as a marker to map the topology of the entire T-domain open channel by chemically attaching 6-His tags to cysteine residues engineered into interior regions of the protein (Senzel *et al.*, 2000). Even from an interior position, the 6-His tag blocked the channel in a voltage-dependent fashion, thus revealing which side of the membrane it was on. Such experiments showed that helices 1 to 4 of the T domain, along with the A fragment, are on the *trans* side when the channel is open. H5 spans the membrane, H6/7 are on the *cis* side and H8 and H9 each span the membrane. Evidently, then, the DT channel is formed from three transmembrane segments (but, see below). It should be noted that there is some evidence, in vesicle systems, that H1 may form a fourth transmembrane segment (Malenbaum *et al.*, 1998; D'Silva and Lala, 2000); however these systems lack the ability to isolate open channels, as noted above. If the translocated regions are merely passengers in the process, then we must conclude that H5–9 alone are capable of translocating hundreds of residues across the bilayer.

The minimal DT channel structure

Before discussing the evidence that the DT channel is a monomer, we will take up the topic of its structure. Perhaps the most astonishing finding is that the channel can be formed by only H8 and H9 (Silverman *et al.*, 1994). (However, as just discussed, the channel normally has a third transmembrane segment.) In a deletion mutant study of channels in planar bilayers, Silverman *et al.* showed that a 61 amino acid stretch corresponding to only slightly more than H8 and H9 formed channels with conductance properties (single channel conductance and reversal potential) almost identical to those of whole T-domain channels. The smallest peptide channels were less active channel formers than wild-type and had altered voltage dependence, but the conductance pathway was evidently the same. Since the smallest channel-forming fragment of T domain was still linked to the A fragment (to improve solubility), one cannot rule out the possibility that part of the A fragment may have, anomalously, contributed to the channel structure, but if so, it would have to have influenced the conductance in exactly the same manner as H5. In any case, the H8–9 hairpin is, at the very least, the most important part of the channel structure.

If such a limited part of the T domain forms the channel, many of its residues should be exposed to the lumen, rather than to the membrane interior. Residues lining a channel can often be identified by engineering a series of unique cysteine mutations into the channel-forming domain, and probing with a small, water-soluble, sulfhydryl-reactive reagent such as a methanethiosulfonate (MTS) derivative, and monitoring effects on the channel (Akabas *et al.*, 1992). This method was refined by Huynh *et al.* (1997), who detected the reaction of MTS with T-domain cysteine mutants by an abrupt change in single-channel conductance. They constructed 49 single cysteine mutants, spanning H8 and H9, and found 23 that reacted and 26 that did not. In order for a reaction to be observed, two conditions must be satisfied: the introduced cysteine must be accessible to the MTS reagent, and the cysteine must be in a suitable location so that its modification by MTS has an effect on the channel conductance. This could be either a direct effect, such as sterically narrowing the lumen or changing the charge distribution in or near the pore, or an indirect effect, such as altering the channel structure. Thus, the failure to observe a reaction at a particular residue is not conclusive evidence for its inaccessibility, but, when a reaction is observed, the residue is definitely accessible.

If H8 and H9 are still alpha-helices in the open channel structure, then the pattern of accessibility of the introduced cysteines should have a periodicity of roughly 3.6 residues: for instance, 1 accessible residue followed by 2 inaccessible residues. However, Huynh and his colleagues found several straight runs of 4 or more reacting (and non-reacting) residues in both H8 and H9. Without belaboring the details, the pattern of MTS susceptibility found does not support a helical hairpin structure for H8/9 in the open channel, nor does it support a beta structure (which would result in every other residue reacting). It is unlikely that this represents any unsuspected defect in the technique, since it has been used successfully to detect both alpha-helical and beta-strand structure in other channels (Akabas *et al.*, 1994; Xu and Akabas, 1996; Benson *et al.*, 1998). Conceivably, this result could be explained by a multimeric channel structure, with different monomers having different faces of their helices exposed to the pore, but, as we shall see shortly, DT appears to be a monomer. Another possibility is that the reactive regions are mobile loops, and expose different residues at different times. This would seem to conflict with the well-defined single-channel conductance, however, unless the movement takes place on such a fast time scale that it is averaged out in the electrical recordings. Spectroscopic studies, on the other hand, have found alpha-helical structure in this region of the channel (Oh *et al.*, 1999b), but all such studies suffer from an inability to isolate the channel in its open state. The apparently anomalous secondary structure of these segments in the open channel is presumably linked to the manifestly incomprehensible structure of the channel itself.

Is the DT channel a monomer?

Several persuasive results argue that the DT channel formed in planar bilayers is indeed a monomer. If the cysteine-substituted channel were a symmetric *n*-mer, it should have *n* accessible cysteines, and hence one would expect to see *n* reactions per channel. In all 23 sites, in H8 and H9 where a reaction was detected, however, only a single step change in conductance was seen. One can imagine that even in a multimeric channel structure, the first reaction might block succeeding reactions by, for instance, sterically or electrostatically preventing reagent from reaching the other cysteines, although this seems unlikely, given the large diameter of the DT channel (Hoch *et al.*, 1985). However, even a small, uncharged reagent (methyl-MTS) produced only a single reaction at the sites where it was tested, and reaction with a positively charged MTS reagent blocks subsequent reaction with a negatively charged one. Again, we should note that there is no apparent problem with the basic experiment that can explain away this untoward result, since, where it has been applied to channels known to be multimeric, multiple reaction steps are indeed seen, as with the channel formed by anthrax toxin (Benson *et al.*, 1998).

Another result with MTS reagents independently suggests a monomeric structure. Most of the specific conductance changes seen are most simply interpreted as arising from a charge effect in the lumen – adding positive charge decreases conductance and adding negative charge increases it. For example, mutating D352–D352C (thus removing its negative charge) reduced its conductance (under conditions where K^+ was the major charge carrier), and reacting D352C with MTS-ES^-, which restores a negative charge at that site, restored the wild-type conductance (Mindell *et al.*, 1994). Likewise, reacting D352C with MTS-EA^+, which results in a side chain very similar to that of lysine, reduced its conductance to a value mimicked by the point mutant D352K. If the channel were an n-mer that, for whatever reason, reacted with only a single MTS molecule, one would not expect the conductance of the reacted channel to be equivalent to that of the corresponding mutant, which has, of course, n-point mutations at residue 352. The simplest inference from the data, then, is that $n = 1$. Finally, a series of experiments designed to detect DT multimeric channels by engineering conductance and gating mutants whose properties would allow monomers and multimers to be easily distinguished, found none (Finkelstein and Gordon, pers. commun.). The cumulative weight of these results forces us to conclude that DT, like colicin, forms monomeric channels.

Are DT and colicin channels functionally identical?

Considering the results discussed above, it is clear that the T domain of DT and the C domain of colicin are alike in numerous ways, despite the lack of sequence homology. It is not yet clear whether they are essentially identical, although that cannot be ruled out at this point. They form channels with similar conductance properties, and both are capable of translocating large pieces of protein at their N-terminal ends. They both appear to do this as monomers, and without the assistance of any other protein. There are, however, some differences, which may or may not turn out to be trivial. For example, the pH and voltage dependences are different, but it has been difficult to characterize these parameters quantitatively and thereby compare them in the two toxins. Also, whereas colicin C domain contains a single hydrophobic hairpin in its crystal structure, formed from H8 and H9, T domain contains two – one formed by H8 and H9 and the other by H5 and H6/7. However, H5/H6–7 does not form a hydrophobic hairpin in the open DT channel, and thus this difference between DT and colicin may not be important.

The recent results with DT discussed above have revealed the structure of the open channel in greater detail than is known for colicin. For the DT channel, the conductance pathway is formed from H8 and H9 with the possible involvement of the third transmembrane segment, H5. For colicin, H8 and H9 (note the convenient convergence of H numbers) form a pair of transmembrane segments (that have not been reported to form or not form the conductance pathway), while H6/7 and H1 each form transmembrane segments, for a total of four. However, H1 is not required for the formation of a channel, although its presence affects the conductance, at least in colicin Ia. Recall also that H8 and H9 of DT are apparently not alpha helices in the open channel, and there is no evidence that colicin is any different in this regard. Figure 7.6 compares the inferred topology of the DT channel (Figure 7.6C) to that of the colicin C domain (Figures 7.6A and B). Results outlined above suggest that colicins E1, Ia and A require their C-terminal 5 or 6 alpha helices in order to form quasi-normal channels, which contrasts sharply with the DT result (only H8 and H9), but this may reflect a difference in insertion mechanism rather than in the structure of the channel [which perhaps manifests itself in the power dependence of the dose–response curve for DT (>1) vs colicin (=1)].

Despite the considerable progress that has been made in characterizing these two systems, two fundamental questions remain: how does so small a piece of protein form a large, monomeric channel, and how does it translocate so much highly charged protein across a lipid bilayer? Our inclination is to suppose that these two phenomena are related, and that both are due to a strong interaction between the protein and the lipid, in which lipid molecules are recruited to form part of the pathway for both protein and ion transport.

Notes

1 Abbreviations: 6-His, six histidine; CTE2, colicin E2 C-terminal domain; CT-S,M,L,XL, colicin Ia C-terminal fragment – small, medium, large, extra large; DOPG, dioleoyl phosphatidyl glycerol; DT, diphtheria toxin; EPR, electron paramagnetic resonance; FLAG epitope, DYKDDDDK; FRET, fluorescence resonance energy transfer; H, Helix; ImmE2, colicin E2 immunity protein; MTS, methanethiosulfonate; NAD, nicotinamide adenine dinucleotide; pS, picosiemens; TMS, trans membrane segment

2 Although this is not the standard term for the N-terminal domain, we use the designation "N domain" to avoid confusion with certain other domains discussed in this chapter.

3 The H6/7 segment is about 38 amino acids long – almost long enough to cross twice as alpha-helices; but it is constrained to cross only once, by the requirement from biotinylation experiments that Y541 (of colicin Ia) be on the *trans* side and that the loop between H7 and H8 (G577) be on the *cis* side (Jakes *et al.*, 1999). H6 was chosen over H7 to form the transmembrane segment in the figure because it is more amphipathic.

4 The diameters of TEA, the translocated colicin helices and the stoppers were determined by visual inspection of the structures, viewed in projection onto a plane perpendicular to the longest axis of the structure; hence, the numbers given can be interpreted as the minimal diameter of a long, right circular cylindrical pore that would let the molecule pass through. If the flexibility of the molecules were taken into account, of course, the estimated diameter would be somewhat smaller. The apparent diameters of the four colicin helices are maximized by this approach, because the side chains from all along the helix are all projected onto the plane. In three dimensions, one would expect the helices to pack together much more closely than shown here, since the side chains of the different helices do not, generally, coincide.

References

Akabas, M. H., Stauffer, D. A., Xu, M. and Karlin, A. (1992) Acetylcholine receptor channel structure probed in cysteine-substitution mutants. *Science*, **258**, 307–310.

Akabas, M. H., Kaufmann, C., Cook, T. A. and Archdeacon, P. (1994) Amino acid residues lining the chloride channel of the cystic fibrosis transmembrane conductance regulator. *J. Biol. Chem.*, **269**, 14865–14868.

Baty, D., Lakey, J., Pattus, F. and Lazdunski, C. (1990) A 136-amino-acid-residue COOH-terminal fragment of colicin A is endowed with ionophoric activity. *Eur. J. Biochem.*, **189**, 409–413.

Bennett, M. J. and Eisenberg, D. (1994) Refined structure of monomeric diphtheria toxin at 2.3 Å resolution. *Protein Sci.*, **3**, 1464–1475.

Benson, E. L., Huynh, P. D., Finkelstein, A. and Collier, R. J. (1998) Identification of residues lining the anthrax protective antigen channel. *Biochemistry*, **37**, 3941–3948.

Berman, H. M., Westbrook, J., Feng, Z., Gilliland, G., Bhat, T. N. and Weissig, H. *et al.* (2000) The Protein Data Bank. *Nucleic Acids Res.*, **28**, 235–242.

Blaustein, R. O. and Finkelstein, A. (1990) Voltage-dependent block of anthrax toxin channels in planar phospholipid bilayer membranes by symmetric tetraalkylammonium ions: effects on macroscopic conductance. *J. General Physiol.*, **96**, 905–919.

Blaustein, R. O., Germann, W. J., Finkelstein, A. and DasGupta, B. R. (1987) The N-terminal half of the heavy chain of botulinum type A neurotoxin forms channels in planar phospholipid bilayers. *FEBS Lett.*, **226**, 115–120.

Blaustein, R. O., Koehler, T. M., Collier, R. J. and Finkelstein, A. (1989) Anthrax toxin: channel-forming activity of protective antigen in planar phospholipid bilayers. *Proc. Natl. Acad. Sci. USA*, **86**, 2209–2213.

Bontems, F., Gilquin, B., Roumestand, C., Menez, A. and Toma, F. (1992) Analysis of side-chain organization on a refined model of charybdotoxin: structural and functional implications. *Biochemistry*, **31**, 7756–7764.

Bruggemann, E. P. and Kayalar, C. (1986) Determination of the molecularity of the colicin E1 channel by stopped-flow ion flux kinetics. *Proc. Natl. Acad. Sci. USA*, **83**, 4273–4276.

Bullock, J. O. and Kolen, E. R. (1995) Ion selectivity of colicin E1: III. Anion permeability. *J. Membr. Biol.*, **144**, 131–145.

Bullock, J. O., Kolen, E. R. and Shear, J. L. (1992) Ion selectivity of colicin E1: II. permeability to organic cations. *J. Membr. Biol.*, **128**, 1–16.

Cassia-Moura, R., Popescu, A., Lima, J. R., Andrade, C. A., Ventura, L. S. and Lima, K. S. *et al.* (2000) The dynamic activation of colicin Ia channels in planar bilayer lipid membrane. *J. Theor. Biol.*, **206**, 235–241.

Cavard, D., Sauve, P., Heitz, F., Pattus, F., Martinez, C. and Dijkman, R. *et al.* (1988) Hydrodynamic properties of colicin A. Existence of a high-affinity lipid-binding site and oligomerization at acid pH. *Eur. J. Biochem.*, **172**, 507–512.

Choe, S., Bennett, M. J., Fujii, G., Curmi, P. M. G., Kantardjieff, K. A., Collier, R. J. and Eisenberg, D. (1992) The crystal structure of diphtheria toxin. *Nature*, **357**, 216–222.

Cleveland, M. V., Slatin, S., Finkelstein, A. and Levinthal, C. (1983) Structure–function relationships for a voltage-dependent ion channel: properties of COOH-terminal fragments of colicin E1. *Proc. Natl. Acad. Sci. USA*, **80**, 3706–3710.

Cole, S. T., Saint-Joanis, B. and Pugsley, A. P. (1985) Molecular characterisation of the colicin E2 operon and identification of its products. *Mol. Gen. Genet.*, **198**, 465–472.

Collarini, M., Amblard, G., Lazdunski, C. and Pattus, F. (1987) Gating processes of channels induced by colicin A, its C-terminal fragment and colicin E1, in planar lipid bilayers. *Eur. Biophys. J.*, **14**, 147–153.

Cramer, W. A., Cohen, F. S., Merrill, A. R. and Song, H. Y. (1990) Structure and dynamics of the colicin E1 channel. *Mol. Microbiol.*, **4**, 519–526.

Cramer, W. A., Heymann, J. B., Schendel, S. L., Deriy, B. N., Cohen, F. S. and Elkins, P. A. *et al.* (1995) Structure–function of the channel-forming colicins. *Ann. Rev. Biophys. Biomol. Struct.*, **24**, 611–641.

D'Silva, P. R. and Lala, A. K. (2000) Organization of diphtheria toxin in membranes: a hydrophobic photolabeling study. *J. Biol. Chem.*, **275**, 27500.

Dankert, J. R., Uratani, Y., Grabau, C., Cramer, W. A. and Hermodson, M. (1982) On a domain structure of colicin E1. A COOH-terminal peptide fragment active in membrane depolarization. *J. Biol. Chem.*, **257**, 3857–3863.

Duché, D., Parker, M. W., Gonzalez-Mañas, J.-M., Pattus, F. and Baty, D. (1994) Uncoupled steps of the colicin A pore formation demonstrated by disulfide bond engineering. *J. Biol. Chem.*, **269**, 6332–6339.

Duché, D., Corda, Y., Geli, V. and Baty, D. (1999) Integration of the colicin A pore-forming domain into the cytoplasmic membrane of Escherichia coli. *J. Mol. Biol.*, **285**, 1965–1975.

Elkins, P., Bunker, A., Cramer, W. A. and Stauffacher, C. V. (1997) A mechanism for toxin insertion into membranes is suggested by the crystal structure of the channel-forming domain of colicin E1. *Structure*, **5**, 443–458.

Engelman, D. M. and Steitz, T. A. (1981) The spontaneous insertion of proteins into and across membranes: the helical hairpin hypothesis. *Cell*, **23**, 411–422.

Espesset, D., Duché, D., Baty, D. and Geli, V. (1996) The channel domain of colicin A is inhibited by its immunity protein through direct interaction in the Escherichia coli inner membrane. *EMBO J.*, **15**, 2356–2364.

Falnes, P. O. and Sandvig, K. (2000) Penetration of protein toxins into cells. *Curr. Opin. Cell Biol.*, **12**, 407–413.

Geli, V., Baty, D., Pattus, F. and Lazdunski, C. (1989) Topology and function of the integral membrane protein conferring immunity to colicin A. *Mol. Microbiol.*, **3**, 679–687.

Geli, V. and Lazdunski, C. (1992) An alpha-helical hydrophobic hairpin as a specific determinant in protein-protein interaction occurring in Escherichia coli colicin A and B immunity systems. *J. Bacteriol.*, **174**, 6432–6437.

Hille, B. (1992) *Ionic Channels of Excitable Membranes*, 2nd edn Sinauer Associates, Inc., Sunderland, Mass.

Hoch, D. H., Romero-Mira, M., Ehrlich, B. E., Finkelstein, A., DasGupta, B. R. and Simpson, L. L. (1985) Channels formed by botulinum, tetanus, and diphtheria toxins in planar lipid bilayers: relevance to translocation of proteins across membranes. *Proc. Natl. Acad. Sci. USA*, **82**, 1692–1696.

Holm, L. and Sander, C. (1993) Globin fold in a bacterial toxin. *Nature*, **361**, 309.

Huynh, P. D., Cui, C., Zhan, H., Oh, K. J., Collier, R. J. and Finkelstein, A. (1997) Probing the structure of the diphtheria toxin channel. Reactivity in planar lipid bilayer membranes of cysteine-substituted mutant channels with methanethiosulfonate derivatives. *J. Gen. Physiol.*, **110**, 229–242.

Jacob, F., Siminovitch, L. and Wollman, E. (1952) Sur la biosynthèse d'une colicine et son mode d'action. *Ann. Pasteur Institute*, **83**, 295–315.

Jakes, K. S., Kienker, P. K., Slatin, S. L. and Finkelstein, A. (1998) Translocation of inserted foreign epitopes by a channel-forming protein. *Proc. Natl. Acad. Sci. USA*, **95**, 4321–4326.

Jakes, K. S., Kienker, P. K. and Finkelstein, A. (1999) Channel-forming colicins: translocation (and other deviant behaviour) associated with colicin Ia channel gating. *Q. Rev. Biophys.*, **32**, 189–205.

Jakes, K. S., Kienker, P. K., Blaustein, R. O., Finkelstein, A. and Miller, C. (2001) Sizing the protein translocation pathway of the colicin Ia channel. *Biophys. J.*, **80**, 495a.

James, R., Kleanthous, C. and Moore, G. R. (1996) The biology of E colicins: paradigms and paradoxes. *Microbiology*, **142**, 1569–1580.

Johnson, A. E. and van Waes, M. A. (1999) The translocon: a dynamic gateway at the ER membrane. *Annu. Rev. Cell Dev. Biol.*, **15**, 799–842.

Kagan, B. L., Finkelstein, A. and Colombini, M. (1981) Diphtheria toxin fragment forms large pores in phospholipid bilayer membranes. *Proc. Natl. Acad. Sci. USA*, **78**, 4950–4954.

Kienker, P. K., Qiu, X.-Q., Slatin, S. L., Finkelstein, A. and Jakes, K. S. (1997) Transmembrane insertion of the colicin Ia hydrophobic hairpin. *J. Membr. Biol.*, **157**, 27–37.

Kienker, P. K., Jakes, K. S. and Finkelstein, A. (2000) Protein translocation across planar bilayers by the colicin Ia channel- forming domain: where will it end? *J. Gen. Physiol.*, **116**, 587–597.

Krasilnikov, O. V., Da Cruz, J. B., Yuldasheva, L. N., Varanda, W. A. and Nogueira, R. A. (1998) A novel approach to study the geometry of the water lumen of ion channels: colicin Ia channels in planar lipid bilayers. *J. Membr. Biol.*, **161**, 83–92.

Kuhlmann, U. C., Pommer, A. J., Moore, G. R., James, R. and Kleanthous, C. (2000) Specificity in protein–protein interactions: the structural basis for dual recognition in endonuclease colicin-immunity protein complexes. *J. Mol. Biol.*, **301**, 1163–1178.

Lakey, J. H., Baty, D. and Pattus, F. (1991) Fluorescence energy transfer distance measurements using site-directed single cysteine mutants: the membrane insertion of colicin A. *J. Mol. Biol.*, **218**, 639–653.

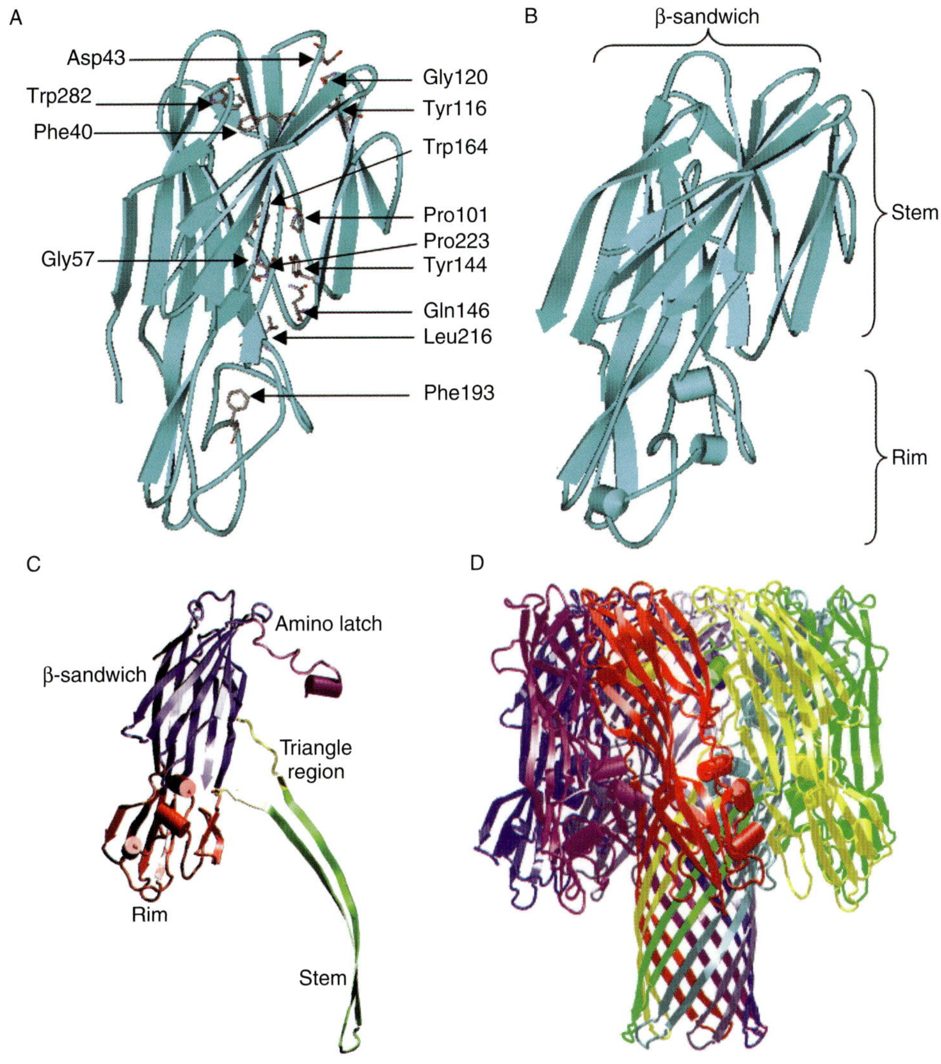

Colour Plate I Comparison of the three-dimensional structures of leucotoxin water-soluble monomers with that of α-toxin protomer and heptameric pore. (See Figure 1.2, p. 7.)

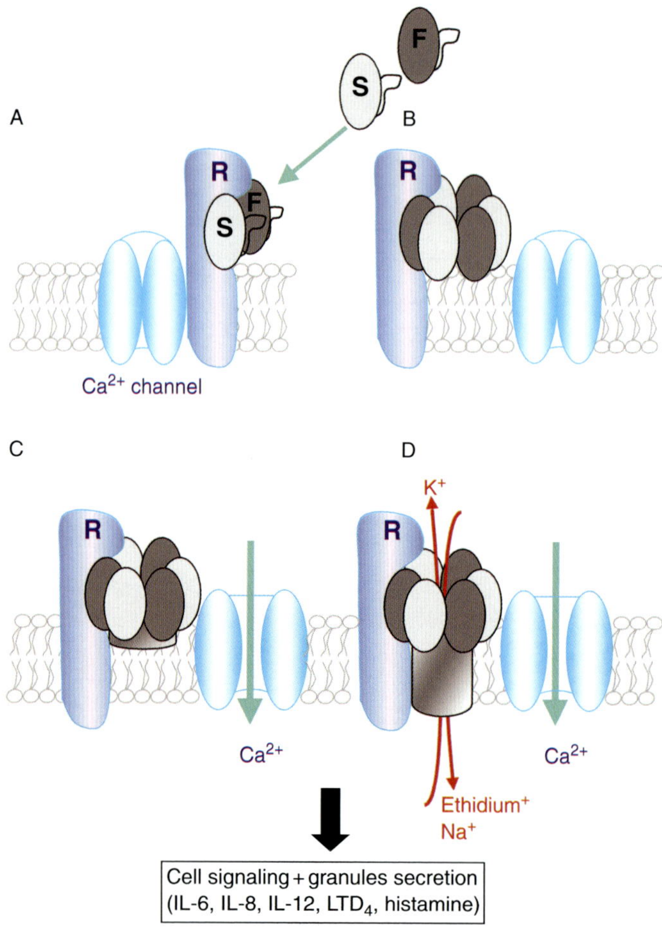

Colour Plate II Scheme of action of the staphylococcal bicomponent leucotoxins. (See Figure 1.4, p. 10.)

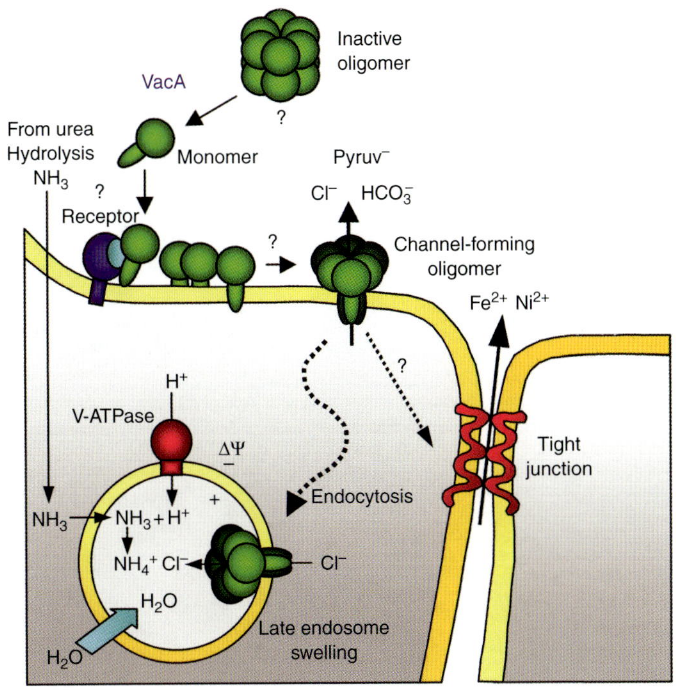

Colour Plate III The current mechanistic model for VacA-induced cell vacuolation. (See Figure 4.3, p. 67.)

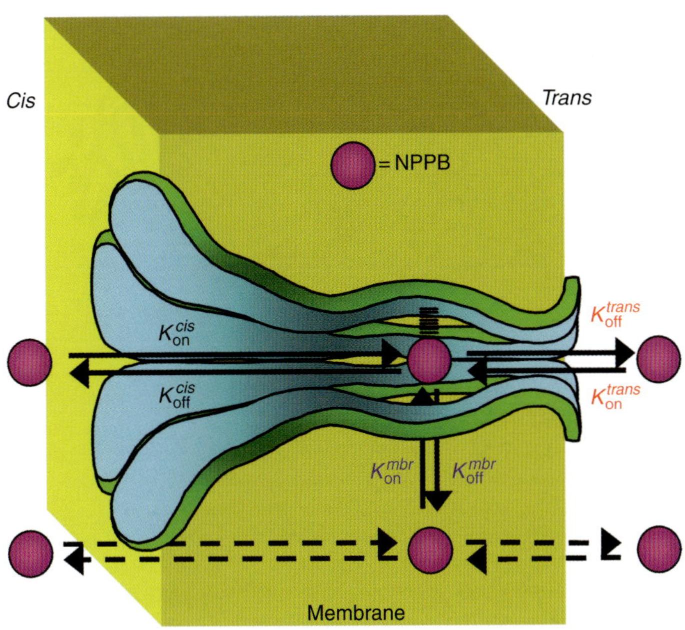

Colour Plate IV A model of VacA inhibition by NPPB. (See Figure 4.6, p. 71.)

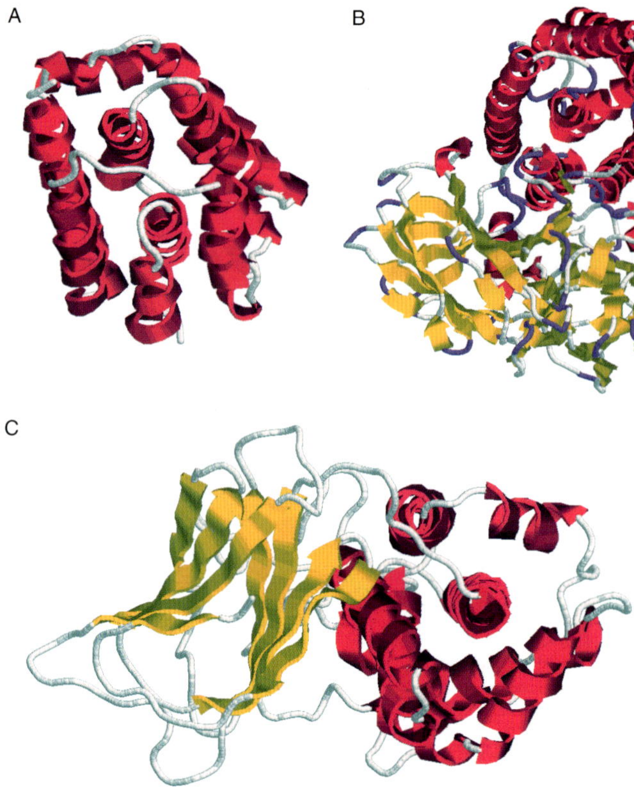

Colour Plate V A cartoon representation of the helical bundle structures of the pore-forming domains from different toxin families. (See Figure 5.1, p. 77.)

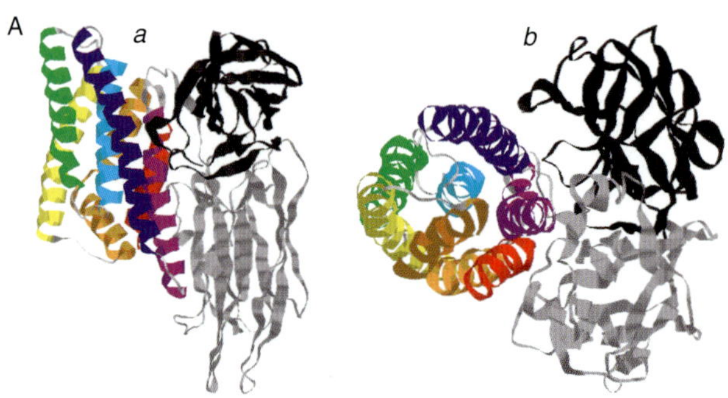

B α1 X–NH–KDVIQKGISVVGDLLGV–COOH
α2 X–NH–GGALVSFYTNFLNT IWPSEDPWKAMENVEALM–COOH
α3 X–NH–DYAKNKALAELQGLQNNVEDYVSALSSWQK–COOH
α4 X–NH–HASQGRIRELFSQAESHFRNSMPSFA–COOH
α5 X–NH–FLTTYAQAANTHLFLLKDAQIYG–COOH
Z
α6 X–NH–KEDIAEFYKRQLKLTQEYTDHCVKWYNVGLDKL–COOH
α7 X–NH–YESWVNFNRYRREMTLTVLDLIALF–COOH

Colour Plate VI Structure of *B. thuringiensis* δ-endotoxin. (See Figure 5.2, p. 79.)

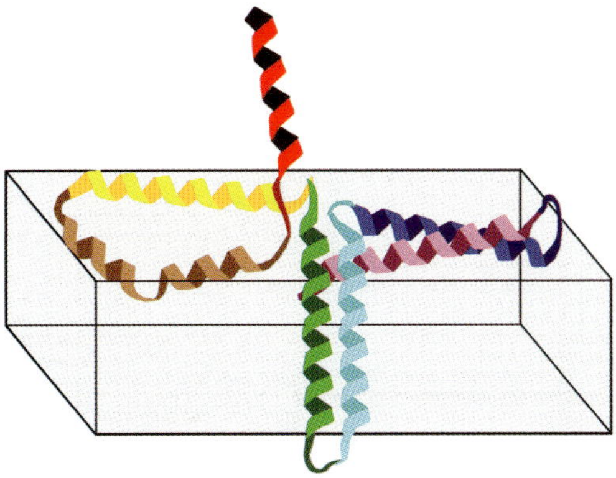

Colour Plate VII The Umbrella model – a schematic presentation of the δ-endotoxin interaction with the phospholipid membranes. (See Figure 5.3, p. 81.)

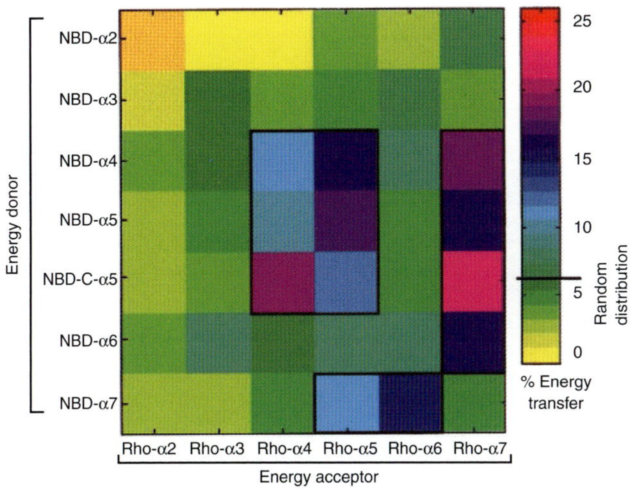

Colour Plate VIII Fluorescence energy transfer between NBD-labeled peptides (donors) and Rho-labeled peptides (acceptors). (See Figure 5.5, p. 83.)

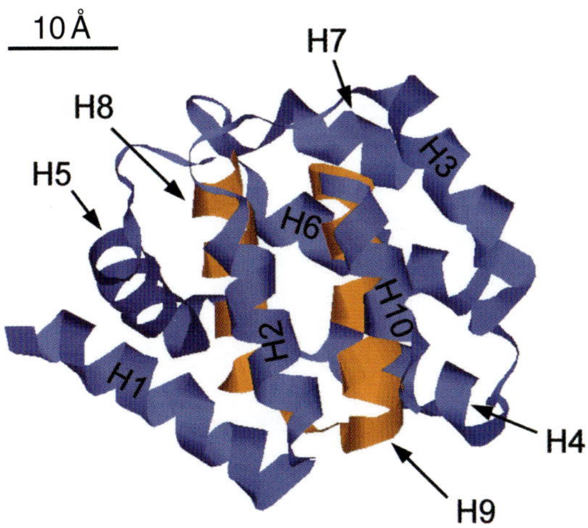

Colour Plate IX The crystal structure of colicin Ia, truncated to show only the C domain portion. (See Figure 7.1, p. 104.)

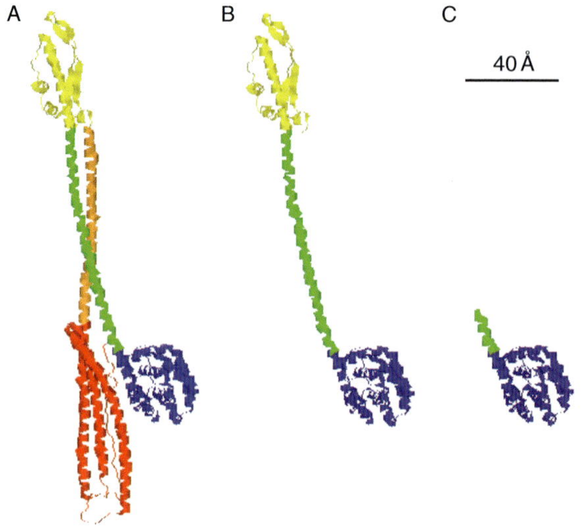

Colour Plate X Whole colicin Ia and two of the C-terminal fragments. (See Figure 7.4, p. 112.)

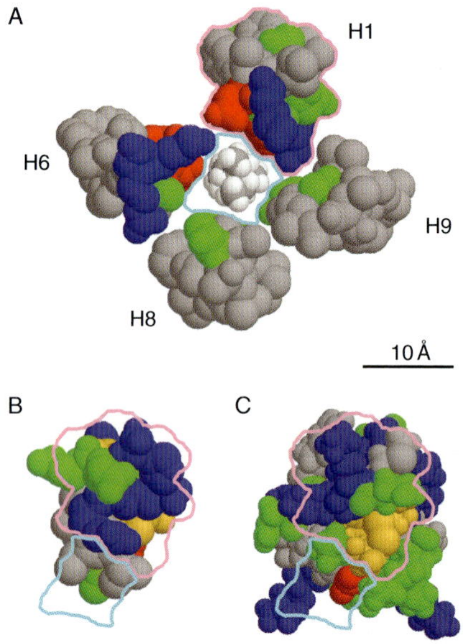

Colour Plate XI Comparison of pathways for ion conduction, "normal" protein translocation and translocation of N-terminal "stoppers" by colicin Ia and its C domain. (See Figure 7.10, p. 120.)

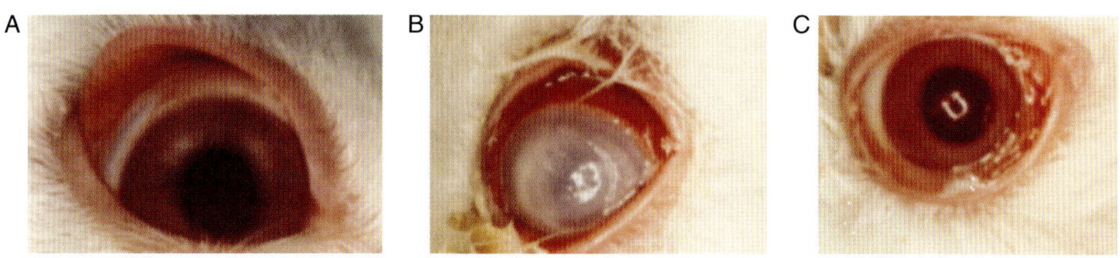

Colour Plate XII Treatment of a *Pseudomonas aeruginosa* experimental keratitis in rabbit with CA(1–7)M(2–9). (See Figure 11.4, p. 234.)

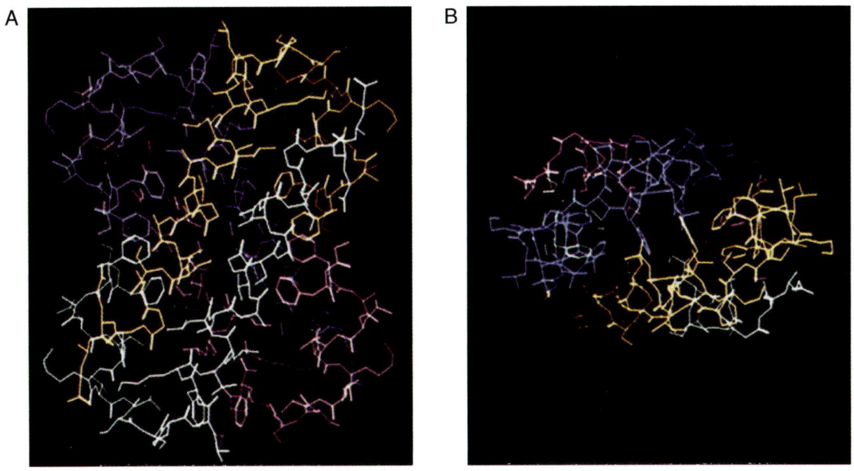

Colour Plate XIII Computerized graphic views of pardaxin tetramers. (See Figure 10.3, p. 184.)

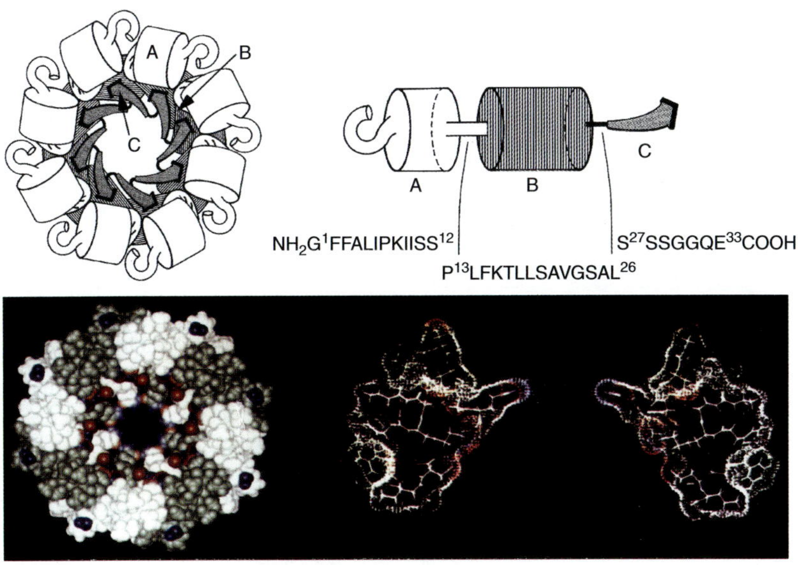

Colour Plate XIV Pardaxin pore model. (See Figure 10.7, p. 190.)

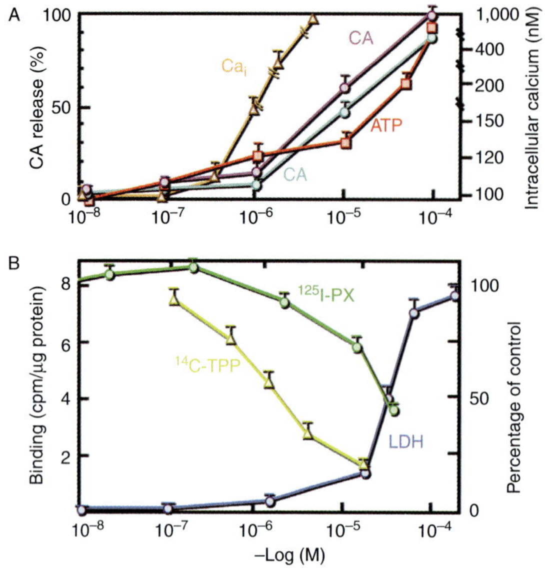

Colour Plate XV Pardaxin-induced catecholamines and ATP release, increase in intracellular calcium binding and neurotoxicity. (See Figure 10.9, p. 194.)

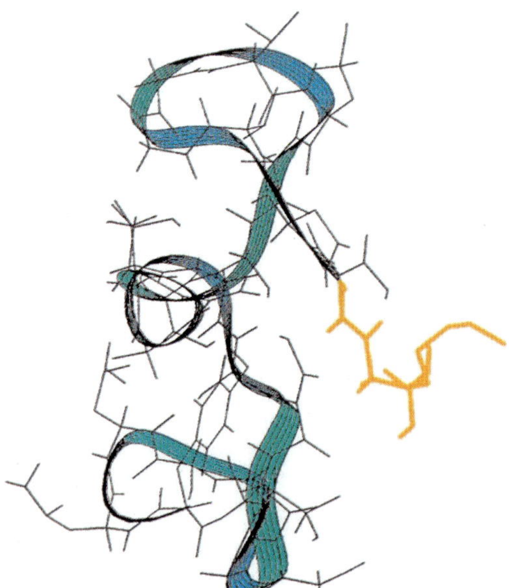

Colour Plate XVI Three-dimensional structure of $SP_{25}A$. The structure is obtained by distance geometry calculations from NMR data of syringopeptin in D_2O at pH 3.6. In yellow it is evidenced the fatty acid chain. (See Figure 13.3, p. 275.)

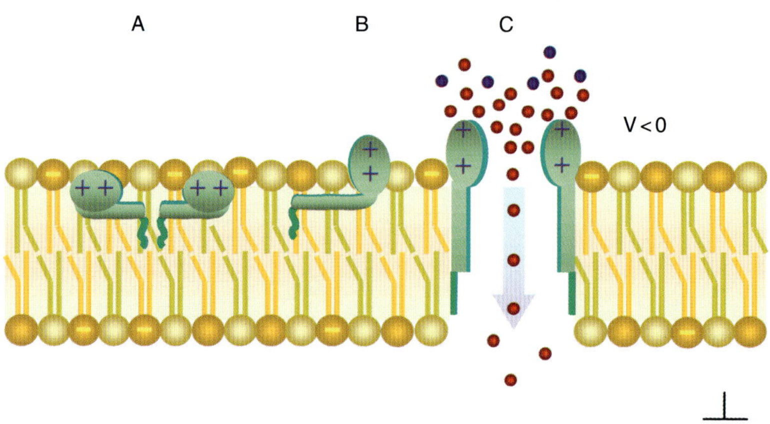

Colour Plate XVII Schematic model of the mechanism of pore formation by $SP_{25}A$. (See Figure 13.9, p. 286.)

Lakey, J. H., Duché, D., Gonzalez-Mañas, J.-M., Baty, D. and Pattus, F. (1993) Fluorescence energy transfer distance measurements: the hydrophobic helical hairpin of colicin A in the membrane bound state. *J. Mol. Biol.*, **230**, 1055–1067.

Lazdunski, C. J., Baty, D., Geli, V., Cavard, D., Morlon, J. and Lloubes, R. *et al.* (1988) The membrane channel-forming colicin A: synthesis, secretion, structure, action and immunity. *Biochim. Biophys. Acta*, **947**, 445–464.

Lear, J. D., Wasserman, Z. R. and DeGrado, W. F. (1988) Synthetic amphiphilic peptide models for protein ion channels. *Science*, **240**, 1177–1181.

Levinthal, F., Todd, A. P., Hubbell, W. L. and Levinthal, C. (1991) A single tryptic fragment of colicin E1 can form an ion channel: stoichiometry confirms kinetics. *Proteins: Structure, Function and Genetics*, **11**, 254–262.

Lin, J. H. and Baumgaertner, A. (2000) Stability of a melittin pore in a lipid bilayer: a molecular dynamics study. *Biophys. J.*, **78**, 1714–1724.

Lindeberg, M., Zakharov, S. D. and Cramer, W. A. (2000) Unfolding pathway of the colicin E1 channel protein on a membrane surface. *J. Mol. Biol.*, **295**, 679–692.

Liu, Q. R., Crozel, V., Levinthal, F., Slatin, S., Finkelstein, A. and Levinthal, C. (1986) A very short peptide makes a voltage-dependent ion channel: the critical length of the channel domain of colicin E1. *Proteins: Structure, Function and Genetics*, **1**, 218–229.

Madshus, I. H. and Stenmark, H. (1992) Entry of ADP-ribosylating toxins into cells. *Curr. Top. Microbiol. Immunol.*, **175**, 1–26.

Malenbaum, S. E., Collier, R. J. and London, E. (1998) Membrane topography of the T domain of diphtheria toxin probed with single tryptophan mutants. *Biochemistry*, **37**, 17915–17922.

Mankovich, J. A., Hsu, C. H. and Konisky, J. (1986) DNA and amino acid sequence analysis of structural and immunity genes of colicins Ia and Ib. *J. Bacteriol.*, **168**, 228–236.

Martinez, M. C., Lazdunski, C. and Pattus, F. (1983) Isolation, molecular and functional properties of the C-terminal domain of colicin A. *EMBO J.*, **2**, 1501–1507.

Matsuzaki, K. (1998) Magainins as paradigm for the mode of action of pore forming polypeptides. *Biochim. Biophys. Acta*, **1376**, 391–400.

Merrill, A. R. and Cramer, W. A. (1990) Identification of a voltage-responsive segment of the potential-gated colicin E1 ion channel. *Biochemistry*, **29**, 8529–8534.

Mindell, J., Silverman, J. A., Collier, R. J. and Finkelstein, A. (1994) Structure function relationships in diphtheria toxin channels: II. A residue responsible for the channel's dependence on *trans* pH. *J. Membr. Biol.*, **137**, 29–44.

Morlon, J., Lloubes, R., Varenne, S., Chartier, M. and Lazdunski, C. (1983) Complete nucleotide sequence of the structural gene for colicin A, a gene translated at a non-uniform rate. *J. Mol. Biol.*, **170**, 271–285.

Muchmore, S. W., Sattler, M., Liang, H., Meadows, R. P., Harlan, J. E., Yoon, H. S. *et al.* (1996) X-ray and NMR structure of human Bcl-xL, an inhibitor of programmed cell death. *Nature*, **381**, 335–341.

Nardi, A., Slatin, S. L., Baty, D. and Duché, D. (2001) The C-terminal half of the colicin A pore-forming domain is active in vivo and in vitro. *J. Mol. Biol.*, **307**, 1293–1303.

Nogueira, R. A. and Varanda, W. A. (1988) Gating properties of channels formed by Colicin Ia in planar lipid bilayer membranes. *J. Membr. Biol.*, **105**, 143–153.

Oh, K. J., Senzel, L., Collier, R. J. and Finkelstein, A. (1999a) Translocation of the catalytic domain of diphtheria toxin across planar phospholipid bilayers by its own T domain. *Proc. Natl. Acad. Sci. USA*, **96**, 8467–8470.

Oh, K. J., Zhan, H., Cui, C., Altenbach, C., Hubbell, W. L. and Collier, R. J. (1999b) Conformation of the diphtheria toxin T domain in membranes: a site-directed spin-labeling study of the TH8 helix and TL5 loop. *Biochemistry*, **38**, 10336–10343.

Ohno, S., Ohno-Iwashita, Y., Suzuki, K. and Imahori, K. (1977) Purification and characterization of active component and active fragment of colicin E3. *J. Biochem.*, **82**, 1045–1053.

Parker, M. W., Pattus, F., Tucker, A. D. and Tsernoglou, D. (1989) Structure of the membrane-pore-forming fragment of colicin A. *Nature*, **337**, 93–96.

Parker, M. W., Tucker, A. D., Tsernoglou, D. and Pattus, F. (1990) Insights into membrane insertion based on studies of colicins. *Trends Biochem. Sci.*, **15**, 126–129.

Parker, M. W. and Pattus, F. (1993) Rendering a membrane protein soluble in water: a common packing motif in bacterial protein toxins. *Trends Biochem. Sci.*, **18**, 391–395.

Pattus, F., Cavard, D., Verger, R., Lazdunski, C., Rosenbusch, J. P. and Schirmer, T. (1982) Formation of voltage dependent pores by colicin A. *Toxicon*, **20**, 205–206.

Pease, J. H. and Wemmer, D. E. (1988) Solution structure of apamin determined by nuclear magnetic resonance and distance geometry. *Biochemistry*, **27**, 8491–8498.

Peterson, A. A. and Cramer, W. A. (1987) Voltage-dependent, monomeric channel activity of colicin E1 in artificial membrane vesicles. *J. Membr. Biol.*, **99**, 197–204.

Pfanner, N. and Geissler, A. (2001) Versatility of the mitochondrial protein import machinery. *Nat. Rev. Mol. Cell Biol.*, **2**, 339–349.

Pugsley, A. P. (1984) The ins and outs of colicins. Part II. Lethal action, immunity and ecological implications. *Microbiol. Sci.*, **1**, 203–205.

Qiu, X.-Q., Jakes, K. S., Kienker, P. K., Finkelstein, A. and Slatin, S. L. (1996) Major transmembrane movement associated with colicin Ia channel gating. *J. Gen. Physiol.*, **107**, 313–328.

Raymond, L., Slatin, S. L. and Finkelstein, A. (1985) Channels formed by colicin E1 in planar lipid bilayers are large and exhibit pH-dependent ion selectivity. *J. Membr. Biol.*, **84**, 173–181.

Raymond, L., Slatin, S. L., Finkelstein, A., Liu, Q. R. and Levinthal, C. (1986) Gating of a voltage-dependent channel (colicin E1) in planar lipid bilayers: translocation of regions outside the channel-forming region. *J. Membr. Biol.*, **92**, 255–268.

Riley, M. A. and Gordon, D. M. (1999) The ecological role of bacteriocins in bacterial competition. *Trends Microbiol.*, **7**, 129–133.

Schein, S. J., Kagan, B. L. and Finkelstein, A. (1978) Colicin K acts by forming voltage-dependent channels in phospholipid bilayer membranes. *Nature*, **276**, 159–163.

Schwartz, S. A. and Helinski, D. R. (1971) Purification and characterization of colicin E1. *J. Biol. Chem.*, **246**, 6318–6327.

Senzel, L., Huynh, P. D., Jakes, K. S., Collier, R. J. and Finkelstein, A. (1998) The diphtheria toxin channel-forming T domain translocates its own NH_2-terminal region across planar bilayers. *J. Gen. Physiol*, **112**, 317–324.

Senzel, L., Gordon, M., Blaustein, R. O., Oh, K. J., Collier, R. J. and Finkelstein, A. (2000) Topography of diphtheria toxin's T domain in the open channel state. *J. Gen. Physiol.*, **115**, 421–434.

Shannon, J. L. and Fernandez, R. C. (1999) The C-terminal domain of the Bordetella pertussis autotransporter BrkA forms a pore in lipid bilayer membranes. *J. Bacteriol.*, **181**, 5838–5842.

Silverman, J. A., Mindell, J. A., Zhan, H., Finkelstein, A. and Collier, R. J. (1994) Structure-function relationships in diphtheria toxin channels: determining a minimal channel-forming domain. *J. Membr. Biol.*, **137**, 17–28.

Slatin, S. L., Raymond, L. and Finkelstein, A. (1986a) Gating of a voltage-dependent channel (colicin E1) in planar lipid bilayers: the role of protein translocation. *J. Membr. Biol.*, **92**, 247–254.

Slatin, S., Raymond, L. and Finkelstein, A. (1986b) Gating of a voltage-dependent channel (colicin E1) in planar lipid bilayers: the role of protein translocation. *J. Membr. Biol.*, **92**, 247–254.

Slatin, S. L. (1988) Colicin E1 in planar lipid bilayers. *Int. J. Biochem.*, **20**, 737–744.

Slatin, S. L., Qiu, X.-Q., Jakes, K. S. and Finkelstein, A. (1994) Identification of a translocated protein segment in a voltage-dependent channel. *Nature*, **371**, 158–161.

Song, H. Y. and Cramer, W. A. (1991) Membrane Topography of ColE1 Gene Products: the immunity protein. *J. Bacteriol.*, **173**, 2935–2943.

Song, H. Y., Cohen, F. S. and Cramer, W. A. (1991) Membrane topography of ColE1 gene products: the hydrophobic anchor of the colicin E1 channel is a helical hairpin. *J. Bacteriol.*, **173**, 2927–2934.

Steer, B. A. and Merrill, A. R. (1997) Characterization of an unfolding intermediate and kinetic analysis of guanidine hydrochloride-induced denaturation of the colicin E1 channel peptide. *Biochemistry*, **36**, 3037–3046.

Steer, B. A., DiNardo, A. A. and Merrill, A. R. (1999) Colicin E1 forms a dimer after urea-induced unfolding. *J. Biochem.*, **340 (Pt 3)**, 631–638.

Tory, M. C. and Merrill, A. R. (1999) Adventures in membrane protein topology. A study of the membrane-bound state of colicin E1. *J. Biol. Chem.*, **274**, 24539–24549.

Vetter, I. R., Parker, M. W., Tucker, A. D., Lakey, J. H., Pattus, F. and Tsernoglou, D. (1998) Crystal structure of a colicin N fragment suggests a model for toxicity. *Structure*, **6**, 863–874.

Wendt, L. (1970) Mechanism of colicin action: early events. *J. Bacteriol.*, **104**, 1236–1241.

Wiener, M., Freymann, D., Ghosh, P. and Stroud, R. M. (1997) Crystal structure of colicin Ia. *Nature*, **385**, 461–464.

Xu, M. and Akabas, M. H. (1996) Identification of channel-lining residues in the M2 membrane-spanning segment of the GABA(A) receptor alpha1 subunit. *J. Gen. Physiol.*, **107**, 195–205.

Yamada, M., Ebina, Y., Miyata, T., Nakazawa, T. and Nakazawa, A. (1982) Nucleotide sequence of the structural gene for colicin E1 and predicted structure of the protein. *Proc. Natl. Acad. Sci. USA*, **79**, 2827–2831.

Yamamoto, K., Nishida, K.-J., Beppu, T. and Arima, K. (1978) Tryptic digest of colicin E2 and its active fragment. *J. Biochem.*, **83**, 827–834.

Zakharov, S. D., Lindeberg, M., Griko, Y., Salamon, Z., Tollin, G., Prendergast, F. G. and Cramer, W. A. (1998) Membrane-bound state of the colicin E1 channel domain as an extended two-dimensional helical array. *Proc. Natl. Acad. Sci. USA*, **95**, 4282–4287.

Zakharov, S. D., Lindeberg, M. and Cramer, W. A. (1999) Kinetic description of structural changes linked to membrane import of the colicin E1 channel protein. *Biochemistry*, **38**, 11325–11332.

8 Actinoporins, pore-forming toxins of sea anemones (Actiniaria)

Gregor Anderluh and Peter Maček

Introduction

Sea anemones (Actiniaria) produce four different classes of cytolytic polypeptides (Anderluh and Maček, 2002). One class, comprising basic ~20 kDa proteins, detected as lethal hemolysins or cytolysins and inhibited by sphingomyelin, was named actinoporins (Kem, 1988). These toxins, forming rectified cation-selective pores in lipid membranes, have been found as isotoxins in more than 22 genera of the fam. Actiniidae and Stichodactylidae (reviewed by Turk, 1991; Maček, 1992). The majority of their biochemical and functional characteristics have been established from the studies of equinatoxins EqtI, EqtII and EqtIII from *Actinia equina* (Ferlan and Lebez, 1974; Maček and Lebez, 1988), tenebrosins TN-A, TN-B and TN-C from A. *tenebrosa* (Norton *et al.*, 1990), sticholysins StI and StII of *Stichodactyla* (formerly *Stoichactis helinthus* (Bernheimer and Avigad, 1976; Kem and Dunn, 1988; Lanio *et al.*, 2001)) (see Table 8.1), magnificalysins HMgI, HmgII of *Heteractis magnifica* (Khoo *et al.*, 1993), and the toxins RTX-A, RTX-S, and RTX-G from *Radianthus macrodactylus* (Monastyrnaya *et al.*, 1999 and refs. therein). Some typical features of the pore-forming activity of actinoporins are presented in Table 8.1 and Figure 8.1. Regardless of the lipid composition of planar lipid membrane, conductances of ~100–400–pS for single actinoporin pores, negative reversal potentials, and molecularity of 3–4 have been shown to be characteristic (Table 8.1). Figure 8.1A illustrates the formation of single pores by the typical actinoporin, EqtII, and its mutated variant EqtIIK77C (lysine 77 replaced by cysteine) in a planar lipid membrane, together with binding to SUV and the consequent release of the vesicle-entrapped marker calcein.

Here, we review briefly the typical biochemical, biophysical and pharmacological features of actinoporins. We have focused on recent advances in studies of their binding to lipids and pore formation in the light of the recently determined crystal structure of equinatoxin II, the first three-dimensional (3-D) structure to be solved of a pore-forming toxin from a eukaryotic organism.

Biological activity of actinoporins

Biological role

Sea anemones, like other Cnidarians, use specially designed stinging organelles (nematocysts or cnidocysts) to deliver their venom. However, it is not known if actinoporins reside in mature nematocysts. Nor is completely elucidated the mechanism by which actinoporins are massively released into the surrounding water upon mechanical stimulation of sea anemones (Senčič and Maček, 1990; Meinardi *et al.*, 1994). Actinoporins have been

Table 8.1 Representatives of Actiniidae and Stichodactylidae actinoporins and their pore-forming characteristics

Family species	*Toxin*	M_r	*pI*	*Planar lipid membrane*	*Pore conductance*[a] *(pS)*	U_{rev} *(mV)*[b]	N_{agg}[c]
Actinidae							
Actinia equina	EqTII[e,f]	19,815	10.5	POPE[k]	23 (50 $CaCl_2$, +40 mV)	−16 (136/50 $CaCl_2$)	n.d.
				POPC/PE 4/1[l]	225 (100 KCl, +40 mV)	−38 (50/500 NaCl)	3–4
Stichodactylidae							
Stichodactyla helianthus[d]	StI[g]	19,391	>9	POPC/PE/SM 1/1/0.1[m]	413 (100 KCl, −40 mV)	−27 (20/200 NaCl)	3
	StII[h,i]	19,283	9.5	PC/Chol 1/2[n]	~100 (500 KCl, −100 mV)	−32 (10/100 NaCl)	3
	(C-III)[j]			PC/Chol/SM ~1/0.5/0.1[o]	~200 (100 KCl, +50 mV)	57 (10/100 K_2SO_4)[r]	4
Radianthus macrodactylus	RTX-A[p]	~20,000	9.8	1-Eru-Gly[q]	75 (100 KCl, +40 mV)	−26 (100/1000 KCl)	4

Notes

a In brackets: mM concentration of electrolyte, holding potential.
b U_{rev}, reversal potential, in brackets: *cis/trans* mM concentrations of electrolyte.
c N_{agg}, number of aggregating monomers in pore.
d Syn.: *Stoichactis helianthus*.
e Maček and Lebez (1988).
f variant EqTII Pro81 is identical to variant Ser177 TN-C (Simpson *et al.*, 1990).
g Lanio *et al.* (2001).
h De los Rios *et al.* (1998).
i StII is identical to cytolysin III, C-III (Blumenthal and Kem, 1983; Huerta *et al.*, 2001).
j Kem and Dunn (1988).
k Zorec *et al.* (1990).
l Belmonte *et al.* (1993).
m Tejuca *et al.* (1996).
n Michaels (1979).
o Varanda and Finkelstein (1980).
p Monastyrnaya *et al.* (1999).
q Rudnev *et al.* (1984).
r K^+-diffusion potential.
n.d. not determined.

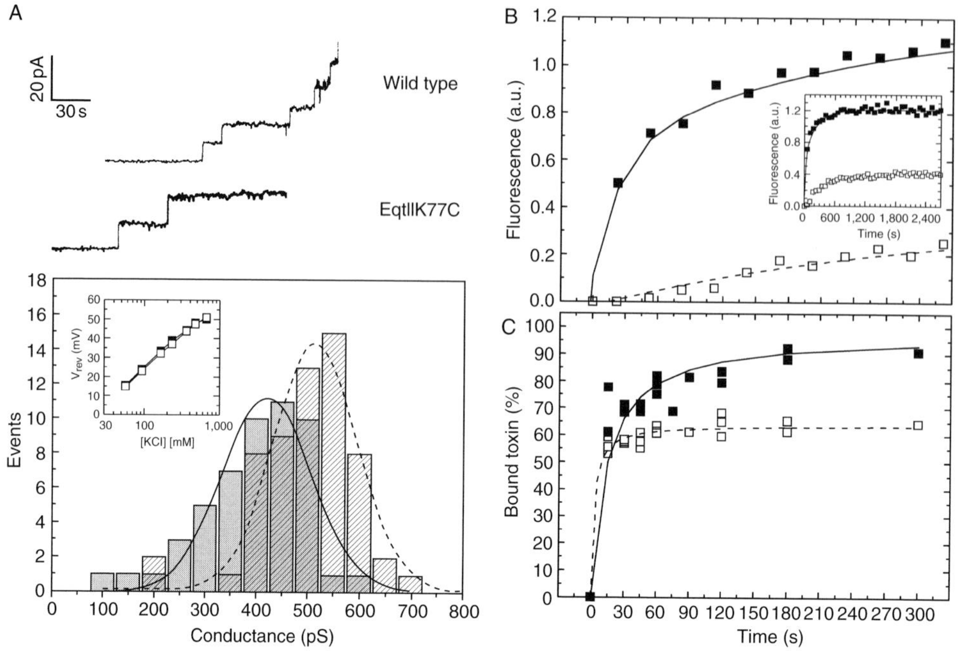

Figure 8.1 Pore-forming activity of wild-type equinatoxin II and mutant EqtIIK77C. (A) Upper traces indicate increased ionic flow due to opening of single pores by wild-type equinatoxin II and mutant EqtIIK77C in planar lipid membrane made of diphytanoylphosphatidylcholine and sphingomyelin (molar ratio 20 : 1). A Gaussian distribution of the conductances of the wild-type pores (solid line) and that of the mutant toxin (dashed line) are shown below. Inset: reversal potential determined under different concentration gradients of KCl for the wild-type (closed squares) and mutated toxin (open squares). (B) Release of fluorescent marker calcein from SUV (SM/POPC, 3 : 1 molar ratio) elicited by 10 μg/ml wild-type EqtII (closed squares) and EqtIIK77C (open squares). The inset shows the marker release on the longer time scale. (C) Binding of the wild-type and mutated toxin was determined in the same conditions as in (B). At a varied time, erythrocytes were added to determine hemolysis by free toxin. A dose-response calibration curve was used to estimate bound toxin. (Reproduced from Anderluh *et al.*, 2000b with permission.)

reported as highly toxic to fish and crustaceans, which may be the natural prey of sea anemones (Mebs, 1994; Giese *et al.*, 1996). In addition to their role in preying, it has been suggested that actinoporins could act, when released in water, as efficient repellents against potential predators (Maček, 1992).

Pharmacology

Actinoporins evoke a wide diversity of pathological effects depending on dose and pharmacological system. The high lethality of actinoporins in mammals (*i.v.* LD_{50} in mice ranges from 20 to 300 μg/kg (Turk, 1991; Maček, 1992; Monastyrnaya *et al.*, 1999)) is ascribed to respiratory arrest, coronary vasospasm and cardiotoxicity (Sket *et al.*, 1974; Ho *et al.*, 1987; Bunc *et al.*, 1999; Drevenšek *et al.*, 2000b). At low doses, tenebrosins were reported as cardiostimulating (Thomson *et al.*, 1987; Galettis and Norton, 1990). A direct vasoconstrictory effect of equinatoxin II on coronary artery and the involvement of L-type calcium channels have been demonstrated (Drevenšek *et al.*, 2000a).

Recent reports imply that, apart from exerting pore-forming activity, actinoporins selectively modify certain cell functions; they inhibit synaptosomal uptake of GABA and choline (Khoo *et al.*, 1995), stimulate exocytosis of glutamate and catecholamines (Migues *et al.*, 1999; Ales *et al.*, 2000), and induce degranulation of thrombocytes and granulocytes (Bunc *et al.*, 1994; Šuput *et al.*, 2001).

Yet, most studies have pointed to a leit-motif of actinoporin activity, that is, a rather non-specific permeabilization of cell membranes. This primary event may induce a variety of deleterious effects in cells typical for other pore-forming toxins. In fact, cell permeabilization by actinoporins may subsequently be manifested as a gross change of whole cell and organelle morphology, cell fragmentation (Batista *et al.*, 1990; Batista and Jezernik, 1992), and cell swelling (Zorec *et al.*, 1990; Meunier *et al.*, 2000). As little as $\sim 10^{-11}$ M equinatoxin II stimulates spontaneous beating of isolated ventricular myocardiocytes and induces retraction of their elongated shape (Maček *et al.*, 1994) while a somewhat higher toxin concentration stimulates platelet aggregation (Teng *et al.*, 1988). Most of the effects are the consequence of increased permeability of the toxin treated cell membranes for low molecular weight solutes including inorganic electrolytes, due to opening of *de novo* pores in the lipid bilayer. In erythrocytes treated with actinoporins, a fast K^+-efflux and delayed hemoglobin release was demonstrated (Ivanov *et al.*, 1987; Maček *et al.*, 1994), characteristic of a colloid-osmotic type of cell lysis. However, even at low concentrations, actinoporins may appear cytotoxic by mediating a rapid entry of extracellular Ca^{2+} into the cytosol, which may trigger a variety of uncontrolled intracellular processes (Zorec *et al.*, 1990).

Structure of actinoporins and pore formation

Structure

Primary structure

Equinatoxins and magnificalysins are translated as preproproteins with a signal sequence followed by a propeptide with a KR recognition site for a subtilisin-like endopeptidase (Anderluh *et al.*, 1996, 2000b; Wang *et al.*, 2000). Mature actinoporins, comprising 175–179 residues, form a well-conserved protein family (Figure 8.2) unrelated to other known proteins. It is characteristic that they all lack cysteine. Isotoxins isolated from one sea anemone species are multigene products and share more than 80% of identical residues and up to 95% similar ones. At least five different isotoxins have been identified at the protein or gene level (Anderluh *et al.*, 1999b) in the case of equinatoxins. Even the most sequentially distant toxins are still more than 60% identical in primary structure. Thus, EqtII shares 64% identical residues with StI and Magn and is more than 80% similar to both (see Figure 8.2).

Actinoporins comprise a highly conserved stretch at the N-terminal end, which is predicted to be an amphiphilic α-helix (underlined in Figure 8.2) similar to melittin and SV5 fusion peptide (Belmonte *et al.*, 1994). The existence of this particular helix has been recently confirmed by an X-ray diffraction study (Athanasiadis *et al.*, 2001) (see also Figure 8.2). Two other remarkable features of the primary structures are a well-conserved tryptophan-rich region in the middle of the sequence and an RGD motif found in all known actinoporin sequences (doubly underlined in Figure 8.2). The tryptophan-rich

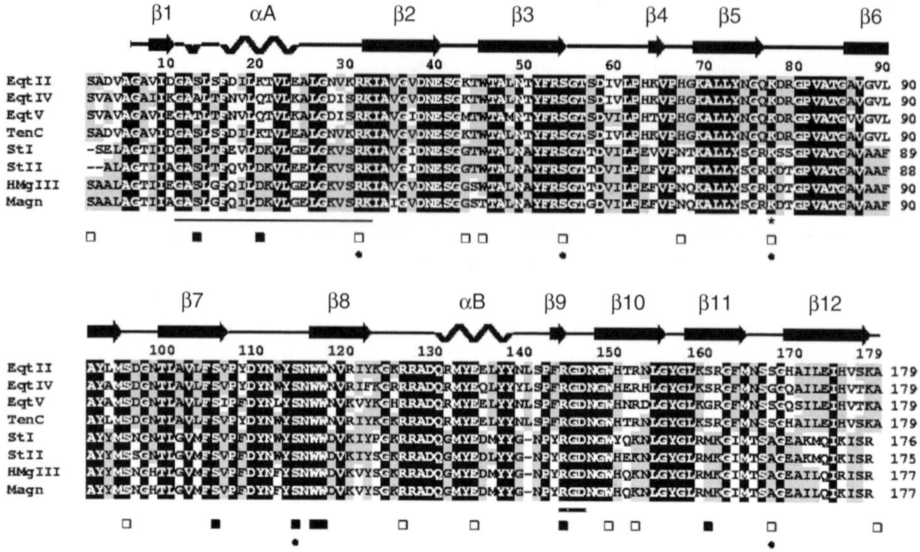

Figure 8.2 Alignment of actinoporin amino acid sequences. The secondary structure elements identified in the EqtII crystal structure are indicated above the alignment. The sequence numbering is according to EqtII. Identical residues in all sequences are shown black, more than 50% of identities are shown as black on the shaded background. Gaps are presented dashed. The N-terminal region similar to melittin is underlined (Belmonte *et al.*, 1994). The conserved RGD motif is doubly underlined. Residues shown to be transferred to a more apolar environment after monomer binding to lipid bilayer are marked by black squares, those exposed to the solvent are shown with open squares. Black circles present residues whose mutation to cysteine results in a considerable loss of hemolytic activity. K77 implicated in oligomerization is denoted by an asterisk. Shown are sequences of EqtII (GeneBank Acc.No. U41661), EqtIV (AF057028), EqtV (U51900), TenC (SwissProt Acc.No. P17723), StI (P81662), StII (GeneBank Acc.No. AJ009931), magnificalysin III (HMgIII) (AF170706), and magnificalysin (Magn) (Samejima *et al.*, 2000).

motif (112 WYSNWW 117 in EqtII) is the second structural element shared with other pore-forming toxins. Specifically, a similar tryptophan motif was found in a highly conserved undecapeptide, that is, 458 ECTGLAWEWWRDV 470 of perfringolysin, of the family of cholesterol-dependent cytolysins from Gram-positive bacteria. This region interacts with the lipid membrane interface (Heuck *et al.*, 2000). Similarly, the tryptophan-rich stretch and residues nearby promote the binding of actinoporins to the membrane as discussed in the section on "structure of actinoporins and pore formation."

Secondary structure

In actinoporins, β-sheet structure is dominant as determined with CD and FT–IR spectroscopy. StI and StII contain 44–50% β-sheet, 18–20% β-turn, 12–15% α-helix and 19–22% random coil as found by FT–IR spectroscopy of their water-soluble forms (Menestrina *et al.*, 1999). These results are in good agreement with the 3-D structural data for EqtII, 47.5% β-sheet, 9.5% α-helix and 47.5% random coil and β-turns (Zhang *et al.*, 2000; Athanasiadis *et al.*, 2001). The toxins were also examined for changes in secondary structure occurring on binding to the lipid membrane. According to analysis of FT–IR spectra, there is slightly increased β-sheet (56–58%) and α-helix (18–20%) at the expense of random coil (9–10%) and β-turn (15%). Similar results were obtained for EqtII using CD spectroscopy (Poklar *et al.*, 1997) and FT–IR spectroscopy (Anderluh *et al.*, 2000a).

Studies of protein conformational stability and pathway of unfolding may provide clues to its mechanism of action. It is reasonable to assume that the transition of a water-soluble to a lipid-inserted form of the actinoporin molecule will involve specific rearrangements of the 3-D structure. The actinoporin secondary structure is relatively stable over a wide range of pH and temperature despite the lack of disulphide bonds. Studies of unfolding revealed that decreasing pH below 1.1, increasing it above 10 or temperatures above 70°C lead to formation of a stable molten-globule state of EqTII and StII (Malavašič *et al.*, 1996; Poklar *et al.*, 1997; Mancheno *et al.*, 2001; Poklar *et al.*, 2001). Its characteristic is a slightly higher content of α-helix as estimated from CD spectra. A similar conformational state with preserved secondary, but disordered tertiary structure was observed when EqtII was bound to DOPG vesicles and heated above 60°C (Poklar *et al.*, 1999).

In addition to the acid-induced molten-globule, a second transition of actinoporins tertiary structure was detected in the pH range 7–10 (Mancheno *et al.*, 2001; Poklar *et al.*, 2001). This broad transition leads to a partially unfolded state and is tightly correlated with enhanced actinoporin activity in natural membranes over the same pH range (see for example, Maček and Lebez, 1981; Doyle and Kem, 1989; Doyle *et al.*, 1989). It was demonstrated for cytolysin C-III, that increasing pH above 7 and up to ~10, markedly promotes the rate of hemolysis, that is, pore-formation, though the extent of binding of the radiolabeled toxin to the cell membranes remains constant (Doyle and Kem, 1989). In artificial lipid membranes, however, an additional permeabilization optimum around pH 4 was observed with both EqtII and StI (Belmonte *et al.*, 1993; Tejuca *et al.*, 1996). The third transition triggered above pH 11 results in complete loss of the tertiary structure, increased non-native helical structure, and protein precipitation (Poklar *et al.*, 2001). The pH data suggest that either acid- or base-induced partial unfolding of actinoporins enhance pore-formation. It is suggestive that the acid-induced transition could occur in the lysosomes in the case of eventual endocytosis of actinoporin molecules adsorbed to cell membrane. This, however, remains to be clarified.

It was also found that a medium with an increased ionic strength enhances StI hemolytic activity (Alvarez *et al.*, 1998). This effect correlates with facilitation of partial unfolding of actinoporins in the presence of increased concentrations of salts (see also Mancheno *et al.*, 2001; Poklar *et al.*, 2001) and could explain the nonspecific stimulatory effect of >10 mM Ca^{2+}, Mg^{2+}, Ba^{2+} and some other divalent cations on EqtII induced hemolysis (Maček and Lebez, 1981), and the changed intrinsic tryptophan fluorescence of EqtII (Maček, unpublished). Moreover, a subtle conformational change of magnificalysins and EqtII, which results in an increased hemolytic activity, could also be induced by 8-anilino-1-naphtalene-sulphonate (ANS), a well-known molten-globule fluorescent probe (Khoo *et al.*, 1997).

In summary, there is ample evidence that transition of the water-soluble actinoporin molecule into a pore form is not accompanied by significant rearrangement of secondary structure, but that certain changes of tertiary structure are involved.

Tertiary structure

The crystal structure of the water-soluble form of EqtII was recently determined at 1.9 Å resolution (Athanasiadis *et al.*, 2001). A model of the 3-D structure to 15 Å resolution has been obtained for StII by crystallization on the surface of a lipid monolayer (Martín-Benito *et al.*, 2000). EqtII (PDB accession number 1IAZ) consists of 174 residues, lacking the first five residues at the N-terminus. It is a single domain protein with dimensions of 42 × 28 × 32 Å (Figure 8.3). It is composed of a tightly folded core of twelve β-sheet strands and two short α-helices A and B, one on each side of the β-sandwich. The arrangement of structural elements exhibits virtually internal two-fold symmetry, and the molecule can be visualized as two halves, each consisting of six β-strands and one α-helix. The C-terminal α-helix B is bound to the body of the molecule by two neighboring β-strands. The first 30 residues at the N-terminus encompassing amphiphilic α-helix A are the only part which can be displaced from the body of the molecule without disrupting the fold of the hydrophobic core. Two β-hairpin loops, K77 to G85 and V106 to N115, are particularly long and protrude from the surface of the molecule. The second loop and its neighboring β8-strand contain the tryptophan-rich region. This part of the molecule is at the bottom in Figure 8.3. Loops and the C-terminus form a flat surface on the other side of the molecule at the top in Figure 8.3 left. The residues H67, S95, R126, R152 and A179, which were shown by cysteine-scanning mutagenesis not to be involved either in membrane binding or oligomerization, are located on this surface (Anderluh *et al.*, 1999a). The RGD motif is exposed on the loop connecting strands β-9 and β-10 and is close to the aromatic-rich surface. Positive charges are distributed mainly on the top and bottom of the molecule, while negative charges are located mainly on the two helices (two and three negatively charged residues in helices A and B, respectively). There are two prominent positive charge clusters on the molecule. One, composed of K30, R31, K32 and K77, lies close to the aromatic-rich cluster. The other is represented by five positively charged residues on the loop connecting strand β-8 with helix B.

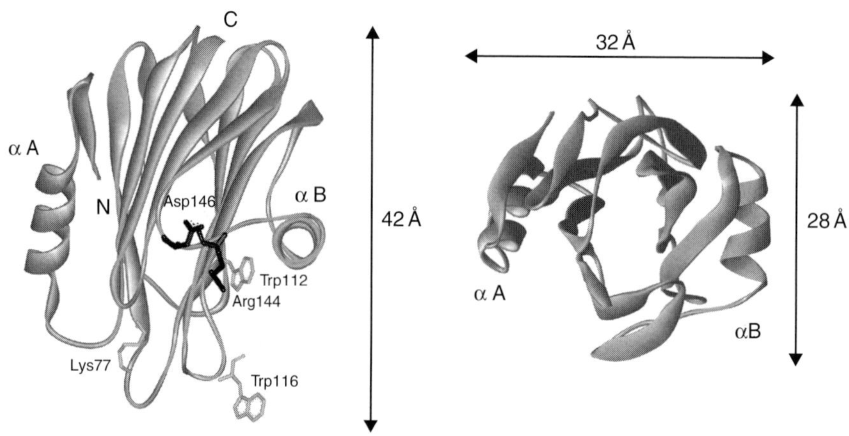

Figure 8.3 Three-dimensional crystal structure of equinatoxin II from *A. actinia*. The 3-D structure of EqtII (PDB accession number 1IAZ) is presented as a solid ribbon model. Numbers and arrows illustrate the molecular dimensions. Left: the bottom of the molecule with tryptophans 112 and 116 is suggested to interact with lipid bilayer. Lysine 77 is essential in oligomerization. Shown are arginine 144 and aspartate 146 of the RGD motif. The protein termini are designated by N and C. The N-terminal helix A (αA) is approximately perpendicular to helix B (αB). Right: view from the top showing the β-sandwich structure of the molecule with two virtually symmetrical halves, each one composed of an α-helix and six β-strands (see Figure 8.1). (Adapted from Athanasiadis *et al.*, 2001.)

Two-dimensional crystals of StII grown on a phosphatidylcholine monolayer, and visualized by high-resolution electron microscopy, exhibit a unit cell with two tetrameric motifs. Within the tetramer, the monomers are in two different orientations, one with its longest axis parallel to the membrane plane, the other at an angle of 60° (Martín-Benito *et al.*, 2000). The adsorbed monomer is glove-shaped with two larger and one smaller lobe. The longest dimension of the monomer is about 51 Å, which is considerably larger than EqtII. One possibility could be that the StII molecule when adsorbed on the membrane dissociates into two halves about its longest axis. It is also noteworthy that a tetrameric structure of StII could be detected in solution (De los Rios *et al.*, 1999) while for EqtII only a low portion of dimers has been observed (Belmonte *et al.*, 1993). At present, the significance of the solution tetrameric structure of StII for insertion and pore-formation remains unclear (Martín-Benito *et al.*, 2000).

Binding to lipids and model membrane permeabilization

Although early competition and binding studies suggested sphingomyelin to be a specific membrane acceptor for an actinoporin from *S. helianthus* (Bernheimer and Avigad, 1976; Linder *et al.*, 1977), later studies provided evidence that actinoporins possess a considerable affinity for a number of lipidic compounds forming either mono- and bilayer membranes or lipoproteins.

Membrane lipid extracts dispersed in water and serum lipoproteins are both very efficient inhibitors of the actinoporin lethal and cytolytic activity. EqTII was found to be co-precipitated with serum lipoproteins (Turk *et al.*, 1989; Narat *et al.*, 1994).

Lipid monolayer experiments proved that actinoporins are surface-active proteins which interact with a variety of pure lipids or their mixtures. It was shown for the cytolysin C-III from *S. helianthus* that it increases the surface pressure at air : water or heptane : water interfaces. This effect of C-III was only slightly greater in SM and DPPC monolayers (Doyle *et al.*, 1989). A more detailed study of equinatoxin II behavior in SM, PC and PC/SM monolayers indicated that when conditions are favorable for insertion, that is, when the initial surface pressure is lower than the critical one, the affinity of EqtII for SM is higher than for PC, and in the PC/SM mixtures, the insertion is governed by the SM content (Caaveiro *et al.*, 2001). This promoting effect of sphingomyelin is even more pronounced in vesicle and planar lipid membrane bilayers, however, the exact role of sphingomyelin is uncertain in either of the model systems.

Currently, there are different interpretations of the role of lipid composition in penetrations of membranes by actinoporins. This is not surprising when faced with the complexity of the pore-forming process, because each consecutive step could be influenced by a single physico-chemical characteristic of the lipid molecule, a combination of several of their properties, or by physical parameters arising from the collective nature of lipids in membranes. Available data suggests that features of lipid head groups themselves may modulate pore-forming activity but do not appear to be essential for actinoporin binding (see Table 8.1 for examples and references). Hence, in contrast to sphingomyelin, its phosphono analogue does not operate as an efficient actinoporin acceptor molecule (Meinardi *et al.*, 1995). EqtII binds to pure PC vesicles in a reversible manner without inserting into the bilayer while SM membranes are penetrated irreversibly by the toxin (Caaveiro *et al.*, 2001). Anionic PG is more efficient in binding and triggering structural changes of EqTII than zwitterionic PC, and moreover, the toxin insertion into the lipid bilayer coincides with a gel to liquid–crystalline phase transition (Poklar *et al.*, 1999). This is in accordance with previous observations that either lipids with unsaturated fatty acid (Michaels, 1979) or free unsaturated fatty acids or alcohols partitioned in bilayers promote permeabilization of PC/SM vesicles by EqTII (Maček *et al.*, 1997). It has been also suggested that the curvature of lipid vesicles may have a role, which was deduced from a lower permeabilization rate of SUV by EqtII as compared to LUV of the same lipid composition (Maček *et al.*, 1994). Moreover, a series of anionic lipids, in decreasing order PA ~ CL > PI ~ PS ~ PG, but also cationic SA, and PE when combined with PC and SM, significantly enhance the rate and extent of permeabilization of LUV by StI. The boosting effect on pore formation is due to increased bilayer negative curvature by the supplemented lipids as suggested by Alvarez *et al.* (2001). Based on the observation that StI markedly increased the flip-flop movement of membrane lipids, the authors also suggested involvement of the inverse hexagonal lipid phase (H_{II}) in pore formation by sticholysin I.

Another significant observation is that membranes made of binary or ternary mixtures of lipids with SM included are more susceptible to actinoporins (Tejuca *et al.*, 1996; De los Rios *et al.*, 1998; Alvarez *et al.*, 2001; Caaveiro *et al.*, 2001). However, adding cholesterol to PC also markedly increases vesicle permeabilization (De los Rios *et al.*, 1998). It is noteworthy that both lipids, SM and Chol, may have a bimodal effect on promoting toxin induced membrane leakage. At mole fractions greater than ~0.3–0.5, the permeabilization efficiency may again decline (Caaveiro *et al.*, 2001).

In summary, there is evidence that actinoporin binding to lipids is governed by both electrostatic and hydrophobic interactions. In addition, the collective behavior of lipids in a bilayer and its physical determinants may modulate the protein activity. Hence, fatty acyl-chain ordering parameter and membrane thickness (Michaels, 1979; Rudnev *et al.*, 1984; Maček *et al.*, 1997), gel-to-crystalline phase transition of lipid (Poklar *et al.*, 1999), and occurrence of lipid microdomains (De los Rios *et al.*, 1998), and negative membrane curvature (Alvarez *et al.*, 2001) have been suggested to modify toxin insertion and pore-formation. It has become obvious that lipid bilayer permeabilization by an actinoporin is not dependent on the chemical nature of one exclusive lipid species. It is rather influenced by several modalities of the bilayer such as thickness, fatty acid acyl-chain ordering, structural defects related to lipid microdomain boundaries, and membrane electrostatic phenomena (i.e. membrane surface and dipole potentials, and transmembrane potential). Thus, lateral microheterogeneity of the membrane, induced by the lipid phase transition or by mixing two polar lipids or adding cholesterol to a polar lipid, may be one reason for the observed effects, as discussed elsewhere in general terms for the penetration of bilayers by matter (Mouritsen *et al.*, 1995; White and Wimley, 1998) and as suggested specifically for StII (De los Rios *et al.*, 1998).

Molecular mechanism of pore formation

In general, formation of protein pore is a multi-step process involving transformation of a soluble protein to a membrane-inserted aggregate (Gouaux, 1997). In the case of actinoporins, it has been demonstrated that the initial binding of monomers to the membrane, which is fast and takes place in seconds, is followed by oligomerization and formation of functional pores (Michaels, 1979; Doyle and Kem, 1989; Belmonte *et al.*, 1993; Maček *et al.*, 1995; Anderluh *et al.*, 1997; De los Rios *et al.*, 1998; Malovrh *et al.*, 2000). Although the existence of StII tetramers in solution makes molecularity of actinoporin binding less clear, the majority of evidence supports the accumulation of monomers on the membrane up to a critical concentration before permeabilization starts (Tejuca *et al.*, 1996; Caaveiro *et al.*, 2001).

In the erythrocyte membrane, an actinoporin pore consists of 3–4 molecules and is considerably less stable than that produced by staphylococcal α-toxin. In the membrane, equinatoxin II pores are identified as oligomers by chemical cross-linking (Belmonte *et al.*, 1993). Assuming trimerization of StI on the membrane, a kinetic model resolving binding and pore formation in lipid vesicles by an actinoporin was established by Tejuca *et al.* (1996). An intact actinoporin pore in both cell and artificial lipid membranes appears to be 2.0–2.2 nm in inner diameter as determined by osmotic protection experiments (Belmonte *et al.*, 1993). It is independent of toxin concentration as reported for StI pores made on large unilamellar vesicles (Tejuca *et al.*, 2001). Somewhat smaller pores, ~1 nm in inner diameter, were reported for StII on lipid vesicles. The estimate was obtained by measuring leakage of a series of solutes of differing sizes (De los Rios *et al.*, 1998).

Structural and functional studies using recombinant actinoporins have been performed to identify parts of the molecule involved in particular steps of the process (Anderluh *et al.*, 1997, 1998, 1999a, 2000a; Malovrh *et al.*, 2000). The first contact of the molecule with the lipid membrane is promoted by the region around the exposed aromatic patch at the bottom of the molecule shown in Figure 8.3. Chemical modification studies of the exposed aromatic residues have pointed to their role in cytolytic and lethal activity (Turk *et al.*, 1989; Turk *et al.*, 1992; Khoo *et al.*, 1997; Campos *et al.*, 1999). Studies of intrinsic tryptophan fluorescence of EqtII and quenching by water-soluble and membrane embedded quenchers suggested that approximately three of the five residues are transferred to the lipid membrane interphase (Maček *et al.*, 1995). Point mutations of EqtII showed that tryptophans 116 and 117, part of the tryptophan-rich region, which is also important for proper toxin conformation, interact with lipids (Malovrh *et al.*, 2000). The tryptophan-rich region is a part of the exposed aromatic patch on the bottom of the molecule (Figure 8.3). It contains two tryptophans (of 3–5 present in actinoporin primary structures) and six tyrosines. The side chain of W116, pointing downwards to the aromatic cluster, is exposed and is responsible for the fluorescence shifts observed upon association with lipid membranes. The side chain of W117 is oriented upwards and is tightly packed within the core of the molecule, and is important for the native fold. In addition, W112 is on the tip of one of the largest protruding loops and completely exposed (Figure 8.3), and it very probably interacts with the lipid surface. Three more residues were shown to interact with the membrane and map to the same area on the 3-D structure. S105 and S114 are part of the tryptophan rich region, while R144 is a part of the RGD motif close to that region (Figure 8.3). Its side chain points downward, in the close vicinity of the aromatic cluster (Anderluh *et al.*, 1999a; Athanasiadis *et al.*, 2001). On the other side of the cluster, around residue 77 and towards helix A (Figure 8.3), are three residues, whose mutation to cysteine results in abolition of hemolytic activity (Figure 8.2; Anderluh *et al.*, 1999a). The side chains of R31, K77 and S167 stabilize the conformation of broad loops, indicating that this part of the molecule should remain rigid during the binding and pore-formation.

The attachment of EqtII, with exposed aromatic residues, to the surface of the membrane appears similar to that of cholesterol-dependent toxins from Gram-positive bacteria. Tryptophan rich domain 4 of perfringolysin, which is structurally similar to EqtII (Athanasiadis *et al.*, 2001), has the role of binding the toxin to the membrane and providing the environment for structural change. This change involves the transition of α-helices from other domains to antiparallel β-sheets which, in the final phase, penetrate the lipid bilayer (Heuck *et al.*, 2000).

The role of the N-terminus of EqtII has been studied by truncation mutagenesis. N-truncation of the recombinant toxin by 5 and 10 amino acid residues results in a decrease of hemolytic activity by 11% and 69%, respectively while deletion of the 33-residue long N-terminal part including helix A (see Figures 8.2 and 8.3) results in an insoluble product (Anderluh *et al.*, 1997). The avidity of the active truncated mutants for sphingomyelin vesicles and erythrocytes was actually augmented. On the other hand, fusion of a 6-histidine tag to the N-terminus of StII does not affect hemolytic activity (De los Rios *et al.*, 2000), and only fusion with larger polypeptides or ubiquitin decreased the activity (Wang *et al.*, 2000). This is also consistent with the finding of Anderluh *et al.* (1999a) that the initial residues of the EqtII N-terminus do not insert into the membrane. Two other mutants from the N-terminus, S13 which precedes helix A and K20, within helix A, were shown to interact with the membrane lipid–water interface. Altogether these facts confirm the N-terminal part containing helix A as being important in pore-formation. However, its interplay with other parts of the toxin molecule, in either lipid binding, toxin aggregation, and/or pore oligomerization in the membrane cannot be ignored.

Lysine 77 located at the bottom of the molecule was found essential for toxin aggregation. Replacement of this residue with cysteine considerably decreased the rate of pore-formation but not toxin binding to either planar lipid membranes or SUV (Figure 8.1). Chemical reversion of cysteine 77 to positively charged cysteinyl-ethylamine with bromoethylamine completely restored the activity but reaction with neutral iodoacetamide or negatively charged iodoacetic acid did not (Anderluh *et al.*, 2000a).

In conclusion, pore-forming toxins are classified into two groups with regard to the structural elements used to penetrate the lipid bilayer (Gouaux, 1997). All predominantly α-helical toxins including colicins employ a cluster of helices to form the transmembrane pore. The second group are toxins rich in β-sheet, that is, *S. aureus* α-hemolysin and cholesterol-dependent toxins (Van der Goot, 2001), and penetrate membranes by β-strands organized in a transmembrane barrel. Actinoporins appear to be different. Their binding is promoted by the surface aromatic patch, but pore-formation appears to require participation of the N-terminal part with involvement of an amphiphilic α-helix.

Concluding remarks

Sea anemone actinoporins appear to be extremely efficient pore-forming toxins as compared to other members of this class isolated from other organisms (Harvey, 1990; Menestrina *et al.*, 1994). Their pore-forming activity in both natural and artificial lipid membranes is well documented. There is, however, a certain discrepancy between the doses required to produce pore-formation and particular pharmacological effects. Specifically, an apparent association constant of $13 \times 10^3\,M^{-1}$ determined for StI in SM/PC LUV (Tejuca *et al.*, 1996) and of $51 \times 10^3\,M^{-1}$ for EqtII in SM/PC SUV (Maček *et al.*, 1995) or partition coefficients of $7.2 \times 10^5 - 2 \times 10^5$ for EqtII for SM/PC LUV (Caaveiro *et al.*, 2001) have been determined. At the same time, concentrations of EqtII higher than $2.3 \times 10^{-7}\,M$ produced pores in bovine lactotrophs (Zorec *et al.*, 1990). On the other hand, much lower concentrations of actinoporins, about $10^{-12}\,M$, suffice to damage membranes of a variety of cells (see for example Batista *et al.*, 1987; Maček *et al.* 1994). This implies that in natural membranes, binding and/or pore-formation may be enhanced by an additional mechanism. In this respect, the conserved RGD or some other motif may be involved. A systematic study using cells with receptors, such as for example integrins, for extracellular proteins bearing the RGD motif is necessary.

At present, a reliable model of the 3-D structure of the actinoporin transmembrane pore cannot be predicted. An innovative model was proposed, suggesting a toroidal pore, in which the lumen of the pore would be created by the N-terminal helix of StI and membrane lipids (Alvarez *et al.*, 2001). The solved 3-D crystal structure of equinatoxin II should provide an impetus to further detailed studies of the mechanism of pore-formation by actinoporins. Moreover, it is also an important tool for construction of recombinant immuno- or mitotoxins targeted at plasmalemma or intracellular membranes of tumor cells or parasites (see for example, Avila *et al.*, 1988; Pederzolli *et al.*, 1995; Panchal, 1998; Tejuca *et al.*, 1999; Anderluh and Menestrina, 2001).

References

Ales, E., Gabilan, N. H., Cano-Abad, M. F., Garcia, A. G. and Lopez, M. G. (2000) The sea anemone toxin Bc2 induces continuous or transient exocytosis, in the presence of sustained levels of high Ca^{2+} in chromaffin cells. *J. Biol. Chem.*, **275**, 37488–37495.

Alvarez, C., Lanio, M. E., Tejuca, M., Martinez, D., Pazos, F., Campos, A. M., *et al.* (1998) The role of ionic strength on the enhancement of the hemolytic activity of sticholysin I, a cytolysin from *Stichodactyla helianthus*. *Toxicon*, **36**, 165–178.

Alvarez, C., Dalla Serra, M., Potrich, C., Bernhart, I., Tejuca, M., Martinez, D., *et al.* (2001) Effects of lipid composition on the membrane permeabilization by sticholysin I and II, two cytolysins of the sea anemone. *Stichodactyla helianthus*. *Biophys. J.*, **80**, 2761–2774.

Anderluh, G. and Maček P. (2002) Cytolytic peptide and protein toxins from sea anemones (Anthozoa: Actiniaria). *Toxicon*, **40**, 111–124.

Anderluh, G. and Menestrina, G. (2001) Pore-forming proteins from sea anemones and the construction of immunotoxins for selective killing of harmful cells. In *Bio-organic Compounds: Chemistry and Biomedical Applications*, edited by M. Fingerman and R. Nagabhushanam, Enfield (NH), USA Science Publishers, Inc., pp. 131–148.

Anderluh, G., Pungerčar, J., Štrukelj, B., Maček, P. and Gubenšek, F. (1996) Cloning, sequencing and expression of equinatoxin II. *Biochem. Biophys. Res. Commun.*, **220**, 437–442.

Anderluh, G., Pungerčar, J., Križaj, I., Štrukelj, B., Gubenšek, F. and Maček, P. (1997) N-terminal truncation mutagenesis of equinatoxin II, a pore- forming protein from the sea anemone *Actinia equina*. *Protein Eng.*, **10**, 751–755.

Anderluh, G., Barlič, A., Križaj, I., Menestrina, G., Gubenšek, F. and Maček, P. (1998) Avidin-FITC topological studies with three cysteine mutants of equinatoxin II, a sea anemone pore-forming protein. *Biochem. Biophys. Res. Commun.*, **242**, 187–190.

Anderluh, G., Barlič, A., Podlesek, Z., Maček, P., Pungerčar, J., Gubenšek, F., *et al.* (1999a) Cysteine-scanning mutagenesis of an eukarytic pore-forming toxin from sea anemone. *Eur. J. Biochem.*, **263**, 128–136.

Anderluh, G., Križaj, I., Štrukelj, B., Gubenšek, F., Maček, P. and Pungerčar, J. (1999b) Equinatoxins, pore-forming proteins from the sea anemone *Actinia equina*, belong to a multigene family. *Toxicon*, **37**, 1391–1401.

Anderluh, G., Barlič, A., Potrich, C., Maček, P. and Menestrina, G. (2000a) Lysine 77 is a key residue in aggregation of equinatoxin II, a pore-forming toxin from sea anemone *Actinia equina*. *J. Membrane Biol.*, **173**, 47–55.

Anderluh, G., Podlesek, Z. and Maček, P. (2000b) A common motif in proparts of Cnidarian toxins and nematocyst collagens and its putative role. *Biochim. Biophys. Acta*, **1476**, 372–376.

Athanasiadis, A., Anderluh, G., Maček, P. and Turk, D. (2001) Crystal structure of the soluble form of equinatoxin II, a pore-forming toxin from the sea anemone *Actinia equina*. *Structure*, **9**, 341–346.

Avila, A. D., Mateo de Acosta, C. and Lage, A. (1988) A new immunotoxin built by linking a hemolytic toxin to a monoclonal antibody specific for immature T lymphocytes. *Int. J. Cancer*, **142**, 568–571.

Batista, U. and Jezernik, K. (1992) Morphological changes of V-79 cells after equinatoxin II treatment. *Cell Biol. Int. Rep.*, **16**, 115–123.

Batista, U., Jezernik, K., Maček, P. and Sedmak, B. (1987) Morphological evidence of cytotoxic and cytolytic activity of equinatoxin II. *Period. Biol.*, **89**, 347–348.

Batista, U., Maček, P. and Sedmak, B. (1990) The cytotoxic and cytolytic activity of equinatoxin II from the sea anemone *Actinia equina*. *Cell Biol. Int. Rep.*, **14**, 1013–1024.

Belmonte, G., Pederzolli, C., Maček, P. and Menestrina, G. (1993) Pore formation by the sea anemone cytolysin equinatoxin II in red blood cells and model lipid membranes. *J. Membrane Biol.*, **131**, 11–22.

Belmonte, G., Menestrina, G., Pederzolli, C., Križaj, I., Gubenšek, F., Turk, T., *et al.* (1994) Primary and secondary structure of a pore-forming toxin from the sea anemone, *Actinia equina* L., and its association with lipid vesicles. *Biochim. Biophys. Acta*, **1192**, 197–204.

Bernheimer, A. W. and Avigad, L. S. (1976) Properties of a toxin from the sea anemone *Stoichactis helianthus*, including specific binding to sphingomyelin. *Proc. Natl. Acad. Sci. USA*, **73**, 467–471.

Blumenthal, K. M. and Kem, W. R. (1983) Primary structure of *Stoichactis helianthus* cytolysin III. *J. Biol. Chem.*, **258**, 5574–5581.

Bunc, M., Frangež, R., Horvat, I., Turk, T. and Šuput, D. (1994) Effects of equinatoxins in vivo. Possible role of degranulation of thrombocytes and granulocytes. *Ann. NY. Acad. Sci.*, **710**, 162–167.

Bunc, M., Drevenšek, G., Budihna, M. and Šuput, D. (1999) Effects of equinatoxin II from *Actinia equina* (L.) on isolated rat heart: the role of direct cardiotoxic effects in equinatoxin II lethality. *Toxicon*, **37**, 109–123.

Caaveiro, J. M., Echabe, I., Gutiérrez-Aguirre, I., Nieva, J. L., Arrondo, J. L. R. and Gonzáles-Manas, J. M. (2001) Differential interaction of equinatoxin II with model membranes in response to lipid composition. *Biophys. J.*, **80**, 1343–1353.

Campos, A. M., Lissi, E. A., Vergara, C., Lanio, M. E., Alvarez, C., Pazos, I., *et al.* (1999) Kinetics and mechanism of St I modification by peroxyl radicals. *J. Prot. Chem.*, **18**, 297–306.

De los Rios, V., Mancheno, J. M., Lanio, M. E., Onaderra, M. and Gavilanes, J. G. (1998) Mechanism of the leakage induced on lipid model membranes by the hemolytic protein sticholysin II from the sea anemone *Stichodactyla helianthus*. *Eur. J. Biochem.*, **252**, 284–289.

De los Rios, V., Mancheno, J. M., Del Pozo, A. M., Alfonso, C., Rivas, G., Onaderra, M., *et al.* (1999) Sticholysin II, a cytolysin from the sea anemone *Stichodactyla helianthus*, is a monomer-tetramer associating protein. *FEBS Lett.*, **455**, 27–30.

De los Rios, V., Onaderra, M., Martínez-Ruiz, A., Lacadena, J., Mancheno, J. M., Del Pozo, A. M., *et al.* (2000) Overproduction in *Escherichia coli* and purification of the hemolytic protein sticholysin II from the sea anemone *Stichodactyla helianthus*. *Prot. Expr. Purif.*, **18**, 71–76.

Doyle, J. W. and Kem, W. R. (1989) Binding of a radiolabeled sea anemone cytolysin to erythrocyte membranes. *Biochim. Biophys. Acta*, **987**, 181–186.

Doyle, J. W., Kem, W. R. and Vilallonga, F. A. (1989) Interfacial activity of an ion channel-generating protein cytolysin from the sea anemone *Stichodactyla helianthus*. *Toxicon*, **27**, 465–471.

Drevenšek, G., Budihna, M. V., Šuput, D. and Bunc, M. (2000a) Nicardipine dose dependently reduces the effect of equinatoxin II on coronary flow in isolated porcine heart. *Pflugers Arch.*, **440**, R145–R146.

Drevenšek, G., Bunc, M., Budihna, M. V. and Šuput, D. (2000b) Lowering of the coronary flow in isolated rat heart by equinatoxin II depends upon extracellular Ca^{2+} concentration. *Pflugers Arch.*, **439**, R150–R151.

Ferlan, I. and Lebez, D. (1974) Equinatoxin, a lethal protein from *Actinia equina*.- I. Purification and characterization. *Toxicon*, **12**, 57–61.

Galettis, P. and Norton, R. S. (1990) Biochemical and pharmacological studies of action of tenebrosin-C, a cardiac stimulatory and haemolytic protein from the sea anemone *Actinia tenebrosa*. *Toxicon*, **28**, 695–706.

Giese, C., Mebs, D. and Werding, B. (1996) Resistance and vulnerability of crustaceans to cytolytic sea anemone toxins. *Toxicon*, **34**, 955–958.

Gouaux, E. (1997) Channel-forming toxins: tales of transformation. *Curr. Opin. Struc. Biol.*, **7**, 566–573.

Harvey, H. L. (1990) Cytolytic toxins. In *Handbook of Toxinology*, edited by W. T. Shier and D. Mebs, New York: Marcel Dekker, pp. 1–66.

Heuck, A. P., Hotze, E. M., Tweten, R. K. and Johnson, A. E. (2000) Mechanism of membrane insertion of a multimeric beta-barrel protein. Perfringolysin O creates a pore using ordered and coupled conformational changes. *Mol. Cell*, **6**, 1233–1242.

Ho, C. L., Ko, J. L., Lue, H. M., Lee, C. Y. and Ferlan, I. (1987) Effects of equinatoxin on the guinea-pig atrium. *Toxicon*, **25**, 659–664.

Huerta, V., Morera, V., Guanche, Y., Chinea, G., Gonzáles, L. J., Betancourt, L., *et al.* (2001) Primary structure of two cytolysin isoforms from *Stichodactyla helianthus* differing in their hemolytic activity. *Toxicon*, **39**, 1253–1256.

Ivanov, A. S., Molnar, A. A., Kozlovskaya, E. P. and Monastyrnaya, M. M. (1987) The action of toxin from *Radianthus macrodactylus* on biological and model membrane permeability. *Biol. Membr.*, **4**, 243–248 (in Russian).

Kem, W. R. (1988) Sea anemone toxins: Structure and action. In *The Biology of Nematocysts*, edited by D. A. Hessinger and H. M. Lenhoff, pp. 375–405. San Diego: Academic Press, Inc.

Kem, W. R. and Dunn, B. M. (1988) Separation and characterization of four different amino acid sequence variants of sea anemone (*Stichodactyla helianthus*) protein cytolysin. *Toxicon*, **26**, 997–1008.

Khoo, K. S., Kam, W. K., Khoo, H. E., Gopalakrishnakone, P. and Chung, M. C. M. (1993) Purification and partial characterization of two cytolysins from a tropical sea anemone *Heteractis magnifica*. *Toxicon*, **31**, 1567–1579.

Khoo, H. E., Lim, J. P. C. and Tan, C. H. (1995) Effects of sea anemone (*Heteractis magnifica* and *Actinia equina*) cytolysins on synaptosomal uptake of GABA and choline. *Toxicon*, **33**, 1365–1371.

Khoo, H. E., Fong, C. L., Yuen, R. and Chen, D. S. (1997) Stimulation of haemolytic activity of sea anemone cytolysins by 8-anilino-1-naphthalenesulphonate. *Biochem. Biophys. Res. Commun.*, **232**, 422–426.

Lanio, M. E., Morera, V., Alvarez, C., Tejuca, M., Gomez, T., Pazos, F., *et al.* (2001) Purification and characterization of two hemolysins from *Stichodactyla helianthus*. *Toxicon*, **39**, 187–194.

Linder, R., Bernheimer, A. W. and Kim, K.-S. (1977) Interaction between sphingomyelin and a cytolysin from the sea anemone *Stoichactis helianthus*. *Biochim. Biophys. Acta*, **467**, 290–300.

Maček, P. (1992) Polypeptide cytolytic toxins from sea anemones (Actiniaria). *FEMS Microbiol. Immunol.*, **105**, 121–130.

Maček, P. and Lebez, D. (1981) Kinetics of hemolysis induced by equinatoxin, a cytolytic toxin from the sea anemone *Actinia equina*. Effect of some ions and pH. *Toxicon*, **19**, 233–240.

Maček, P. and Lebez, D. (1988) Isolation and characterization of three lethal and hemolytic toxins from the sea anemone *Actinia equina* L. *Toxicon*, **26**, 441–451.

Maček, P., Belmonte, G., Pederzolli, C. and Menestrina, G. (1994) Mechanism of action of equinatoxin II, a cytolysin from the sea anemone *Actinia equina* L, belonging to the family of actinoporins. *Toxicology*, **87**, 205–227.

Maček, P., Zecchini, M., Pederzolli, C., Dalla Serra, M. and Menestrina, G. (1995) Intrinsic tryptophan fluorescence of equinatoxin II, a pore-forming polypeptide from the sea anemone *Actinia equina* L, monitors its interaction with lipid membranes. *Eur. J. Biochem.*, **234**, 329–335.

Maček, P., Zecchini, M., Stanek, K. and Menestrina, G. (1997) Effect of membrane-partitioned n-alcohols and fatty acids on pore-forming activity of a sea anemone toxin. *Eur. Biophys. J.*, **25**, 155–162.

Malavašič, M., Poklar, N., Maček, P. and Vesnaver, G. (1996) Fluorescence studies of the effect of pH, guanidine hydrochloride and urea on equinatoxin II conformation. *Biochim. Biophys. Acta: Bio-Membranes*, **1280**, 65–72.

Malovrh, P., Barlič, A., Podlesek, Z., Maček, P., Menestrina, G. and Anderluh, G. (2000) Structure-function studies of tryptophan mutants of equinatoxin II, a sea anemone pore-forming protein. *Biochem. J.*, **346**, 223–232.

Mancheno, J. M., De los Rios, V., Del Pozo, A. M., Lanio, M. E., Onaderra, M. and Gavilanes, J. G. (2001) Partially folded states of the cytolytic protein sticholysin II. *Biochim. Biophys. Acta*, **1545**, 122–131.

Martín-Benito, J., Gavilanes, F., De los, Rios., Mancheno, J. M., Fernández, J. J. and Gavilanes, J. G. (2000) Two-dimensional crystallization on lipid monolayers and three-dimensional structure of sticholysin II, a cytolysin from the sea anemone *Stichodactyla helianthus*. *Biophys. J.*, **78**, 3186–3194.

Mebs, D. (1994) Anemonefish symbiosis: vulnerability and resistance of fish to the toxin of the sea anemone. *Toxicon*, **32**, 1059–1068.

Meinardi, E., Azcurra, J. M., Florin-Christensen, M. and Florin-Christensen, J. (1994) Coelenterolysin: a hemolytic polypeptide associated with the coelenteric fluid of sea anemones. *Comp. Biochem. Physiol.*, **109B**, 153–161.

Meinardi, E., Florin-Christensen, M., Paratcha, G., Azcurra, J. M. and Florin-Christensen, J. (1995) The molecular basis of the self non self selectivity of a coelenterate toxin. *Biochem. Biophys. Res. Commun.*, **216**, 348–354.

Menestrina, G., Cabiaux, V. and Tejuca, M. (1999) Secondary structure of sea anemone cytolysins in soluble and membrane bound form by infrared spectroscopy. *Biochem. Biophys. Res. Commun.*, **254**, 174–180.

Menestrina, G., Schiavo, G. and Montecucco, C. (1994) Molecular mechanism of action of bacterial toxins. *Mol. Asp. Med.*, **15**, 79–193.

Meunier, F. A., Frangež, R., Benoit, E., Ouanounou, G., Rouzaire-Dubois, B., Šuput, D., *et al.* (2000) Ca^{2+} and Na^{+} contribute to the swelling of differentiated neuroblastoma cells induced by equinatoxin-II. *Toxicon*, **38**, 1547–1560.

Michaels, D. W. (1979) Membrane damage by a toxin from the sea anemone *Stoichactis helianthus*. I. Formation of transmembrane channels in lipid bilayers. *Biochim. Biophys. Acta*, **555**, 67–78.

Migues, P. V., Leal, R. B., Mantovani, M., Nicolau, M. and Gabilan, N. H. (1999) Synaptosomal glutamate release induced by the fraction Bc2 from the venom of the sea anemone *Bunodosoma caissarum*. *NeuroReport*, **10**, 67–70.

Monastyrnaya, M., Zykova, T. and Kozlovskaya, E. P. (1999) Isolation and characterization of high-molecular cytolytic toxins from the sea anemone *Radianthus macrodactylus*. *Biorganic Chem.*(in Russian), **25**, 733–741.

Mouritsen, O. G., Jorgensen, K. and Honger, T. (1995) Permeability of lipid bilayers near the phase transition. In *Permeability and Stability of Lipid Bilayers*, edited by E. A. Disalvo and S. A. Simon, 137–160. London: CRC Press.

Narat, M., Maček, P., Kotnik, V. and Sedmak, B. (1994) The humoral and cellular immune response to a lipid attenuated pore-forming toxin from the sea anemone *Actinia equina* L. *Toxicon*, **32**, 65–71.

Norton, R. S., Bobek, G., Ivanov, J. O., Thomson, M., Beer, E. F., Moritz, R. L., *et al.* (1990) Purification and characterization of proteins with cardiac stimulatory and hemolytic activity from the sea anemone *Actinia tenebrosa*. *Toxicon*, **28**, 29–41.

Panchal, R. G. (1998) Novel therapeutic strategies to selectively kill cancer cells. *Biochem. Pharmacol.*, **55**, 247–252.

Pederzolli, C., Belmonte, G., Dalla Sera, M., Maček, P. and Menestrina, G. (1995) Biochemical and cytotoxic properties of conjugates of transferrin with equinatoxin II, a cytolysin from a sea anemone. *Bioconjugate Chem.*, **6**, 166–173.

Poklar, N., Lah, J., Salobir, M., Maček, P. and Vesnaver, G. (1997) pH and temperature-induced molten globule-like denatured states of equinatoxin II: a study by UV-melting, DSC, far- and near-UV CD spectroscopy, and ANS fluorescence. *Biochemistry*, **36**, 14345–14352.

Poklar, N., Fritz, J., Maček, P., Vesnaver, G. and Chalikian, T. V. (1999) Interaction of the pore-forming protein equinatoxin II with model lipid membranes: a calorimetric and spectroscopic study. *Biochemistry*, **38**, 14999–15008.

Poklar, N., Völker, J., Anderluh, G., Maček, P. and Chalikian, T. V. (2001) Acid- and base-induced conformational transitions of equinatoxin II. *Biophys. Chem.*, **90**, 103–121.

Rudnev, V. S., Likhatskaya, G. N., Kozlovskaya, E. P., Monastyrnaya, M. M. and Elyakov, G. B. (1984) Effect of hemolysin from the sea anemone *Radianthus macrodactylus* on lipid membrane conductivity. *Biol. Membr.* (in Russian), **10**, 1019–1024.

Samejima, Y., Yanagisawa, M., Aoki-Tomomatsu, Y., Iwasaki, E., Ando, J. and Mebs, D. (2000) Amino acid sequence studies on cytolytic toxins from sea anemone *Heteractis magnifica, Entacmaea quadricolor* and *Stichodactyla mertensii* (Anthozoa). *Toxicon*, **38**, 259–264.

Senčič, L. and Maček, P. (1990) New method for isolation of venom from the sea anemone *Actinia cari*. Purification and characterization of cytolytic toxins. *Comp. Biochem. Physiol.*, **97B**, 687–693.

Simpson, R. J., Reid, G. E., Moritz, R. L., Morton, C. and Norton, R. S. (1990) Complete amino acid sequence of tenebrosin-C, a cardiac stimulatory and haemolytic protein from the sea anemone *Actinia tenebrosa*. *Eur. J. Biochem.*, **190**, 319–328.

Sket, D., Drašlar, K., Ferlan, I. and Lebez, D. (1974) Equinatoxin, a lethal protein from *Actinia equina*-II. Pathophysiological action. *Toxicon*, **12**, 63–68.

Šuput, D., Frangež, R., Bunc, M. (2001) Cardiovascular effects of equinatoxin III from the sea anemone *Actinia equina* (L.). *Toxicon*, **39**, 1421–1427.

Tejuca, M., Dalla Serra, M., Ferreras, M., Lanio, M. E. and Menestrina, G. (1996) Mechanism of membrane permeabilization by sticholysin I, a cytolysin isolated from the venom of the sea anemone *Stichodactyla helianthus*. *Biochemistry*, **35**, 14947–14957.

Tejuca, M., Anderluh, G., Maček, P., Marcet, R., Torres, D., Sarracent, J., *et al.* (1999) Antiparasite activity of sea-anemone cytolysins on *Giardia duodenalis* and specific targeting with anti-*Giardia* antibodies. *Int. J. Parasitol.*, **29**, 489–498.

Tejuca, M., Dalla Serra, M., Alvarez, C. Potrich, C. and Menestrina, G. (2001) Sizing the radius of the pore formed in erythrocytes and lipid vesicles by the toxin sticholysin I from the sea anemone *Stichodactyla helianthus*. *J. Membr. Biol.*, **183**, 125–135.

Teng, C. M., Lee, L. G., Lee, C. Y. and Ferlan, I. (1988) Platelet aggregation induced by equinatoxin. *Thromb. Res.*, **52**, 401–411.

Thomson, M., Moritz, R. L., Simpson, R. J. and Norton, R. S. (1987) Tenebrosin-A, a new cardiostimulant protein from the Australian sea anemone *Actinia tenebrosa*. *Biochem. Int.*, **15**, 711–718.

Turk, T. (1991) Cytolytic toxins from sea anemones. *J. Toxicol. Toxin Rev.*, **10**, 223–262.

Turk, T., Maček, P. and Gubenšek, F. (1989) Chemical modification of equinatoxin II, a lethal and cytolytic toxin from the sea anemone. *Actinia equina* L. *Toxicon*, **27**, 375–384.

Turk, T., Maček, P. and Gubenšek, F. (1992) The role of tryptophan in structural and functional properties of equinatoxin II. *Biochim. Biophys. Acta*, **1119**, 1–4.

Van der Goot, G. (2001) *Pore-Forming Toxins*, 1st edn. Springer-Verlag: Berlin.

Varanda, W. and Finkelstein, A. (1980) Ion and nonelectrolyte permeability properties of channels formed in planar lipid bilayer membranes by the cytolytic toxin from the sea anemone, *Stoichactis helianthus*. *J. Membrane Biol.*, **55**, 203–211.

Wang, Y., Chua, K. L. and Khoo, H. E. (2000) A new cytolysin from the sea anemone, *Heteractis magnifica*: isolation, cDNA cloning and functional expression. *Biochim. Biophys. Acta*, **1478**, 9–18.

White, S. H. and Wimley, W. C. (1998) Hydrophobic interactions of peptides with membrane interfaces. *Biochim. Biophys. Acta: Rev. Biomembr.*, **1376**, 339–352.

Zhang, W., Hinds, M. G., Anderluh, G., Hansen, P. E. and Norton, R. S. (2000) Letter to the editor: sequence-specific resonance assignements of the potent cytolysin equinatoxin II. *J. Biomol. NMR*, **18**, 281–282.

Zorec, R., Tester, M., Maček, P. and Mason, W. T. (1990) Cytotoxicity of equinatoxin II from the sea anemone *Actinia equina* involves ion channel formation and an increase in intracellular calcium activity. *J. Membrane Biol.*, **118**, 243–249.

Part 2

Pore-forming peptides

9 Structural and charge requirements for antimicrobial peptide insertion into biological and model membranes

Hong Xia Zhao, Andrea C. Rinaldi, Anna Rufo, Argante Bozzi, Paavo K. J. Kinnunen and Antonio Di Giulio

Introduction

The cohabitation of microorganisms and multicellular organisms on this planet is sometimes problematic. This stems from the fact that the living tissues of the latter are indeed a rich source of nutrients for legions of microbial pathogens, either bacteria, protozoa, or fungi. To overcome this unfortunate situation, multicellular organisms along all the evolutionary scale, from plants to mammals, produce a plethora of peptides with antibiotic activity (Boman, 1995; Ganz and Lehrer, 1998; Garcia-Olmedo *et al.*, 1998; Otvos, 2000), a vital arsenal to combat the endless chemical warfare against an invasive enemy.

When animals are concerned, antimicrobial peptides play an important role in the first, rapidly-reacting line of defense, the so-called "innate immunity," opposed by the host organism against potential pathogens characterized by fast growing kinetics (Barra *et al.*, 1998; Hancock and Chapple, 1999). To serve this scope, antimicrobial peptides are produced where the initial encounter between microbial pathogens and their host takes place, usually the skin and moist epithelial surfaces, where these molecules are synthesized, and excreted by specialized cells and glands (Simmaco *et al.*, 1998). In addition, antimicrobial peptides are stored in granules and vesicles of mobile phagocytic cells, ready to be released directly in the area of microbial invasion (Ganz and Lehrer, 1997). On the other side, the activation and deployment of the humoral and cellular immune responses (antibodies and antigen-recognizing cytotoxic lymphocytes) that form the "adaptive immunity," may require several days, and have therefore a minimal impact on the prevention of onset of infections. Whereas adaptive immunity appeared late during evolution and is found only in higher vertebrates, innate immunity is an ancient "discovery" of multicellular organisms, spurred by the basic need to resist pathogenic invaders.

The effectors of the innate immune response, including antimicrobial peptides, do not possess two of the main properties of the classical immune system: memory, necessary to prevent reinfection by microbes previously encountered by the host, and high specificity. Indeed, antimicrobial peptides are rather unspecific in their activity. Many, but not all, have an unusually broad spectrum of activity against both Gram-positive and -negative bacteria, protozoa, yeasts, fungi and even viruses, but have remarkably low toxicity against normal mammalian cells. Nevertheless, antimicrobial peptides do share with the factors of adaptive immunity other important features, like that of being inducible by bacteria or their products (Diamond *et al.*, 1996; Mangoni *et al.*, 2001) (although they can also be, in some instances, constitutively expressed in healthy tissues), and the ability to distinguish self from infectious non-self.

Crafted by a long evolution in a remarkable diversity of structures, antimicrobial peptides from higher organisms are gene-encoded and synthesized by ribosomes, and the resulting primary translation product is normally processed further to the mature active peptide. Usually 12–50 amino acids in length, the many determined sequences of antimicrobial peptides show no clear homology, but share peculiar characteristics such as a large percentage of hydrophobic residues (usually around 50%), and an excess of basic amino acids (essentially arginine and lysine and, to a minor extent, histidine) over acidic amino acids, which gives a cationic character to the entire molecule with a net positive charge of +2 to +9 (Hwang and Vogel, 1998).

Most of the antimicrobial peptides are composed of L-amino acids, but there are several examples of coexistence of L- and D-amino acids in the same sequence, this feature giving rise to unique structures and properties (Mignogna *et al.*, 1998). Finally, some antimicrobial peptides have an irregular amino acid composition, showing a sequence which is rich in one or more specific amino acids that are usually not abundant in peptides or proteins. Relevant examples include the tryptophan-rich tritrpticin and indolicidin, histatins, a family of small peptides found in human saliva which are very rich in histidin and many proline-rich cathelicidins, members of large family of antimicrobial peptides (Epand and Vogel, 1999). Antimicrobial peptides have a variety of secondary structures, including well-defined α-helix, β-sheet (stabilized by 2–4 disulfide bridges), and both β-sheet and α-helix structures when in a membrane-like environment, whereas they can be random-structured in free solution. Smaller groups of peptides are composed by cyclic molecules with thioether groups in the ring, peptides terminating in an amino alcohol (peptaibols), macrocyclic cystine-knotted peptides and loop peptides formed by a single disulfide bridge (Hwang and Vogel, 1998; Epand and Vogel, 1999).

In contrast to the vast array of different structural features of antimicrobial peptides, all present an amphipathic character, with one surface being highly hydrophilic and the other hydrophobic. Stemming from this shared property, and according to several studies, a unifying vision of their primary mechanism of action has emerged and it is now widely accepted that most of them exert their lethal effect by perturbing the integrity of cell membranes of microorganisms (Boman *et al.*, 1994; McElhaney and Prenner, 1999). In addition to many experimental evidences that support this view, indications that antimicrobial peptides act on cell membranes can be derived by a purely speculative process. So, if one considers the diversity of the microbial targets which are susceptible of attack by peptides, it is apparent that the only feature they share is that they are all surrounded by a membrane. As expected for a factor of innate immunity, the interaction with membranes would be unspecific, and only driven by the hydrophobicity and cationic character of the peptide on one side, and by the physicochemical properties of the membrane itself on the other. Thus, the basis for necessary discrimination between host self and pathogen non-self cells seems to be the lipid composition of the target membrane, and the existence of a large transmembrane electrical potential in peptide susceptible organisms (Oren and Shai, 1998; Hancock and Chapple, 1999). A clear demonstration that antimicrobial peptides do not interact with chiral receptor molecules embedded in the lipid matrix of a membrane, but that instead direct peptide–lipid interactions play a major role in their function, came from the finding that analogs of peptides composed entirely of D-amino acids exhibited the same antimicrobial activity as their all-L parent molecules (Wade *et al.*, 1990). The existence of such a physical mechanism of action, based on cationic and hydrophobic interactions with selected membrane lipids, would also account for the high speed of action shown by most antimicrobial peptides when challenged in vitro with target microbes. Another advantage for the host of investing in antimicrobial peptides as defense weapons, is that, given the non-specificity of their mechanism, it is not easy for microbial pathogens to develop resistant mutants to overcome peptide intervention.

Although all evidences indicate that antimicrobial peptides act by permeabilizing the cell membranes of microorganisms, it must be stressed at this point that this is not the end of the story. Indeed, increasing evidence is accumulating that antimicrobial peptides might also act through a variety of mechanisms other than disruption of membrane integrity, and that their role in host defense goes well beyond direct killing of microorganisms. For example, stimulation of host defense-mechanisms by antimicrobial peptides is becoming apparent in several cases. Among these, dermaseptin, which was found to stimulate the release of myeloperoxidase and the production of reactive oxygen species by polymorphonuclear leukocytes (Ammar *et al.*, 1998), and defensins, that can enhance the recruitment of neutrophils to the infected tissue (Welling *et al.*, 1998), thus increasing the ability of host to resist local infections. Receptor-mediated signaling activities of some peptides have also been reported (Hoffman *et al.*, 1999; Yang *et al.*, 1999), and the idea that many antimicrobial peptides actually do act by entering the selected cells and binding to some intracellular targets such as DNA, interfering with their metabolic function, is rapidly growing among researchers. Another emerging view is that many peptides could act synergistically with other host molecules with antimicrobial activity to kill microbes. Positive cooperativity has been reported between peptides and lysozyme, and between different antimicrobial peptides (McCafferty *et al.*, 1999; Hancock and Scott, 2000). More in general, it is becoming clear that there exist a certain degree of coupling between the innate and adaptive immune systems, and that antimicrobial peptides have a large impact on the quality and effectiveness of immune and inflammatory responses. Although the present scenario is far from complete, excellent reviews are available that highlight recent examples of these intriguing themes (Hancock and Diamond, 2000; Hancock and Scott, 2000; Scott and Hancock, 2000).

The purpose of this review is to describe the recent progresses made in understanding the mechanisms of interaction of cationic peptides with biological membranes and artificial lipid bilayers on one side, and with lipid monolayers on the other one. In particular, we will focus on the structural and charge requirements of peptides for activity and on how this depends on the physicochemical properties of target membrane. Since the primary literature of interest for our topic is bewilderingly vast and often redundant, reference is made in many instances to recent, comprehensive reviews that may represent a suitable gateway to a more-in-depth perusal of specific areas.

Our discussion will be restricted to ribosomally synthesized peptides (and their synthetic analogs) with antimicrobial activity produced by higher eukaryotes, and will not address the important area of antimicrobial peptides produced by bacteria, either ribosomally or through non-ribosomal biosynthetic means, which often contain unusual amino acids that require extensive post-translational modifications. Recent advances in the latter field are covered by a number of good review articles, to which the interested reader is addressed (Sahl, 1994; Konz and Marahiel, 1999; Jack and Jung, 2000; Moffitt and Neilan, 2000; Sablon *et al.*, 2000).

Diversity of antimicrobial peptides and their synthetic analogs

As briefly outlined above, antimicrobial peptides are extremely common in nature, and it can be reasonably said that they have been found almost in each animal, plant or microorganism where they have been sought. The number of those already known exceeds several hundreds (a good collection of sequences, besides a wealth of other useful information, can be found at http://www.bbcm.univ.trieste.it/~tossi/pag1.htm), and many are discovered and characterized each year. Tigerinins, for example, are an entire novel class of very small

(11–12 residues), peculiar antimicrobial peptides recently isolated from the skin of the Indian frog *Rana tigerina* (Sai *et al.*, 2001). A comprehensive list and discussion of all known antimicrobial peptides falls beyond the scope of this review, and we will restrict ourselves to delineate some general features of these molecules and their synthetic analogs.

Antimicrobial peptides show a remarkable variety in their primary structure, but may usually be grouped into families on the basis of sequence similarity. It is not rare that many different peptides, sometimes belonging to distinct families, can be recovered from a single eukaryotic genus or even a single species. Anuran amphibians (toad and frogs), with their peptide-rich skins, are a good example (Simmaco *et al.*, 1998). The existence of such peptide assemblies on a single organism has been interpreted as a requirement for wide-spectrum antibacterial protection, that would be better achieved through the contemporary presence of many peptides with slightly different structures and thus antibiotic specificity (Simmaco *et al.*, 1998).

Diversity of antimicrobial peptides is not limited to their sequences, but extends to secondary structure. The application of solution- and solid-state NMR methods has greatly extended, in the last years, our knowledge of the three-dimensional structures of many peptides, both in solution and membrane bound, the latter information being particularly important for our understanding of peptide functioning (Bechinger, 1999; Bechinger *et al.*, 1999). These progresses have allowed the identification of several major structural classes: α-helical, the first to be identified and characterized; β-sheet peptides, single β-hairpins composed of about 20 residues and with 1–2 disulfide linkages; cystine-rich peptides, longer than the precedents and comprising 2–3 β-strands stabilized by several disulfide bridges; peptides with an unusual composition of regular amino acids. Table 9.1 describes the main structural features of some well-studied antimicrobial peptides originating from a wide range of sources, selected as representative examples of their structural class. This classification follows that proposed by Hwang and Vogel (1998), but we want to make clear that the features distinguishing one class from the other are sometimes not so sharply delineated, as in the case of the β-sheet and cystine-rich peptides, and that other grouping schemes have been reported (Hancock and Lehrer, 1998). In addition, there are smaller groups of peptides, some of them cited in the introduction to this book chapter, which cannot be comprised in any of the major structural classes defined above.

As previously reported, the diversity of folding patterns of antimicrobial peptides is counterbalanced by a surprising convergence to a common three-dimensional configuration, with opposing hydrophobic and polar/cationic faces. The hydrophilic/cationic face is believed to initiate peptide interaction with the negatively charged headgroups of membrane lipids, while the hydrophobic side would permit the peptides to enter the membrane core. Recently, the existence of another type of common three-dimensional configuration for antimicrobial peptides has been claimed, being a sort of "cationic double-wing structure with two regions of positive charge bracketing a hydrophobic core" (Hancock and Scott, 2000).

Starting from the reference templates offered by many naturally occurring antimicrobial peptides, researchers have focused their efforts on the structure–function relationships of synthetic analogs, mostly of α-helical peptides. In recent years, this approach has been extensively exploited, thanks to the facile gateway to custom peptide synthesis offered by modern automated solid-phase methods, to produce analogs with modified primary and/or secondary structure. A plethora of sequence analogs of cecropins, magainins and melittin, only to quote some examples involving well known peptides, have been synthesized and tested as for their antimicrobial and/or cytotoxic properties (Andreu *et al.*, 1985; Blondelle and Houghten, 1991; Maloy and Kari, 1995). Peptides that differ at one or more sequence

Table 9.1 Main characteristics of some representative antimicrobial peptides

Class	*Peptide*	*Sequence*	*Source*
α-helical	Melittin	GIGAVLKVLTTGLPALISWIKRKRQQ-Am	*Apis mellifera* (honey bee venom)
	Magainin 2	GIGKFLHSAKKFGKAFVGEIMNS	*Xenopus laevis* (clawed toad skin)
	Temporin A	FLPLIGRVLSGIL-Am	*Rana temporaria* (European red frog skin)
	Cecropin A	KWKLFKKIEKVGQNIRDGIKAGPAVAVVGQATQIAK-Am	*Hyalophora cecropia* (moth)
β-sheet	Protegrin-1	$RGGRLC_1YC_2RRRFC_1VC_2VGR$-Am	Porcine leukocytes
	Tachyplesin-1	$KWC_1FRVC_2YRGIC_2YRRC_1R$-Am	*Tachypleus gigas* (Asian horseshoe crab)
	Polyphemusin-1	$RRWC_1FRVC_2YRGFC_2YRKC_1R$-Am	*Limulus polyphemus* (Atlantic horseshoe crab)
	Androctonin	$RSVC_1RQIKIC_2RRRGGC_2YYKC_1TNRPY$	*Androctonus australis* (scorpion hemolymph)
Cystine-rich	Drosomycin	$DC_1LSGRYKGPC_2AVWDNETC_3RRVC_4KEEGRSSGHC_2SPSLKC_3WC_4EGC_1$	*Drosophila melanogaster* (fruit fly)
	Hamster defensin (HANP-1)	$VTC_1FC_2RRRGC_3ASRERHIGYC_2RFGNTIYRLC_3C_1RR$	Hamster neutrophils
	Human β-defensin 1	$DHYNC_1VSSGGQC_2LYSAC_3PIFTKIQGTC_2YRGKAKC_1C_3K$	Several human tissues
	Sapecin B	$LTC_1EIDRSLC_2LLHC_3RLKGYLRAYC_1SQQKVC_2RC_3VQ$	*Sarcophaga peregrina* (flesh fly)
Unusual composition	Indolicidin	ILPWKWPWWPWRR-Am	Bovine neutrophils
	Histatin-5	DSHAKRHHGYKRKFHEKHHSHRGY	Human saliva
	Bactenecin-5	RFRPPIRRPPIRPPFYPPFRPPIRPPIFPPIRPPFRPPLGPFP	bovine neutrophils
	PR-39	RRRPRPPYLPRPRPPPFFPPRLPPRIPPGFPPRFPPRFP-Am	Pig intestine

Note
-Am indicates an amidated C-terminus; subscript numbers represent amino acids that are joined by disulfide bridges.

positions, are shortened or contain deletions, contain all-D amino acids, or are hybrids of two different natural peptides (like many cecropin–melittin peptides), all of these have been designed, synthesized and extensively studied by many research groups worldwide.

Temporins represent another recent example of peptides isolated from a natural source that have been the object of synthetic analogs studies. Short (10–13 residues long) peptides isolated from the skin of *Rana temporaria*, a red frog found in many regions of Central Europe, temporins are among the smallest of the antimicrobial peptides isolated so far from amphibian skin, and in general from animal sources (Simmaco *et al.*, 1996). Temporins were found to be active against Gram-positive bacteria and *Candida albicans*, but to be generally non-toxic to human erythrocytes (Simmaco *et al.*, 1996; Mangoni *et al.*, 2000). Very recent data, acquired by means of several biophysical techniques, started to shed light on the modes of interaction of temporins with model lipid membranes (Rinaldi *et al.*, 2001, Zhao *et al.*, 2001b), suggesting that they are likely to be membrane active peptides also in vivo. Aiming at improving the activity of a member of temporin family, namely temporin A, against clinically important bacterial strains, synthetic analogs were synthesized and tested against *Staphylococcus aureus* and vancomycin-resistant enterococcal strains (Harjunpää *et al.*, 1999; Wade *et al.*, 2000). Results obtained showed, among other things, that not only the hydrophobic/hydrophilic character of a given residue is important in retaining antibacterial activity, but also that the size of the amino acid side chain matters. In the case of temporin A, indeed, residues with bulky hydrophobic side chains (such as isoleucine and/or leucine) must be present at positions 5 and 12, and substitutions of alanine for the same positions were totally devoid of bactericidal activity (Wade *et al.*, 2000).

Instead of working on synthetic analogs of naturally occurring peptides *tout court*, other investigators analyzed their sequences and extracted sequence regions that may contribute to either antimicrobial and hemolytic activity, and thereafter synthesized peptide analogs according to these patterns. A remarkable example of this approach is the extensive work performed by Sitaram and colleagues on the synthetic peptides, and their analogs, corresponding to various putative α-helical segments of seminal plasmin (SPLN), a 47-residue protein isolated from bovine seminal plasma (reviewed in Sitaram and Nagaraj, 1999). The synthesized SPLN fragments, varying in length and sequence, showed variable biological activity compared to the entire natural peptide, with altered bactericidal specificity and hemolytic activity. In particular, the 13-residue peptide SPF (PKLLETFLSKWIG), corresponding to the most hydrophobic segment of SPLN, was selected for further study and several substitution, addition and omission analogs (including a couple of cyclic ones) prepared, modulating charged, polar and hydrophobic residues. The overall peptide charge and position of charges, hydrophobicity, hydrophobic moment and conformation were all found to affect SPF biological activity to some extent (Bikshapathy *et al.*, 1999a,b).

Alternative methodologies have been explored for studying artificial antimicrobial peptides. One is the use of conformationally defined combinatorial libraries, successfully applied to the development of novel α-helical antimicrobial peptides (Blondelle and Houghten, 1996). Another method is the design of synthetic antimicrobial peptides starting from the comparison of the sequences of known, naturally occurring peptides. Following this methodology, Tossi and coworkers analyzed the sequences (considering only the first 20 residues) of 85 peptides from various natural sources, and calculated the frequency for each residue at each position (Tossi *et al.*, 1997). From these data, authors obtained a sequence template that directed the synthesis of four α-helical peptides, each displaying a potent antimicrobial activity against selected Gram-positive and -negative bacteria (Tossi *et al.*, 1997).

Many other studies have systematically investigated, by means of *de novo* and *ad hoc* synthesized peptides, the role of peptide structural properties in membrane interactions. A short synopsis of these investigations is offered below.

Biophysical parameters and activity of α-helical peptides

As seen in the previous section, antimicrobial peptides are divided into some groups on the basis of their structure and activity spectrum (Hwang and Vogel, 1998). The larger group includes molecules containing positive residues along their length, so creating an amphipathic condition: these peptides form amphipathic α-helices when interacting with lipid bilayers. Aimed to explain peptide selectivity and activity on model and biological membranes, several authors have focused on some biophysical parameters of selected peptides, which have been analyzed by a range of experimental techniques (Blondelle *et al.*, 1999; Dathe and Wieprecht, 1999; Epand and Vogel, 1999).

The first key parameter is helicity, which characterizes both smaller peptides and peptide molecules with more complex structures (Kourie and Shortouse, 2000). Many studies performed using analogs of natural peptides, such as a work on a synthetic peptide composed of leucine and lysine residues designed by Cornut and coworkers (Cornut *et al.*, 1994), or the several works on magainin 2 analogs (Chen *et al.*, 1988; Ohsaki *et al.*, 1992), emphasized the importance of the α-helix structure in conferring cytotoxicity to the peptide of choice. More recently, titration calorimetry studies confirmed that the α-helix formation is a strong driving force for the binding to the membranes (Wieprecht *et al.*, 1999). The role played by this parameter modulating the peptide–membrane interaction has been further explored by several investigations. For example, Wieprecht and colleagues reported that the substitution of two neighboring amino acids in magainin by their D-enantiomers, causing a decrease in helicity but no modification in other structural parameters, reduces the activity against neutral and moderately negatively charged membranes, whereas the effects seen on the permeability of highly negatively charged membranes are less significant (Wieprecht *et al.*, 1996). Other data obtained using the amphipathic model peptide KLAL (KLALKLALKALKAAKLA–NH_2) confirmed that the degree of helicity plays a negligible role in the binding between a given peptide and a very negatively charged bilayer, whereas a strong modulation is observed on membranes with reduced negative charge (Dathe *et al.*, 1996).

All these results seem in agreement with the effects observed on erythrocytes, because a decrease in helicity of KLAL analogs correlates with a remarkable reduction in hemolytic activity, whereas both Gram-negative and -positive bacteria show to be not much sensitive to a trouble in this parameter (Dathe *et al.*, 1996). Nevertheless, more recent data concerning some synthetic small peptides (10 amino acid long) asserted the necessity to widen these considerations, since helicity seems to condition the antimicrobial activity of these peptides in a significant way (Oh *et al.*, 1999).

Peptide charge is another key factor. Most cytotoxic peptides are positively charged thanks to the presence of lysine and arginine residues (and, to a lesser extent, histidine) in their sequences. The net charge of these molecules usually ranges from +2 to +9 and can vary with pH as a result of the ionization state of various residues. The role of this parameter for the peptide–membrane interaction is documented by many studies on both natural peptides and synthetic analogs. First of all, native melittin and magainin interact with membranes in a manner revealing their different structures. Melittin, characterized by a hydrophobic N-terminus and a charged C-terminus, interacts with both neutral and negatively charged bilayers, showing both hemolytic and bactericidal activity. Magainin does not have a hydrophobic region along its chain, and preferentially interacts with negatively charged bilayers and bacteria (Dathe and Wieprecht, 1999). These observations are in agreement with many results obtained on synthetic peptides. Dhople and Nagaraj suggest that the substitution of some aspartic acids by lysines, makes an α-helical segment of δ-toxin a potent antimicrobial and hemolytic peptide (Dhople and Nagaraj, 1995), and Matsuzaki and colleagues found that a loss of positive charges reduces the antimicrobial activity of magainin, whereas the hemolytic activity is less affected (Matsuzaki *et al.*, 1997).

Nevertheless, these electrostatic interactions may inhibit the permeabilizing efficiency of a given peptide, when they overcome some limit. In fact, an increased negative membrane charge can strongly bind cationic peptide molecules and hamper their penetration into the membrane, as seen for KLAL model peptide (Dathe *et al.*, 1996), magainin 2 amide (Wieprecht *et al.*, 1997a) and melittin (Benachir and Lafleur, 1995), but also an increase in positive peptide charge can produce the same effect (Matsuzaki *et al.*, 1997).

Hydrophobicity is a measure of the ability of a peptide to move from an aqueous to a hydrophobic phase, so providing other data on the feature of the interaction between a lipid bilayer and a given peptide. This parameter measures the degree of peptide affinity for the lipid acyl chains at the core of model and biological membranes. Most data are in agreement to associate an increase in hydrophobicity with an enhanced permeabilizing activity on zwitterionic bilayers, and therefore on erythrocytes. This is evident in the studies done on analogs of magainin 2 amide (Dathe *et al.*, 1997; Wieprecht *et al.*, 1997a). The permeabilizing effects on lipid vesicles bearing high content of zwitterionic phosphatidylcholine increase due to some substitutions in the peptide sequence which enhance its hydrophobicity, while other structural parameters remain constant, and the hemolytic activity shows the same correlation. Moreover, these investigations also emphasize that hydrophobicity plays a secondary role in the interaction between peptides and negatively charged bilayers. At the same time, antimicrobial activity is not much influenced by this parameter, and besides it seems to present a negative correlation with the antibacterial specificity. All these observations have been confirmed by other studies on brevinin 1E amide (Kwon *et al.*, 1998) and on some synthetic amphiphilic α-helical peptides (Kiyota *et al.*, 1996).

Hydrophobic moment is a measure of the asymmetry of peptide hydrophobicity and it is defined as the vector sum of the hydrophobicity of each amino acid in a given protein or peptide (Eisenberg, 1984). The numeric value of the hydrophobic moment is small for the helices where the residues are evenly distributed and large for the helices which present most hydrophobic residues on one face and most hydrophilic residues on the other face.

Some investigations reported by Wieprecht and colleagues suggest that the hydrophobic moment influences much more the activity on neutral model membranes than on negatively charged bilayers (Dathe *et al.*, 1997; Wieprecht *et al.*, 1997c). An increase of this parameter in magainin 2 analogs causes an enhanced permeabilizing efficiency on phosphatidylcholine-rich bilayers and on bacteria and erythrocytes, whereas the negatively charged model membranes are less influenced. This fact determines two results: first, hydrophobic interactions with acyl chains of lipids dominate the activity on neutral membranes and second, this parameter influences in a negative manner the antimicrobial specificity. Nevertheless, this last point of view is not in agreement with more recent data obtained by a C-terminal fragment of melittin (Subbalakshmi *et al.*, 1999). Few substitutions done to enhance the hydrophobic moment of this peptide, and consequently the extent of its hydrophobic face, cause a more remarkable increase in antibacterial activity compared to hemolytic activity, thus providing important directions to the design of small peptides with selective antimicrobial activity.

Last, but not least, we should finally consider a biophysical parameter that expresses the distribution of hydrophilic and hydrophobic residues along an α-helix, that is, the angle subtended by the hydrophilic or hydrophobic helix face. This was found to be a very important factor in the overall peptide–bilayer interaction, because it emphasizes the differences between binding and permeabilization. In the last years, this aspect was elucidated by several investigations on model peptides designed from magainin 2 amide which underlined that the value of the hydrophilic angle (subtended by the positively charged helix face) correlates inversely with peptide–membrane affinity and membrane permeabilization (Wieprecht *et al.*, 1997b). These synthetic peptides, having a different polar angle but no other structural modification, show that the binding is favored by a large angle and by a negatively charged bilayer, whereas the permeabilization is enhanced by a small angle and by a low negative membrane charge. Moreover both antimicrobial and hemolytic activity present a positive correlation with an increase of the angle. Nevertheless, these and other data show that this same increase causes a decrease in specificity against bacteria compared to erythrocytes (Dathe *et al.*, 1997; Wieprecht *et al.*, 1997b). These observations are partly confirmed by more recent investigations on two model peptides, θp100 and θp180, which differ only in their polar angles (Uematsu and Matsuzaki, 2000). Indeed, the binding affinity of θp180 (with a larger positive angle) for a negatively charged bilayer, is only slightly larger than that of θp100, whereas the latter is more effective in permeabilizing the same bilayer, thus confirming the positive correlation between a small polar angle and the ability of a given peptide to penetrate in a lipid bilayer to form a pore. Instead, the data on hemolytic and antimicrobial activity in this case do not seem completely in agreement with the previous observations (Uematsu and Matsuzaki, 2000), and that emphasizes the complexity of peptide interaction with a biological membrane.

In conclusion, we can assert that each peptide structural parameter here analyzed is of some importance in modulating the interaction of the peptide itself with lipid membranes under different conditions, and that a given balance between these factors is most certainly crucial in tuning the modes of membrane activity and selectivity of antimicrobial peptides.

Membrane phospholipids and model membranes

The antimicrobial peptides are considered to kill bacteria by permeabilizing and/or disrupting bacterial membranes and they are found to target the lipid matrix rather than the proteins (Bessalle *et al.*, 1990; Wade *et al.*, 1990; Maget-Dana, 1999). These peptides are generally able to interact with the negatively charged headgroups of the bacterial membrane phospholipids and the incorporation of these "aliens" molecules causes alteration in the bilayer leading to membrane perturbation (Wu *et al.*, 1999). It is reported that not only the nature of the peptide but also the characteristics of cell membrane, as well as the metabolic states of the cells, determine the antimicrobial mechanism of a peptide (Liang and Kim, 1999). Nevertheless the influence of the membrane structure of a given organism sensible to the action of the peptides, remains an unsolved question. Although the majority of the antimicrobial peptides have high toxicity in bacterial membranes without harming differentiated mammalian cells, some of them show effects also on the latter either inducing lysis in erythrocytes and other eukaryotic cells (Cruciani *et al.*, 1991), or stimulating hematopoietic cell funcions in polymorphonuclear leukocytes (Ammar *et al.*, 1998).

Many of these self-defense peptides act on cytoplasmic membrane of microorganisms with a physical mechanism based on cationic and hydrophobic interactions with membrane lipids (for review, see Blondelle *et al.*, 1999) that would account both for the high speed of action shown by most of such peptides when challenged in vitro with target microbes and for the non-specificity of the challenge. In particular, the peptides can bind the polyanionic lipopolysaccharide (LPS), a major component of the outer membrane of Gram-negative bacteria, as well as to other negatively charged lipids which would favor the association of cationic peptides. In fact, phosphatidylglycerol (PG), phosphatidylserine (PS) and cardiolipin (CL) are more abundant in bacterial outer membranes than in the surface of mammalian cells in which zwitterionic lipids, phosphatidylcholine (PC) and sphingomyelin (SM), are predominant (Table 9.2). It is therefore, the electrostatic association between cationic peptide and anionic lipids that alters the physical structure of the membrane favoring its permeability to normally impermeant hydrophobic molecules (Vaara, 1992). Indeed, inhibition of such association by divalent cations is found to reduce the antimicrobial activity of many peptides like defensins (Lehrer *et al.*, 1993) and magainins (Matsuzaki *et al.*, 1999).

The interactions between an antimicrobial peptide and membranes, can be studied either by an in vivo approach using intact cells or, alternatively, by extensive use of the in vitro experiments with isolated or model membranes. However, natural membranes are very complex structures with a great variety of lipids and proteins interactions, thus studies carried out in whole cells allow characterization only of the global membrane damage phenomena. Hence most of our knowledge of the significance of specific lipid–peptide interactions derived from studies on model membranes. Among them liposomes, lipid vesicles of different size and lipid composition, represent the most promising tool. Since the 1980s several methods have been developed for the generation, from a lipid mixture, of closed lipid bilayers. These include, among others, the extrusion through polycarbonate membranes (Olson *et al.*, 1979), reverse phase evaporation (Szoka and Papahadjopoulos, 1978), detergent dialysis (Zumbuehl and Weder, 1981), ethanol injection (Batzri and Korn, 1973), together with the classical rehydration of dried lipid films. Serious drawbacks present in all the vesicle preparations prevent the definition for the liposomes of perfect membrane lipid mimic sytems – for example, injection of lipid ethanol, rehydration and reverse-phase evaporation give rise to vesicles with large size distribution; detergent dialysis leaves residual detergent that can perturb membrane properties and affect peptide action; the extrusion method is considerably time-consuming. The liposomes are commonly classified, on the basis

Table 9.2 Percentage of the main phospholipid components of representative cells

	Zwitterionic lipids			*Negatively charged lipid*		
	PE	PC	SM	PS	PG	CL + lysoPG
Escherichia coli (cytoplasmic membrane)	82	—	—	—	6	12
Salmonella typhimurium	60	—	—	—	33	7
Pseudomonas capacia	82	—	—	—	18	—
Staphylococcus aureus	—	—	—	—	57	43
Candida albicans	70	4	15	11	—	—
Candida neoformans	29	51	—	16	—	4
Erythrocyte (outer layer)	—	50	50	—	—	—

Notes
PE, phopshatidylethanolamine; PC, phosphatidylcholine; SM, sphingomyelin; PS, phosphatidylserine; PG, phosphatidylglycerol; CL, cardiolipin.

of their mean diameter and number of lipid layers (Winterhalter and Lasic, 1993), in: small unilamellar vesicles (SUVs, diameter less than 50 nm), large unilamellar vesicles (LUVs, diameter between 50 and 150 nm), multilamellar vesicles (diameter from 200 to 800 nm), giant liposomes (diameter of 10 μm or more). However, due to the better balance between optical interferences and lipid membrane curvature and packing, the SUVs and LUVs resulted the most suitable models for permeability studies (Eytan, 1982; Matsuzaki *et al.*, 1989; Ladokhin *et al.*, 1997; Kang *et al.*, 1998; El Jastimi *et al.*, 1999; Epand *et al.*, 1999). Using these models was possible to follow the initial steps of peptide–lipid interactions. By EPR measurements were investigated the membrane affinity of the peptide named helix5, both as a function of its structural context and bilayer depth (Russell *et al.*, 1999), as well as the effect on the bilayer fluidity induced on liposomes of various lipid composition by temporins (Rinaldi *et al.*, 2001). Furthermore, by oriented CD, it was possible to assess that magainin adopts an alpha-helical conformation with two distinct orientations when interacting with a lipid bilayer. At low concentrations, magainin is absorbed parallel to the membrane surface; at high concentrations instead, it is inserted into the membrane showing a cooperative effect in the insertion (Ludke *et al.*, 1994). These vesicles were also very useful in fluorometric studies for the correlation between the membrane binding of some temporin family peptides with the induced leakage of liposome entrapped molecules. The results of this study demonstrate that although the lytic activity of temporins is not greatly affected by the membrane composition of liposomes: (i) temporins A and B allow the leakage of large-size molecules from the bacterial cells; (ii) temporin H renders both the outer and inner membrane of bacteria permeable to hydrophobic substances of low molecular mass; and (iii) temporin D, although devoid of antibacterial activity, has a cytotoxic effect on erythrocytes. (Mangoni *et al.*, 2000). In a recent review (Nir and Nieva, 2000) leakage from liposomes, induced by several peptides, are discussed and a pore model is described. In this model, using GALA as inducing peptide, was possible to predict either the persistance of the pores as a function of the peptide–lipid molar ratio and the effect of membrane composition on pore formation. Other aspects of the antimicrobial peptide–lipid interaction, namely the correlation between peptide charge and antimicrobial activity or membrane surface charge and peptide affinity, were also underlined using these membrane models (Dathe *et al.*, 1996). In this study, an amphipathic peptide was used to analyze the process of peptide binding at lipid vesicles of different surface charge and to determine the structure of the lipid-bound peptides using CD spectroscopy. The relationship between peptide helicity, model membrane permeability, and biological activity has been studied and investigation of antibacterial and hemolytic activity were carried out.

In specific cases the peptide–membrane interactions were also studied in the so-called giant liposomes (Thieffry *et al.*,1992; Holopainen *et al.*, 2000). These vesicles which are spherical, closed molecular bilayers and entrap an aqueous compartment (Luisi and Walde, 2000), are good models for studies of undulation, budding, wounding, healing and other manifestations of cell-like behavior induced by membrane-acting reagents (Riquelme *et al.*, 1990; Menger and Lee, 1995; Menger and Angelova, 1998; Menger and Keiper, 1998; Angelova and Tsoneva, 1999; Angelova *et al.*, 1999; Holopainen *et al.*, 2000; Zhao *et al.*, 2001a,b). Due to their large size, giant vesicles can be observed by light microscopy, allowing the direct visualization of processes involving supramolecular chemistry and biophysics (Luisi and Walde, 2000).

Another membrane model is the monomolecular film at the air–water interface or monolayer (Brockman, 1999). This last structure overcome most of the liposomes disadvantages showing at the same time a direct thermodynamic relationship with bilayer membranes (Feng, 1999). The interest of phospholipid monolayers is due to their homogeneity, their stability and their planar geometry, where the lipid molecules have specific orientation (at an air–water interface, the heads of phospholipids point toward the water subphase and the acyl chains toward the air). Also the two dimensional molecular density and the ionic conditions of the subphase can be varied, as well as the temperature and composition. Since the amphipathic character of antimicrobial peptides makes them surface-active products and their biological activity begins at lipid membrane interface, the monolayer technique is particularly suitable to study their physicochemical and biological properties. The monolayer technique has proved to be a very useful approach for the study of the surface activities of peptides at hydrophobic–hydrophilic interfaces and, the interaction of peptides with phospholipids at the air–water interface provides a unique model in understanding the insertion mechanisms of antimicrobial peptide into cell membranes (for a review, see Maget-Dana, 1999).

Penetration of antimicrobial peptides into lipid monolayers

The principle of selectivity and differing levels of activity of antimicrobial peptides with cell membranes are not well understood, nor is the exact mechanism of action. The lipid specificity of antimicrobial peptides has been extensively studied with lipid bilayers as well as lipid monolayers (for reviews, see McElhaney and Prenner, 1999). A large number of antimicrobial peptides are cationic and preferentially bind to anionic lipids (Maget-Dana and Ptak, 1997; Matsuzaki, 1998; Zhang *et al.*, 2000). This can provide a potential mechanism for microbial specificity, since most of the anionic lipids of mammalian membranes are sequestered on the cytoplasmic side of the membrane while they are exposed to external medium with microbial membranes (Op den Kamp, 1979). Penetration of antimicrobial peptide into lipid monolayers can provide the affinity of the peptide for different lipids. The interactions can be measured by two ways. One is keeping the surface pressure constant and observing the increase in area of the film (Seelig, 1987). Another is the most widely used method of maintaining the film area constant and measuring the surface pressure changes upon addition of compounds into the subphase (Pethica, 1955; Wiedmer *et al.*, 1978). Molecules that interact only with the headgroups of monolayer lipids typically induce minor changes in surface pressure. In contrast, insertion into the hydrophobic region of lipid monolayers causes a significant increase in surface pressure. Thus, the degree of surface pressure change ($\Delta\pi$) can be used to resolve whether peptide–membrane interactions include insertion and disturbance of the fatty acyl core of the membrane upon addition of peptides into the subphase.

In a recent study, the lipid specificity of polyphemusin I and its variants was monitored by the extent of the surface pressure change upon addition of these peptides to the subphase of different lipid monolayers (Zhang *et al.*, 2000). These peptides effectively penetrate into hydrophobic portion of negatively charged phospholipid monolayers and the insertion is more effective on PG than on CL monolayers. However, none of the peptides are able to penetrate monolayers composed of neutral lipids such as 1-palmitoyl-2-oleoyl-*sn*-glycero-3-phosphocholine (POPC) or 1-palmitoyl-2-oleoyl-*sn*-glycero-3-phospho-ethanolamine (POPE). A cooperative interaction of these peptides with monolayers containing negatively charged phospholipids is suggested due to sigmoidal increases in surface pressure as a function of peptide concentration. Similar to the lipid specificity of polyphemusin I and its variants, the extent of penetration by defensin A is higher in monolayers made of anionic phospholipids than in monolayers constituted of zwitterionic phospholipids (Maget-Dana and Ptak, 1997). Studies of colicin A insertion into lipid monolayers composed of different phospholipids also show that the peptide interacts preferentially with negatively charged phospholipids although it is able to penetrate into many phospholipid monolayers (Pattus *et al.*, 1983). When protegrin-1, an 18 amino acid peptide isolated from pig leukocytes is injected underneath dipalmitoylphosphatidylglycerol (DPPG), dipalmitoylphosphatidylethanolamine (DPPE) and dipalmitoylphosphatidylcholine (DPPC) films, the peptide readily penetrates into a DPPG monolayer while not penetrating into DPPE and DPPC monolayers at surface pressure of 20 mN/m (Gidalevitz *et al.*, 2000).

The studies mentioned above emphasize the importance of cationic charges of antimicrobial peptides for their preferential binding to bacterial membranes. However, because of the sensitive balance of electrostatic and hydrophobic peptide–membrane interactions, a simple correlation between the amount and distribution of cationic residues and antimicrobial specificity does not exist (Kiyota *et al.*, 1996). Penetration of peptides into phospholipid monolayers residing at the air–water interface not only depends on the charge property of lipids but is also highly dependent on the physical state of lipids, the ionic strength and pH value in the subphase (Hendrickson *et al.*, 1983; Maget-Dana and Ptak, 1997; van Kan *et al.*, 2001). Increased penetration of melittin into DPPC and DPPG monolayers was found at the L-E-L-C phase transition region (Hendrickson *et al.*, 1983). The penetration of defensin A into PS monolayers is decreased by the presence of Ca^{2+} in the subphase, while such an effect is not observed for CL monolayers, suggesting that the nature of the phospholipid polar head group plays a role in the interaction with this peptide (Maget-Dana and Ptak, 1997). Further investigations focused on the effects of peptide sequence, concentration, nature and packing of the phospholipids and the condition of the subphase (pH, salt concentration) on the penetration of clavanins into lipid films (van Kan *et al.*, 2001). Clavanins are histidine-rich, amidated α-helical antimicrobial peptides isolated from the hemocytes of the tunicate *Styela clava*, broadly effective against Gram-positive and -negative bacteria (Lee *et al.*, 1997). Wild-type clavanin A was found to insert into different lipid monolayers in a concentration-dependent manner and readily inserts at initial surface pressures near the physiological range of 30–35 mN/m (Demel *et al.*, 1975), at both pH 5.5 and 7.4. However, the insertion rate was reduced upon increasing pH and the maximal extent of monolayer insertion of clavanin A hardly depended on the pH in the range between 5.5 and 7.7. Considering this pH dependence and the essentially identical insertion behavior into both negatively charged and neutral phospholipids, it can be concluded that the hydrophobic interaction of this peptide with lipids dominates and the state of charge of the histidines does not play a major role in the clavanin–lipid interactions. Consistently, minor effects of ionic strength in the subphase as well as of the primary sequence of the peptides suggest the importance of the hydrophobic forces for insertion of the clavanin derivatives.

Monolayers of peptides and mixtures of peptides and phospholipids

Because of their amphipathic properties, membrane-active peptides can generally be expected to exhibit surface activity. A number of antimicrobial peptides have been found to form stable monolayers at the air–water interface and their interfacial properties have been extensively studied (Fidelio *et al.*, 1982; Tournois *et al.*, 1989; Signor *et al.*, 1994; Wackerbauer *et al.*, 1996; Maget-Dana and Ptak, 1997). Peptides are generally spread by carefully adding a small amount of a concentrated aqueous solution or diluting the aqueous peptide solution with a volatile solvent miscible with water at the air–water interface.

The main parameters characterizing a film state of a given substance on an aqueous subphase are temperature, surface pressure, surface area and the amount of molecules. The observed surface pressure plotted versus the average area per molecule, based on the total amount of peptides added to the interface, results in the conventionally evaluated pressure-area isotherm and are often carried out by means of a Langmuir film balance (Adamson, 1976). The behavior of antimicrobial peptides at interfaces is suggested to be mainly determined by their predominant secondary structures (Maget-Dana *et al.*, 1999). Comparative pressure-area isotherm characteristics of model peptides in α-helical, β-sheet and random coil structures indicate that β-sheet peptides form stable monolayers in a condensed state which collapse at a high pressure, while monolayers of α-helical peptides are more compressible and less stable (Maget-Dana *et al.*, 1999). Defensin A, an antimicrobial peptide isolated from the larvae of the flesh fly *Phormia terranovae*, is only partially unfolded at the air–water interface due to the presence of the three disulfide bridges (Maget-Dana and Ptak, 1997). Monolayers of gramicidins and their mutants at the air–water interface showed a phase transition behavior which might relate to cluster formation, reorientation of molecules, conformational changes or multilayer formation (Tournois *et al.*, 1989). The surface area characteristics were slightly affected when a charged group on either NH_2- or COOH-terminus was coupled or the length of the peptide was increased by two amino acids. However, modification of tryptophans or even replacement of a single tryptophan by phenylalanine caused a drastic change in the surface-area characteristics and a partial loss of the phase transition, suggesting an important role of tryptophan in the interfacial behavior of gramicidin (Tournois *et al.*, 1989). Yet, the surface activity of antimicrobial peptides is affected not only by their amphipathic structure, molecular size and net charge, but also by the composition of lipid monolayers and their physicochemical state, as well as by the conditions of the subphase (for a review, see Maget-Dana, 1999). For example, the interfacial behavior of defensin A is significantly affected by pH and by the ionic strength in the subphase, and a bilayer organization of this peptide film was suggested on acidic subphases (Maget-Dana and Ptak, 1997). A cyclic peptide, bacteriocin AS-48 produced by *Enterococcus faecalis* S-48 has a broad inhibitory spectrum against Gram-positive and-negative bacteria. This peptide forms a stable monolayer at the air–water interface and the interfacial behavior of bacteriocin AS-48 as a function of pH explains its biological activity (Abriouel *et al.*, 2001). A novel approach has been presented to analyze Langmuir monolayer experiments using melittin, offering a definite means of determining the extent of partitioning between a monomolecular film at the air–water interface and its aqueous subphase (Wackerbauer *et al.*, 1996). The data suggest the existence of two structural conversions in the course of an increasing lateral compression, which is corroborated by surface potential measurements reflecting corresponding changes of the effective dipole moment perpendicular to the surface.

To further understand peptide–lipid interactions mixed peptide and phospholipid monolayers have been applied and the quantitative analysis of the molecular interactions in mixed monolayers provide important information on the molecular mechanisms underlying peptide-mediated cell lysis. The driving forces of peptide–lipid association are electrostatic interactions between positively charged residues of the peptides and negatively charged headgroups of the lipids, as well as the hydrophobic interactions between non-polar amino acids and hydrophobic core of the membrane. The hydrophobicities of the peptides are generally too low to effectively associate with zwitterionic phospholipids and thus prevent toxicity against the host cells. In contrast, the positive charges of antimicrobial peptides enable selective binding to bacterial membranes containing negative charges on the outer leaflet of the membrane and thus disruption of the bacterial membranes (for a review, see Matsuzaki, 1999). The interaction of bacteriocin AS-48 with dipalmitoyl phosphatidic acid (DPPA) suggest that the main interaction of this peptide with lipids is of electrostatic nature, since only when the lipid is charged enough and the peptide is unfolded an interaction between them was observed (Abriouel *et al.*, 2001). Maget-Dana and Ptak have studied the defensin A behavior in mixed films with two natural phospholipids, egg lecithin and bovine brain PS (Maget-Dana and Ptak, 1997). The compression isotherm of mixed defensin A–phospholipid monolayers and calculated excess free energy of mixing as a function of monolayer composition suggest that defensin A interacts with phospholipids by forming 1 : 9 complexes (Maget-Dana, 1999). These complexes are not miscible in the lipid phase and induce microheterogeneity in the lipid membrane. These clusters might be related to the ion-channel structures responsible for the biological activity of defensin A. Similarly, formation of melittin-1,2-dimyristoyl-*sn*-glycero-3-phosphocholine (DMPC) complexes with a 1 : 24 stoichiometry was suggested from the area-composition plot of melittin in mixed monolayers while the interaction of melittin with the negatively charged ganglioside GM_1 forms 1 : 3 melittin-GM_1 complexes (Fidelio *et al.*, 1986). The behavior of gramicidins incorporated into lipid monolayers show that these peptides are not miscible with dioleoylphosphatidycholine and their orientation depends on surface pressure, namely lipid packing density (Van Mau *et al.*, 1988). Surface potential measurements suggest the existence of a relationship between the single channel characteristics and the surface potential. Tryptophan residues of gramicidin A and M are indicated to be essential for lowering the lipid surface potential in agreement with the single channel behavior of both peptides (Van Mau *et al.*, 1988). Infrared reflection adsorption spectroscopy (IRRAS) studies of monolayers of gramicidin S mixed with phospholipids indicate that this peptide penetrates more deeply into the zwitterionic lipid monolayers in the liquid-expanded state as compared to the liquid-condensed state (Lewis *et al.*, 1999). The insertion of this peptide depends on the lateral surface pressure and some gramicidin S remains inserted into the negatively charged phospholipid monolayers whereas the peptide is squeezed out of the zwitterionic monolayers at high surface pressure.

Depending on the temperature, the subphase, and the particular phospholipids applied, the mixed peptide–phospholipid monolayers can be homogeneous or display phase separation. This behavior was observed with a membrane-active peptide, bombolitin III, by means of the Langmuir film coupled with epifluorescence microscope (Signor *et al.*, 1994). In well-defined conditions the formation of phase-separated peptide domains at the air–water interface and the effect of their presence on the organization of the lipids were visualized. The bombolitins were found to significantly perturb the structure of phospholipid monolayers, forming phase separated peptide domains.

Conformation and orientation of peptides at the air–water interfaces

Less is known about the conformation and orientation changes of antimicrobial peptides that may be associated with the individual steps of binding to the membrane surface and insertion into the bilayer hydrocarbon core. Several techniques combined with Langmuir film balance experiments may improve and extend the monolayer technique in understanding the mechanism(s) of these peptides interacting with lipid membranes. These techniques include circular dichroism (CD) (Sui *et al.*, 1994; Boncheva *et al.*, 1997), IRRAS (Boncheva *et al.*, 1997), and X-ray (Ludtke *et al.*, 1995), allowing to study the conformation and orientation of the peptides at the interface. CD spectra analysis of transferred dimyristoylphosphatidic acid (DMPA) monolayers showed that inserted melittin contains more α-helix structure than adsorbed melittin, and the peptide undergoes conformational changes upon binding to phospholipid monolayers (Sui *et al.*, 1994). Polarized ATR-FTIR spectroscopy of a monomolecular lipid film formed at the air–water interface of a Langmuir trough provides a useful model to investigate these different steps of lipid–peptide interactions (Benayad *et al.*, 1991). FTIR studies on the interactions of melittin with phospholipid monolayers at different surface pressures show that melittin inserts into lipid monolayers and perturbs the organization of the films (Benayad *et al.*, 1991). The behavior of lytic ideally amphipathic peptides Li*Kj* ($i = 2j$) and LKn ($n = i + j$) was investigated in situ by the monolayer technique combined with PM-IRRAS (Castano *et al.*, 1999). A change in the secondary structure occurs versus peptide length, and the orientation of these peptides at the air–water interface was suggested to be flat. This structure and orientation behavior is independent on the nature of the interface, air–water or DMPC monolayer, and the lateral surface pressure. Lipid perturbation due to these peptide insertions was detected and it indicates disorder of the lipid head groups. The conformation and orientation of gramicidin A were studied with polarization-modulated FTIR spectroscopy and the data show the peptide adopting a $\beta^{6.3}$ helix in a DMPC monolayer (Ulrich and Vogel, 1999). Also an orientation change of this peptide in a lipid monolayer is suggested and this behavior is strongly dependent on the surface pressure.

In conclusion, the monolayer technique has proved to be a peculiar approach for the study of surface activities of antimicrobial peptides. Combined with sensitive analytical and spectroscopic techniques, this approach provides greater insight into the molecular basis of antimicrobial peptide action. The dynamic interactions between the peptides and the membrane lipids are suggested to be key elements in the biological activities of the peptides as well as in the specificity of their activities.

Concluding remarks

The studies summarized above give an idea, although necessarily incomplete and fragmentary, of the considerable efforts made to elucidate the antimicrobial mechanisms of several peptides by investigating their interaction with biological and model membranes. Based on this intense research work, several general theories have been proposed to account for peptide action. The most studied group, as repeatedly stressed above, includes the linear, mostly α-helical peptides, so the models of action so far available have been developed mainly considering the properties of these peptides.

The formation of transmembrane pores was postulated to be a feasible mechanism for membrane permeation by amphipathic α-helical peptides (Boman *et al.*, 1994 and references therein). To form a pore across the bilayer, a bundle of α-helical peptides spans perpendicularly to the membrane, with the peptides' hydrophobic moities interacting with the lipid core and their hydrophilic surfaces pointing inward, lining the resulting aqueous channel (Oren and Shai, 1998). Thus, the pore behaves as a protein complex inserted into the membrane. Such a mechanism for membrane permeation by α-helical peptides is often referred to as "barrel-stave" (Ehrenstein and Lecar, 1977) (Figure 9.1A″,B″,C″). The formation of such transmembrane pores (or channels) leads to dissipation of the proton motive force and disruption of the membrane potential, cell metabolite leakage and ultimately to cell lysis.

Looking closer at the "barrel-stave" model, some interesting details of the proposed mechanism for peptide–lipid interaction can be discerned. First of all, since the α-helical

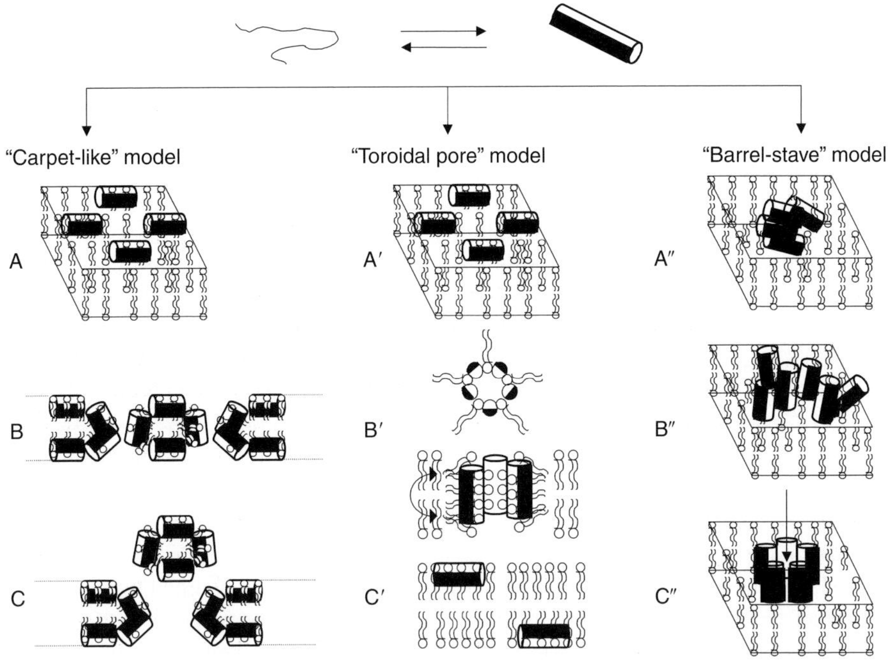

Figure 9.1 Cartoon of the major current hypotheses for antimicrobial peptide action (adapted from Shay, 1999; Matsuzaki, 2001). Antimicrobial peptides, randomly structured in solution, adopt a typical amphiphilic α-helix upon interacting with target membranes, with a face of the helix highly hydrophilic (blank half cylinder) and the other one hydrophobic (solid half cylinder) (upper part). *"Carpet-like" model:* (A) the peptides bind the membrane lying with their main axis parallel to it; when a peptide critic concentration is reached, the peptides rotate and reorient (B): leading to collapse of membrane structure and micellization (C) *"Toroidal pore" model:* (A′) at low concentration the peptides are oriented parallel to the plane of the bilayer; (B′) when a threshold concentration is reached, peptide molecules reorient perpendicular to the membrane and, together with lipid molecules, adopt a transient multi-pore state which leads to irreversible membrane disruption; translocation of peptide molecules into the inner monolayer, also occurs (C′) together with an increase in the flip-flop rate of lipids (double-head arrow in B′). *"Barrel-stave" model:* (A″) on the bilayer plane, peptide monomers get linked to each other through their hydrophilic surfaces; after recruiment of more monomers (B″), a bundle of α-helical peptides crosses the membrane to form a channel with an aqueous interior (C″), which causes dissipation of membrane potential and metabolites leakage.

peptides must cross the entire membrane, penetrating the lipid core of the bilayer, hydrophobic interactions are thought to predominantly drive the process (Oren and Shai, 1998). As an important consequence, the charge of the lipid headgroups is considered in this case to be a factor of minor relevance, with the peptides binding to both zwitterionic and charged phospholipid membranes. Being certainly energetically unfavorable to keep the hydrophilic surface of the peptidic α-helix in direct contact with the hydrophobic lipid acyl chains, peptide insertion into the bilayer is believed to be preceded by the assembly of at least two peptide monomers on the membrane surface, the polar surfaces of the monomers facing each other and thus providing reciprocal shielding from hydrophobic attractions (Figure 9.1A″). This molecular recognition and assembly between surface membrane-bound monomers occurs already at low peptide concentration, and is followed by insertion of the organized monomers into the bilayer core, to form a pore (Figure 9.1B″,C″). Progressive recruitment of additional monomers would then occur, increasing the pore magnitude until a critical size is reached (Shai, 1999). The barrel-stave model is now considered by most authors to hold mainly for cell non-selective lytic peptides, that is, peptides that lyse different microorganisms but also normal mammalian cells, such as erythrocytes (Shai, 1999). Melittin and the member of the family of peptaibols alamethicin, a 20 amino acid fungal peptide produced by *Trichoderma viride*, are well characterized examples of peptides that adopt the transmembrane helical bundle model (Bechinger, 1999).

The so-called "carpet-like" model is another major hypothesis of how antimicrobial peptides can induce the disruption of membrane integrity (Gazit *et al.*, 1995). In this case, peptides are thought to bind the target membrane lying with their main axis parallel to it, covering the bilayer in a carpet-like manner (Figure 9.1A). When a peptide threshold concentration is reached, the peptide monomers are believed to rotate and reorient towards the hydrophobic core of the membrane, leading to collapse of membrane packing and micellization (Shai, 1995) (Figure 9.1B,C).

The carpet model has been used to describe the mechanism of action of the majority of antimicrobial peptides with a cationic net charge, including dermaseptins and cecropins. In this mechanism, electrostatic interactions between the negatively charged lipid headgroups of the target membrane and the positively charged peptide are envisaged to drive the first steps of peptide binding (Shai, 1995; Shai, 1999). Consistently, peptides claimed to follow a carpet-like mechanism were proved to have low affinity to zwitterionic lipids compared to acidic ones. As discussed above, this preferential binding is probably one, although not the only one, of the means by which antimicrobial peptides are able to discriminate between bacteria and normal mammalian cells. An additional point to underline is that the carpet model can explain the mechanism of action of peptides shorter than 23–24 residues that cannot cross the membrane and, therefore, are not easy to reconcile with the barrel-stave mechanism.

More recently, a quite complex model, named "toroidal" or "two-state" model, has been proposed that has the potential to explain the action of both helical and β-sheet peptides (Ludtke *et al.*, 1996; Matsuzaki *et al.*, 1996a). At a low peptide concentration (i.e. at a low molar peptide-to-lipid ratio, P/L), the peptide is oriented parallel to the plane of the bilayer, being adsorbed in a functionally inactive state in the phospholipid headgroup region (Figure 9.1A′). Above a certain threshold P/L value, the peptides reorient perpendicular to the plane of the bilayer and, together with several surrounding lipids, flip inward, adopting a multi-pore state that leads to irreversible membrane disruption (Figure 9.1B′,C′). Importantly, the transition between the two states (inactive/active) of a membrane bound peptide not only depends on the concentration of the peptide itself, but is also modulated by the lipid composition of the bilayer (Huang, 2000). It follows that the composition of the lipid assembly is seen as a truly crucial regulatory parameter that determines the susceptibility of a cell to an antimicrobial peptide, whereas the peptide charge regulates membrane affinity and target cell specificity (Huang, 2000). Binding affinity and lytic activity may thus be not necessarily correlated.

It should be noted how the pore structure in the toroidal model differs dramatically from the classic barrel-stave channel. In fact, the former is described as a "peptide–lipid supramolecular complex pore, where rows of lipids interpose the helices oriented perpendicular to the membrane surface" (Matsuzaki, 1998). The lipid headgroups and the polar face of the amphiphilic helices (in the case of an α-helical peptide) together line the inside of the pore (Figure 9.1B′). Therefore, the peptide monomers are always associated with the polar lipid headgroups rather than with the acyl chains, and the outer and inner leaflets of the bilayer become continuous via the pore-lining lipids (Matsuzaki, 2001). The reported pore magnitude, duration and concentration depend greatly on the peptide studied, but, in general, the multi-pore state formed at high peptide concentration is considered to be rather stable, although individual pores may have a short lifetime (Matsuzaki *et al.*, 1997), and the pores formed allow the free diffusion of ions and even larger molecules (Wu *et al.*, 1999; Huang, 2000). Matsuzaki and coworkers suggested that the formation of the peptide–lipid supramolecular complex leads to a coupled transbilayer transport of lipids and peptides, with an increased flip-flop rate of lipids between leaflets of the membrane and peptide monomers that translocate into the inner monolayer (Matsuzaki *et al.*, 1996a) (Figure 9.1B′,C′). Furthermore, the same authors reported that different antimicrobial peptides, namely magainin and PGLa, a member of the magainin family, can act synergistically to form mixed toroidal, 1 : 1 stoichiometric complexes with increased rate of formation and stability compared to those formed by each peptide alone, resulting in an enhanced membrane disruption activity (Matsuzaki *et al.*, 1998; see also McCafferty *et al.*, 1999).

Magainin has been the peptide of choice for extensive research work that brought to the definition of the toroidal model. However, many other peptides are being recognized to act following this mechanism. The α-helical peptides mastoparan X from *Vespa xanthoptera* and PGLa (Matsuzaki *et al.*, 1996b; Latal *et al.*, 1997), and the β-sheet ones tachyplesin and protegrin-1, are well-known examples (Heller *et al.*, 1998; Matsuzaki, 1999). Thanks to a thorough study of the interaction of selected peptides with model membranes, Hancock and coworkers have subsequently shown that the general toroidal model can be reasonably extended to peptides of remarkably different sizes (10–28 amino acids) and belonging to a range of different structural classes (α-helical, extended, loop, linear β, β-structured loop, cyclic) (Wu *et al.*, 1999). As underlined above, the mechanism of action of small peptides (10–14-monomers) like those used by authors of this review is objectively difficult to rationalize in terms of the barrel-stave model, unless hypothesizing that monomers of short peptides can connect to each other to form a bilayer-spanning channel, as postulated to occur in the case of the α-helical, 13-residues long temporin A and its analogs (Wade *et al.*, 2000).

It has been recently pointed out that the toroidal model and the carpet-like are not necessarily incompatible, if one imagines that the formation of transient pores in the membrane when a local high concentration of peptide is reached (toroidal model), is just an early step before the disruption of the membrane packing and the advent of micellization (carpet-like) (Shai, 1999).

The models and mechanisms outlined above certainly constitute relevant progresses towards our understanding of the mechanism of action of antimicrobial peptides. Striving to provide plausible explanations for the plethora of experimental results produced so far, these hypotheses give us some answers but also leave many open questions. In particular, the partially unsolved problem of understanding exactly how specificity in peptide–lipid interactions is achieved is of crucial importance, since it relates to the target cell specificities exhibited by antimicrobial peptides and to the mechanism of discrimination of self versus non-self. Indeed, as discussed above, behind the already wide array of naturally occurring peptides an army looms formed by hundreds of synthetic variants and analogs derived from these natural peptides that could constitute an original and novel class of antimicrobial drugs with a remarkable clinical potential. Several cationic peptides have already entered clinical trials with good future prospects, testifying how great is the interest for the clinical application of antimicrobial peptides (Hancock and Chapple, 1999; Hancock, 2000). While the future will continue to see investigation on these fascinating biomolecules advancing on several distinguished fronts, it will be the task of biochemists and biophysicists to clarify the details of peptide action and to provide informations to tune the design of improved antimicrobial peptides for clinical use.

References

Abriouel, H., Sanchez-Gonzalez, J., Maqueda, M., Galvez, A., Valdivia, E. and Galvez-Ruiz, J. (2001) Monolayer charateristics of bacteriocin AS-48, pH effect and interactions with dipalmitoyl phosphatidic acid at the air/water interface. *Journal of Colloid and Interface Science*, **233**, 306–312.

Adamson, A. W. (1976) *Physical Chemistry of Surfaces*, 3rd edn. John Wiley & Sons, New York.

Ammar, B., Perianin, A., Mor, A., Sarfati, G., Tissot, M., Nicolas, P., *et al.* (1998) Dermaseptin, a peptide antibiotic, stimulates microbicidal activities of polymorphonuclear leukocytes. *Biochemical and Biophysical Research Communications*, **247**, 870–875.

Andreu, D., Merrifield, R. B., Steiner, H. and Boman, H. G. (1985) N-terminal analogues of cecropin A: synthesis, antibacterial activity, and conformational properties. *Biochemistry*, **24**, 1683–1688.

Angelova, M. I. and Tsoneva, I. (1999) Interactions of DNA with giant liposomes. *Chemistry and Physics of Lipids*, **101**, 123–137.

Angelova, M. I., Hristova, N. and Tsoneva, I. (1999) DNA-induced endocytosis upon local microinjection to giant unilamellar cationic vesicles. *European Biophysical Journal*, **28**, 142–150.

Barra, D., Simmaco, M. and Boman, H. G. (1998) Gene-encoded peptide antibiotics and innate immunity. Do "animalcules" have defence budgets? *FEBS Letters*, **430**, 130–134.

Batzri, S. and Korn, E. D. (1973) Single bilayer liposomes prepared without sonication. *Biochimimica et Biophysica Acta*, **298**, 1015–1020.

Bechinger, B. (1999) The structure, dynamics and orientation of antimicrobial peptides in membranes by multidimensional solid-state NMR spectroscopy. *Biochimica et Biophysica Acta*, **1462**, 157–183.

Bechinger, B., Kinder, R., Helmie, M., Vogt, T. C. B., Harzer, U. and Schinzel, S. (1999) Peptide structural analysis by solid-state NMR spectroscopy. *Biopolymers*, **51**, 174–190.

Benachir, T. and Lafleur, M. (1995) Study of vesicle leakage induced by melittin. *Biochimica et Biophysica Acta*, **1235**, 452–460.

Benayad, A., Benamar, D., Van Mau, N., Page, G. and Heitz, F. (1991) Single channel and monolayer studies of acylated gramicidin A: influence of the length of the alkyl group. *European Biophysical Journal*, **20**, 209–223.

Bessalle, R., Kapitkovsky, A., Gorea, A., Shalit, I. and Fridakin, M. (1990) All-D-magainin: chirality, antimicrobial activity and proteolytic resistance. *FEBS Letters*, **274**, 151–155.

Bikshapathy, E., Sitaram, N. and Nagaraj, R. (1999a) Addition and omission analogs of the 13-residue antibacterial and hemolytic peptide PKLLKTFLSKWIG: structural preferences, model membrane binding and biological activities. *Journal of Peptide Research*, **53**, 47–55.

Bikshapathy, E., Sitaram, N. and Nagaraj, R. (1999b) Effect of introducing p-fluorophenylalanine and multiple tryptophan residues in a 13-residue antibacterial peptide. *Protein and Peptide Letters*, **6**, 67–72.

Blondelle, S. E. and Houghten, R. A. (1991) Hemolytic and antimicrobial activities of the twenty-four individual omission analogues of melittin. *Biochemistry*, **30**, 4671–4678.

Blondelle, S. E. and Houghten, R. A. (1996) Novel antimicrobial compounds identified using synthetic combinatorial library technology. *Trends in Biotechnology*, **14**, 60–65.

Blondelle, S. E., Lohrer K. and Aguilar, M. I. (1999) Lipid-induced conformation and lipid binding properties of cytolytic and antimicrobial peptides: determination and biological specificity. *Biochimica et Biophysica Acta*, **1462**, 89–108.

Boman, H. G. (1995) Peptide antibiotics and their role in innate immunity. *Annual Review of Immunology*, **13**, 61–92.

Boman, H. G., Marsh, J. and Goode, J. A. (eds) (1994) *Antimicrobial Peptides*, Ciba Foundation Symposium 186. John Wiley & Sons, Chichester.

Boncheva, M. and Vogel, H. (1997) Formation of stable polypeptide monolayers at interfaces: controlling molecular conformation and orientation. *Biophysical Journal*, **73**, 1056–1072.

Brockman, H. (1999) Lipid monolayers: why use half of a membrane to characterize protein–membrane interactions? *Current Opinion in Structural Biology*, **9**, 438–443.

Castano, S., Desbat, B., Laguerre, M. and Dufourcq, J. (1999) Structure, orientation, and affinity for interfaces and lipids of ideally amphipathic lytic LiKj ($i=2j$) peptides. *Biochimica et Biophysica Acta*, **1416**, 176–194.

Chen, H.-C., Brown, J. H., Morell, J. L. and Huang, C. M. (1988) Synthetic magainin analogues with improved antimicrobial activity. *FEBS Letters*, **236**, 462–466.

Cornut, I., Buttner, K., Dasseux, J.-L. and Dufourcq, J. (1994) The amphipathic α-helix concept application to the *de novo* design of ideally amphipathic Leu, Lys peptides with hemolytic activity higher than that of melittin. *FEBS Letters*, **349**, 29–33.

Cruciani, R. A. Barker, J. L., Zasloff, M., Chen, H. C. and Colamonici, O. (1991) Antibiotic magainins exert cytolytic activity against transformed cell lines through channel formation. *Proceedings of the National Academy of Sciences USA*, **88**, 3792–3796.

Dathe, M., Schumann, M., Wieprecht, T., Winkler, A., Beyermann, M., Krause, E. *et al.* (1996) Peptide helicity and membrane surface charge modulate the balance of electrostatic and hydrofobic interactions with lipid bilayer and biological membranes. *Biochemistry*, **35**, 12612–12622.

Dathe, M., Wieprecht, T., Nikolenko, H., Handel, L., Maloy, W. L., MacDonald, D. L., *et al.* (1997) Hydrophobicity, hydrophobic moment and angle subtended by charged residues modulate antibacterial and hemolytic activity of amphipathic helical peptides. *FEBS Letters*, **403**, 208–212.

Dathe, M. and Wieprecht, T. (1999) Structural features of helical antimicrobial peptides: their potential to modulate activity on model membranes and biological cells. *Biochimica et Biophysica Acta*, **1462**, 71–87.

Demel, R. A., Geurts van Kessel, W. S. M., Zwaal, R. F. A., Roelofsen, B. and van Deenen, L. L. M. (1975) Relation between various phospholipase actions on human red cell membranes and the interfacial phospholipid pressure in monolayers. *Biochimica et Biophysica Acta*, **406**, 97–107.

Dhople, V. M. and Nagaraj, R. (1995) Generation of analogs having potent antimicrobial and hemolytic activities with minimal changes from an inactive 16-residue peptide corresponding to the helical region of *Staphylococcus aureus* δ-toxin. *Protein Engineering*, **8**, 315–318.

Diamond, G., Russell, J. P. and Bevins, C. L. (1996) Inducible expression of an antibiotic peptide gene in lipopolysaccharide-challenged tracheal epithelial cells. *Proceeding of the National Academy of Science USA*, **93**, 5156–5160.

Ehrenstein, G. and Lecar, H. (1977) Electrically gated ionic channels in lipid bilayers. *Quarterly Reviews of Biophysics*, **10**, 1–34.

Eisenberg, D. (1984) Three-dimensional structure of membrane and surface proteins. *Annual Review of Biochemistry*, **53**, 595–623.

El Jastimi, R., Edwards, K. and Lafleur, M. (1999) Characterization of permeability and morphological perturbations induced by nisin on phosphatidylcholine membranes. *Biophysical Journal*, **77**, 842–852.

Epand, R. M. and Vogel, H. J. (1999) Diversity of antimicrobial peptides and their mechanisms of action. *Biochimica et Biophysica Acta*, **1462**, 11–28.

Epand, R. F., Epand, R. M., Monaco, V., Stoia, S., Formaggio, F., Crisma, M. *et al.* (1999) The antimicrobial peptide trichogin and its interaction with phospholipid membranes. *European Journal of Biochemistry*, **266**, 1021–1028.

Eytan, G. D. (1982) Use of liposomes for reconstitution of biological functions. *Biochimica et Biophysica Acta*, **694**, 185–202.

Feng, S. (1999) Interpretation of mechanochemical properties of lipid bilayer vesicles from the equation of state of pressure–area measurement of the monolayer at the air–water or oil–water interface. *Langmuir*, **15**, 998–1010.

Fidelio, G. D., Maggio, B. and Cumar, F. A. (1982) Interaction of soluble and membrane proteins with monolayers of glycosphingolipids. *Biochemical Journal*, **203**, 717–725.

Fidelio, G. D., Maggio, B. and Cumar, F. A. (1986) Molecular parameters and physical state of neutral glycosphingolipids and gangliosides in monolayers at different temperatures. *Biochimica et Biophysica Acta*, **862**, 49–56.

Ganz, T. and Lehrer, R. I. (1997) Antimicrobial peptides of leukocytes. *Current Opinion in Hematology*, **4**, 53–58.

Ganz, T. and Lehrer, R. I. (1998) Antimicrobial peptides of vertebrates. *Current Opinion in Immunology*, **10**, 41–44.

Garcia-Olmedo, F., Molina, A., Alamillo, J. M. and Rodriguez-Palenzuela, P. (1998) Plant-defense peptides. *Biopolymers*, **47**, 479–491.

Gazit, E., Boman, A., Boman, H. G. and Shai, Y. (1995) Interaction of the mammalian antibacterial peptide cecropin P1 with phospholipid vesicles. *Biochemistry*, **34**, 11479–11488.

Gidalevitz, D., Waring, A. J., Lehrer, R. I. and Lee, K. Y. C. (2000) Penetration of antimicrobial peptide protegrin-1 into various phospholipid monolayers: toward new antibiotic drugs. Book of abstracts, 219th ACS National Meeting, San Francisco, CA, March 26–30. American Chemical Society, Washington DC.

Hancock, R. E. W. and Lehrer, R. (1998) Cationic peptides: a new source of antibiotics. *Trends in Biotechnology*, **16**, 82–88.

Hancock, R. E. W. and Chapple, D. S. (1999) Peptide antibiotics. *Antimicrobial Agents and Chemotherapy*, **43**, 1317–1323.

Hancock, R. E. W. (2000) Cationic antimicrobial peptides: towards clinical applications. *Expert Opinion on Investigational Drugs*, **9**, 1723–1729.

Hancock, R. E. W. and Diamond, G. (2000) The role of cationic antimicrobial peptides in innate host defences. *Trends in Microbiology*, **8**, 402–410.

Hancock, R. E. W. and Scott, M. G. (2000) The role of antimicrobial peptides in animal defenses. *Proceedings of the National Academy of Science USA*, **97**, 8856–8861.

Harjunpää, I., Kuusela, P., Smoluch, M. T., Silberring, J., Lankinen, H. and Wade, D. (1999) Comparison of synthesis and antibacterial activity of temporin A. *FEBS Letters*, **449**, 187–190.

Heller, W. T., Waring, A. J., Lehrer, R. I. and Huang, H. W. (1998) Multiple states of β-sheet peptide protergin in lipid bilayers. *Biochemistry*, **37**, 17331–17338.

Hendrickson, H. S., Fan, P., Kaufman, D. and Kleiner, D. (1983) The effect of a phase transition on penetration of phospholipid monolayers by melittin and glucagon. *Archives of Biochemistry and Biophysics*, **227**, 242–247.

Hoffmann, J. A., Kafatos, F. C., Janeway, C. A. and Ezekowitz, R. A. B. (1999) Phylogenetic perspectives in innate immunity. *Science*, **248**, 1313–1318.

Holopainen, J. M., Angelova, M. I. and Kinnunen, P. K. (2000) Vectorial budding of vesicles by asymmetrical enzymatic formation of ceramide in giant liposomes. *Biophysical Journal*, **78**, 830–838.

Huang, H. W. (2000) Action of antimicrobial peptides: two-state model. *Biochemistry*, **39**, 8347–8352.

Hwang, P. M. and Vogel, H. J. (1998) Structure–function relationships of antimicrobial peptides. *Biochemistry and Cell Biology*, **76**, 235–246.

Jack, R. W. and Jung, G. (2000) Lantibiotics and microcins: polypeptides with unusual chemical diversity. *Current Opinion in Chemical Biology*, **4**, 310–317.

van Kan, E. J. M., van der Bent, A., Demel, R. A. and de Kruijff, B. (2001) Membrane activity of the peptide antibiotic clavanin and the importance of its glycine residues. *Biochemistry*, **40**, 6398–6405.

Kang, J. H., Shin, S. Y., Jang, S. Y., Lee, M. K. and Hahm, K. S. (1998) Release of aqueous contents from phospholipid vesicles induced by cecropin A (1–8)-magainin 2 (1–12) hybrid and its analogues. *Journal of Peptide Research*, **52**, 45–50.

Kiyota, T., Lee, S. and Sugihara, G. (1996) Design and synthesis of amphiphilic α-helical model peptides with systematically varied hydrophobic–hydrophilic balance and their interaction with lipid- and bio-membranes. *Biochemistry*, **35**, 13196–13204.

Konz, D. and Marahiel, M. A. (1999) How do peptide synthetases generate structural diversity? *Chemistry and Biology*, **6**, R39–48.

Kourie, J. I. and Shorthouse, A. A. (2000) Properties of cytotoxic peptide-formed ion channels. *American Journal of Physiology. Cell Physiology*, **278**, C1063–C1087.

Kwon, M.-Y., Hong, S.-Y. and Lee, K.-H. (1998) Structure–activity analysis of brevinin 1E amide, an antimicrobial peptide from *Rana esculenta*. *Biochimica et Biophysica Acta*, **1387**, 239–248.

Ladokhin, A. S., Selsted, M. E. and White, S. H. (1997) Bilayer interactions of indolicidin, a small antimicrobial peptide rich in tryptophan, proline, and basic amino acids. *Biophysical Journal*, **72**, 794–805.

Latal, A., Degovics, G., Epand, R. F., Epand, R. M. and Lohner, K. (1997) Structural aspects of the interaction of peptidyl–glycylleucine–carboxyamide, a highly potent antimicrobial peptide from frog skin, with lipids. *European Journal of Biochemistry*, **248**, 938–946.

Lee, I. H., Cho, Y. and Lehrer, R. I. (1997) Effects of pH and salinity on the antimicrobial properties of clavanins. *Infection and immunity*, **65**, 2898–2903.

Lehrer, R. I., Lichtesnstein, A. K. and Ganz, T. (1993) Defensins: antimicrobial and cytotoxic peptides of mammalian cells. *Annual Review of Immunology*, **11**, 105–128.

Lewis, R. N., Prenner, E. J., Kondejewski, L. H., Flach, C. R., Mendelsohn, R., Hodges, R. S. *et al.* (1999) Fourier transform infrared spectroscopic studies of the interaction of the antimicrobial peptide gramicidin S with lipid micelles and with lipid monolayer and bilayer membranes. *Biochemistry*, **38**, 15193–15203.

Liang, J. F. and Kim, S. C. (1999) Not only the nature of the peptide but also the characteristics of cell membrane determine the antimicrobial mechanism of a peptide. *Journal of Peptide Research*, **53**, 518–522.

Ludtke, S., He, K. and Huang, H. W. (1995) Membrane thinning caused by magainin 2. *Biochemistry*, **34**, 16764–16769.

Ludtke, S. J., He, K., Heller, W. T., Harroun, T. A., Yang, L. and Huang, H. W. (1996) Membrane pores induced by magainin. *Biochemistry*, **35**, 13723–13728.

Luisi, P. L. and Walde, P. (eds) (2000) *Giant Vesicles*. John Wiley & Sons, Chichester.

Maget-Dana, R. (1999) The monolayer technique: a potent tool for studying the interfacial properties of antimicrobial and membrane-lytic peptides and their interactions with lipid membranes. *Biochimica et Biophysica Acta*, **1462**, 109–140.

Maget-Dana, R. and Ptak, M. (1997) Penetration of the insect defensin A into phospholipid monolayers and formation of defensin A complexes. *Biophysical Journal*, **73**, 2527–2533.

Maget-Dana, R., Lelievre, D. and Brack, A. (1999) Surface active properties of amphiphilic sequential isopeptides: comparison between helical- and -sheet conformations. *Biopolymers*, **49**, 415–423.

Maloy, W. L. and Kari, U. P. (1995) Structure–activity studies on magainins and other host defense peptides. *Biopolymers*, **37**, 105–122.

Mangoni, M. L., Rinaldi, A. C., Di Giulio, A., Mignogna, G., Bozzi, A., Barra, D. *et al.* (2000) Structure–function relationship of temporins, small anti-microbial peptides from amphibian skin. *European Journal of Biochemistry*, **267**, 1447–1454.

Mangoni, M. L., Miele, R., Renda, T. G., Barra, D. and Simmaco, M. (2001) The synthesis of antimicrobial peptides in the skin of *Rana esculenta* is stimulated by micoorganisms. *FASEB Journal*, **15**, 1431–1432.

Matsuzaki, K., Harada, M., Handa, T., Funakoshi, S., Fujii, N., Yajima, H. *et al.* (1989) Magainin 1-induced leakage of entrapped calcein out of negatively-charged lipid vesicles. *Biochimica et Biophysica Acta*, **981**, 130–134.

Matsuzaki, K., Murase, O., Fujii, N. and Miyajima, K. (1996a) An antimicrobial peptide, magainin-2, induced rapid flip-flop of phospholipids coupled with pore formation and peptide translocation. *Biochemistry*, **35**, 11361–11368.

Matsuzaki, K., Yoneyama, S., Murase, O. and Miyajima, K. (1996b) Transbilayer transport of ions and lipids coupled with mastoparan X translocation. *Biochemistry*, **35**, 8450–8456.

Matsuzaki, K., Sugishita, K., Harada, M., Fujii, N. and Miyajima, K. (1997). Interactions of an antimicrobial peptide, magainin 2, with outer and inner membranes of Gram-negative bacteria. *Biochimica et Biophysica Acta* **1327**, 119–130.

Matsuzaki, K. (1998) Magainins as paradigm for the mode of action of pore forming polypeptides. *Biochimica et Biophysica Acta* **1376**, 391–400.

Matsuzaki, K., Mitani, Y., Akada, K., Murase, O., Yoneyama, S., Zasloff, M. and Miyajima, K. (1998) Mechanism of synergism between antimicrobial peptides magainin 2 and PGLa. *Biochemistry*, **37**, 15144–15153.

Matsuzaki, K. (1999) Why and how are peptide–lipid interactions utilized for self-defense? Magainins and tachyplesins as archetypes. *Biochimica et Biophysica Acta* **1462**, 1–10.

Matsuzaki, K., Sugishita, K. and Miyajima, K. (1999) Interactions of an antimicrobial peptide, magainin 2, with lipopolysaccharide-containing liposomes as a model for outer membranes of Gram-negative bacteria. *FEBS Letters*, **449**, 221–224.

Matsuzaki, K. (2001) Molecular mechanisms of membrane perturbation by antimicrobial peptides. In *Development of Novel Antimicrobial Agents: Emerging Strategies*, Horizon Scientific Press, Wymondham, UK, pp. 167–181.

McCafferty, D. G., Cudic, P., Yu, M. K., Behenna, D. C. and Kruger, R. (1999) Synergy and duality in peptide antibiotic mechanisms. *Current Opinion in Chemical Biology*, **3**, 672–680.

McElhaney, R. N. and Prenner, E. J. (eds) (1999) The interaction of antimicrobial peptides with model lipid bilayer and biological membranes. *Biochimica et Biophysica Acta*, **1462** (Special Issue), 1–234.

Menger, F. M. and Lee, S. J. (1995) Induced morphological changes in synthetic giant vesicles – Growth, fusion, undulation, excretion, wounding, and healing. *Langmuir*, **11**, 3685–3689.

Menger, F. M. and Angelova, M. I. (1998) Giant vesicles: imitating the cytological processes of cell membranes. *Accounts of Chemical Research*, **31**, 789–797.

Menger, F. M. and Keiper, J. S. (1998) Giant vesicles – micromanipulation of membrane bilayers. *Advanced Materials*, **10**, 888–890.

Mignogna, G., Simmaco, M. and Barra, D. (1998) Occurrence and function of D-amino acid-containing peptides and proteins: antimicrobial peptides. *EXS*, **85**, 29–36.

Moffitt, M. C. and Neilan, B. A. (2000) The expansion of mechanistic and organismic diversity associated with non-ribosomal peptides. *FEMS Microbiology Letters*, **191**, 159–167.

Nir, S. and Nieva, J. L. (2000) Interaction of peptides with liposomes: pore formation and fusion. *Progress in Lipid Research*, **39**, 181–206.

Oh, J. E., Hong, S. Y. and Lee, K.-H. (1999) Structure–activity relationship study: short antimicrobial peptides. *Journal Peptide Research*, **53**, 41–46.

Ohsaki, Y., Gazdar, A. F., Chen, H.-C. and Johnson, B. E. (1992) Antitumor activity of magainin analogues against human lung cancer cell lines. *Cancer Research*, **52**, 3534–3538.

Olson, F., Hunt, C. A., Szoka, F. C., Vail, W. J. and Papahadjopoulos, D. (1979) Preparation of liposomes of defined size distribution by extrusion through polycarbonate membranes. *Biochimica et Biophysica Acta*, **557**, 9–23.

Op den Kamp, J. A. (1979) Lipid asymmetry in membranes. *Annual Review of Biochemistry*, **48**, 47–71.

Oren, Z. and Shai, Y. (1998) Mode of action of linear amphipathic α-helical antimicrobial peptides. *Biopolymers*, **47**, 451–463.

Otvos, L. Jr. (2000) Antibacterial peptides isolated from insects. *Journal of Peptide Science*, **6**, 497–511.

Pattus, F., Martinez, C., Dargent, B., Cavard, D., Verger, R. and Lazdunski, C. (1983) Interaction of colicin A with phospholipid monolayers and liposomes. *Biochemistry*, **22**, 5698–5703.

Pethica, B. A. (1955) Thermodynamics of monolayer penetration at constant area. I. *Transactions of Faraday Society*, **51**, 1402–1411.

Rinaldi, A. C., Di Giulio, A., Liberi, M., Gualtieri, G., Oratore, A., Schininà, M. E. *et al.* (2001) Effects of temporins on molecular dynamics and membrane permeabilization in lipid vesicles. *Journal of Peptide Research*, **58**, 213–220.

Riquelme, G., Lopez, E., Garcia-Segura, L. M., Ferragut, J. A. and Gonzalez-Ros, J. M. (1990) Giant liposomes: a model system in which to obtain patch-clamp recordings of ionic channels. *Biochemistry*, **29**, 11215–11222.

Russell, C. J., Thorgeirsson, T. E. and Shin Y. K. (1999) The membrane affinities of the aliphatic amino acid side chains in an alpha-helical context are independent of membrane immersion depth. *Biochemistry*, **38**, 337–346.

Sablon, E., Contreras, B. and Vandamme, E. (2000) Antimicrobial peptides of lactic acid bacteria: mode of action, genetics and biosynthesis. *Advances in Biochemical and Engineering/Biotechnology*, **68**, 21–60.

Sahl, H. G. (1994) Gene-encoded antibiotics made in bacteria. In *Antimicrobial Peptides*, edited by H. G. Boman, J. Marsh, and J. A Goode, Ciba Foundation Symposium 186, John Wiley & Sons, Chichester, pp. 27–42.

Sai, K. P., Jagannadham, M. V., Vairamani, M., Raju, N. P., Devi, A. S., Nagaraj, R. and Sitaram, N. (2001) Tigerinins: novel antimicrobial peptides from the Indian frog *Rana tigerina*. *Journal of Biological Chemistry*, **276**, 2701–2707.

Scott, M. G. and Hancock, R. E. W. (2000) Cationic antimicrobial peptides and their multifunctional role in the immune system. *Critical Reviews in Immunology*, **20**, 407–431.

Seelig, A. (1987) Local anesthetics and pressure: a comparison of dibucaine binding to lipid monolayers and bilayers. *Biochimica et Biophysica Acta*, **899**, 196–204.

Shai, Y. (1995) Molecular recognition between membrane-spanning polypeptides. *Trends in Biochemical Science*, **20**, 460–465.

Shai, Y. (1999) Mechanism of the binding, insertion and destabilization of phospholipid bilayer membranes by α-helical antimicrobial and cell non-selective membrane-lytic peptides. *Biochimica et Biophysica Acta*, **1462**, 55–70.

Signor, G., Mammi, S., Peggion, E., Ringsdorf, H. and Wagenknecht, A. (1994) Interaction of bombolitin III with phospholipid monolayers and liposomes and effect on the activity of phospholipase A_2. *Biochemistry*, **33**, 6659–6670.

Simmaco, M., Mignogna, G., Canofeni, S., Miele, R., Mangoni, M. L. and Barra, D. (1996) Temporins, novel antimicrobial peptides from the European red frog *Rana temporaria*. *European Journal of Biochemistry*, **242**, 788–792.

Simmaco, M., Mignogna, G. and Barra, D. (1998) Antimicrobial peptides from amphibian skin: what do they tell us? *Biopolymers*, **47**, 435–450.

Sitaram, N. and Nagaraj, R. (1999) Interaction of antimicrobial peptides with biological and model membranes: structural and charge requirements for activity. *Biochimica et Biophysica Acta*, **1462**, 29–54.

Subbalakshmi, C., Nagaraj, R. and Sitaram, N. (1999) Biological activities of C-terminal 15-residue synthetic fragment of melittin: design of an analog with improved antibacterial activity. *FEBS Letters*, **448**, 62–66.

Sui, S. F., Wu, H., Guo, Y. and Chen, K. S. (1994) Conformational changes of melittin upon insertion into phospholipid monolayer and vesicle. *Journal of Biochemistry*, **116**, 482–487.

Szoka, F. Jr. and Papahadjopoulos, D. (1978) Procedure for preparation of liposomes with large internal aqueous space and high capture by reverse-phase evaporation. *Proceedings of the National Academy of Sciences USA*, **75**, 4194–4198.

Thieffry, M., Neyton, J., Pelleschi, M., Fevre, F. and Henry, J. P. (1992) Properties of the mitochondrial peptide-sensitive cationic channel studied in planar bilayers and patches of giant liposomes. *Biophysical Journal*, **63**, 333–339.

Tossi, A., Tarantino, C. and Romeo, D. (1997) Design of synthetic antimicrobial peptides based on sequence analogy and amphipathicity. *European Journal of Biochemistry*, **250**, 549–558.

Tournois, H., Gieles, P., Demel, R., de Gier, J. and de Kruijff, B. (1989) Interfacial properties of gramicidin and gramicidin–lipid mixtures measured with static and dynamic monolayer techniques. *Biophysical Journal*, **55**, 557–569.

Uematsu, N. and Matsuzaki, K. (2000) Polar angle as a determinant of amphipathic α-helix–lipid interaction: a model peptide study. *Biophysical Journal*, **79**, 2075–2083.

Ulrich, W. P. and Vogel, H. (1999) Polarization-modulated FTIR spectroscopy of lipid/gramicidin monolayers at the air/water interface. *Biophysical Journal*, **76**, 1639–1647.

Vaara, M. (1992) Agents that increase the permeability of the outer membrane *Microbiology Reviews*, **56**, 395–411.

Van Mau, N., Trudelle, Y., Daumas, P. and Heize, F. (1988) Mixed monolayers of linear gramicidins and phospholipids. Surface pressure and surface potential studies. *Biophysical Journal*, **54**, 563–567.

Wackerbauer, G., Weis, I. and Schwarz, G. (1996) Preferentially partitioning of melittin into air/water interface: structural and thermodynamic implications. *Biophysical Journal*, **71**, 1422–1427.

Wade, D., Boman, A. Wåhlin, B., Drain, C. M., Andreu, D., Boman, H. G. *et al.* (1990) All-D amino acid-containing channel-forming antibiotic peptides. *Proceedings of the National Academy of Science USA*, **87**, 4761–4765.

Wade, D., Silberring, J., Soliymani, R., Heikkinen, S., Kilpelainen, I., Lankinen, H. and Kuusela, P. (2000) Antibacterial activities of temporin A analogs. *FEBS Letters*, **479**, 6–9.

Welling, M. M., Hiemstra, P. S., van den Barselaar, M. T., Paulusma-Annema, A., Nibbering, P. H., Pauwels, E. K. *et al.* (1998) Antibacterial activity of human neutrophil defensins in experimental infections in mice is accompanied by increased leukocyte accumulation. *Journal of Clinical Investigation*, **15**, 1583–1590.

Wiedmer, T., Brodbeck, U., Zahler, P. and Fulpius, B. W. (1987) Interactions of acetylcholine receptor and acetylcholinesterase with lipid monolayers. *Biochimica et Biophysica Acta*, **506**, 161–172.

Wieprecht, T., Dathe, M., Schümann, M., Krause, E., Beyermann, M. and Bienert, M. (1996) Conformational and functional study of magainin 2 in model membrane environments using the new approach of systematic double-D-amino acid replacement. *Biochemistry*, **35**, 10844–10853.

Wieprecht, T. Dathe, M., Beyermann, M., Krause, E., Maloy, W. L., MacDonald, D. L. *et al.* (1997a) Peptide hydrophobicity controls the activity and selectivity of magainin 2 amide in interaction with membranes. *Biochemistry*, **36**, 6124–6132.

Wieprecht, T., Dathe, M., Epand, R. M., Beyermann, M., Krause, E., Maloy, W. L. *et al.* (1997b) Influence of the angle subtended by the positively charged helix face on the membrane activity of amphipathic, antibacterial peptides. *Biochemistry*, **36**, 12869–12880.

Wieprecht, T., Dathe, M., Krause, E., Beyermann, M., Maloy, W. L., MacDonald, D. L. *et al.* (1997c) Modulation of membrane activity of amphipathic, antibacterial peptides by slight modifications of the hydrophobic moment. *FEBS Letters*, **417**, 135–140.

Wieprecht, T., Beyermann, M. and Seelig, J. (1999) Binding of antibacterial magainin peptides to electrically neutral membranes: thermodynamics and structure. *Biochemistry*, **38**, 10377–10387.

Winterhalter, M. and Lasic, D. D. (1993) Liposome stability and formation: experimental parameters and theories on the size distribution. *Chemistry and Physics of Lipids*, **64**, 35–43.

Wu, M., Maier, E., Benz, R. and Hancock, R. E. W. (1999) Mechanism of interaction of different classes of cationic antimicrobial peptides with planar bilayers and with the cytoplasmic membrane of *Escherichia coli*. *Biochemistry*, **38**, 7235–7242.

Yang, D., Chertov, O., Bykovskaia, S. N., Chen, Q., Buffo, M. J., Shogan, J. *et al.* (1999) ß-defensins: linking innate and adaptive immunity through dendritic and T cell CCR6. *Science*, **286**, 525–528.

Zhang, L. Scott, M. G., Yan, H., Mayer, L. D. and Hancock, R. E. W. (2000) Interaction of polyphemusin I and structural analogs with bacterial membranes, lipopolysaccharide, and lipid monolayers. *Biochemistry*, **39**, 14504–14514.

Zhao, H. X., Mattila, J. P., Holopainen, J. M and Kinnunen, P. K. J. (2001a) Comparison of the membrane association of two antimicrobial peptides, magainin 2 and indolicidin. *Biophysical Journal*, **81**, 2979–2991.

Zhao, H. X., Rinaldi, A. C., Di Giulio, A., Simmaco, M. and Kinnunen, P. K. J. (2001b). Interactions of the antimicrobial peptides temporins with model biomembranes. Comparison of temporin B and temporin L. *Biochemistry*, **41**, 4425–4436.

Zumbuehl, O. and Weder, H. G. (1981) Liposomes of controllable size in the range of 40 to 180 nm by defined dialysis of lipid/detergent mixed micelles. *Biochimica et Biophysica Acta*, **640**, 252–262.

10 Pardaxins – pore-forming neurotoxins as pharmacological tools in dissecting neurotransmitter exocytosis and neurotoxicity

Philip Lazarovici

Introduction

Neurons store and release two major kinds of chemical signals: neurotransmitters and neuropeptides (Kelly, 1993). Both types of signals are stored in secretory vesicles that release their contents by regulated exocytosis when cytosolic calcium levels are raised (Trimble *et al.*, 1991). A few dozen proteins, collectively named SNARE, which function at specific stages during exocytosis, have been identified (Linial, 1997), but detailed knowledge of their mechanism of action is lacking. Neurotoxins have been successfully used as tools to dissect the molecular mechanisms of exocytotic synaptic transmission (Dolly, 1988). Presynaptic calcium channels were detected and studied with specific conotoxins and agatoxins (Garcia *et al.*, 1997), which are their potent and selective blockers. The nerve terminal presynaptic SNARE proteins synaptobrevin, syntaxin and SNAP-25 were identified as essential minimal core components of the exocytotic apparatus because they are proteolytically cleaved by tetanus and botulinum neurotoxins, the most potent inhibitors of exocytosis known to date (Schiavo *et al.*, 1992; Montecucco *et al.*, 1998). Another stimulatory tool of exocytosis is α-latrotoxin, a protein component of black widow spider venom (Grasso, 1988). This toxin is selectively toxic for presynaptic exocytotic machinery and induces massive release of neurotransmitters and neuropeptides (Grasso, 1988; Surkova, 1994). This chapter is devoted to yet another presynaptic neurotoxin, pardaxin, which is similar to α-latrotoxin but, in contrast to the clostridial toxins, is a strong and universal stimulator of neurotransmitter release (Bloch-Shilderman *et al.*, 1997). This toxin belongs to the large group of pore-forming (ionophore) toxins.

Pore-forming toxins act directly on the cell plasma membrane. They make pores in the cell membrane, disturb the ionic homeostasis and normal cell physiology and may ultimately cause cell death (Thelestam and Molby, 1979). Pore-forming toxins of microbial, plant and animal origin represent an abundant, heterogeneous group of chemical structures with various mechanisms of action (Harvey, 1990). Ionophore toxins include polypeptide toxins of different molecular weights and phospholipid selectivity, which aggregate different numbers of toxin monomers to assemble a functional pore in the cell plasma membrane. This produces a variety of pore sizes. According to pore size and the mechanism of cell membrane perforation, we prefer to classify ionophore toxins into four groups (Bloch-Shilderman *et al.*, 1997). The first group, represented by the thiol-activated ionophore toxins from certain Gram-positive bacteria, require a thiol group for binding the toxins to the cholesterol in a eukaryotic cell's plasma membrane (Alouf, 1986). In general, these toxins form aggregates of 70–80 molecules to make large transmembranal pores with a ring- or arch-shaped structure, as seen in electron micrographs (Niedermeyer, 1985). The estimated diameter of such pores is about 15 nm and a single membrane lesion of this size will collapse voltage- and receptor-operated ionic channel activity, resulting in cell death. The second group of ionophore toxins is represented by single chain, low molecular weight polypeptides of 32 (streptolysin S) or 26 (delta toxin, melitin) amino acid residues. They do not form visible, pore-like structures in electron micrographs of intoxicated cells, but cause leakage of intracellular low molecular weight markers, indicative of plasma membrane lesions of about 4–5 nm (Thelestam and Molby, 1979; Buckingham and Duncan, 1983). These lesions formed by toxin oligomers do not display typical pore/channel behavior and may be characterized as causing a surfactant membranal effect, different from the detergent-like membrane solubilization process. The end result is a rapid loss of transmembranal ionic gradients, resulting in the collapse of the signal transduction pathways, loss of ionic homeostasis and cell death. The third group of toxins, represented by *Staphylococcus aureus* alpha-toxin and certain sea anemone toxins, includes proteins that form 4–7 mer pores in the cell membrane, which can be visualized by electron microscopy (Bhakdi and Tranum-Jensen, 1987). These toxins appear to bind selectively to cardiolipin and sphingomyelin. They also interact non-selectively with different lipids in artificial membranes (Menestrina, 1986), forming discrete, non-selective ion channels and small plasma membrane pores, 0.5–2 nm in diameter (Thelestam and Mollby, 1979). At high concentrations and/or following prolonged exposure, this leads to cell lysis. A fourth group of ionophore toxins, which form dimers (gramicidin) or octamers (alamethicin), includes the low molecular weight polypeptide antibiotics that do not appear to bind selectively to any particular phospholipid. These are small (>0.2 nm), ion-permeable channels that may be voltage-dependent and cation-selective (Hall *et al.*, 1981; Anderson, 1984).

The Moses sole fish – a source of pore-forming toxins

Certain fish secrete toxic compounds that repel their predators. One of these, the Red Sea Moses sole (*Pardachirus marmoratus*) (Figure 10.1A,B), exudes a fluid from specialized epithelial glands (Figure 10.1C) into the surrounding ocean water. This secretion repels sharks, and it may function as a naturally occurring weapon of defense against shark predation. The principal factors in the fish-secretion, a family of pore-forming toxins and a family of steroidaminoglycosides, responsible for both its toxicity and noxious effects in sharks, were isolated, identified and named pardaxins (Lazarovici *et al.*, 1986; Shai *et al.*, 1988; Adermann *et al.*, 1998) and mosesins (Tachibana *et al.*, 1986), respectively. The secretion of *P. pavonius* (Peacock sole) in the western Pacific presumably contains similar factors: a family of steroid monoglycosides (pavonins) (Tachibana *et al.*, 1986) and ichthyotoxic pardaxin peptides (Thompson *et al.*, 1986).

Pardachirus marmoratus toxins are secreted by a double row of small, cylindrical, simple acinar glands (Figure 10.1C), located between each of the fin rays from head to tail. The ventral member of the pair of glands releases its secretion from a pore located on the ventral side of the fish, peripheral (lateral) to that of its dorsal partner (Figure 10.1C). Each cylinder contains a central channel (Figure 10.1C–E) that is part of the secretory duct and is lined with interdigitating epithelial cells. The periphery of the cylinder houses a prominent capillary network and the secretory acini lined by thin secretory cells, which surround the acinus, a compartment filled with the secretion. Together, these pairs of secretory units comprise an appreciable portion of the body mass of this small flatfish. A light microscope image through a longitudinally sectioned pair of toxin-secreting glands shows the acini in each gland (Figure 10.1C). The secretory duct lies within the plane of the fin ray, whose articulation can be seen to the left of the image (Figure 10.1C), the articulation separating

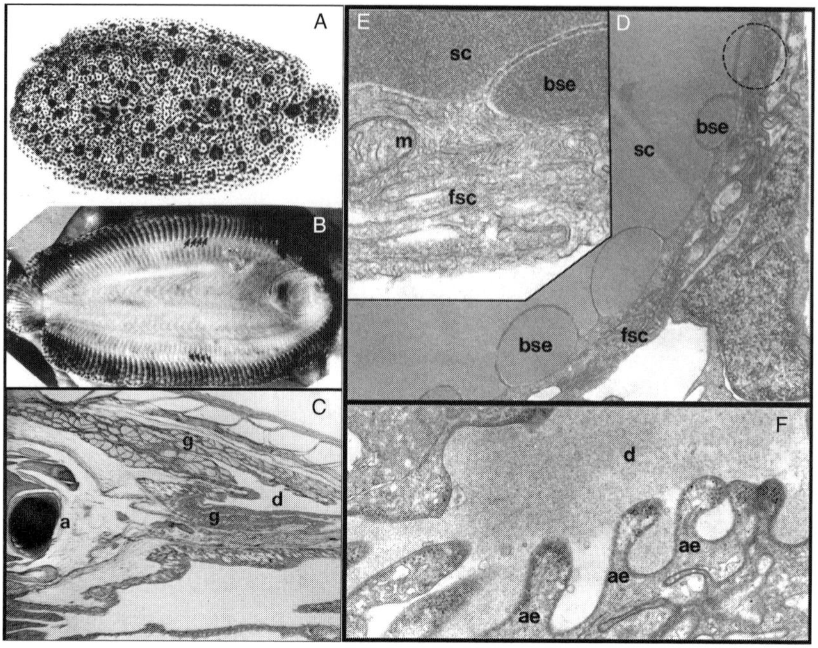

Figure 10.1 *Pardachirus marmoratus* and the morphology of its toxin glands. (A,B)–Lateral views of the fish. Arrows indicate white secretion around the gland openings. (C) Photomicrograph (20×) of the two toxin-secreting glands in sagittal section. The glands (g) are seen right of the ray articulation (a) and are filled with secretion. The clear space within each gland includes the secretory duct(s) (d) which end out of the image to the right. (D) Electron micrograph (30,000×) of the glandular secretory epithelium and a portion of one acinus. Peripheral to the area of moderately electron dense material (sc), which is the toxin secretion(s), is the thin secretory epithelial cell (fsc). Several secretory vesicles (bse) can be seen at its surface, some separated from the contents of the acinar pool only by the thickness of the plasma membrane. (E) Electron micrograph (80,000×) showing secretory cell cytoplasm in detail. The fibrillar nature of the cytoplasmic matrix and the smooth (agranular) appearance of the endoplasmic membranes are evident. Also seen are mitochondria (m) and a portion of one of the plasma membrane elevations, within which is material of almost identical granularity and electron density to that in the acinar pool. No secretory granules are visible in the cytoplasm. (F) This electron micrograph (20,000×) is an image of the cells lining the secretory duct (d). Their rounded apical elevations (ae) project into the secretory space and come into contact with the secretion.

the glands (Lazarovici *et al.*, 1990). An electron micrograph of an acinus shows a large mass of stored secretory material, adjacent to which lies the thin cytoplasm of the secretory cell (Figure 10.1D). The secretion is released into the acinar pool from the secretory cell as globules (Figure 10.1D,E), some of which can be seen raising the plasma membrane. The cytoplasm of an epithelial cell (Figure 10.1E) contains mitochondria, smooth endoplasmic reticulum, and an extensive and dense network of intermediate filaments, probably keratin, but lacks secretory granules. The acinus is surrounded by a system of satellite cells (cells with pleomorphic nuclei) applied closely to them. An electron micrograph of the epithelium lining the secretory duct shows irregular, rounded apical elevations that protrude into the duct space (Figure 10.1E). These cells display a peculiar electron density subadjacent to the plasma membrane, and under that a cytoplasm rich in electron-dense particles with endoplasmic membranes. The other most conspicuous structural feature of these cells is the complex nature of their interdigitation in the region of their lateral boundaries. Many mitochondria with prominent cristae are present in the basal regions of some duct cells. This entire constellation of structures suggests the production of energy to support active ion transport processes (Lazarovici *et al.*, 1990).

Pardaxin sequence, secondary structure and modeling

Pardaxins have been isolated and characterized from the western Pacific Peacock sole, *P. pavonius* (P1, P2 and P3) (Thompson *et al.*, 1986) and the Red Sea Moses sole, *P. marmoratus* (P4) (Lazarovici *et al.*, 1986, 1988; Adermann *et al.*, 1998) (Figure 10.2B). All known pardaxins contain a single-peptide chain composed of 33 amino acids, and are acidic, amphipathic and hydrophobic peptides with a mass of about 3,500 da, and show a high tendency to

A

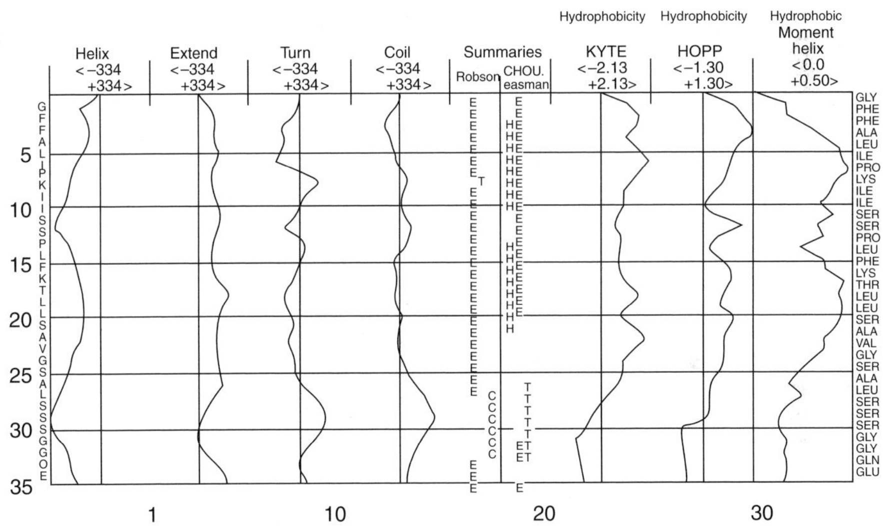

B

P1 GFFA **L** IPKIISSP **L** FKTLLSAVGSALSSSG **E** QE
P2 GFFA **L** IPKIISSP **I** FKTLLSAVGSALSSSG **G** QE
P3 GFFA **F** IPKIISSP **L** FKTLLSAVGSALSSSG **E** QE
P4 GFFA **L** IPKIISSP **L** FKTLLSAVGSALSSSG **G** QE
P5 GFFA **L** IPKIISSP **L** FKTLLSAVGSALSSSG **D** QE

Figure 10.2 Secondary structure analysis and amino acid sequence of pardaxins. (A) The amino acid sequence of pardaxin P4 and the empirical parameters of folding, hydrophobicity and hydrophobic moment using several secondary structure analyses. The 33-residue sequence is presented on both sides of the figure. The computerized secondary structure predictions are based on the method of Garnier *et al.* (1978). H = helix; T = β-turn; C = coil; E = extended. Hydrophobicity values were calculated using computerized programs based on the method of Kyte and Doolittle (1982) or Hopp and Woods (1982). The hydrophobic moment was calculated based on the Eisenberg (1984) scale. (B) Amino acid alignment of pardaxins. P1, P2 and P3 are from *P. pavonius*; P4 and P5 are from *P. marmoratus*. The variable residues are boldfaced.

aggregate in aqueous solution. Their sequences are homologous, differing only at positions 5, 14 or 31 (Figure 10.2B) (Adermann *et al.*, 1998).

Recently, we reported the isolation of a novel pardaxin isoform (P5) from *P. marmoratus* (Adermann *et al.*, 1998). The primary structure of the native peptide was determined by amino acid sequencing, endoproteinase cleavage and mass spectrometry. The replacement of Gly31 of pardaxin P4 by Asp (Figure 10.2B) is the major difference between P4 and P5. The sequence of the novel pardaxin P5 differs by one point mutation from the pardaxin P1 isolated from *P. pavonius* (Figure 10.2B). The substitution of Asp31 by Glu31 would require a mutation of the third base of the Asp31 codon. In addition, the mutation of C to U, the first base of the Leu5 codon of P1, would lead to the sequence of P3, another pardaxin found in *P. pavonius*. In contrast, no single point mutation of any P1, P3 or P5 would result in P2. Pardaxin P2 is genetically closest to P5, differing in position 31 (Asp to Gly:C to A substitution of the first base of the Leu14 codon). We suggest that the pardaxin isoform P4 may provide the mutational gap between P5 and P2. It would be one point mutation away from P5 (Asp31 to Gly31) and another point mutation away from P2 (Leu14 to Ile14). However, the investigation of this evolutionary structure–function relationship of these pore-forming toxins requires further evaluation using other *Pardachirus* species.

The hydrophobic moment plot of pardaxin residue values (Figure 10.2A), based on the consensus hydrophobicity scale of Eisenberg (Eisenberg, 1984), indicates that most of the points fall on or near the peptide's surface region, but that those from its highly hydrophilic carboxy terminal lie in the globular region. This property is typical of surface-seeking, amphiphilic proteins that have large helical hydrophobic moments, such as melittin, *S. aureus* delta-toxin, and other pore-forming toxins. Possible secondary structures of pardaxin were predicted by the Delphi program, using decision constants (DHA = − 75, DHE = 50) favoring α helices over extended segments, and by the Amphi program that identifies possible amphipathic α-helices and β-strands. Delphi predicted that pardaxin segments 1–8 and 16–25 were α-helices and that the remainder might be coils or turns (Figure 10.2A). Amphi predicted that segments 1–12 and 13–28 could form amphipathic α-helices (Figure 10.2A).

Pardaxin is predominantly found in different oligomeric forms in aqueous solutions (Lazarovoci *et al.*, 1986). Approximate three-dimensional models of pardaxin aggregates were developed using computer graphics (Lazarovici *et al.*, 1992). These models then served as starting points for energy refinement calculations to obtain more precise models. In the models presented here (Figure 10.3), pardaxin has two α-helical segments, the amino terminal helix and the carboxy terminal helix, comprising residues 2–10 and 13–27. These segments were assigned helical structures for the following reasons: (i) proline is frequently found in the first position of α-helices and Pro-13 is immediately followed by residues that are frequently observed in α-helices; (ii) initial models indicated that adjacent α-helices pack well next to each other when the C-helix extends to residue 29; and (iii) the segment 11–13 (Ser–Ser–Pro) is quite likely not to be in α-helical conformation and the residues that precede a proline are not usually an α-helical conformation. The initial carboxy terminal helix was created by a program that generates α-helices with the backbone structure and side chain conformations most commonly observed for α-helices in crystal structures (Lazarovici *et al.*, 1990, 1992). This program was not appropriate for an α-helix containing an interior proline. The initial backbone structure, the amino helix, was modeled from the α-helical segment 120–128 in lactate dehydrogenase with the sequence PHE–lys–phe–*Ile*–ILE–PRO–asn–ILE–*Val*, in which most residues were identical or similar to those of the pardaxin segment 2–10, as indicated by capital letters and italics, respectively.

The remaining carboxy terminal residues were assigned phi and psi backbone torsion angles commonly observed in random coil segments, and that extended the Glu33 carboxyl groups into a region postulated to be occupied by water and near Lys6 and Lys16 of adjacent helices. Initially, the conformations of Ser11 and Ser12 were adjusted manually, using computer graphics, so that the amino helices could undergo apparently favorable interactions with the membrane lipid and with other protein segments. Connolly surfaces were added to the structures and monomers were manually docked to form tetrameric structures (Figure 10.3), using the Mogli program on an Evans and Sutherland computer graphics monitor. Carboxy-terminal helices were packed so that Connolly surfaces formed a tight barrier between the inside and outside of the peptide. Most hydrophilic groups not involved in internal hydrogen bonds were positioned to be exposed to water, and most hydrophobic groups were positioned to be in contact with each other. Energies for individual dimers thus created were refined with the CHARMM program, using the adapted basic Newton–Raphson method. A conversion criterion of 0.01 Å was used for the rms gradient during a cycle of minimization. These energy-refined dimers were then used to reconstruct aggregates and a final energy refinement was applied to the complete tetramer. Each monomer was in an identical environment and had the same conformation (Lazarovici *et al.*, 1992). Experimental data indicate that in aqueous solution pardaxin appears to exist primarily as a tetramer and shows concentration-dependent oligomerization (Lazarovici *et al.*, 1986). On this basis, a tetramer (Figure 10.3) was developed, using dimers that have the same packing. The major differences among the monomer conformations of these models involve the carboxy terminus and residues Ser11 and Ser12 that control the relative positions of the amino- and carboxy-helices to one other. In the tetramer the dimers were packed next to each other so that the row of large hydrophobic side chains on the carboxy terminal helices of each dimer were packed next to those of the adjacent dimer (Figure 10.3). This produced a structure similar to that of melittin tetramers. The tetramer structure has two-fold symmetry about each orthogonal axis (Figure 10.3). Helical interactions between carboxy-terminal-helix dimers have "4–4 ridges-into-grooves"-type packing. The amino-terminal helices are positioned so they fit into the hydrophobic groove formed by the carboxy-terminal helices. The amino-termini of two amino-terminal helices approach each other on each side of the tetramer. All hydrophilic side chains are exposed on the surface and most hydrophobic side chains are buried.

Pore-forming activity of pardaxin on artifical membranes

Effects on electrical properties of planar lipid bilayers

In planar lipid bilayers of 2% soybean lecithin (Moran *et al.*, 1984) or cholesterol/phosphatidylcholine (1 : 4, w/w) (Lazarovici *et al.*, 1988), the addition of pardaxin (Figure 10.4) (Lazarovici *et al.*, 1986) at a concentration range of 10^{-6} to 10^{-8} M caused a gradual decrease in membrane resistance after a 5–30 min delay (concentration-dependent), but only when pardaxin was added on the positive side of the membrane. No membrane changes were seen when pardaxin was added to the negative compartment. Consecutive addition of the toxin to the positive compartment of the membrane resulted, under voltage clamp, in a rapid drop in membrane resistance, each time to a lower steady-state level (Moran *et al.*, 1984; Shi *et al.*, 1995). The relationship between resistance and pardaxin concentration was linear and the slope indicated that an aggregate of eight pardaxin molecules is needed to form an active cation-conductive pore (Moran *et al.*, 1984; Shi *et al.*, 1995).

A B

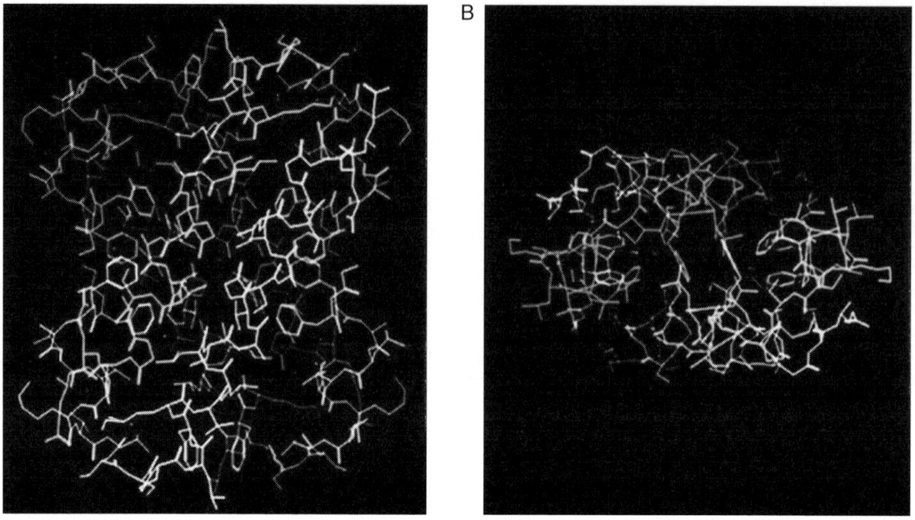

Figure 10.3 Computerized graphic views of pardaxin tetramers (A) The front central portions of the green and yellow monomers are the long amphipathic α-helices that cross at an angle of 15° predicted by "3–4 ridges-into-grooves" helix packing, as in the channel model. Helices of the green and yellow dimer cross those of the blue and purple dimer at an angle of 40°, consistent with "4–4 ridges-into-grooves" helix packing. This places the two residue-wide hydrophobic strip on each dimer next to each other. The N-helices of each monomer fit into hydrophobic grooves and interact with the hydrophobic strips on each side of the tetramer. (B) The tetramer viewed from the top. The transition segment between the N-helix and the C-helix is shown in yellow and blue and the C-terminus is shown in green and purple. The hydrophobic strips are in the center and the N-helices are on each side. The tetramer shows two-fold symmetry about each of the three axes. (See Colour Plate XIII.)

The synthetic amino terminal decapeptide of pardaxin did not affect membrane resistance, suggesting that the channel-forming activity requires more of the intact pardaxin molecule than the hydrophobic amino-terminal 10 residue sequence (Lazarovici *et al.*, 1988). Cross-linked pardaxin is not able to form channels, emphasizing the importance of a native structure for channel conductance. In the presence of pardaxin, when the potential is negative with respect to the side without the toxin, the current–voltage relationship is linear, as in the non-treated planar bilayer (Figure 10.4C). However, if the potential is made positive with respect to the *trans* side of the membrane, rectification is observed (Figure 10.4C): above a certain voltage, the conductance increases. The voltage-dependent (non-ohmic) conductance change seen in the pardaxin-treated planar bilayer is most simply interpreted as a transition of pardaxin pores from a low state of conduction to a highly conductive state. Planar bilayer lipid membranes treated with low concentrations of pardaxin (10^{-8}–10^{-9} M) show current fluctuations under voltage-clamp conditions corresponding to the opening and closing of individual channels (Figure 10.4b). The fluctuations are a common phenomenon at low concentrations and the frequency of these events is dependent on pardaxin concentration and the applied potential (Figure 10.4b). During the fluctuations, the currents tend to assume certain values, which are dependent on the applied voltage.

The stepwise increments in current grow as a function of positive potential (Figure 10.4B, +40 mV). At zero-voltage, there are no current fluctuations (Figure 10.4B, 0 mV) and at −10 mV the channels open in the opposite direction (Figure 10.4B, −10 mV). Thus, pardaxin interacts with the planar bilayer to form hydrophilic pores. The rectification part of the current–voltage curve (Figure 10.4C) (Moran *et al.*, 1984), and the observation that the frequency of opening of a single channel increases sharply above the threshold potential are reminescent of the behavior of many pore-forming toxins that show voltage dependency. The available data indicate that pardaxin pores exhibit poor cation selectivity (Shi *et al.*, 1995),

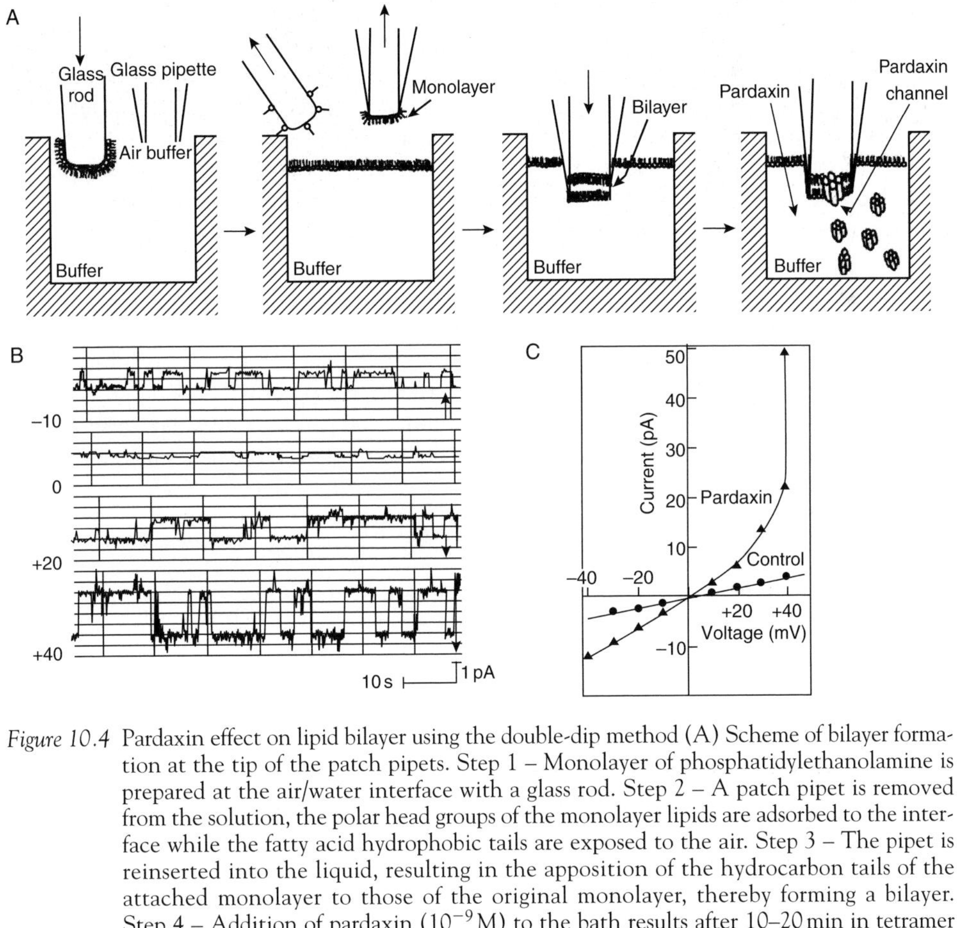

Figure 10.4 Pardaxin effect on lipid bilayer using the double-dip method (A) Scheme of bilayer formation at the tip of the patch pipets. Step 1 – Monolayer of phosphatidylethanolamine is prepared at the air/water interface with a glass rod. Step 2 – A patch pipet is removed from the solution, the polar head groups of the monolayer lipids are adsorbed to the interface while the fatty acid hydrophobic tails are exposed to the air. Step 3 – The pipet is reinserted into the liquid, resulting in the apposition of the hydrocarbon tails of the attached monolayer to those of the original monolayer, thereby forming a bilayer. Step 4 – Addition of pardaxin (10^{-9} M) to the bath results after 10–20 min in tetramer insertion into the bilayer and pore formation, as measured by the current fluctuations depicted in b at a positive potential of 100 mV. The single channel conductance, estimated from the amplitude of the current steps divided by the applied voltage, was in the range of 10 pS. (B) Single channel fluctuations of the planar bilayer induced by pardaxin (5×10^{-8} M) in 5 mM Hepes-Na containing 400 mM $RbCl_2$, pH 7.4 Pardaxin was added to the *cis* side. Current fluctuations were measured at the designated potentials. Arrows represent channel openings. (C) Current–voltage relationship of the membrane composed of phosphatidylcholine/cholesterol (4 : 1; w/w) in the absence (control) or presence (pardaxin) of toxin (10^{-7} M). Pardaxin was applied on the *cis* (positive) side of the membrane at a positive potential of +20 mV and the changes of bilayer resistance were measured at different potentials set by voltage clamp.

suggesting also that the pores may be large. Pardaxin effects are concentration-dependent; with time, the bilayers break and the time of breakage of the planar lipid bilayer is reduced as the concentration increases. This instability of the membrane, similar to liposome lysis (Lazarovici *et al.*, 1986) or an increase in the permeability of biological membranes, is most probably due to formation of large pardaxin/phospholipid micellar structures, resulting in membrane collapse (Lazarovici *et al.*, 1986).

Single channel recordings and selectivity of pardaxin channels on planar bilayers of phosphatidylethanolamine at the tip of patch pipets

To learn about the ion selectivities of the pardaxin channels, their behavior in lipid bilayers was examined. Bilayers of phosphatidylethanolamine were formed at the tips of micropipets (Figure 10.4A) by the double-dip method (Shi *et al.*, 1995). After the bilayer was found to be stable, pardaxin was added to the solution on one side, its potential was set at +20 mV and the currents across the bilayer in the voltage-clamp mode were monitored. At low concentrations of pardaxin (10^{-9}–10^{-8} M) single channel events (Figure 10.4B) usually appeared within 10–20 min; at higher toxin concentrations the latencies were shorter. A potential gradient was required for the toxin molecules to insert and/or express channel activity into the membrane; little or no activity was found in the absence of a potential gradient.

To determine the ionic selectivity of the pardaxin channels, various ion substitutions were performed and the bionic reversal potential, that is, the potential at which the current across a bilayer with many open pardaxin channels changed sign, was determined. The relative permeabilities of the ions could then be determined from the general equation:

$$E_{rev} = \mathrm{RT}\ (P_{C_1}[C_1] + P_{A_1}\ [A_1]\ \mathrm{n}F\ (P_{C_2}[C_2] + P_{A_2}[A_2])$$

where C_1, C_2, A_1 and A_2 are the concentrations of cations and anions on sides one and two. In practice, it was found that Tris was an impermeant cation and HEPES an impermeant anion. Therefore, salts of these compounds were used. For monovalent cations the selectivity sequence was (Figure 10.5A):

$$Tl^+ > Cs^+ > Rb^+ > K^+\ NH_4^+ > Li+ > Na^+$$

which, except for Li^+, is the same as that of the relative hydrated sizes. For monovalent anions the sequence was (Figure 10.5B):

$$I^- > NO_3^- > Br^- > Cl^- > Cl^- > ClO_4^- > SCN^- > HCOO^-$$

which is quite different from that of the relative hydrated sizes. Therefore, the mechanisms governing the permeabilites of cations and anions appear to be different (Shi *et al.*, 1995). To measure the relative permeabilities of K^+ and Cl^-, that is, P_{Cl}/P_K, the reversal potential between 50 mM KCl and 50 mM NaCl was measured. P_{Cl}/P_K, calculated using the equation above and E_{rev}, was 0.78. The experimental value of P_{Na}/P_K was 0.43, and P_{Cl}/P_{Na} was 1.8. Thus, based on these values, pardaxin pores showed only a modest selectivity among these small anions and cations (Shi *et al.*, 1995), compared with many ionic channels in the plasma membranes of eukaryotic cells.

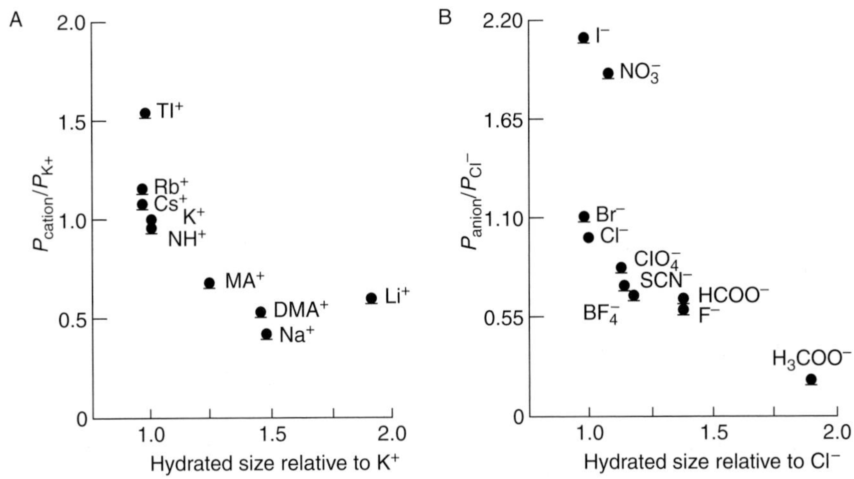

Figure 10.5 Relationship between P_{cation}/P_K and hydrated cation size. (A) Relationship between P_{anion}/P_{Cl} and hydrated anion size. (B) Each point indicates the mean ± SD of 3–8 experiments.

Fluorometric analysis of pore formation in unilamellar liposomes by pardaxin

A valinomycin-mediated diffusion potential is produced across the membrane of unilamellar liposomes (Figure 10.6A) in the presence of a K^+ concentration gradient (Figure 10.6C); the potential can be monitored by the fluorescence of a voltage-sensitive dye (Loew *et al.*, 1985). Addition of 7.5×10^{-12} M gramicidin (Anderson, 1984) causes the immediate collapse of the membrane potential in a fraction of the liposome population, as shown by the recorded recovery in fluorescence (Figure 10.6C). Pardaxin at a concentraion of 5×10^{-10} M showed a similar rate of fluorescence recovery. The partial recovery suggests that this action of pardaxin is mediated through pores, and not by detergent lysis of the liposomes. The same behavior was observed in sodium and choline solutions, indicating that the pores are non-selective. The molar concentration of the pores can be estimated by using gramicidin as a standard, or directly from the product of the fractional fluorescence recovery and the liposome concentration (Loew *et al.*, 1985). The methods give comparable results and allow quantitative estimation of pore-forming activity. For pardaxin the concentration dependence of the pore-forming activity can be treated with Poisson statistics and fits a model in which the pore is composed of between four and twelve monomers (Loew *et al.*, 1985). Electron microscopic examination of unilamellar liposomes (Figure 10.6A, inset) incubated for 30 min with pardaxin at 10^{-9}–10^{-10} M did not reveal any significant lysis, swelling or other morphological changes (Lazarovici *et al.*, 1986). However, at concentrations higher than 10^{-7} M, lysis of the large liposomes and disruption or aggregation of the small liposomes was observed (Lazarovici *et al.*, 1986, 1988).

^{86}Rb flux through gramicidin- and pardaxin-induced pores in multilamellar liposomes

The presence of ion-selective pores in multilamellar liposomes (Figure 10.6B) was investigated by measuring the uptake of radioactive ions against a large chemical gradient of the

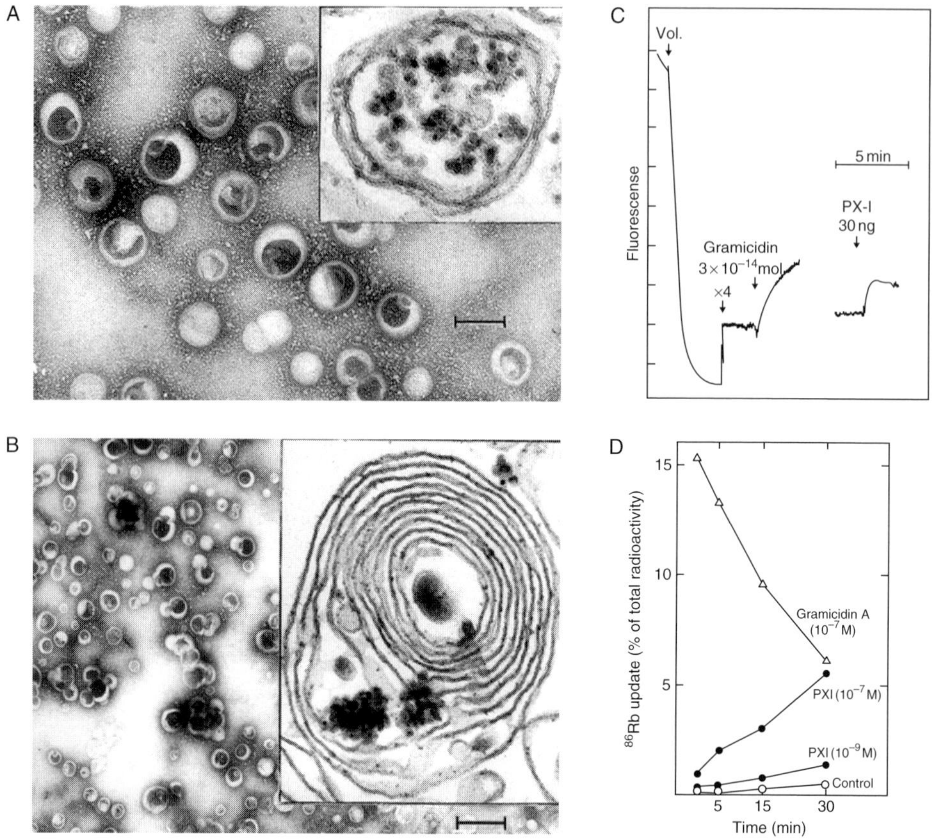

Figure 10.6 Pore activity of pardaxin in large unilamellar (A,C) and multilamellar (B,D) liposomes visualized by negative staining (A,B) and transmission electron microscopy (inset of A,B). (A) Typical views of negatively stained, Sephacryl S-200 gel permeation fractions of large, unilamellar vesicles. Bar = 0.1 μm. (C) Fluorometric analysis of pardaxin and gramicidin pores. The two traces are records from separate experiments. The first trace shows the partial recovery of fluorescence due to incorporation of gramicidin into a small fraction of the vesicle population. The second trace was recorded after the diffusion potential had already been established at a gain identical to that established before adding gramicidin in the first trace. Arrows show introduction of toxin. (B) Typical view of negatively stained, multilamellar vesicles. Bar = 0.05 μm. (D) ^{86}Rubidium uptake in multilamellar liposomes, against a chemical gradient induced by pardaxin and gramicidin. The data (% of total radioactivity) represent the mean of triplicates in which liposomes were diluted in Tris–EDTA-imidazole buffer, pH 7.2, containing the different compounds. The uptake of control liposomes and the standard errors did not exceed 0.5% of the total radioactivity.

same unlabeled ion (Lazarovici *et al.*, 1988). Under these conditions, an electrical diffusion potential was formed spontaneously, as a consequence of a K^+ gradient; the magnitude of the potential was determined by the relative permeabilities of the liposomal membrane to K^+, Cl^- and Tris. In those vesicles containing pardaxin pores the K^+ permeability was greater than that of Cl^- and Tris, producing a K^+ diffusion potential. Since ^{86}Rb usually behaves similarly to K^+, it should tend to equilibrate according to the membrane potential, and to accumulate in vesicles containing the pardaxin pores. The effects of gramicidin and pardaxin on the time-course of ^{86}Rb uptake into soybean phospholipid multilamellar liposomes containing 150 mM KCl in the presence of a large potassium gradient are shown in Figure 10.6D. Gramicidin at 10^{-7} M caused, within 1 min of incubation, significant uptake of ^{86}Rb. Pardaxin (10^{-7} M), however, had no measurable effect at 1 min. However, there was considerable ^{86}Rb uptake after a 30-min incubation. Compared with gramicidin, pardaxin showed much slower ^{86}Rb uptake (Lazarovici *et al.*, 1988). It is possible that this attentuated uptake is a reflection of the time required for the pardaxin molecules to form an active channel or is the result of some complex behavior of the channel, resulting in both conductance of ^{86}Rb and dissipation of the gradient. However, in the absence of a K^+ gradient or pardaxin, the uptake of ^{86}Rb proved to be insignificant.

Modeling pardaxin pores

Pore-forming toxins have protein sequences consistent with the formation of an amphipathic α-helix (Guy and Raghunathan, 1989). An amphipathic α-helix is defined as an asymmetric distribution of polar and apolar residues between the opposite sides of the helix. Transmembrane portions of many ion channels and ionophore toxins appear to be comprised primarily of amphipathic α-helices packed in parallelly or antiparallelly, and spanning a lipid bilayer membrane with the hydrophilic faces of the α-helix lining a central pore and the hydrophobic surfaces interacting with the hydrophobic phospholipid chains of the membrane (Guy and Conti, 1990; Raghunathan *et al.*, 1990). The pardaxins, 33-amino acid residue monomer sequences, can be divided into three regions (Figure 10.6). The amino-terminal region A (Gly1-Ser12), with the exception of Lys8, is highly hydrophobic and the Lys8–Pro13 segment is of a helical nature (Shai *et al.*, 1990; Barrow *et al.*, 1992; Lazarovici *et al.*, 1992). Region B (Ile14–Leu26) is an amphipathic α-helix (Zagorsky *et al.*, 1991; Barrow *et al.*, 1992; Lazarovici *et al.*, 1992). The non-polar residues (Ile14, Phe15, Leu18, Leu19, Ala21, Val22, Ala25, Leu26) reside on the face opposite the polar or charged residues (Lys16, Thr17, Ser20, Ser24) (Zagorsky *et al.*, 1991). Pro13 causes a bend between the two helices (Zagorsky *et al.*, 1991; Lazarovici *et al.*, 1992). The carboxy-terminal region C (Ser27–Glu33) appears to exist in an extended coil or β-sheet conformation, as evident from NMR data (Zagorsky *et al.*, 1991) and molecular modeling (Lazarovici *et al.*, 1992). Using synthetic pardaxin analogs, the basic structural activity requirements for pore formation have been identified (Shai *et al.*, 1990, 1991; Barrow *et al.*, 1992; Rapaport and Shai, 1992; Shai, 1994): (1) Pardaxin binds strongly to the membrane via the hydrophobic amino-terminal region A; (2) Pardaxin penetrates into the hydrophobic phase of the membrane, as detected by fluorescence energy transfer (Rapaport and Shai, 1992) or proteolysis and HPLC analysis (Lazarovici *et al.*, 1992); (3) Pardaxin monomers self-associate in its membrane-bound state with parallel orientation of the monomers within the aggregate (Rapaport and Shai, 1992); (4) The pardaxin B region contains the minimal requirements for amphipathicity and maintenance of the two α-helix segments (Barrow *et al.*, 1992); (5) The pardaxin carboxy-terminal C region is crucial for pore activity and is the pore-binding segment (Shai *et al.*, 1990); (6) Upon association of pardaxin with the membrane, rearrangement of the secondary structure occurs, resulting in increased α-helicity (Lazarovici *et al.*, 1988, 1992); (7) An oligomer of eight pardaxin monomers is required for an active pore (Loew *et al.*, 1985). Based on these data, three-dimensional models of pardaxin pores were developed (Figure 10.6) with the aid of computer graphics (Guy and Raghunathan, 1989; Lazarovici *et al.*, 1992).

Using the criteria described in the modeling of the pardaxin tetramer, the pardaxin pore was conceived as a cylinder of eight parallel monomers in which eight stranded β-barrels formed by residues 28–33 (region C, Figure 10.7) are surrounded by eight amphipathic α-helices formed by residues 13–26 (region B, Figure 10.7). Residues 1–10 (region A, Figure 10.7) are postulated to form a ring of α-helices around the outer entrance of the pore (Lazarovici *et al.*, 1992). The shape of the pore is an hourglass with a narrow selectivity filter of 4.8 Å diameter, and there are no free charges inside the pore lining (Figure 10.7) (Lazarovici *et al.*, 1992). This is consistent with the finding that these channels transport both cations and anions (Shi *et al.*, 1995). This model also conforms to structure-activity studies (Shai, 1994) and is similar to other models developed for melittin, *S. aureus* delta toxin, magainins and other ionophore toxins (Guy and Raghunathan, 1989; Raghunathan *et al.*, 1990). The relation between structure and function at the single amino acid residue level and the understanding of ion binding and movement through putative ionic channels or ionophore toxin pores is nonetheless not available. Pardaxin and other pore-forming toxins should continue to serve as general models for transmembrane pores as their small size and availability make them ideal tools for biophysical investigations.

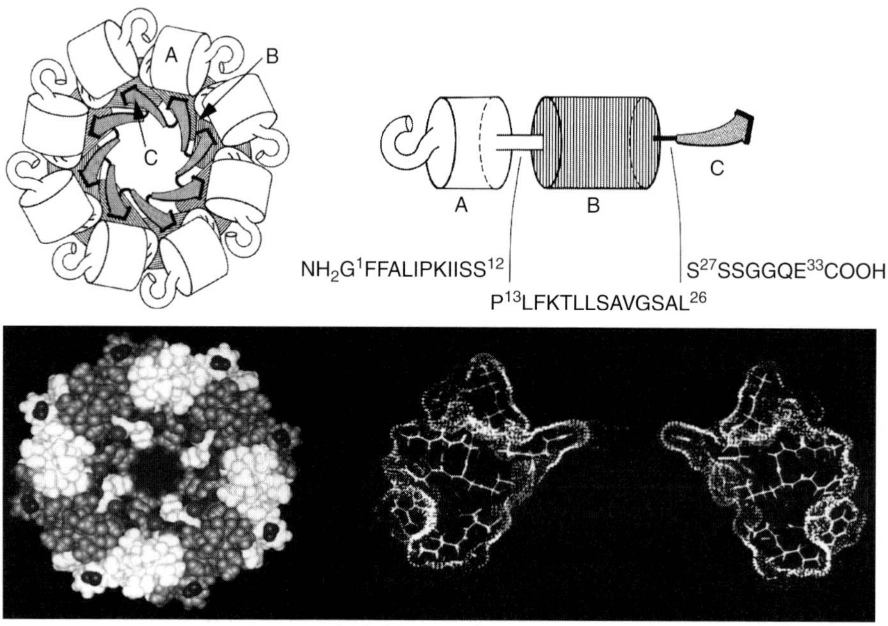

Figure 10.7 Pardaxin pore model. Upper part: A schematic view of pardaxin secondary structure and pore. The pardaxin pore is composed of eight monomers (left). Each monomer contains three regions (right). (A) residues 1–12 from the amino terminal containing one helix: (B) residues 13–26 containing an amphipathic α-helix. (C) residues 27–32 at the carboxy terminal involved in ion permeation. Lower part: Computerized models of pardaxin pores. Left – Space-filled rendering of the pardaxin channel looking down the channel from the *cis* side. Charged oxygens are shown in red, noncharged oxygens in brown, charged nitrogens in blue and noncharged nitrogens in light blue. The ring of hydrogen bonds formed by Gln32 at the narrow region and salt bridges formed between Gln33 and Lys8 and Lys 16 at the entrance to the channel are shown. Every other monomer is coloured green or white. Right – Cross-section of the channel showing the molecular surface. The hourglass shape of the channel with the narrow region formed by Gln32 is seen. (See Colour Plate XIV.)

Presynaptic effects of pardaxin on the neuromuscular junction and on synaptosomes

Pardaxin increases the frequency of the spontaneous release of transmitter quanta at the neuromuscular junction (Renner *et al.*, 1987) by up to 100-fold (Figure 10.8A) and also increases the quantal content (Figure 10.8C) of the evoked end plate potential. The increase in the frequency of the spontaneous release of transmitter quanta is dose-dependent and influenced by temperature. The increase in m.e.p.p. frequency (Figure 10.8A) is not accompanied by a change in amplitude (Figure 10.8B), suggesting that postsynaptic effects are absent. The delay (5–10 min) between pardaxin application to the bath and the start of its effects (Figure 10.8A–C) is indicative of slow kinetics of insertion and/or action on the nerve terminal. The presynaptic effects of pardaxin were reduced when Ca^{2+} in the Ringer solution was replaced by Mg^{2+} or Sr^{2+}, indicating that the toxin effect on m.e.p.p. frequency is at least partly dependent on external Ca^{2+}. The synthetic amino-terminal

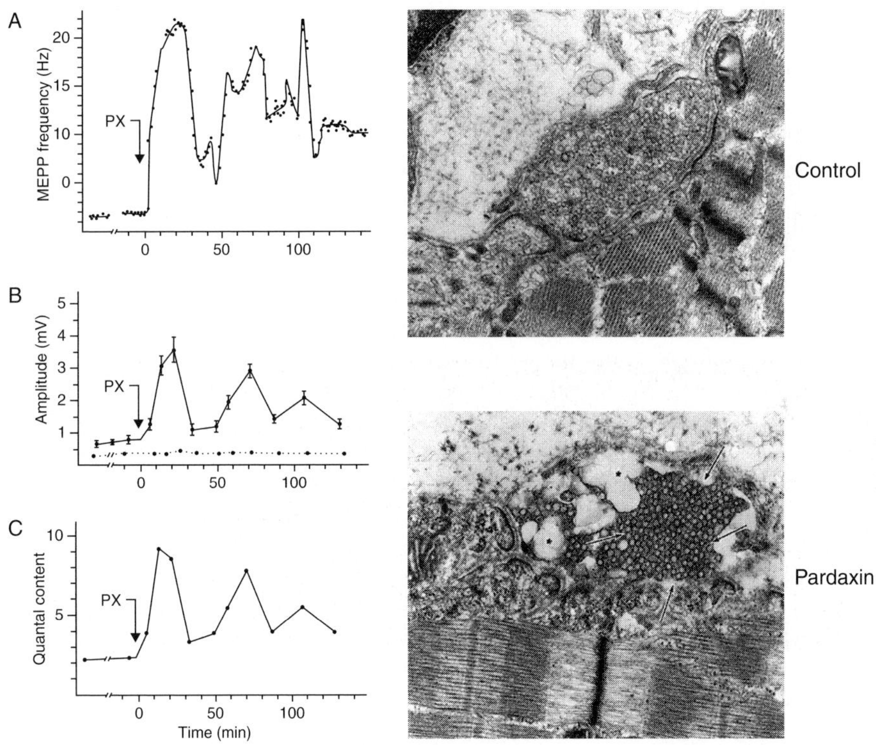

Figure 10.8 Effect of pardaxin on the release of acetylcholine and the morphology of the neuromuscular junction. Left: Effect of a single application of pardaxin (PX) (0.5 μg/ml) on the release of acetylcholine at a single end-plate in the sartorius muscle of the frog. The muscle was equilibrated in low-calcium Ringer solution (0.6 mM Ca; 4.2 mM Mg) at room temperature. At time 0 the toxin was added to the bath. The abscissa for all three graphs: time in min. Ordinates: A, frequency of the MEPPs in Hz; B, amplitude of the EPPs (—) and of the MEPPs (······) in mV; C, quantal content of the EPPs. Right: Cross-section (control) and longitudinal section (PX) through control nerve terminal or terminal treated with 3×10^{-7} M pardaxin for 30 min at 21°C. Arrows indicate vesicular cluster. Magnification ×53,000.

decapeptide of pardaxin was ineffective in promoting neurotransmitter release at up to 10^{-5} M, indicative of the importance of an intact pardaxin molecule structure for the presynaptic effect. The specificity of the presynaptic effects of pardaxin was tested in a series of experiments (Renner *et al.*, 1987). The effects of lanthanum ions (1 mM) at the end of the experiment varied, depending on the previous presynaptic effects of the toxin. At low concentrations of pardaxin, lanthanum caused the usual extremely high-frequency burst of m.e.p.p.s before the frequency fell to zero (Figure 10.8). If the toxin levels showed a strong neurotoxic effect or a fairly large effect over a long period of time (more than 1 h), the addition of lanthanum produced only a rapid decline in the frequency of m.e.p.p.s to zero. The possible effects of pardaxin on nerve conduction were examined on the isolated *N. ischiadicus* nerve of the frog. The amplitude of the extracellularly recorded action potential of the whole nerve did not change, even after the application of 10^{-5} M pardaxin for more than 2 h. Comparative experiments were performed to determine whether the channel-forming ionophore gramicidin has presynaptic effects. Gramicidin, at a concentration range of 1–5 μg/ml, failed to increase the frequency of the m.e.p.p.s significantly; the main effect was a loss of the resting potential, which reduced the amplitude of the m.e.p.p.s to undetectable values.

The effects of pardaxin on the ultrastructure of the nerve terminal at the sartorius neuromuscular junction are also shown in Figure 10.8. Electron microscopy shows that the nerve terminal is normally separated from the muscle by a synaptic cleft (~450 Å) containing an electron-dense basement membrane. The dense region in the presynaptic membrane is considered to be the active zone of neurotransmitter release. The synaptic vesicles appear to be of uniform size (~500 Å) and shape (Figure 10.8, control). Finger-like projections of Schwann cells are seen. After treatment with 3×10^{-7} M pardaxin, there was much aggregation of cholinergic vesicles inside the terminal (Figure 10.8, pardaxin) as well as empty areas. These histological changes are similar to the earlier findings showing clumping of acetylcholine vesicles after prolonged incubation of neuromuscular junctions in buffer containing a high Ca^{2+} concentration (Renner *et al.*, 1987).

The effect of pardaxin on neurotransmitter release was also tested using purely cholinergic synaptosomes of *Torpedo marmorata* electric organ (Arribas *et al.*, 1993). Pardaxin elicited the release of acetylcholine with a biphasic dose-dependency. At low concentrations (up to 3×10^{-7} M), the release was calcium-dependent and synaptosomal structure was well preserved, as revealed by electron microscopy and measurements of released lactate dehydrogenase activity. At concentrations of 3×10^{-7}–10^{-5} M, the pardaxin-induced release of acetylcholine was independent of extracellular calcium, and released synaptosomal lactate dehydrogenase activity decreased, indicating a synaptosomal membrane perturbation. Electron microscopy of 10^{-6} M pardaxin-treated synaptosomes revealed nerve terminals depleted of synaptic vesicles and containing cisternae. At higher toxin concentrations ($\geq 10^{-5}$ M), there were striking effects on synaptosomal morphology and released lactate dehydrogenase activity, suggesting a membrane lytic effect. We concluded that, at low concentrations, this neurotoxin is a promising tool for investigating mechanisms of neurotransmitter release in the nervous system and that the excitatory presynaptic effect measured in vivo can be reproduced in vitro using synaptosomal preparations.

PC12 pheochromocytoma cells: a neuronal model of pardaxin-induced neurotransmitter release

The clonal PC12 cell line is now widely used in various studies on neurons and adrenal gland chromaffin cells (Greene and Tischler, 1976). Regulated exocytosis of catecholamines has been investigated largely in bovine adrenal chromaffin cells (Burgoyne, 1991) and in rat PC12 cells (Ahnert-Hilger *et al.*, 1985). The secretory vesicles of these cells, the chromaffin granules, similar to synaptic vesicles, store dopamine, norepinephrine, ATP and various proteins. The essential role of calcium in catecholamine secretion from chromaffin cells has been well established (Burgoyne, 1991). Cytosolic calcium ($[Ca]_i$) can be increased due to membrane depolarization, opening of receptor-operated channels or release of calcium from intracellular stores (Burgoyne, 1991). The concentration of ($[Ca]_i$) is strictly regulated, and it is considered that the increase in calcium concentration within the microdomain of the active exocytotic zone allows vesicles to fuse and release their catecholamine content. Therefore, we have been using the PC12 cell model to investigate the mechanism by which pardaxin induces neurotransmitter exocytosis, with emphasis on both calcium-dependent and calcium-independent processes (Lazarovici, 1994).

The relative high potency and specificity of the presynaptic effects of pardaxin suggest that the toxin may induce its biological effect via a specific interaction with some component(s) of the presynaptic membrane. To explore this possibility and to elucidate the relationships between binding and neurotransmitter release, we iodinated pardaxin. Radioreceptor assays, measuring the binding of $[^{125}I]$pardaxin (Lazarovici *et al.*, 1988) to pheochromocytoma PC12 cells, did not reveal any receptor-mediated binding phenomena such as saturability, specificity or competition. Using PC12 cells, the maximal binding of pardaxin occurred between 10^{-10} and 10^{-8} M (Figure 10.9B). However, the binding of pardaxin was reduced only by a high concentration of unlabeled pardaxin (greater than 10^{-6} M), suggesting that this phenomenon did not represent binding to specific receptors on the cell surface but, most probably, reflected a more complex physiochemical interaction of the toxin with the cells. At high concentrations of pardaxin (10^{-8}–10^{-4} M) there was a decrease in the binding of pardaxin and the PC12 cell cultures were depolarized, as reflected by a reduction in the uptake of the radioactive permeant cation triphenylphosphonium (Figure 10.9B). The most plausible interpretation of the binding data is that the pardaxin-induced depolarization of the cell membrane (10^{-8}–10^{-4} M) decreases the cell-associated $[^{125}I]$pardaxin. Indeed, at this range of pardaxin concentration the PC12 cell membrane was found to be permeable to both ^{86}Rb and ^{45}Ca influx. Pardaxin at concentrations of 10^{-8}–10^{-4} M stimulated the release of ATP and catecholamines in the absence or presence of millimolar concentrations of calcium in Krebs–Ringer buffer (Figure 10.9A and Lazarovici and Lelkes, 1992). Maximal release was found at concentrations of 10^{-5}–10^{-4} M (Figure 10.9A), at which pardaxin pore formation significantly changed cell membrane permeability. Measurements of intracellular calcium using Fura-2-loaded cells indicated a strong influx of extracellular calcium at pardaxin concentrations of 10^{-7}–10^{-5} M (Figure 10.9A, and Lazarovici and Lelkes, 1992). At concentrations higher than 10^{-5} M, pardaxin induced a strong necrotic, cytotoxic effect, reflected by leakage of lactate dehydrogenase (LDH) (Figure 10.9B), followed by cell disintegration (Bloch-Shilderman *et al.*, 2002a). In conclusion, to avoid cytotoxic effects, the pharmacological experiments with pardaxin should be performed at concentrations of 10^{-7}–10^{-5} M.

Characterization of pardaxin-induced neurotransmitter release: the role of calcium

The ability of pardaxin versus other secretagogues to induce dopamine release from PC12 cells in the presence or absence of extracellular calcium ($[Ca]_0$) was previously investigated in detail (Abu-Raya *et al.*, 1999). Pardaxin stimulated dopamine release under both conditions. Pardaxin-induced dopamine release is an exocytotic process, as measured by

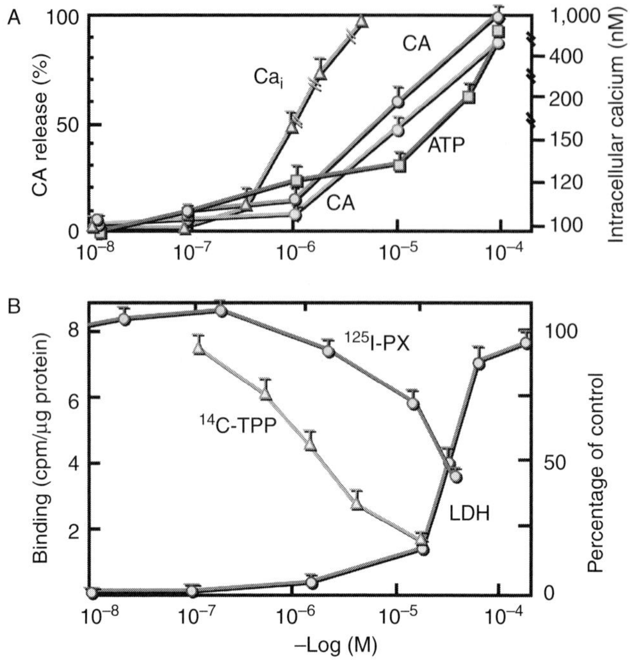

Figure 10.9 Pardaxin-induced catecholamines and ATP release, increase in intracellular calcium binding and neurotoxicity. (A) Relationship between pardaxin-induced catecholamine (CA) (cyan, purple) and ATP release (red). Levels of intracellular calcium (Ca) (mustard) were measured using Fura-2-loaded cells at the indicated pardaxin concentrations. Each value is the average of three culture wells. S.D. was in the range of $\pm$ 10%. CA release in the presence (purple) or absence (cyan) (and 1 mM EGTA) of $[Ca^{++}]_0$ was performed for 15 min at 37°C. The values represent the average of triplicate wells with a S.D. $\pm$ 2%. ATP release was assayed by the luciferin/luciferase method after toxin incubation for 15 min at 37°C with cell cultures. (B) Pardaxin binding, cytotoxicity and [^{3}H]TPP uptake by PC12 cells. In binding experiments (green), PC12 cell monolayers were incubated for 1 h at 37°C in choline medium (140 mM choline chloride, 5 mM KCl, 0.8 mM $MgSO_4$, 1.8 mM $CaCl_2$, 10 mM glucose, 25 mM Hepes and 10 mM Tris, pH 7.4) with ^{125}I-pardaxin (3×10^{-10}M) in the presence of increasing concentrations of unlabeled toxin. The binding reaction was terminated by removing the incubation reaction mixture and washing the cells twice with binding buffer containing 0.2% ovalbumin, before solubilization with 0.5 N NaOH and counting of cell-associated radioactivity. Neurotoxicity was estimated in a 45 min assay, at 37°C, by measuring lactate dehydrogenase (LDH, blue); the cell membrane potential was estimated by following the uptake of [^{14}C]triphenylphosphonium bromide (^{14}C-TPP, yellow). Uptake for 45 min at 37°C was initiated by adding 0.1 µCi to cell monolayers in 2 ml choline buffer and was allowed to proceed in the presence or absence of the indicated pardaxin concentrations. After removal of the radioactive reagents and washing with buffer, cell-associated radioactivity was measured. (See Colour Plate XV.)

a vesicle fusion-based ELISA reflecting the exposure of the vesicular SNARE protein synaptotagmin to the cell surface (Bloch-Shilderman *et al.*, 2001). Pardaxin, in contrast to KCl, carbachol and bradykinin, induced dopamine release from PC12 cells in the absence of $[Ca]_0$ (Abu-Raya *et al.*, 1999), similar to pardaxin-stimulated, calcium-independent catecholamine release from bovine adrenal medullary chromaffin cells (Lazarovici and Lelkes, 1992). To clarify the effect of pardaxin on intracellular calcium, PC12 cells loaded with acetoxymethyl ester of Fura-2 (Fura-2-AM) were treated with pardaxin in the presence or absence of extracellular calcium ($[Ca]_0$). There was a gradual and sustained increase in cytosolic calcium in calcium-containing medium following the addition of pardaxin. In the absence of extracellular calcium, pardaxin did not augment $[Ca]_i$. Since pardaxin did not increase $[Ca]_i$ in calcium-free medium we concluded that pardaxin did not release calcium from intracellular stores. The influx of calcium elicited by pardaxin in the presence of extracellular calcium should enhance dopamine release. To test this possibility, we determined the effect of 1,2-bis(2-aminophenoxy) ethane N,N,N′,N′-tetra-acetic acid (BAPTA-AM), a membrane-permeant chelator of cytosolic calcium. In the presence of $[Ca]_0$, BAPTA (10 μM) inhibited KCl-induced dopamine release by 63% and pardaxin-induced dopamine release by 40%. However, pardaxin-induced dopamine release in the absence of $[Ca]_0$ was not affected by BAPTA treatment. These data suggest that the calcium influx is partially involved in pardaxin-induced dopamine release and that mobilization of calcium from intracellular stores is not the mechanism involved in the pardaxin stimulation (Abu-Raya *et al.*, 1999).

Pardaxin stimulation of the arachidonic acid cascade in PC12 cells

As the triggering of neurotransmitter release induced by pardaxin could be due to stimulation of phospholipases, we explored this possibility by treating PC12 cultures labeled with arachidonic acid ($[^3H]$AA) with several concentrations of pardaxin, in the presence or absence of extracellular calcium (Abu-Raya *et al.*, 1998). In the presence of extracellular calcium, 10^{-6} M and 10^{-5} M pardaxin stimulated AA release 2.3- and 10-fold vs that in the control, respectively (Figure 10.10C). In the absence of extracellular calcium, 10^{-6} M and 10^{-5} M pardaxin stimulated AA release 1.6- and 7.1-fold vs the control, respectively (Figure 10.10C). Stimulation of AA release by pardaxin was detected after 5 min of incubation, whereas maximal stimulation was measured after 30 min of incubation. The same time course of AA release was observed in the presence or absence of extracellular calcium. The effect of pardaxin on prostaglandin E_2 (PGE_2) (an arachidonic acid metabolite produced by the cyclooxygenase pathway) release in the presence or absence of extracellular calcium was also determined (Abu-Raya *et al.*, 1998). Treatment of the cultured cells with 5×10^{-5} M and 1.5×10^{-6} M pardaxin for 30 min in calcium-containing medium stimulated PGE_2 release by five- and thirteen-fold over that of the control, respectively (Figure 10.10D). In the absence of extracellular calcium, PGE_2 release in the presence of pardaxin was between 50 and 70% of that obtained in the presence of extracellular calcium (Figure 10.10D). Similar to native pardaxin, synthetic pardaxin stimulated PGE_2 release about three-fold more than in the control. To verify the selective effect of pardaxin on PGE_2 release, three pardaxin structural analogs lacking a helical structure were tested and proved to be ineffective in stimulating PGE_2 release (Abu-Raya *et al.*, 1998). To verify that AA release induced by pardaxin is mediated mainly by phospholipase A_2 (PLA_2) activation, the effect of pardaxin on the ^{32}P-phospholipid content of PC12 cells was examined (Figure 10.10A and B, and Abu-Raya *et al.*, 2000). The levels of the major phospholipid species phosphatidylethanolamine (PE), phosphatidylinositol (PI) and phosphatidylserine (PS) did not change after pardaxin treatment (Abu-Raya *et al.*, 1998). These data indicate that phospholipase C activation is not involved in the pardaxin-induced arachidonic acid cascade. In addition, synthesis of phosphatidic acid (PA) was not observed. Since pardaxin did

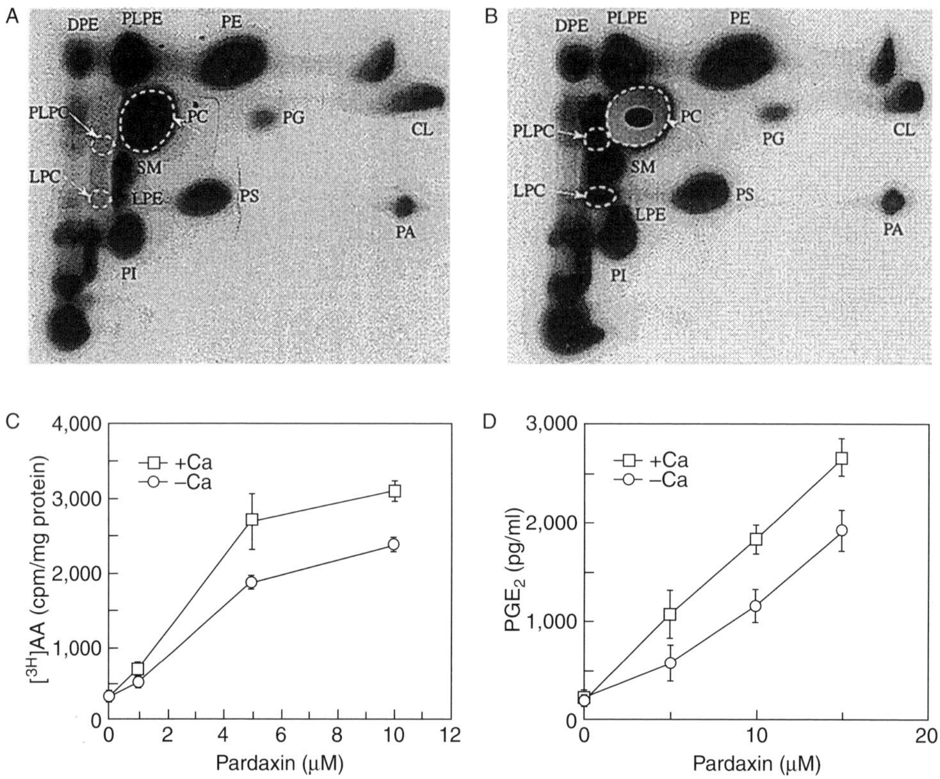

Figure 10.10 Pardaxin stimulation of the arachidonic acid pathway. (A,B) Thin layer chromatography (TLC) of [^{32}P] orthophosphate incorporated into the major phospholipids in PC12 cells before (A) or after (B) pardaxin treatment. PC12 cells (2×10^6 cells/dish) were labeled with [^{32}P] orthophosphate, washed and treated with 5 μM pardaxin for 15 min at 37°C. The phospholipids were extracted with n-propanol and separated by TLC. The plates were exposed for autoradiography and scanned. PE – phosphatidylethanolamine; PS – phosphatydylserine; PC – phosphatidylcholine; CL – cholesterol; PI- phosphatidylinositol; LPC – lysophosphatidylcholine; PLPC – phospholysophosphatidylcholine; PA – phosphatidic acid. (C) Dose-dependent stimulation of [^{3}H]AA release from PC12 cells by pardaxin in the presence (+Ca) or absence (−Ca) of $[Ca]_0$ (+ 1 mM EGTA). Cells labeled with [^{3}H]AA were incubated with the toxin for 30 min. (D) Dose-dependent stimulation of PGE_2 release from PC12 cells by pardaxin in the presence (+Ca) or absence (−Ca) of $[Ca]_0$ (+1 mM EGTA). PC12 cells were incubated with pardaxin at the appropriate concentrations at 37°C for 30 min, and PGE_2 was determined in the culture medium.

not lead to PA generation, it appears also that phospholipase D (PLD) is not activated and/or involved in pardaxin-induced AA release. However, a reduction of 30% in the phosphatidylcholine (PC) level was observed after pardaxin treatment, which was accompanied by an approximately 40% increase in lysophosphatidylcholine (LPC) synthesis (Figure 10.10B), indicating PLA_2 activation. As pardaxin affects the PC level, but not PS or PE, these findings suggest that PC is one of the major phospholipid sources utilized by pardaxin for AA mobilization in PC12 cells via stimulation of phospholipases A_2 (Abu-Raya *et al.*, 1998, 2000).

Characterization of pardaxin-induced neurotransmitter release: the role of arachidonic acid and eicosanoids

Since AA and eicosanoids may act as intracellular second messengers and affect synaptic transmission, we examined the effect of AA cascade inhibitors on pardaxin-induced dopamine release in PC12 cells (Abu-Raya *et al.*, 1999). Following treatment with indomethacin (a cyclooxygenase inhibitor), there was an increase of about 50% in dopamine release in response to pardaxin in the presence or absence of $[Ca]_0$. At 10^{-5} M pardaxin, the indomethacin effect was not significant. These results suggest that the cyclooxygenase pathway is not involved in pardaxin-induced dopamine release.

To determine the involvement of the lipoxygenase pathway in pardaxin-induced dopamine release, several lipoxygenase inhibitors were tested (Abu-Raya *et al.*, 1999; Bloch-Shilderman *et al.*, 2002b). Esculetin, 2-(12-hydroxydodeca-5,10-diynyl)-3,5, 6-trimethyl-1,4-benzoquinone (AA861) and nordihydroguaiaretic acid (NDGA) inhibited pardaxin-induced dopamine release by about 50% in calcium-containing medium. A parallel inhibition of pardaxin-induced 5-hydroxyeicosatetranoic acid (5-HETE) release by these compounds was also observed. In the absence of extracellular calcium, NDGA and AA861 inhibited pardaxin-induced dopamine release by about 90% (Abu-Raya *et al.*, 1999). To further confirm the involvement of the arachidonic acid cascade in pardaxin-induced dopamine release, we tested the effect of two PLA_2 inhibitors. 4-Bromophenacyl bromide (BPB) and mepacrine inhibited by 47% and 60%, respectively, pardaxin-induced dopamine release in the presence of $[Ca]_0$, and by 73% and 87%, respectively, in the absence of $[Ca]_0$. In parallel cultures, BPB and mepacrine inhibited by 75–90% pardaxin-induced AA release independently of $[Ca]_0$ (Abu-Raya *et al.*, 1999; Bloch-Shilderman *et al.*, 2002b). These results suggest the involvement of PLA_2 activation in neurotransmitter release by pardaxin (Abu-Raya *et al.*, 1999; Bloch-Shilderman *et al.*, 2002b).

Although the role of calcium in neurotransmitter release has been widely investigated, the intracellular signaling pathways whereby eicosanoids activate neurotransmitter exocytosis in the absence of calcium are unknown. It would be interesting to know whether AA metabolites act as intracellular second messengers or in an autocrine fashion. The possibility that the generation of lipid products such as eicosanoids directly regulates the SNARE-mediated synaptic vesicle fusion process, independently of calcium, also merits careful examination. Pardaxin can provide a new pharmacological tool for clarifying these issues.

Pardaxin stimulation of MAPKs in PC12 cells

One of the groups of enzymes under intense investigation as part of the effort to understand signaling cascades that are activated by a number of extracellular stimuli is the extracellular signal-regulated protein kinase (ERK), also referred to as the mitogen-activated protein kinase (MAPK) superfamily of enzymes. This large family of protein kinases includes over a dozen members that participate in many eukaryotic regulatory pathways (Blenis, 1993). These enzymes comprise at least three parallel, yet interwoven, signal transduction cascades that are differentially regulated in response to mitogens, growth factors, cytokines and various forms of stress (Force and Bonventre, 1998). Each cascade consists of a minimum of three enzymes activated in series: a MAP kinase/ERK kinase or MEKK (a MEK activator), a MAP kinase/ERK kinase or MEK (a MAP kinase activator), and a MAP kinase/ERK homolog (Blenis, 1993).

The first and best-studied enzymes of the ERK/MAPK superfamily are part of in the classical mitogen-activated (MAPK) subfamily pathway. The c-Jun N-terminal protein kinase

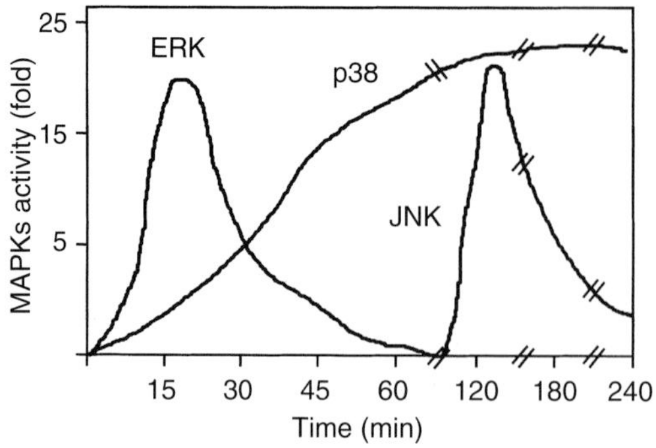

Figure 10.11 Kinetics of pardaxin activation of MAPKs (fold activation over the control) of ERK, p38 and JNK enzymes in PC12 cells after pardaxin (5×10^{-6} M) treatment.

(JNK) and p38 enzymes comprise two additional subfamilies, whose members are potently activated by a variety of stress-related stimuli. The JNK enzymes, also referred to as stress-activated protein kinases (SAPKs), are activated by a variety of stress-related insults (e.g. heat shock, osmotic imbalance, endotoxin, UV and protein synthesis inhibition) and cytokines (Gupta *et al.*, 1996). The p38 kinase, also referred to as high osmolarity glycerol response kinase (HOG), is activated by a variety of stress-related stimuli, as well as cytokines and insulin (Molnar *et al.*, 1997).

To test the possible activation of ERK/MAPK superfamily cascades, PC12 cells were incubated with pardaxin for different periods of time. The cells were lysed and the resulting lysates were subjected to immunoblot analysis using anti-phosphorylated ERK, p38 and JNK antibodies (Bloch-Shilderman *et al.*, 2001). Phosphorylated ERK was detectable within 30 min and disappeared within 60 min after addition of pardaxin (Figure 10.11). Pardaxin stimulation of ERK phosphorylation activity is a specific effect since pretreatment of PC12 cells with 2-(2-amino-3-methoxyphenyl-4H-1-benzopyran-4-one) (PD 98059), a selective MEK1 inhibitor, completely blocks pardaxin stimulation of ERK (Bloch-Shilderman *et al.*, 2001). We also observed that the phosphorylated ERK is translocated to the nucleus in PC12 cells treated for 20 min with pardaxin (Bloch-Shilderman *et al.*, 2001). The rapid stimulation of MAPK in PC12 cells by pardaxin is correlated to the time course of pardaxin-induced arachidonic acid and dopamine release. On the other hand, activation of stress-kinases p38 and JNK by pardaxin entails much longer kinetics (Figure 10.11). This finding and the observation that pardaxin stimulates p38 and JNK activation to an extent similar to that of anisomycin, H_2O_2, palytoxin, hyperosmotic medium etc., support the central role of these kinases in pardaxin-induced neurotoxicity (Bloch-Shilderman *et al.*, 2002a). Since it has been demonstrated that certain A_2 phospholipases are activated by ERK phosphorylation (Lin *et al.*, 1993), it is likely that ERK phosphorylation activity is required in the pardaxin-induced arachidonic acid and dopamine release from PC12 cells. This aspect is now the major focus of our research.

Signal transduction mechanisms of pardaxin pores

Binding experiments with ionophore toxins in eukaryotic cells indicated the absence of specific glycoprotein receptors, and that pore formation is initiated by binding of the toxin to different lipids. This rule applies to pardaxin as well. As summarized schematically in Figure 10.12, pardaxin at concentrations of 10^{-11}–10^{-7} M forms voltage-gated pores that are permeable to cations in artificial membranes (Moran *et al.*, 1984; Lazarovici *et al.*, 1988). At concentrations higher than 10^{-5} M it acts as a lytic agent on eukaryotic cells (Figure 10.12). It is very tempting to propose that all of these effects of pardaxin reflect its concentration-dependent oligomerization. We would like to suggest the following working model for pardaxin oligomerization and pore formation: (1) In aqueous buffers pardaxin monomers are organized as a tetramer or octamer in which the α-helix polar residues are exposed to water, and the hydrophobic residues are located inside the tetramer. (2) The interaction of pardaxin with an artificial or a biological membrane involves polar or charged forces that trigger the dissociation of the pardaxin tetramers and insertion of the monomers into the lipid bilayer. (3) In the lipid bilayer, a structural reaggregation of pardaxin monomers takes place in which the charged side chains interact, and the hydrophobic residues are externally oriented in the pardaxin tetramer, octamer or higher oligomeric form. The quaternary structures schematically presented in the molecular model of Figure 10.7 predict a hydrophilic interior that allows the passage of ions. Most probably, each oligomeric form is characterized by different levels of conductance. (4) The transmembrane voltage opens the pores formed by the pardaxin molecules and/or affects pardaxin oligomers supposedly stabilized by dipole–dipole interactions of monomer helices.

Although the precise pharmacological effects of pore-forming toxins are largely unkown and many fundamental questions remain unanswered, the past several years have greatly increased our understanding of pardaxin actions on signal transduction pathways (summarized in Figure 10.13). Upon exposure of cells to pardaxin, they bind and organize to form conductive pores within the cell membrane (step 1, Figure 10.13). Due to poor selectivity, these pores transport both monovalent and divalent cations into the cell (Fink *et al.*, 1989; Lazarovici and Lelkes, 1992; Nikodijevic *et al.*, 1992; Abu-Raya *et al.*, 1993). The increase in the intracellular Na^+ concentration depolarizes the cell which, in turn, opens voltage-sensitive

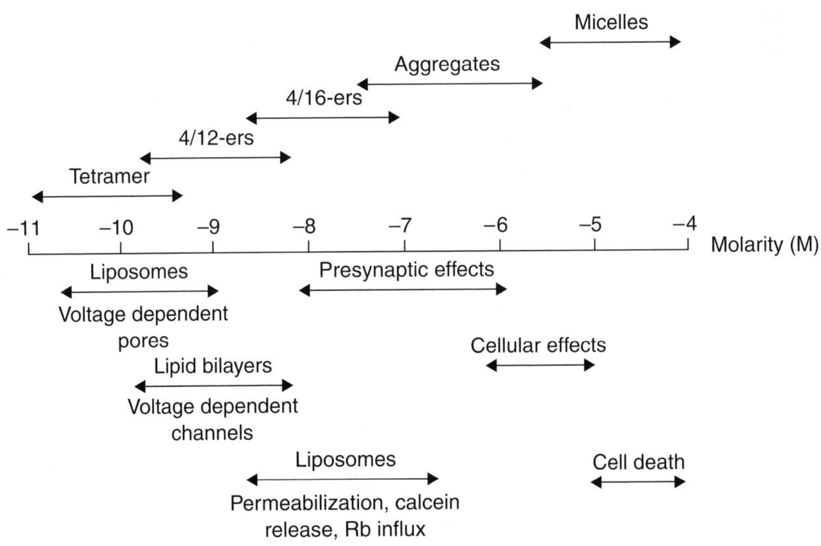

Figure 10.12 Schematic representation of pardaxin oligomerization, pore formation, lysis, presynaptic and neurotoxic effects as a function of toxin concentration.

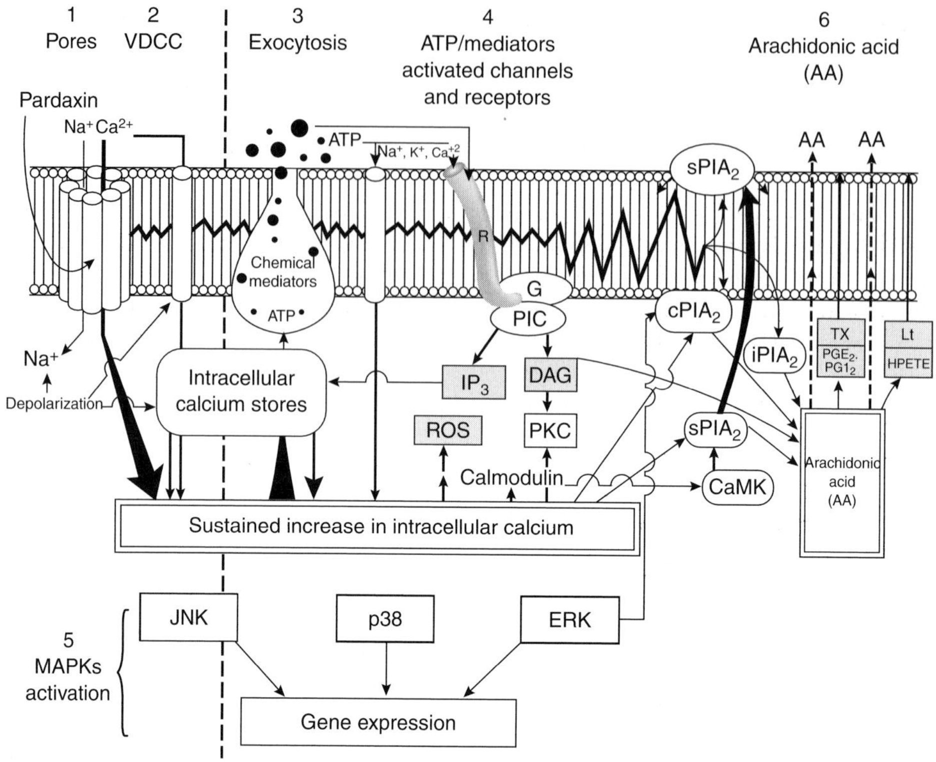

Figure 10.13 Model of pardaxin signal transduction pathways. (1) Pore formation in the plasma membrane. Influx of Na^+ and Ca^{2+}. (2) Depolarization opens voltage-dependent calcium channels (VDCC), resulting in Ca^{2+} influx and Ca^{2+} release from intracellular stores. (3) The elevation in intracellular calcium induces secretion of ATP and chemical mediators. (4) ATP and chemical mediators activate non-selective cation channels and receptors (R) coupled by G proteins to phospholipase C (PLC) to increase intracellular calcium in a second wave by influx, depolarization and/or release from intracellular stores. The increase in intracellular calcium activates calmodulin, protein kinase C (PKC) and calcium–calmodulin (CaM) kinases and is involved in the generation of radical oxygen species (ROS). (5) Activation of the MAPK family with different kinetics resulting in modulation of gene expression. (6) Release of arachidonic acid (AA) from membrane phospholipids occurs via several pathways: I. activation of cytosolic PLA_2 ($cPLA_2$) by calcium, protein-kinases and/or receptor-mediated signals; II. Activation of secretory PLA_2 ($sPLA_2$) by calcium and/or protein phosphorylation signals; III. Activation of $iPLA_2$; after release, free AA may be incorporated into the membrane phospholipids or metabolized by one of the three pathways, generating platelet activating factor (PAF), cyclooxygenated to form prostaglandins (PG), prostacyclin (PGI_2) and thromboxane (TX), or lypooxygenated to form hydroperoxyeicosatetreanoic acids (HPETE), which undergo conversion into leukotrienes (Lt). The model is based on studies performed with pardaxin and other pore-forming toxins, using pheochromocytoma PC12 cell cultures. Na^+, Ca^{2+}, K^+ – sodium, calcium and potassium ions, respectively; IP_3 – inositol (1, 4, 5) P_3; arrows indicate influx of ions or directional flow of activation. Arrow width indicates the intensity of the signal.

calcium channels (VDCC, step 2, Figure 10.13), resulting in the influx of extracellular calcium into the cell (Nikodijevic *et al.*, 1992). Pardaxin pores cause a rapid or gradual, progressive increase in intracellular calcium (Lazarovici and Lelkes, 1992; Abu-Raya *et al.*, 1993). The influx of calcium from pardaxin pores and VDCC (Nikodijevic *et al.*, 1992) causes a sustained rise in the concentrations of intracellular calcium. A rapid consequence of the increase in intracellular calcium is the exocytotic secretion of a variety of chemical mediators, hormones or neurotransmitters, depending on cell type or pore-forming toxin (step 3, Figure 10.13). Pardaxin was found to induce exocytotic secretion of acetylcholine (Renner *et al.*, 1987; Arribas *et al.*, 1993), serotonin and norepinephrine (Wang and Friedman, 1986), dopamine and adrenocorticotropic hormone (Bloch-Shilderman *et al.*, 1996). As secretory granules of cells contain high concentrations of ATP, it is also released by the action of pardaxin (Lazarovici and Lelkes, 1992) when added to the extracellular media (step 3, Figure 10.13). From this moment, in either a paracrine or autocrine fashion, the chemical mediators, hormones and/or ATP released may activate their respective receptors and/or ionic channels (step 4, Figure 10.13), thus amplifying and propagating the initial pore-forming toxin signals. For example, external ATP triggers a rapid and short-lived increase in internal calcium concentration by activating homeostatic channels permeable to Na^+, Ca^{2+} and K^+ (step 4, Figure 10.13). The extracellular ATP also activates purinergic receptors that increase the intracellular calcium level through influx or release from IP_3 (inositol 1,4,5-triphosphate) – sensitive intracellular calcium stores, through activation of phospholipase C (step 4, Figure 10.13; deSouza *et al.*, 1995). The picture might be more complex, since ATP can be hydrolyzed to adenosine, which could then activate additional signal transduction pathways through adenosine A_2 receptors. Release of chemical mediators such as bradykinin induces phospholipase C activation, resulting in synthesis of the second messengers, IP_3 and diacylglycerol (DAG), and the subsequent mobilization of calcium from internal stores to increase the cytoplasmic calcium (step 4, Figure 10.13). Indeed, *S. aureus* α-toxin was found to stimulate phospholipase C (Fink *et al.*, 1989). In many cell types, receptor-mediated IP_3 formation and calcium release from intracellular stores is followed by a calcium influx across the plasma membrane (Putney *et al.*, 1989), required to refill depleted stores and provide enough calcium to produce important metabolic changes. These sequential, progressive four steps of pore-forming toxin action (Figure 10.13) may cause a self-maintained, sustained increase in intracellular calcium due to its influx from the extracellular medium and its release from intracellular stores. This, in turn, probably initiates, in the short term, an array of cellular changes in metabolism, secretion, muscle contraction, gene expression, cell cycle and cross-talk between different signal transduction pathways, depending on cell type. Most probably, a chronic increase in intracellular calcium (calcium overload) to millimolar concentrations may be the lethal insult causing necrotic cell death (Orrenius *et al.*, 1989). This results in the influx of water and extracellular ions, cell swelling and rupture, typical of ionophore toxin action in general, and pardaxin in particular (Shier, 1985; Bloch-Shilderman *et al.*, 2002a). The physiologic effects of μM elevations in the intracellular free calcium concentration are transduced by many effector systems, such as protein kinases and phosphatases. Calcium and DAG will stimulate protein kinase C (Figure 10.13), while a sustained increase in intracellular calcium saturates calmodulin and strongly activates both calcium–calmodulin (CaM) kinases (Figure 10.13), such as myosin light chain kinase (MLCK), and multifunctional CaM kinases, for example, CaM kinase I, II (Shulman, 1993). In the pheochromocytoma PC12 cells, multiple signals converge on CaM kinases, including a ligand gate calcium influx, via purinergic and nicotinic cholinergic receptors and IP_3-mediated release of intracellular calcium through muscarinic and bradykinin receptors (MacNicol and Schulman, 1992). Cross-talk between PKC and CaM kinases, as well as other calmodulin-dependent processes, culminate in phosphorylation changes in multiple substrates localized in the nucleus, cytoskeleton, plasma membrane and cytosol (Figure 10.13).

Ras-MAPK is well characterized as a major signal transduction pathway activated by tyrosine kinase receptors such as NGF-tyrosine kinase A receptor (Kaplan and Stephens, 1994). Cross-talk between MAPK pathways and Ca^{2+} pathways has been demonstrated in numerous cell types (Ely *et al.*, 1990; Chao *et al.*, 1992; Rosen *et al.*, 1994; Clapman, 1995), but the precise mechanism by which Ca^{2+} mediates the activation of ERK is unknown. Since pardaxin did not activate protein kinase C (Bloch-Shilderman *et al.*, 1996) and we do not have any evidence that pardaxin directly stimulates tyrosine kinases, we assume that pardaxin stimulation of MAPKS (Bloch-Shilderman *et al.*, 2001) (step 5, Figure 10.13) reflects cross-talk between Ca^{2+} pathway(s) and the MAPK cassette. Indeed, evidence for this concept was provided by molecular studies (Hawes *et al.*, 1995; Lopez-Ilasaca, 1998) on G protein-coupled receptors. Activation of such receptors induces disassociation of the G protein into $\beta\gamma$- and α-subunits. The $\beta\gamma$-subunits then stimulate ERK activation, most probably increasing intracellular Ca^{2+} levels and activating protein kinase C. In other studies with PC12 cells it was reported that the influx of Ca^{2+} ions, resulting from depolarization, leads to Ras-dependent ERK activation (Rosen *et al.*, 1994; Rusanescu *et al.*, 1995). It was suggested that the protein tyrosine kinase PYK2, which is activated by an increase in intracellular Ca^{2+} and by protein kinase C, leads to activation of ERK (Lev *et al.*, 1995). The possibility that pardaxin stimulates PYK2 kinase, and as a result activate ERKs, requires further investigation since PYK2 may be a candidate kinase functioning upstream of MEK/MAPK. An important function of ERKs is to regulate gene expression. Upon activation, ERKs are translocated to the nucleus where they phosphorylate transcription factors (Blenis, 1993). We found that ERKs translocated to the nucleus following pardaxin activation (Bloch-Shilderman *et al.*, 2001). Therefore, transcription factors phosphorylated by ERKs may be of significance in the action of pardaxin (Figure 10.13). Indeed, experiments in our laboratory show that pardaxin stimulates c-fos early gene expression (unpublished data). This probably requires both Ca^{2+} and the MAPK pathway, as recently reported for depolarization-induced c-fos activation in PC12 cells (Lee *et al.*, 2000).

Another major cell signaling effect of pardaxin is the stimulation of the arachidonic acid cascade (Abu-Raya *et al.*, 1993), initiated via activation of phospholipase A_2 (PLA_2, phosphatide 2-acylhydrolase) enzymes. PLA_2 action releases polyunsaturated arachidonic fatty acid (AA), which is subsequently converted into a family of biologically active metabolites, collectively known as "eicosanoids" (step 6, Figure 10.13; Piomelli, 1993). Some of the enzymatic pathways involved in AA release and metabolism are depicted in Figure 10.13. In PC12 cells, streptolysin toxin, *S. aureus* α- and δ-toxins, *S. helianthus* toxin, alamethicin and pardaxin strongly stimulated both PLA_2 cyclooxygenase and lipoxygenase pathways, resulting in the formation of prostaglandin E_2 (PGE_2), thromboxane B_2 (TXB_2) and 5-HETE (Fink *et al.*, 1989; Abu-Raya *et al.*, 1993; Bloch-Shilderman *et al.*, 1996). It has also been shown that *S. aureus* α-toxin strongly stimulated AA pathways with subsequent formation of prostacyclin PGI_2 in endothelial cells (Suttorp *et al.*, 1985), leukotriene B4 in polymorphonuclear leukocytes (Suttorp *et al.*, 1987) and platelet-activating factor (PAF) in pulmonary artery endothelial cells (Suttorp *et al.*, 1992). The generation of eicosanoids by pore-forming toxins might explain a wide spectrum of pathological effects such as endotoxic shock, pulmonary hypertension, blood coagulation, inflammation and edema (Seeger *et al.*, 1984; Bhakdi *et al.*, 1988; Snyder, 1990). The initial step in eicosanoid generation is the hydrolysis of AA from the sn-2 position (where AA is most often esterified) of the glycerol backbone of plasma membrane phospholipids by the activities of the two major groups of PLA_2. The first group, cytosolic, high molecular weight (70–110 kDa) AA-specific enzyme ($cPLA_2$), is activated by ERK phosphorylation and then translocated to the plasma membrane in response to rising nanomolar levels of calcium upon receptor-stimulation, and is regulated by PKC-dependent phosphorylations (Piomelli, 1993). $cPLA_2$s are activated by pardaxin (Bloch-Shilderman *et al.*, 2002b), most probably amplifying the receptor-dependent hydrolysis of AA-containing phospholipids during steps 4 and 6 (Figure 10.13). The second group, low molecular weight, secreted and membrane-bound enzymes ($sPLA_2$, 14 kDa), non-selectively hydrolyze AA-containing phospholipids and are regulated by glucocorticoids. These $sPLA_2$s are involved in phospholipid remodeling, particularly important in preserving plasma membrane integrity, and are secreted in large quantities at inflammatory sites, leading to a marked increase in the release of AA and the generation of eicosanoids (Piomelli, 1993). Since ionophore toxins disturb cell membrane organization, as reflected by changes in phase-transition properties and phospholipid aggregation (Lelkes and Lazarovici, 1988), they might recruit and activate $sPLA_2$s used to repair the damaged membrane. Recently, using relatively selective inhibitors for PLA_2s, we demonstrated that all these isoenzymes are stimulated by pardaxin under physiological conditions (Bloch-Shilderman *et al.*, 2002b), but only calcium-independent PLA_2 ($iPLA_2$) is stimulated by pardaxin under calcium-free conditions (Bloch-Shilderman *et al.*, 2002b).

Alternatively, AA release by pore-forming toxins may be initiated by the activation of phosphoinositide-specific PLC (Fink *et al.*, 1989), producing 1,2-diacylglycerol (DAG). This intermediate has been shown to activate PKC, which in turn reportedly activates PLA_2 (Parker *et al.*, 1987). In addition, DAG may be hydrolyzed by a DAG-lipase to yield AA and monoacylglycerol. This PLC–DAG lipase signal transduction pathway was found to be induced by bradykinin (Allen *et al.*, 1992) and it was proposed that it behaves as a second chemical mediator of pore-forming toxins (Figure 10.13). This is not characteristic of pardaxin. AA and eicosanoids may act as intracellular second messengers, modifying the activities of phospholipases, protein-kinases, G-proteins, adenylate and guanylate cyclase as well as ion channels. Like local chemical mediators, they may be released outside the intoxicated cells and act on neighboring cells by binding to high-affinity membrane receptors, triggering new cascades of cell signaling. Our signal transduction model (Figure 10.13) of pore-forming toxin activity in general, and of pardaxin in particular, reflects complex, amplified, self-generating cascades of cellular signals propagated by intracellular calcium and the arachidonic acid cascade. At high toxic concentrations or following prolonged exposure, pardaxin induces necrotic cell death by several pathological mechanisms: excess arachidonic acid, production and release of eicosanoids, calcium overload, oxygen radical formation, activation of stress kinases JNK and p38, deregulation of gene expression, collapse of ionic homeostasis and lysis of plasma membranes (Bloch-Shilderman *et al.*, 2002a). Identification and characterization of pore-forming toxin molecular mechanisms of action in different cells are fields of active research with major unresolved questions. An understanding of pore-forming toxin actions will lay the foundation for efficient therapies and provide new pharmacological concepts in cell signaling.

Conclusions

Pore-forming toxins such as pardaxin belong to a large group of proteins of bacterial, plant and animal origin that alter cell plasma membrane permeability. Although pore-forming toxins appear to have little structural homology, they share a common requirement – the need to insert and form pores in plasma cell membranes. The conversion from a water-soluble toxin to a membrane pore requires a large change in conformation and organization, achieved by new packaging of the hydrophobic, amphipathic α-helices and assembly in the plasma membrane in oligomeric, poorly selective channels. At subcytotoxic concentrations, these small lesions in the plasma membrane induce a sustained increase in intracellular calcium and activation of the arachidonic acid pathways. These crucial events are further amplified by cascades of cellular signaling, due to the release of local mediators and/or hormones and modifications of intracellular enzymes, ion channels and receptor activities. The end result is a lethal hit to the signal transduction machinery of the attacked cell, resulting in cell death. At high concentrations and/or chronic exposure of the cells to ionophore toxins, rapid cytotoxicity will ensue. However, ionophore toxins at subcytotoxic concentrations and under defined experimental conditions could be used as pharmacological tools to study exocytosis, to permeabilize cells, to change the intracellular milieu, to induce neurotransmitter release, to activate the arachidonic acid cascade, to investigate mechanisms of cell death and cytoprotection and for drug delivery. The emerging diversity of signal transduction pathways on the one hand, and the multiplicity of action of pore-forming toxins on these signaling cascades and cellular physiology on the other, might reveal potential sites of action for novel therapeutic agents.

Pardaxins are polypeptide pore-forming marine neurotoxins that induce massive release of neurotransmitters from neurons. Using PC12 pheochromocytoma cells as a neuronal model for studying exocytosis, we found evidence that pardaxin stimulates dopamine release independently of extracellular calcium. Our results also indicate that pardaxin stimulates the release of arachidonic acid and eicosanoids, independently of calcium. Pardaxin stimulates the ERK superfamily of enzymes, with different kinetics. It is conceivable that pardaxin activation of ERK is involved in pardaxin stimulation of arachidonic acid and dopamine release. In addition, activation of stress-kinases JNK and p38 may be involved in pardaxin-induced neurotoxicity. Elucidation of the neuronal signal transduction pathways affected by pardaxin could provide a new understanding of synaptic transmission and new targets for therapy of patients affected by neurotoxins.

References

Abu-Raya, S., Trembovler, V., Shohami, E. and Lazarovici, P. (1993) Cytolysins increase intracellular calcium and induce eicosanoid release by pheochromocytoma PC12 cell cultures. *Nat. Toxins*, **1**, 263–270.

Abu-Raya, S., Bloch-Shilderman, E., Shohami, E., Trembovler, V., Shai, Y., Weidenfeld, J. *et al.* (1998) Pardaxin, a new pharmacological tool to stimulate the arachidonic acid cascade in PC12 cells. *J. Pharmacol. Exp. Ther.*, **287**, 889–896.

Abu-Raya, S., Bloch-Shilderman, E., Lelkes, P. I., Trembovler, V., Shohami, E., Gutman, Y. *et al.* (1999) Characterization of pardaxin-induced dopamine release from PC12 cells: The role of calcium and eicosanoids. *J. Pharmacol. Exp. Ther.*, **288**, 399–406.

Abu-Raya, S., Bloch-Shilderman, E., Jiang, H., Adermann, K., Schaefer, E. M., Goldin, E. *et al.* (2000) Cellular signaling in PC12 cells affected by pardaxin. In *Natural and Selected Synthetic Toxins – Biological Implications*, edited by T. A. Tu and W. Gaffield, pp. 22–42. Washington, DC: American Chemical Society, ACS Symposium Series, vol. 745.

Adermann, K., Manfred, R., Paul, E., Lazarovici, P., Hochman, J. and Wellhoner, H. (1998) Isolation, characterization and synthesis of a novel pardaxin isoform. *FEBS Lett.*, **435**, 173–177.

Ahnert-Hilger, G., Bhakdi, S. and Gratzl, M. (1985) Minimal requirements for exocytosis: a study using PC12 cells permeabilized with Staphylococcal-toxin. *J. Biol. Chem.*, **260**, 12730–12734.

Allen, A. C., Gammon, C. M., Ousley, A. H., McCarthy, K. D. and Morell, P. (1992) Bradykinin stimulates arachidonic acid release through the sequential actions of an Sn-1 diacylglycerol lipase and a monocyl glycerol lipase. *J. Neurochem.*, **58**, 1130–1139.

Alouf, J. E. (1986) Streptococcal toxins (streptolysin O, streptolysin S, erythrogenic toxins). In *Pharmacology of Bacterial Toxins*, edited by F. Dorner and J. Drews, p. 199. Oxford: Pergamon Press.

Andersen, D. S. (1984) Gramicidin channels. *Ann. Rev. Physiol.*, **46**, 531–548.

Arribas, M., Blasi, J., Lazarovici, P. and Marsal, J. (1993) Calcium-dependent and independent acetylcholine release from electric organ synaptosomes by pardaxin: evidence of a biphasic action of an excitatory neurotoxin. *J. Neurochem.*, **60**, 552–558.

Barrow, C. J., Nakanishi, K. and Tachibana, K. (1992) Structure and activity studies on pardaxin and analogues using model membranes of phosphatidylcholine. *Biochem. Biophys. Acta.*, **1112**, 235–240.

Bhakdi, S. and Tranum-Jensen, J. (1987) Damage to mammalian cells by proteins that form transmembrane pores. *Rev. Physiol. Biochem. Pharmacol.*, **107**, 147–223.

Bhakdi, S., Muhly, M., Mannhardt, U., Hugo, F., Klapettek, K., Muller-Eckhardt, C. *et al.* (1988) Staphylococcal alpha-toxin promotes blood coagulation via attack on human platelets. *J. Exp. Med.*, **168**, 527–542.

Blenis, J. (1993) Signal transduction via the MAP kinases: proceed at your own RSK. *Proc. Natl. Acad. Sci., USA*, **90**, 5889–5892.

Bloch-Shilderman, E., Abu-Raya, S., Rasouly, D., Furman, O., Trembovler, V., Shavit, D. *et al.* (1996) The role of calcium, protein kinase C, pertussis toxin substrates and eicosanoids on pardaxin-induced dopamine release from PC12 cells. In *Biochemical Aspects of Marine Pharmacology*, edited by P. Lazarovici, M. Spira and E. Zlotkin, pp. 158–174. Ft Collins, Co.: Alaken Inc.

Bloch-Shilderman, E., Abu-Raya, S. and Lazarovici, P. (1997) Ionophore polypeptide toxins and signal transduction. In *Toxins and Signal Transduction*, edited by Y. Gutman and P. Lazarovici, pp. 211–232. Amsterdam, The Netherlands: Harwood Academic Publishers.

Bloch-Shilderman, E., Jiang, H., Abu-Raya, S., Linial, M. and Lazarovici, P. (2001) Involvement of extracellular signal-regulated kinase (ERK) in pardaxin-induced dopamine release from PC12 cells. *J. Pharmacol. Exp. Ther.*, **296**, 704–711.

Bloch-Shilderman, E., Jiang, H. and Lazarovici, P. (2002a) Pardaxin, an ionophore neurotoxin, induces PC12 cell death: activation of stress kinases and production of reactive oxygen species. *J. Nat. Toxins*, **11**, 71–85.

Bloch-Shilderman, E., Abu-Raya, S., Trembovler, V., Boschwitz, H., Gruzman, A. *et al.* (2002b) Pardaxin stimulation of phospholipases A2 and their involvement in exocytosis in PC12 cells. *J. Pharmacol. Exp. Ther.*, **301**, 953–962.

Buckingham, L. and Duncan, J. L. (1983) Approximate dimensions of membrane lesions produced by Streptolysin S and Streptolysin O. *Biochem. Biophys. Acta*, **729**, 115–122.

Burgoyne, R. D. (1991) Control of exocytosis in adrenal chromaffin cells. *Biochem. Biophys. Acta.*, **1071**, 174–202.

Chao, T. S., Byron, K. L., Lee, K. M., Villereal, M. and Rosner, M. R. (1992) Activation of MAP kinases by calcium-dependent and calcium-independent pathways. Stimulation by thapsigargin and epidermal growth factor. *J. Biol. Chem.*, **267**, 19876–19883.

Clapman, D. E., (1995) Calcium signaling. *Cell*, **80**, 259–268.

De Souza, L. R., Moore, H., Raha, S. and Reed, J. K. (1995) Purine and pyrimidine nucleotides activate distinct signalling pathways in PC12 cells. *J. Neurosci. Res.*, **41**, 753–763.

Dolly, J. O. (1988) *Neurotoxins in Neurochemistry*. Chichester, England: Ellis Horwood.

Eisenberg, D. (1984) Three-dimensional structure of membrane and surface proteins. *Ann. Rev. Biochem.*, **53**, 595–623.

Ely, C. M., Oddie, K. M., Litz, J. S., Rossomando, A. J., Kanner, S. C., Sturgill, T. W. *et al.* (1990) A 42-kD tyrosine kinase substrate linked to chromaffin cell secretion exhibits as associated MAP kinase activity and is highly related to a 42-kDa mitogen-stimulated protein in fibroblasts. *J. Cell Biol.*, **110**, 731–742.

Fink, D., Contreras, M. L., Lelkes, P. I. and Lazarovici, P. (1989) *Staphylococcus aureus* α-toxin activates phospholipases and induces a Ca^{2+} influx in PC12 cells. *Cell. Signal.*, **4**, 387–393.

Force, T. and Bonventre, J. V. (1998) Growth factors and mitogen-activated protein kinases. *Hypertension*, **31**, 152–161.

Garcia, A. G., Albillos, A., Gandia, L., Lopez, M. G., Michelena, P. and Montiel, C. (1997) W-toxins, calcium channels and neurosecretion. In *Toxins and Signal Transduction*, edited by Y. Gutman and P. Lazarovici, pp. 155–209. Amsterdam, The Netherlands: Harwood Academic Publishers.

Garnier, J., Osguthorpe, D. J. and Robson, B. (1978) Analysis of the accuracy and implications of simple methods for predicting the secondary structure of globular proteins. *J. Mol. Biol.*, **120**, 97–120.

Grasso, A. (1988) α-Latrotoxin as a tool for studying ionic channels and transmitter release process. In *Neurotoxins in Neurochemistry*, edited by O. J. Dolly, pp. 67–78. Chichester, England: J. Wiley and Sons.

Greene, L. A. and Tischler, A. S. (1976) Establishment of a noradrenergic clonal line of rat adrenal pheochromocytoma cells which respond to nerve growth factor. *Proc. Natl. Acad. Sci. USA*, **73**, 2424–2428.

Gupta, S., Barrett, T., Whitmarsch, A. J., Caranagh, J., Sluss, H. K., Derijard, B. *et al.* (1996) Selective interaction of JNK protein kinase isoforms with transcription factors. *EMBO J.*, **15**, 2760–2770.

Guy, H. R. and Conti, F. (1990) Pursuing the structure and function of voltage-gated channels. *TINS*, **13**, 201–206.

Guy, H. R. and Raghunathan, G. (1989) Structural models of membrane insertion and channel formation by antiparallel-helical membrane peptides. In *Transport Through Membranes: Carriers, Channels and Pumps*, edited by E. Pullman, pp. 369–379. Jerusalem: Magnes Found.

Hall, J. E., Vodyanoy, I., Balasubramanian, J. M. and Marshall, G. R. (1981) Alamethicin. A rich model for channel behaviour. *Biophys. J.*, **45**, 233–247.

Harvey, A. L. (1990) Cytolytic toxins. In *Handbook of Toxinology*, edited by W. T. Shier and D. Mebs, pp. 1–66. New York: Marcel Dekker, Inc.

Hawes, B. E, Biesen, T. V., Koch, W. J., Luttrell, L. M. and Lefkowitz, R. J. (1995) Distinct pathways of G_i- and G_q-mediated mitogen-activated protein kinase activation. *J. Biol. Chem.*, **270**, 17148–17153.

Hopp, T. P. and Woods, K. R. (1982) A computer program for predicting protein antigenic determinants. *Mol. Immunol.*, **20**, 483–489.

Kaplan, D. R. and Stephens, R. M. (1994) Neurotrophin signal transduction by the trk receptor. *J. Neurobiol.*, **25**, 1404–1417.

Kelly, R. B. (1993) Storage and release of neurotransmitters. *Cell*, **72**, 43–53.

Kyte, J. and Doolittle, R. (1982) A simple method for displaying the hydropathic character of a protein. *J. Mol. Biol.*, **157**, 105–132.

Lazarovici, P. (1994) Challenging catecholamine exocytosis with pardaxin, an excitatory ionophore fish toxin. *J. Toxicol. Toxin Rev.*, **13**, 45–63.

Lazarovici, P. and Lelkes, P. I. (1992) Pardaxin induces exocytosis in bovine adrenal medullary chromaffin cells independent of calcium. *J. Pharmacol. Exp. Ther.*, **263**, 1317–1326.

Lazarovici, P., Primor, N. and Loew, L. M. (1986) Purification and pore forming activity of two hydrophobic polypeptides from the secretion of the Red Sea Moses sole (*Pardachirus marmoratus*). *J. Biol. Chem.*, **261**, 16704–16713.

Lazarovici, P., Primor, N., Caratsch, C. G., Munz, K., Lelkes, P., Loew, L. M. *et al.* (1988) Action on artificial and neuronal membranes of pardaxin, a new presynaptic excitatory polypeptide neurotoxin with ionophore activity. In *Neurotoxins in Neurochemistry*, edited by O. J. Dolly, pp. 219–240. Chichester: Ellis Horwood.

Lazarovici, P., Primor, N., Gennaro, J., Fox, J., Shai, Y., Lelkes, P. I., *et al.* (1990) Origin, chemistry and mechanisms of action of a repellent, presynaptic excitatory, ionophore polypeptide. In *Marine Toxins: Origin, Structure and Molecular Pharmacology* (ACS Symposium Series), edited by S. Hall and G. Strichartz, pp., 347–364. Washington, D.C.: American Chemical Society.

Lazarovici, P., Edwards, C., Raghunathan, G. and Guy, H. R. (1992) Secondary structure, permeability and molecular modeling of pardaxin pores. *J. Nat. Toxins*, **1**, 1–15.

Lee, S. A., Park, J. K., Kang, E. K., Bae, H. R., Bae, K. W. and Park, H. T. (2000) Calmodulin-dependent activation of p38 and p42/44 mitogen-activated protein kinases contributes to c-fos expression by calcium in PC12 cells: Modulation by nitric oxide. *Brain Res. Mol. Brain Res.*, **75**, 16–24.

Lelkes, P. I. and Lazarovici, P. (1988) Pardaxin induces aggregation, but not fusion of phosphatidylserine vesicles. *FEBS Lett.*, **242**, 161–166.

Lev, S., Moreno, H., Martinez, R., Canoll, P., Peles, E., Musacchio, J. M. *et al.* (1995) Protein tyrosine kinase PYK2 involved in Ca^{2+}-induced regulation of ion channel and MAP kinase functions. *Nature (Lond)*, **376**, 737–745.

Lin, L. L., Wartman, M., Lin, A. Y., Knopf, J. L., Seth, A. and Davis, R. J. (1993) $cPLA_2$ is phosphorylated and activated by MAP kinase. *Cell*, **72**, 269–278.

Linial, M. (1997) SNARE proteins – why so many? Why so few? *J. Neurochem.*, **69**, 1781–1792.

Loew, L. M., Benson, L., Lazarovici, P. and Rosenberg, I. (1985) Fluorimetric analysis of transferable membrane pores. *Biochemistry*, **24**, 2101–2104.

Lopez-Ilasaca, M. (1998) Signaling from G-protein-coupled receptors to mitogen-activated protein (MAP)-kinase cascades. *Biochem. Pharmacol.*, **56**, 269–277.

MacNicol, M. and Shulman, H. (1992) Multiple Ca^{2+} signaling pathways converge on CaM kinase in PC12 cells. *FEBS Lett.*, **304**, 237–240.

Menestrina, G. (1986) Ionic channels formed by *Staphylococcus aureus* alpha-toxin: voltage-dependent inhibition by divalent and trivalent cations. *J. Memb. Biol.*, **90**, 177–190.

Molnar, A., Theodoras, A. M., Zon, L. I. and Kyriakis, J. M. (1997) Cdc42Hs, but not Rac1, inhibits serum-stimulated cell cycle progression at G1/S through a mechanism requiring p38/RK. *J. Biol. Chem.*, **272**, 13229–13253.

Montecucco, C., Pellizzari, R., Rossetto, O., Schiavo, G., Tonello, F. and Washbourne, F. (1998) Clostridial neurotoxins as enzymes: structure and function. In *Secretory Systems and Toxins*, edited by M. Linial, A. Grasso and P. Lazarovici, pp. 315–331, Amsterdam, The Netherlands: Harwood Academic Publishers.

Moran, A., Korchak, Z., Moran, N. and Primor, N. (1984) Surfactant and channel-forming activities of the Moses sole toxin. In *Toxins, Drugs and Pollutants in Marine Animals*, edited by L. Bolis, J. Zadunaisky and R. Gilles, pp. 13–25. Berlin: Springer Verlag.

Niedermeyer, W. (1985) Interaction of streptolysin O with biomembranes: kinetic and morphological studies on erythrocyte membranes. *Toxicon.*, **23**, 425–439.

Nikodijevic, B., Nikodijevic, D. and Lazarovici, P. (1992) Pardaxin-stimulated calcium uptake in PC12 cells is blocked by cadmium and is not mediated by L-type calcium channels. *J. Basic Clin. Physiol. Pharmacol.*, **3**, 359–370.

Orrenius, S., McConkey, D. J., Bellomo, G. and Nicotera, P. (1989) Role of Ca^{2+} in toxic cell killing. *Trends Pharmacol. Sci.*, **10**, 281–285.

Parker, J., Daniel, L. W. and White, M. (1987) Evidence of protein kinase C involvement in phorbol diester-stimulated arachidonic acid release and prostaglandin synthesis. *J. Biol. Chem.*, **262**, 5385–5393.

Piomelli, D. (1993) Arachidonic acid in cell signalling. *Curr. Opin. Cell. Biol.*, **5**, 274–280.

Putney, J. W. Jr, Takemura, H., Hughes, A. R., Horstman, D. A. and Thastrup, O. (1989) How does inositol phosphate regulate calcium signaling? *FASEB*, **7**, 1899–1905.

Raghunathan, G., Seetharamulu, P., Brooks, B. and Guy, H. R. (1990) Models of delta-hemolysin membrane channels and crystal structures. *Proteins: Structure, Function and Genetics*, **8**, 213–225.

Rapaport, D. and Shai, Y. (1992) Aggregation and organization of pardaxin in phospholipid membranes. *J. Biol. Chem.*, **267**, 6502–6509.

Renner, P., Caratsch, C. E., Waser, P. G., Lazarovici, P. and Primor, N. (1987) Presynaptic effects of the pardaxins, polypeptides isolated from the gland secretion of the flatfish *Pardachirus marmoratus*. *Neuroscience*, **23**, 319–325.

Rosen, L. B., Ginty, D. D., Weber, M. J. and Greenberg, M. E. (1994) Membrane depolarization and calcium influx stimulates MEK and MAP kinases via activation of Ras. *Neuron*, **12**, 1207–1221.

Rusanescu, G., Qi, H., Thomas, S. M., Brugge, J. S. and Halegoua, S. (1995) Calcium influx induces neurite growth through a Src-Ras signaling cassette. *Neuron*, **15**, 1415–1425.

Schiavo, G., Benfenati, F., Poulain, B., Rossetto, O., Polverino de Laureto P., DasGupta, B. R. *et al.* (1992) Tetanus and botulinum-B neurotoxins block neurotransmitter release by proteolytic cleavage of synaptobrevin. *Nature*, **359**, 832–835.

Seeger, W., Bauer, M. and Bhakdi, S. (1984) Staphylococcal alpha-toxin elicits hypertension in isolated rabbit lungs due to stimulation of the arachidonic acid cascade. *J. Clin. Invest.*, **74**, 849–858.

Shai, Y. (1994) Pardaxin: channel formation by shark repellant peptide from fish. *Toxicology*, **87**, 109–129.

Shai, Y., Fox, J., Caratsch, C., Shih, Y., Edwards, C. and Lazarovici, P. (1988) Sequencing and synthesis of pardaxin, a polypeptide from the Red Sea Moses sole with ionophore activity. *FEBS Lett.*, **242**, 161–166.

Shai, Y., Nach, D. and Yanovsky, A. (1990) Channel formation properties of synthetic pardaxin and analogues. *J. Biol. Chem.*, **265**, 20202–20209.

Shai, Y., Hadari, Y. R. and Finkels, A. (1991) pH-dependent pore formation properties of pardaxin analogues. *J. Biol. Chem.* **266**, 22346–22354.

Shi, J. L., Edwards, C. and Lazarovici, P. (1995) Ion selectivity of the channels formed by pardaxin, an ionophore, in bilayer membranes. *Nat. Toxins*, **3**, 151–155.

Shulman, H. (1993) The multifunctional Ca^{2+}/calmodulin-dependent protein kinases. *Curr. Opin. Cell Biol.*, **5**, 247–253.

Shier, W. T. (1985) The final steps to toxic cell death. *J. Toxicol. Toxin Rev.*, **4**, 191–249.

Snyder, F. (1990) Platelet-activating factor and related acetylated lipids as potent biologically active cellular mediators. *Am. J. Physiol.*, **259**, C697–C708.

Surkova, I. (1994) Can exocytosis inudced by α-latrotoxin be explained solely by its channel-forming activity? *Ann. N.Y. Acad. Sci.*, **710**, 48–64.

Suttorp, N., Seeger, W., Dewein, E., Bhakdi, S. and Roka, L. (1985) Staphylococcal alpha-toxin induced PGI_2 production in endothelial cells: role of calcium. *Am. J. Physiol.*, **248**, C127–C134.

Suttorp, N., Seeger, W., Zucker-Raimann, J., Roka, L. and Bhakdi, S. (1987) Mechanism of leukotriene generation in polymorphonuclear leukocytes by staphylococcal alpha-toxin. *Infect. Immun.*, **55**, 104–110.

Suttorp, N., Buerke, M. and Tannert-Otto, S. (1992) Stimulation of PAF-synthesis in pulmonary artery endothelial cells by *Staphylococcus aureus* α-toxin. *Trombosis Res.*, **67**, 243–252.

Tachibana, K., Sakaitanai, M. and Nakanishi, K. (1986) Pavonins: shark-repelling ichthyotoxins from the defense secretion of the Pacific sole. *Science*, **226**, 703–705.

Thelestam, M. and Molby, R. (1979) Classification of microbial, plant and animal cytolysins based on their membrane damaging effects on human fibroblasts. *Biochem. Biophys. Acta*, **557**, 156–169.

Thompson, S. A., Tachibana, K., Nakanishi, K. and Kubota, I. (1986) Mellitin-like peptide from the shark-repelling defense secretion of the sole *Pardachirus pavoninus*. *Science*, **233**, 341–343.

Trimble, W. S., Linial, M. and Scheller, R. H. (1991) Cellular and molecular biology of the presynaptic nerve terminal. *Ann. Rev. Neurosci.*, **14**, 93–122.

Wang, H. Y. and Friedman, E. (1986) Increased 5-hydroxytryptamine and norepinephrine release from rat brain slices by the Red Sea flatfish toxin pardaxin. *J. Neurochem.*, **47**, 656–658.

Zagorski, M. G., Norma, D. G., Barrow, C. J., Iwashita, T., Tachibana, K. and Patel, D. J. (1991) Solution structure of pardaxin P-2. *Biochemistry*, **30**, 8009–8017.

11 Cecropin–melittin hybrid peptides as versatile templates in the development of membrane-active antibiotic agents[1]

Luis Rivas and David Andreu

Cecropins, together with melittin and magainins, are the most widely used templates to design synthetic structures useful as models of membrane-active peptides and candidate antimicrobial agents. The different strategies used for this purpose, from simple residue substitution to design of chimeric sequences which recombine well defined regions of the parent peptides, are reviewed. The impact of these modifications on antibiotic activity upon microorganisms or tumoral cells, as well as the perspectives for their clinical use, are discussed.

Introduction

Eucaryotic antibiotic peptides (EAPs) are an important component of the non-adaptive immune system. Their deterrent role against invading pathogens has been convincingly demonstrated for animals lacking antigen-specific immunity, such as insects (Otvos, 2000), as well as for higher eucaryotes, including man, despite the coalescence of antigen-specific immunity and the coexistence of several antibiotic peptides with overlapping functions (for reviews, see Hancock and Lehrer, 1998; Ganz and Lehrer, 1999; Hancock and Chappel, 1999; Hancock, 2000; Hancock and Diamond, 2000; Hancock and Scott, 2000). Also, their function in the control of local microbiological fauna is a largely unexplored field that deserves further investigation (reviewed by Boman, 1998, 2000).

Eucaryotic antibiotic peptides can also be viewed as a therapeutic alternative to the worldwide spread of multiresistance against classical antibiotics. This claim is based on the wide range of EAP-susceptible pathogens, and on the lack – or very difficult induction – of resistance to EAPs, which target essential, well conserved physiological systems of the pathogen, such as plasma membrane and cellular homeostasis. On the other hand, several instances of EPA action on intracellular targets have been recently reported (Boman *et al.*, 1993; Park *et al.*, 1998; Castle *et al.*, 1999; Wu *et al.*, 1999; Otvos *et al.*, 2000; Helmershorst *et al.*, 2001; Kim *et al.*, 2001; Kragol *et al.*, 2001), with important secondary effects other than a direct killing effect on the pathogen (Hancock and Diamond, 2000).

Aside from C-terminal amidation and intramolecular disulfide bridges, EAPs undergo minimal postranslational modifications (reviewed by Andreu and Rivas, 1998). This, in addition to their relatively small size, has from early days made EAPs ideal targets for chemical synthesis (Andreu *et al.*, 1983; Andreu and Rivas, 2001) and subsequent SAR studies, aimed to expand the range of susceptible organisms, to delimit minimal active structures and thus define potential drug candidates. Genetic coding of EAPs, in contrast with non-ribosomally-produced peptide antibiotics, allows their production through recombinant techniques, as well as the development of EAP-expressing transgenic animals and plants with improved resistance against pathogenic challenge. Preliminary accounts of EAP-based gene therapy to fight infections have also appeared (Bals *et al.*, 1999).

Hybrid eucaryotic linear antibiotic peptides

Linear antimicrobial peptides

Initially, EAPs were classified taxonomically, according to zoological source. This criterium was soon found inadequate, as usually more than one EAP was present in a given species, and peptides with similar structures were found in widely different organisms. The more rational classification of EAPs in five structural groups (Boman, 1995) remains largely valid, with the addition of a newly discovered class of polycylic structures (head-to-tail and multiple internal disulfide), represented so far by one peptide, the θ-defensin found in primates (Tang *et al.*, 1999).

Within the class of linear peptides, the most populated subdivision is comprised by structures tending to adopt α-helical conformation, including insect (melittin, bombolittin) or fish toxins (pardaxins) endowed with antimicrobial activity, as well as strict EAPs from insects (cecropins, sarcotoxins, andropins, ceratotoxins), tunicates (clavanins, styelins), amphibians (PGLa, magainins, bombinins, dermaseptins, caerins, citropins), fishes (hagfish peptide, pleurocidin, misgurin) or mammals (cecropin P1, LL-37, BAMP-27, BAMP-28). Recently, a cecropin-like peptide was detected as amino terminal sequence of the ribosomal protein L_1 in the bacteria *Helicobacter pylori* (Pütsep *et al.*, 1999). The reader is referred to an exhaustive and regularly updated database (http://www.bbcm.univ.trieste.it/~tossi/pag1.html) for references, as well as to other chapters of this book describing the requirements for EAP activity on membranes. The present chapter will focus mainly on the development of synthetic chimeric antibiotic peptides derived from naturally occurring, α-helical EAPs.

Concept and design of hybrid sequences

A hybrid – or chimeric – peptide (Figure 11.1) can be broadly defined as made up by at least two fragments from different peptides (sequence hybridization), or from the same peptide (self-hybridization) (Table 11.3) but rearranged in a non-native fashion (e.g. reversed N- and C-terminal domains). Retro versions, in which the entire sequence is inverted (i.e. the C-terminus becomes the N-terminus and so on) (Table 11.5), might be viewed as an extreme form of hybridization but will not be discussed as such here (see, however, under *Stereoisomers* below). Nature provides some examples of what might be considered chimeric antibiotic peptides. For instance, for θ-defensin a biosynthesis mechanism has been sketched (Tang *et al.*, 1999), involving head-to-tail ligation of two α-defensin-related nonapeptides. Also, scorpine (Conde *et al.*, 2000) has an N-terminal domain close to cecropins and a C-terminal domain related to defensins.

As to synthetic versions of chimeric antimicrobial peptides, those incorporating cecropin segments are by far the best documented and will be dealt with in greater detail in the following pages (see also Tables 11.1–11.5). Cecropins have been mainly hybridized with other cecropin sequences, as well as with melittin or magainin and exceptionally with bombinin (Shin *et al.*, 1997b). Other α-helical chimeras recently reported involve combination of fragments from (fish EAPs) pleurocidin and misgurin, ceratotoxin (insect) and dermaseptin (amphibian), some of them with substantial improvement in antimicrobial

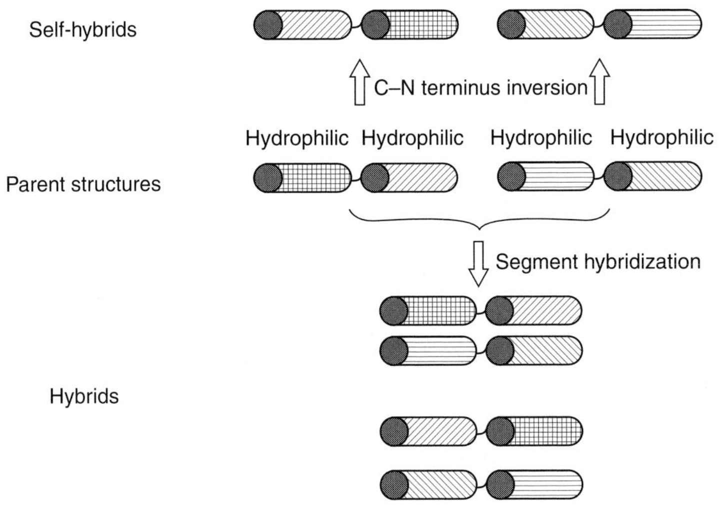

Figure 11.1 Design approaches to hybrid peptides. In agreement with most linear α-helical EAP structures, two chemically distinct regions in each parent structure are defined and involved in the hybridization process.

activity against human or fish pathogenic bacteria relative to the parent peptides (Jia *et al.*, 2000).

Historically, a first attempt of hybridization involved the N-terminal regions of cecropin B and sarcotoxin IA, which differ only in a pair of consecutive residues (Li *et al.*, 1988). The synthetic hybrids, which preserved their own C-termini, did not show improved antibiotic activity. A more successful attempt involved the hybrid peptide CA(1–11) CD(12–37)[2] (Fink *et al.*, 1989a), which was over 30 times more active than parental cecropin D on *Pseudomonas aeruginosa*, *Micrococcus luteus* and *Bacillus subtilis*. The criteria underlying the design of this hybrid were to increase the charge of the N-terminus (from 3+ in cecropin D to 7+ in cecropin A) while preserving structural motifs common to both peptides. Sequence inversion in cecropin A, as in CA(25–37)CA(1–13), provided completely inactive hybrids (Boman *et al.*, 1989). More recently, Chen *et al.* (1997) has designed cecropin B hybrids such as *CB-1* [CB(1–24)CB(1–10)] or *CB-2* [CB(1–24)-GP-CB(1–10)], with the hydrophylic N-terminal moiety replacing the hydrophobic C-terminus of the native molecule. These hybrids are in general less active than the parent peptide on liposomes or bacteria but are more effective in killing cancer cells. Another cecropin B inverted hybrid, *CB-3* [CB(25–35)CB(11–24)CB(25–35)] (Wang *et al.*, 1998), with hydrophobic helices at both N- and C-termini, was inactive as antibiotic but provided some insights on the mechanisms of lipid bilayer lysis (Table 11.4) (Srisailam *et al.*, 2001).

Other experiments in sequence hybridization have focused on melittin, which like the cecropins consists of two α-helical segments but arranged in opposite order (hydrophobic N-terminus, hydrophylic C-terminus). In the M(16–26)M(1–13) hybrid, the location of the two helical segments was reversed, to make it cecropin-like. This resulted in a lower bactericidal activity, but also decreased the strong hemolytic effect of melittin (Boman *et al.*, 1989). Melittin inversion hybrids have been revisited (Juvvadi *et al.*, 1996b): one analog, M(21–26)M(1–20), shows an interesting antibacterial profile.

Table 11.1 Parental peptides and their properties

Peptide (abbreviation) species	*Sequence*	*Antibiotic activity**						*Comments*
		G^+	G^-	*F*	*P*	*T*	*C*	
Cecropin A (CA) *Hyalophora cecropia* (American silk moth)	KWKLFKKIEKVGQNIRDGIIKAGPAVA VVGQATQIAK-NH_2	+	++	+	+	+	–	First cecropin to be chemically synthesized (Andreu *et al.*, 1993)
Cecropin B (CB) *H. cecropia*	KWKVFKKIEKMGRNIRNGIVKAGPAIA VLGEAKAILS-NH_2	+	++	+	+	+	–	Antibiotic activity higher than cecropin A. (Steiner *et al.*, 1981)
Cecropin D (CD) *H. cecropia*	WNPFKELEKVGQRVRDAVISAGPAVAT VAQATALAK-NH_2	–	+	n.d.	n.d.	n.d.	–	The less basic and active of *Hyalophora* cecropins (Hultmark *et al.*, 1982; Fink *et al.*, 1989a)
Sarcotoxin IA (SC) *Sarcophaga peregrina* (flesh fly)	GWLKKIGKKIERVGQHTRDATIQGLGI AQQAANVAATAR-NH_2	+	+	n.d.	n.d.	n.d.	–	Cecropin from Diptera (Okada and Natori, 1985)
Cecropin P1 (CP) *Sus scrofa* (pig)	SWLSKTAKKLENSAKKRISEGIAIAIQG GPR-COOH	–	+	+	+	n.d.	–	Forms a continuous α-helix (Lee *et al.*, 1989)
Melittin (M) *Apis mellifera* (honey bee)	GIGAILKVLSTGLPALISWIKRKRQE-NH_2	+	+	+	+	+	+	Main toxin from bee venom (Habermann *et al.*, 1972)
Magainin 2 (MA) *Xenopus laevis* (African clawed frog)	GIGKFLHSAKKFGKAFVGEIMNS-NH_2	+	+	+	+	+	–	One of the first amphibian antibiotic peptides described. Lead for a large number of analogs (Zasloff, 1987)
Bombinin *Bombina variegata* (Yellow bellied-toad)	GIGGALLSAAKVGLKGLAKGLAEHFA-COOH	++	+	+	n.d.	n.d.	–	(Simmaco *et al.*, 1991; Shin *et al.*, 1997b)

Notes
* G^+ – Gram positive; G^- – Gram negative; F – Fungicidal; P – Parasiticidal; T – Tumoricidal; C – Cytotoxic and/or hemolytic.
n.d. Not determined.

Table 11.2 Relevant analogs from parental peptides

Peptide	*Sequence*	*Comments and references*
Abp3	KWKVFKKIEKMGRNLREGLVKAGPAIAVLGEAKAL-NH_2	Improved antibiotic activity respect to cecropin B. NMR structure determined (Xia *et al.*, 1998)
SB-37	MPKWKFKKIEKVGRNIRNGIVKAGPAIAVLGEAKALG	Cecropin B analog with improved antibiotic activity (Jaynes *et al.*, 1993)
Shiva-1	MPRWRLFRRIDRVGKQIKQGILRAGPAIALVGDARAVG	Cecropin B analog. It induces resistance against potato wilt (Jaynes *et al.*, 1993)
MB-38	HQPKWKVFKKIEKVGRNIRNGIVKAGPAIAVLGEAKALG	Cecropin B analog with N-terminus modified to improved secretion. $Met^{11} \rightarrow Val$ substitution decreases its susceptibility to plant proteinases (Owens and Heutte, 1997)
Cycled melittin	CGIGAVLKVLTTGLPALISWIKRKRQQC (cyclized, C to C)	Devoid of most of the hemolytic activity inherent to natural melittin (Unger *et al.*, 2001)
Hecate-1, DC-1	FALALKALKKALKKLKKALKKAL	Strongly modified peptides used as comparison with cecropin A
DC-2	FAKKLAKKLKKLAKKLAKLALAL	High parasiticidal activity (Barr *et al.*, 1995)

Table 11.3 Selected sequences of melittin and cecropin analogs produced by self-transposition, or cecropin–cecropin and self-cecropin hybridizations

Peptide	*Sequence*[a]	*Comments and references*
1 *Transposed analogs*		
CA(25–37)CA(1–22)	AVAVVGQATQIAKKWKLFKKIEKVGQNIRDGIIKA–NH_2	Transposed cecropin A analog preserving most of its microbicidal activity (Wade *et al.*, 1992)
CA(25–37)CA(1–13)	AVAVVGQATQIAKKWKLFKKIEKVGQ–NH_2	Transposed analog of cecropin A, devoid of relevant antibiotic activity (Wade *et al.*, 1992; Boman *et al.*, 1989)
M(16–26)M(1–13)	PALISWIKRKQQVLKVLTTGL–NH_2	Transposed melittin endowed with antibiotic but not hemolytic activity (Juvvadi *et al.*, 1996b)
MCFA	GLKKLISWIKRAAQQG–NH_2	Transposition of positive charges from the C- to the N-terminal end of melittin abrogates hemolysis (Subbalakshmi *et al.*, 1999)
2 *Hybrids among cecropins*		
L^3K^4-CB	KWLKKVFKKIEKMGRNIRNGIVKAGPAIAVLGE AKAL–NH_2	Mimicking of sarcotoxin IA by insertion of the differential dipeptide into cecropin A sequence. Substantial loss of antibiotic activity (Li *et al.*, 1988)
CA(1–11)CD(12–37)	KWKLFKKIEKVGQRVRDAVISAGPAVATVAQAT ALAK–NH_2	First hybrid between different cecropins with improvement of bactericidal activity relative to cecropin D (Fink *et al.*, 1989a)
3 *Cecropin B self hybrids*		
CB-1 = CB(1–25)CB(1–10)	KWKVFKKIEKMGRNIRNGIVKAGPKWKVFKKIEK–NH_2	Improvement of tumoricidal activity respect to cecropin B. Studies on hinge flexibility (Chen *et al.*, 1997; Wang *et al.*, 1998)
CB-2 = CB(1–25)-GP-CB(1–10)	KWKVFKKIEKMGRNIRNGIVKAGPGPKWKVFK KIEK–NH_2	
CB-3 = CB(25–35)CB(11–24)CB(25–35)	AIAVLGEAKALMGRNIRGIVKAGPAIAVLGEAKA L–NH_2	

Note
a Underlining of some sequences stretches is intended to differentiate both parental sequences.

Table 11.4 Selected cecropin–non-cecropin hybrids

Peptide[a]	*Sequence*[b]	*Comments and references*
1 *Long (>15 residues) cecropin A–melittin hybrids*		
CA(1–8)M(1–18); CEME; MBI-27	KWKLFKKIGIGAVLKVLTTGLPALIS–NH_2	One of the most active cecropin A-melittin hybrids (Wade *et al.*, 1990)
CEMA, OLP-8, MBI-28	KWKLFKKIGIGAVLKVLTTGLPALKLTK–NH_2	CA(1–8)M(1–8) analog with improved antiendotoxic activity (Gough *et al.*, 1996)
MsrA	MALEHMKWKLFKKIGIGAVLKVLYTGLPALKLTK–NH_2	CEMA analog optimized for expression and antibiotic activity in Plants (Osusky *et al.*, 2000)
CP29	KWKSFIKKLTTAVKKVLTTGLPALIS–NH_2	Cp29 and CP26 improved CEMA amphipathicity. V analogs were obtained from a CP26 variant. The non-bent analog V681n has an increased hemolytic activity, whereas V68pp with two prolines has poor antibacterial activity (Scott *et al.*, 1999a; Zhang *et al.*, 1999)
CP26	KWKSFIKKLTSAAKKVVTTAKPLISS–NH_2	
V681n	KWKSFLKTFKSAVKTVLHTALKAISS–NH_2	
V68pp	KWKSFLKTFKSPARTVLYTALKPISS–NH_2	
490lin	KWKSFIKLEKVLKPGGLLSNIVTSL–NH_2	Comparison between linear and lactam-cycled analogs of CA(1–8)M(1–8) attempted to stabilize α-helix in aqueous solutions, but did not lead to enhanced antibiotic activity (Houston *et al.*, 1998)
490lac	KWKSFIKLEKVLKPGGLLSNIVTSL–NH_2	
CA(1–13)M(1–13)	KWKLFKKIEKVGQGIGAVLKVLTTGL–NH_2	The first cecropin A–melittin hybrid (Wade *et al.*, 1990)
M(1–13)CA(1–13)	GIGAVLKVLTTGLKWKLFKKIEKVGQ–NH_2	Improved bactericidal activity and poor hemolytic activity (Wade *et al.*, 1992)

(Continued)

Table 11.4 (Continued)

Peptide[a]	*Sequence*[b]	*Comments and references*
2 *Shortened (≤15 residues) cecropin A–melittin hybrids*		
CA(1–7)M(2–9)	KWKLFKKIGAVLKVL–NH_2	One of the most active shortened cecropin A–melittin hybrids (Andreu *et al.*, 1992)
$FA^{\alpha1}$- CA(1–7)M(2–9)	FA- KWKLFKKIGAVLKVL–NH_2	Fatty acid acylated CA(1–7)M(2–9) analogs. Acylation at K^1 but not at K^7 increased antibiotic activity (Chicharro *et al.*, 2001)
$FA^{\varepsilon1}$- CA(1–7)M(2–9)°°	KWKLFKKIGAVLKVL–NH_2 (FA on K[1])	
$FA^{\varepsilon7}$- CA(1–7)M(2–9)	KWKLFKKIGAVLKVL–NH_2 (FA on K[7])	
CAM135	GWRLIKKILRVFKGL–NH_2	The most active analog obtained from CA(1–7)M(2–9) by QSAR. Twice as active as the initial peptide (Mee *et al.*, 1997)
CA(1–7)M(5–9)	KWKLFKKVLKVL–NH_2	Bactericidal activity tested in vivo (Nos-Barberá *et al.*, 1997)
CA(2–7)M(6–9)	WKLFKKILKVL–NH_2	One of the shortest cecropin A–melittin analogs with antibiotic activity (Cavallarin *et al.*, 1998)
3 *Other cecropin hybrids*		
CB(1–13)M(1–13)	KWKVFKKIEKMGRGIGAVLKVLTTGL–NH_2	The only cecropin B–melittin described, improved Gram-positive activity without hemolytic activity (Wade *et al.*, 1992)
CA(1–8)MA(2–12)	KWKLFKKIGIGKFLHSAKKF–NH_2	Lead peptide of cecropin A–magainin 2 hybrids. High tumoricidal and antibacterial activity, but less toxic than the equivalent CA(1–8)M(2–12) (Shin *et al.*, 1997a; Wang *et al.*, 1998; Oh *et al.*, 2000a)
CA(1–8)BO(1–12)	KWKLFKKIGIGGALLSAAKV–NH_2	Cecropin A–bombinin hybrid, with antifungal activity and not hemolytic (Shin *et al.*, 1997b)

Notes
a Alternative names were included.
b Unmodified cecropin fragments are underlined.
FA – Fatty acid acylation at NH_2.

Table 11.5 Selected sequences of enantio, retro and retroenantio peptides

Peptide	*Sequence*[a,b]	*Comments and references*
1 *Examples of some EAPs stereoisomers*		
All-L-CA(1–7)M(2–9)	KWKLFKKIGAVLKVL–NH_2	Merrifield *et al.*, 1995b; Vunnam *et al.*, 1998
All-D-CA(1–7)M(2–9)	kwklfkkigavlkvl–NH_2	
Retro-CA(1–7)M(2–9)	LVKLVAGIKKFLKWK–NH_2	
Retroenantio-CA(1–7)M(2–9)	lvklvagikkflkwk–NH_2	
[D]-$V^{5,8}$, I^{17}, K^{21}-melittin	GIGAvLKvLTTGLPALiSWIkRKRQQ–NH_2	Diastereomer of melittin. Substantial loss of α-helix content and hemolytic, but not bactericidal activity (Oren and Shai, 1997; Sharon *et al.*, 1999)
[D]-K^{11}, F^{12}-magainin2	GIGKFLHSAKkfGKAFVGEIMNS–NH_2	Magainin diastereomer with strong loss of α-helix content (Wieprecht *et al.*, 1996)
2 *Other peptides whose enantio-, retro- and retroenantiomers antibiotic activities have been explored*		
All-D-, retro- and retroenantio-CA(1–13)M(1–13)	KWKLFKKIEKVGQGIGAVLKVLTTGL–NH_2	Decreased hemolytic, preserved bactericidal activity (Wade *et al.*, 1990; Merrifield *et al.*, 1995a; Vunnam *et al.*, 1998)
Retro-melittin M(26–1)	qqrkrkiwsilaplgttlvklvagig–NH_2	No dramatic changes in activities (Juvvadi *et al.*, 1996b)
M(20–1)M(21–26)	IWSILAPLGTTLVKLVAGIGQQKRK–NH_2	Partial retro-melittin self hybrid endowed with good antibacterial activity with large decrease in hemolysis (Juvvadi *et al.*, 1996b)
M(26–21)M(1–20)	QQKRKGIGAVLKVLTTGLPALISWI–NH_2	Slightly improved bactericidal, less hemolytic activity relative to melittin (Juvvadi *et al.*, 1996b)
M(1–20)-*all*-D(21–26)	GIGAVLKVLTTGLPALISWIkrkqq–NH_2	Semi-enantio analog, lower antiviral activity than melittin (Wachinger *et al.*, 1992)
All-L- and retro-Ac-CP1(1–10)M(2–9)	Ac–SWLSKTAKKGAVLKVL–NH_2	Cecropin P1-melittin retro-analog, less active than the *all*-L-form against bacteria (Juvvadi *et al.*, 1999)

Notes
a Small letters stand for D-amino acid.
b Underlined sequences aim to differentiate the two parental sequences. Ac-Acetylated.

A very successful approach by the Merrifield and Boman groups involved cecropin A-melittin hybrids (reviewed by Merrifield *et al.*, 1994) such as CA(1–13)M(1–13) and CA(1–8)M(1–18), where the cationic N-terminus of cecropin A is followed by the hydrophobic N-terminal helix of melittin. This hybridization endowed the peptides with strong activity on Gram-positive bacteria, against which parent cecropin A is fairly inactive, while eliminating or greatly reducing the hemolytic activity associated with the cationic C-terminus of melittin (Boman *et al.*, 1989; Wade *et al.*, 1990, 1992). Taking CA(1–8)M(1–18) as lead, one of our laboratories has shown that substantial size reduction to 15 (Andreu *et al.*, 1992), 12 (Nos-Barberá *et al.*, 1998) or even 11 residues (Cavallarin *et al.*, 1998) can be achieved without seriously compromising antimicrobial activity. Further attempts to improve the activity of cecropin A-melittin hybrids include a CA(1–8)M(1–18) analog with an extended, cationic C-terminus (Piers *et al.*, 1994), or the use of QSAR based on combinatorial search algorithms (Mee *et al.*, 1997) as design principles; in either case no appreciable improvement has been found. Further modification of cecropin A-melittin hybrids has included an N-terminally extended, transgenic potato version (Osusky *et al.*, 2000) providing broad resistance to phytopathogens and fatty acid-acylation (Chicharro *et al.*, 2001) at the N-terminus (α- or ε-amino groups of Lys1) of CA(1–7)M(2–9) conferring improved activity on *Leishmania* (Table 11.4).

Cecropin–magainin hybrids such as CA(1–8)M(1–12) have also been developed (Shin *et al.*, 1997a; Kang *et al.*, 1998) and shown to possess interesting antitumor activity. Finally, porcine cecropin P1-melittin hybrids have also been prepared (Juvvadi *et al.*, 1999), with interesting activities for peptides such as Ac-CP(1–10)M(2–9).

Structural considerations

Parent structures

Conformational studies in solution by NMR of cecropins (Holak *et al.*, 1988), magainins (Gesell *et al.*, 1997) or melittin (Bazzo *et al.*, 1988) suggested a structural pattern of two α-helix segments separated by a short flexible (hinge) sequence. In the cecropins, the N-terminal and C-terminal helices were amphipathic and hydrophobic, respectively, whereas in melittin and magainin the order was reversed.

Additional structural work on related analogs confirms the general pattern. Thus, a cecropin B analog (CB, $I^{15} \rightarrow L$, $Q^{17} \rightarrow E$, $I^{19} \rightarrow L$) [*Abp3*] (Xia *et al.*, 1998) also displays an amphipathic N-terminal helix (Phe^{5}–Lys^{21}), a hinge (Ala^{22}–Gly–Pro–Ala^{25}) and a C-terminal hydrophobic helix (Ile^{26}–Leu^{35}). The NMR structure of sarcotoxin IA, a cecropin from the *Sarcophaga peregrina* fly, has been determined in 80% methanol (Iwai *et al.*, 1993). Again, the N-terminal amphipathic (Leu^{3}–Gln^{23}) and C-terminal hydrophobic helices (Ala^{28}–Ala^{38}) are connected by a hinge (Gly^{24}–Ile^{27}), with considerable similarity to cecropin A. Porcine cecropin P1 (Sipos *et al.*, 1992) in 15–30% HFIP follows a similar trend: an amphipathic N-terminal helix (Ser^{4}–Ile^{19}) is followed by a "hinge" sequence (Glu^{20}–Gly^{21}) that can provide a certain degree of flexibility (depending on the ionization state of the Glu residue) and by a short (1–2 turn) hydrophobic helix.

Structural studies on hybrid peptides

CA(1–13)M(1–13), one of the most active cecropin A-melittin hybrids, was the first to be examined by NMR (Sipos *et al.*, 1991). It showed the cecropin-like helix-hinge-helix

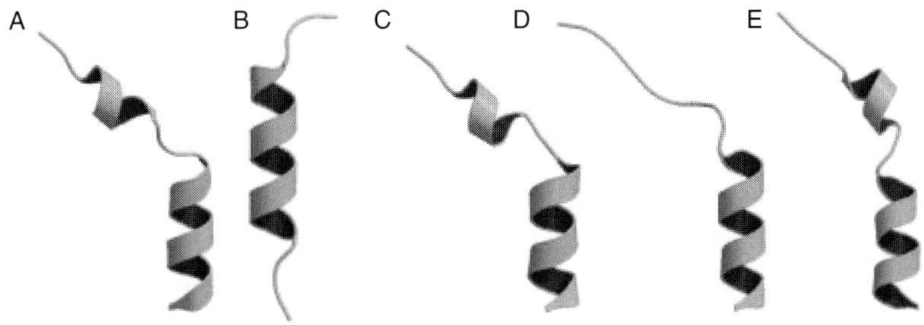

Figure 11.2 Ribbon diagrams depicting NMR solution structures of CA(1–8)MA(1–12) hybrid peptides. (A) CA(1–8)MA(1–12). (B) CA(1–8)MA(1–12), with G^9,I^{10}, G^{11} deletions. (C) CA(1–8)MA(1–12), G^9,$I^{10}G^{11} \rightarrow$ P. (D) CA(1–8)MA(1–12), $W^2 \rightarrow$ A. (E) CA(1–8)MA(1–12), $W^2 \rightarrow$ L. (Reprinted with permission from Oh *et al.* (2000a), *Biochemistry*, **39**, 11855–11864. Copyright 2000. American Chemical Society.)

pattern, the short N-terminal (2–3 helix turns, Ala^4–Gly^{12}) segment followed by a Gln–Gly–Ile hinge and by a hydrophobic C-terminal helix (Gly^{16}–Leu^{26}) that reproduces quite well the structure of the parental sequence. In contrast, the N-terminal region shows several differences with cecropin. Hybrids with a shorter cecropin A moiety, such as CA(1–8)MA(1–12) (Oh *et al.*, 2000a) predictably display a shorter amphipathic helix (see Figure 11.2).

Notably, in this short stretch replacement of Trp^2 (not actually part of the α-helix) by Leu causes a complete destructuration of the helix, whereas Ala replacement preserves it (Oh *et al.*, 2000a). Similar findings had been earlier described (by circular dichroism) for cecropin A analogs with $Trp^2 \rightarrow$ Glu and $Trp^2 \rightarrow$ Phe replacements, respectively (Andreu *et al.*, 1985). On the other hand, in the even shorter CA(1–7)M(2–9) (Andreu *et al.*, 1992; Fernández *et al.*, 1994), removal of the Gly^1 residue from the melittin moiety abrogates the hinge, causing the amphipathic and hydrophobic helical sections to coalesce into a single helix spanning most of the sequence. Several long-range NOEs observed for this peptide were incompatible with a single α-helical *monomeric* structure. Instead, the NOE constraints could be fitted into a model of aggregation involving antiparallel packing of neighboring helices (Fernández *et al.*, 1996). The tendency to aggregate seems to be an inherited feature from melittin and may enhance antibiotic activity, either by facilitating peptide incorporation into the membrane in large quantities or by promoting membrane disruption. Other active hybrids of the same length, such as CA(1–7)M(4–11) or CA(1–7)M(5–12), follow similar trends (Andreu *et al.*, 1992).

Finally, in the CB(1–25)CB(1–10) [*CB-1*] hybrid (Srisailam *et al.*, 2000), the overall cecropin features are preserved: the peptide displays a well defined α-helix spanning residues 3–21, followed by a helix-breaking Gly–Pro hinge and again by a short (amphipathic, not hydrophobic) helical stretch. In contrast CB(25–35)CB(11–24)CB(25–35) [*CB-3*] adopts a crumpled structure with four segments forming α-helix (Srisailam *et al.*, 2001).

Key structural features for hybrids

The preceding conformational data, along with other additional structure-activity results, allow to define some general structural principles for cecropin-hybrid peptides.

Chain length. As membrane-active peptides often exert their lethal activity through formation of membrane-spanning pores, a minimal length of 20 amino acids was initially proposed for a peptide adopting a standard α-helix to traverse the membrane (20 residues; 1.5 Å/residue = 30 Å). In tune with this, deletion of seven [M(7–26), Habermann and Koballek, 1970] or eight N-terminal residues [M (8–25), Dawson *et al.*, 1978] from melittin abrogated hemolytic activity. Also, Scott *et al.* (1999b) found that Gram-negative bactericidal activity of a set of CA(1–8)M(1–18) analogs was related to the length of the chain.

That general wisdom came into question by the fact that much shorter hybrids, such as 15-residue CA(1–7)M(2–9) (Andreu *et al.*, 1992), or even shorter peptides (Nos-Barberá *et al.*, 1997; Cavallarin *et al.*, 1998) all containing at least seven residues from cecropin A and four from melittin were still very potent antimicrobials. More recently M(12–26), a melittin fragment with 5–6-fold less antibacterial activity relative to the parent molecule, but also considerably (300-fold) less hemolytic activity, could be engineered into a fully active 15-residue analog by appropriate replacements [$Pro^{14} \rightarrow Lys$; $Ala^{15} \rightarrow Lys$; $Lys^{23} \rightarrow Ala$; $Arg^{24} \rightarrow Ala$], with preservation of its non-hemolytic character (Subbalakshmi *et al.*, 1999). These results, diverging from the originally accepted view, have been interpreted on the basis of peptide–membrane interaction (see the Section on "Studies on model membranes").

Although for most cecropin hybrids the usual arrangement involves placing the cationic N-terminal sequence of cecropin A at the N-terminus followed by the N-terminal hydrophobic part of melittin, other orientations have also been tested, with various results. Thus, inversion of the consensus arrangement, as in CA(25–37)CA(1–13) or CA(25–37) CA(1–22), involved almost total loss of bactericidal activity (Boman *et al.*, 1989 Wade *et al.*, 1992). On the other hand, the melittin inversion hybrid M(16–26)M(1–13) is devoid of hemolytic activity, but preserves almost completely the bactericidal activity (Wade *et al.*, 1992).

Hinge. In the context of antimicrobial peptides, a hinge has been defined as a flexible sequence that separates two helical domains, as in native cecropins (see above). This was originally considered an essential feature for antimicrobial action (Fink *et al.*, 1989a,b), as it would allow independent electrostatic binding of the N-terminal cationic region to phospholipid head groups on the membrane, while the hydrophobic portion could be inserted in the membrane to form a pore. This view has gradually come into question. Thus, using individual deletion analogs Blondelle and Houghten (1991b) found that deletions inside the hinge had a much less dramatic effect than inside the helical domains. Similarly, for cecropin A-melittin hybrids, flexibility is not essential, nor requires a strong helix-breaking sequence such as Gly–Pro, but can be simply achieved by having two nearby Gly residues in a short stretch, as in CA(1–13)M(1–13) (Sipos *et al.*, 1991), or even a single residue, as suggested for Ser^{11} in CA(1–10)M(18–26) (Wade *et al.*, 1992). Deletion of Pro or Gly–Pro sequences increased bactericidal and hemolytic activity for cecropin A–melittin analogs (Wade *et al.*, 1992; Zhang *et al.*, 1999), whereas inserting a hinge within a helical region was deleterious, as with the Pro replacements within [CP26] analogs (Zhang *et al.*, 1999), with cecropin A–magainin hybrids (Shin *et al.*, 2000; Oh *et al.*, 2000a), or with the additional Gly–Gly sequence of [CP201] (Friedrich *et al.*, 1999). A similar trend had earlier been found in parental peptides such as cecropin A (Andreu *et al.*, 1985) or cecropin P1 (Gazit *et al.*, 1995).

Sometimes flexibility requirements are more strict. A case in point is the cecropin A–magainin hybrid CA(1–8)MA(1–12), for which deletion of the Gly–Ile–Gly hinge gave rise to an helical structure extending almost through the entire sequence, but causing partial bactericidal and almost total tumoricidal activity losses, respectively (Shin *et al.*, 2000;

Oh *et al.*, 2000a). Replacing the Gly–Ile–Gly hinge by Pro produced a recovery of bactericidal and tumoricidal activity, but when flexibility was improved by substitution with the Gly–Pro–Gly tripeptide, a significant drop of all activities was noted. On the other hand, insertion of a Gly–Pro dipeptide after the constitutive Gly^{23}–Pro^{24} of the self-hybrid CB(1–24)CB(1–10) [*CB-1*] caused a decrease in tumoricidal but not bactericidal activity (Chen *et al.*, 1997).

Helicity and amphipathicity. Most of the peptides mentioned in this chapter (cecropins, magainins, melittin and their corresponding hybrids), are unstructured and monomeric in aqueous buffers (Steiner, 1982; Gazit *et al.*, 1994, 1995; Silvestro and Axelsen, 2000); scarce exceptions to this rule arise from peptide aggregation at high concentrations and/or non-physiological conditions (Terwilliger *et al.*, 1982a,b; Andreu *et al.*, 1992; Fernández *et al.*, 1994). The prevalent view for a number of years has been to associate peptide antimicrobial (or, for that matter, membrane) activity with the tendency to adopt an amphipathic α-helical conformation when bound to the membrane. As discussed below, this concept has been critically reevaluated in recent years.

The conventional view is supported by experiments such as systematic single residue deletions in melittin: those within the two helical regions caused substantial drops in hemolytic activity while those at the hinge or C-terminus were less severe (Blondelle and Houghten, 1991a,b). Bactericidal activity in other peptides also followed this general trend. Thus, for cecropin A, disruption of the N-terminal amphipathic helix by Pro insertion at position 8 caused practical loss of bactericidal activity except for *E. coli* (Andreu *et al.*, 1985). In tune with these observations, attempts to improve peptide activity have involved substitution analogs designed to enhance the helical character and/or amphipathicity at certain points while trying to preserve overall charge, length and hydrophobicity (Shin *et al.*, 1997a; Friedrich *et al.*, 1999). This approach has not produced substantial improvements. An even more drastic approach at stabilizing the helical conformation involved introducing an internal lactam clamp in the form of a covalent bond between Glu and Lys side chains located one helix turn (roughly 4 residues) away from each other (Houston *et al.*, 1998). Although the resulting peptides showed increased α-helical contents in aqueous media, the conformational rigidity imposed by the lactam bridge was detrimental to antibacterial activity.

A further consideration regarding helicity concerns terminal amino and carboxyl groups and their influence on the helix macrodipole. At physiological pH, these charged (NH_3^+ and COO^-) groups oppose the dipole (which has a $N^{\delta+} \rightarrow C^{\delta-}$ direction). Therefore, end group capping contributes to helix stability and may result in increased activity. The fact that a number of helical antimicrobial peptides (e.g. bombolittin, cecropins, melittin, PGLa, clavanin) are C-terminally amidated seems to reinforce this view. Thus, C-amidation of cecropin P1, which is non-amidated in the native form, increases both antibacterial and hemolytic activity (Lee *et al.*, 1989), while the carboxylic forms of cecropin A (Li *et al.*, 1988) or CA(1–13)M(1–13) (Vunnam *et al.*, 1998) show decreased activity. It must be noted that, in addition to helix stabilization, amidation may also increase resistance to carboxypeptidase degradation. The effect of reducing the positive charge the N-terminus is illustrated by CA(1–8)M(1–18), which adopts substantial levels of α-helix (vs aperiodic conformation in aqueous media) in the presence of heparin, presumably by neutralization of the Lys charge by the highly anionic oligosaccharide heparin (Rivas, unpublished circular dichroism data). Similar results were obtained for melittin analogs with colominic (polysialic) acid (Pérez-Payá *et al.*, 1995), or for magainin (Matsuzaki *et al.*, 1999), cecropin A (De Lucca *et al.*, 1995) and cecropin B (Bland *et al.*, 2001) with LPS. Acetylation of

CA(1–7)M(2–9) or CA(1–13)M(1–13) increases the helical content to levels comparable with those obtained in the presence of structure-inducing solvents such as HFIP (Vunnam *et al.*, 1998; Chicharro *et al.*, 2001). On the other hand, for cecropin P1–melittin analogs, acetylation decreased while succinylation increased the activity of the peptides (Juvvadi *et al.*, 1999).

While these data are all consistent with the conventional view that helicity and activity are related, recent findings suggest that an α-helix may not be a strict requirement for antibiotic activity. For instance, discrepancies in α-helix content are observed for membranes compared to solvents, since membrane interactions are anisotropic. Thus, for magainin, not only the helical content but also the strength of the helix varies depending on the system used (Wieprecht *et al.*, 1996). Even more compelling evidence comes from work at two independent laboratories showing that melittin (Oren and Shai, 1997) or magainin (Wieprecht *et al.*, 1996). Diastereoisomers (incorporating D-amino acids in helical regions of the parent molecule) can have lower overall helicity and hemolytic activity and yet retain most of their antibacterial activity.

Charge. Again, the general view has been that a certain level of positive charge is needed to attain specific microbicidal activity. For instance, cecropin D is less cationic and less active than cecropin A or B, and an improvement was found in the more cationic hybrid CA(1–11)CD(12–37) (Fink *et al.*, 1989a). Succinylation of CA(1–2)M(2–9) completely abolished activity (Fernández *et al.*, 1994), and similar relationships were derived for magainins (Besalle *et al.*, 1992; Matsuzaki *et al.*, 1997a). Improved permeabilization of the outer membrane was obtained upon C-terminal extension of CA(1–8)M(1–18) with Lys residues (Piers *et al.*, 1994), as well as improved antitumoral activity for the highly cationic CA(1–8)M(1–12), $T^{18} \rightarrow K$, $T^{19} \rightarrow K$ analogs (Shin *et al.*, 1997a). Again, for CB(1–25)CB(1–10)] [*CB-1*] or CB(1–25)-GP-CB(1–10) [*CB-2*] the presence of a strong cationic sequence at the C-terminus improved antitumoral, though not antibacterial activity (Chen *et al.*, 1997). Nevertheless, an excess of positive charged led to a loss of either efficacy (Matsuzaki *et al.*, 1997a) or even selectivity (Dathe *et al.*, 2001) for magainin analogs. Positive charge, however, is only a partial requirement, since in addition to cationic character a hydrophobic moiety is equally needed for antibiotic effect. Thus, in cecropin A–melittin hybrids, where most of the positive charge is accumulated at the N-terminus, deletion of the C-terminal hydrophobic section, as in the truncated CA(1–12) or CA(1–8) analogs, produced inactive molecules (Fink *et al.*, 1989b, Sipos *et al.*, 1991; Díaz-Achirica *et al.*, 1998).

Tryptophan residues. Tryptophan appears in many antibiotic peptides (cecropins, dermaseptins, enbocin, indolicidin, lycotoxin, melittin, pleurocidin, styelin) in a frequency much higher than its natural abundance in proteins. In most cases, the Trp residue is located at or very near the N-terminus, while in peptides with a reverse arrangement of amphipathic/hydrophobic moieties (e.g. melittin) it is found at the C-terminal region.

Although a Trp in position 2 is a conserved feature in many insect cecropins as well as in cecropin P1, its role in membrane permeabilization is still rather puzzling. NMR analysis shows it does not integrate within the N-terminal amphipathic helix (Holak *et al.*, 1988), which may explain why [D-Trp^2] cecropin A is fully active (Li *et al.*, 1988). Also, in cecropin A, Trp formylation, removal, or replacement by non-aromatic residues causes a substantial decrease of antibacterial activity (Merrifield *et al.*, 1982; Andreu *et al.*, 1985). Similar observations have been made for some cecropin A-melittin hybrids (Scott *et al.*, 1999a,b). Again, in CA(1–8)MA(1–12), a $Trp^2 \rightarrow Ala$ replacement caused substantial loss of α-helix and loss of antibiotic activity, which were however restored in a Leu^2 analog (Oh *et al.*, 2000a). Other Trp-related observations within the cecropin A family are the inactivity of peptides with two consecutive Trp residues (Scott *et al.*, 1999b), and the possibility to exchange Lys^1 and Trp^2 positions without appreciable loss of activity (D. Andreu, unpublished).

For melittin, remarkably, deletion of Trp^{19} leads to loss of hemolytic activity (Blondelle and Houghten, 1991a). On the other hand, when a Trp residue is "walked" along the sequence (Blondelle *et al.*, 1993), significant decreases in hemolytic activities are observed, particularly when replacing Leu residues at positions 9, 13, 16 and, to a lesser extent, 6. In contrast, placing the Trp on the hydrophilic face of the amphipathic α-helix results in increased hemolysis.

Conjectures about the special role of tryptophan in antimicrobial activity have assigned it an anchoring role into the membrane, following the initial electrostatic interaction between the cationic groups of peptide and the anionic heads of phospholipids. Insertion of the Trp indole ring into membranes at the vicinity of glycerol units has been described (Yau *et al.*, 1998); upon insertion, the flat aromatic ring may further hamper acyl chain mobility, as well as intervene in structure-based interactions with other membrane components. In melittin, the Trp indole ring has been located immediately under the phospholipid head in acidic vesicles (Ghosh *et al.*, 1997). This is in agreement with the quenching of tryptophan fluorescence of cecropin A–melittin hybrids by external quenchers (Zhang *et al.*, 1999).

Moreover, in a theoretical model of a cecropin A pore, a key role for the Trp residue in the stabilization of peptide dimers was postulated, by fitting it into the hydrophobic pocket formed by Phe^5 and Ile^8 of an adjacent monomer (Durrell *et al.*, 1992).

Stereoisomers. The enantiomeric versions of cecropin A, melittin, magainin2 and the CA(1–13)M(1–13) and CA(1–8)M(1–18) hybrids (Wade *et al.*, 1990) served a double purpose: (i) to make the peptides resistant to proteolysis, and (ii) to ascertain whether chirality was relevant for the interaction of the peptides with their targets. The finding that the enantiomers performed equally or better than their *all*-L analogs on both model membranes and pathogens ruled out the involvement of a stereospecific (protein, oligosaccharide) receptor in the mechanism of action of these peptides. Similar conclusions have been drawn for a set of cecropin A–melittin hybrid pentadecapeptides (Merrifield *et al.*, 1995b), for CA(1–8)M(1–18) on *Leishmania* (Díaz-Achirica *et al.*, 1998), or for the *all*-D-cecropin B (De Lucca *et al.*, 2000; Bland *et al.*, 2001). Exceptions to this behavior are the differential binding of L- or D-cecropin B to LPS, a chiral molecule (Bland *et al.*, 2001), or the decreased activity of *all*-D forms of CA(1–13)M(1–13) and CA(1–8)M(1–18) against *Plasmodium falciparum* merozoites (Wade *et al.*, 1990).

Further insights into the topological requirements of EAP action were obtained from the retro (reversed amino acid sequence) and retroenantio (retro made up of D-amino acids) versions of CA(1–13)M(1–13) and CA(1–7)M(2–9) (Merrifield *et al.*, 1995a,b). The topological relationship between native (*all*-L) and retroenantio forms becomes evident if the latter is drawn in fully extended conformation and rotated 180° in the plane: the residues are arranged in identical order and the side chains adopt the same spatial orientation; on the other hand, the end groups and the direction of peptide bonds (CO(NH vs NH(CO)) are opposite.

All CA(1–13)M(1–13) analogs displayed the same activity for a set of different bacteria, with the exception of *Bacillus subtilis*, for which both retro and retroenantio hybrids were less active, suggesting that both sequence and amide-bond direction must be preserved for activity. A similar result and interpretation were found for the retro and retroenantio isomers of the shorter CA(1–7)M(2–9) hybrid on *Pseudomonas aeruginosa*. Remarkably, for *Staphylococcus aureus*, the retro and retroenantio isomers of CA(1–7)M(2–9) were also inactive, but the corresponding acetyl derivatives retained activity, implying that the location of the $-NH_3^+$ group is important for activity on this particular organism (Merrifield *et al.*, 1995a). More recently, cecropin P1-melittin hybrids CP(1–9)M(2–8) and CP(1–9)M(1–8) have been investigated, and a strong requirement for helix and dipole direction has been found (Juvvadi *et al.*, 1999). The subtle differences encountered in these studies suggest that sequence and dipole effects must be considered for each peptide on an individual basis.

As mentioned above, diastereomeric versions of EAPs such as magainin and melittin, with reduced levels of α-helix but preserving the overall charge and the nature of their respective side chains have been investigated. Destruction of helicity decreases the hemolytic activity of melittin but not its antibiotic activity (Oren and Shai, 1997). The NMR structure of a melittin diastereomer with D-residues at Val^5, Ile^{17} and Lys^{21} shows that helix formation is restricted to the C-terminal but not the N-terminal region of the molecule, and only in the presence of trifluoroethanol or of acidic but not neutral phospholipids (Sharon *et al.*, 1999). Although similar studies for cecropin–melittin hybrids are not yet available, it seems reasonable to conclude that the N-terminal moiety of melittin included in cecropin–melittin hybrids is related to cytolytic but not to antibacterial activity. One can also predict that diastereomeric modification of cecropin–melittin hybrids will presumably result in analogs with lower cytotoxicity.

Other structural features. Our laboratories have recently explored the effect of acylation of CA(1–7)M(2–9) at positions 1(N^α and N^ε) or 7 (N^ε) with saturated fatty acids (Chicharro *et al.*, 2001). Whereas acylation at either amino group of the N-terminal Lys improves antimicrobial and, to a lesser extent, hemolytic activities, acylation at position 7 impairs helix formation and is highly detrimental for antibiotic activity. The length of the fatty acid chain has some influence on the activity of the lipopeptide, with maximum values observed for lauroyl (C_{12}) or myristoyl (C_{14}) groups, further elongation leading to significant aggregation and partial loss of activity.

Cyclization of magainin and melittin by disulfide bridge formation between N- and C-terminal Cys residues has been explored (Unger *et al.*, 2001). Whereas cyclization reduced hemolytic activity of both peptides, antibacterial potency only decreased for magainin variation in α-helical content of the cyclic vs linear peptide form only decreased by 25%.

Studies on model membranes

Background

Given their intrinsic simplicity, interactions between EAPs and model membranes, either planar or liposomes, are a good testing ground for the determination of parameters involved in the permeabilization and killing mechanisms, avoiding the complexities inherent to living systems. Obviously, such simplification often hampers extrapolation from in vitro to in vivo contexts (Andreu *et al.*, 1983; Friedrich *et al.*, 1999; Wu *et al.*, 1999; Zhang *et al.*, 1999; Dathe *et al.*, 2001, among others). Compilation of results from different groups is further hindered by the variety of experimental parameters (lipid composition, lipid : peptide ratio) and by the different techniques used to monitor the corresponding effects (Shai, 1999).

Compared to classical membrane pore-forming peptides such as δ-hemolysin, EAPs possess some distinctive features, as follows:

1 For EAPs, "pore" is defined in a functional sense, as any dynamic structure (e.g. structural membrane defect) induced exclusively by the peptide and which causes release of

soluble molecules (Matsuzaki, 1998). Possible variations in stoichiometry may give rise to transient structures which might go unnoticed from a macroscopic point of view.

2 EAPs achieve membrane permeabilization at higher peptide : lipid ratios (*c*.1 : 100) than other pore-forming peptides, for which ratios as low as 1 : 10,000 can induce permeabilization.

3 Pore formation always requires the presence of phospholipids that act not only as catalyzer for peptide structuration, but also as a physical part of the pore.

The various cecropin hybrids described in the previous sections can differ widely from their parent peptides in terms of membrane interaction. For better comparison of the different models proposed for these peptides, their membrane interaction will be discussed as a gradual process, as in a series of snapshots.

Steps in peptide–membrane interaction

Membrane binding

Peptide–membrane binding is thermodynamically governed by three factors: (i) electrostatic interaction between the anionic phospholipid head groups and the cationic peptide; (ii) α-helix formation; and (iii) hydrophobic interactions between segments or specific residues within the peptide sequence and the hydrocarbon core of the membrane (Jacobs and White, 1989; Wieprecht *et al.*, 1996; Ladokhin and White, 1999).

Electrostatic interaction

Soon after the discovery of EAPs, their specificity was tracked down to a preferential interaction with organisms with a high content in acidic phospholipids at the outer leaflet of the membrane (Zasloff, 1987; Steiner *et al.*, 1988). Accordingly, increasing the positive charge of the peptide or the content in acidic phosholipids improves membrane interaction for cecropin A (Andreu *et al.*, 1985), CA(1–11)-CD(12–37) (Fink *et al.*, 1989b), magainin (Matsuzaki *et al.*, 1991, 1997a; Besalle *et al.*, 1992), self-hybrids of cecropin B (Wang *et al.*, 1998) or analogs of CA(1–8)MA(1–12) (Kang *et al.*, 1998). Particularly instructive in this respect was the work of Matsuzaki *et al.* (1991), showing that binding of magainin 2 to vesicles with a fixed percentage of different acidic phospholipids correlated with the zeta potential of the membrane. Other experiments had shown that inclusion of cationic lipids abrogated binding of cecropin A to liposomes (Steiner *et al.*, 1988), or that increased ionic strength hampered membrane interaction (Matsuzaki *et al.*, 1991). As found for peptides, chirality of the phospholipid head group (e.g. use of L- or D-POPS) did not influence the permeabilization process (Matsuzaki *et al.*, 1998).

α-helix formation

α-helix induction is thermodynamically favored: formation of internal hydrogen bonds makes up for the entropy loss involved in peptide structuration and binding to the membrane, plus the energetic cost of burying polar peptide bonds into an hydrophobic environment (White and Wimley, 1998; Ladokhin and White 1999; Wieprecht *et al.*, 1999b). On the other hand, the contribution of helix formation to the overall free energy of binding depends both on the peptide (i.e. it accounts for 50% in magainin whereas for melittin it is down to 24–30%; Wieprecht *et al.*, 1999a; Unger *et al.*, 2001) and on the environment (i.e. cecropin A can be induced into an α-helix without insertion in the membrane (Silvestro and Axelxen, 2000).

The question has been raised whether acidic phospholipids are indeed essential for α-helix formation. Although binding to pure zwitterionic vesicles is difficult to monitor (Matsuzaki *et al.*, 1991; Mchaourab *et al.*, 1994; Mancheño *et al.*, 1996; Wang *et al.*, 1998), helix formation in the presence of neutral vesicles has been conclusively demonstrated for cecropin A (Steiner *et al.*, 1988; Silvestro and Axelsen, 2000), cecropin A–melittin hybrids (Fernández *et al.*, 1994; Mancheño *et al.*, 1996), melittin (Dempsey, 1990), magainin 2 (Wieprecht *et al.*, 1996), cecropin B self-hybrids (Wang *et al.*, 1998) and hydrophobic analogs of CA(1–12) MA (1–12) (Kang *et al.*, 1998). While the amount of acidic phospholipids did not determine the helical content of a peptide, it did show a favorable effect on the kinetics of helix formation (Matsuzaki *et al.*, 1991; Wang *et al.*, 1998). Helical content in neutral lipid environments was also relatively independent of the charge of the peptide (Wieprecht *et al.*, 1996), although full charge reversal, such as that resulting from complete succinylation of CA(1–7)M(2–9), prevented helix formation in the presence of neutral liposomes (Fernández *et al.*, 1994).

Recent work by two different groups seems to confirm that α-helix formation takes place at the membrane interphase before further penetration of the peptide into the membrane. Thus, cecropin A appears to be already structured as α-helix at high pressure conditions that avoid peptide penetration into monolayers (Silvestro and Axelxen, 2000). Under these extreme conditions, the hydrophobicity of the head of PC seems to be sufficient to trigger helix formation. Also, cecropin B self-hybrids undergo an α-helix formation in a single kinetic step in stopped-flow circular dichroism experiments (Wang *et al.*, 1998).

Hydrophobic interactions

Several observations point to requirements for membrane binding other than electrostatic attraction and helix formation. Peptide permeabilization of pure neutral (zwitterionic) vesicles (devoid of electrostatic interactions) has been demonstrated not only for melittin but also to a variable extent for cecropin A (Steiner *et al.*, 1988; Silvestro and Axelxen, 2000), magainin (Wieprecht *et al.*, 1999a), cecropin A–melittin hybrid peptides (Fernández *et al.*, 1994; Mancheño *et al.*, 1996) or cecropin B self hybrids analogs (Wang *et al.*, 1998). This behavior must be related to the observation that partition coefficients of EAPs for neutral phospholipids, though lower (10- to 60-fold) than for anionic phospholipids (Nakajima *et al.*, 1987; Christensen *et al.*, 1988; Gazit *et al.*, 1994, 1995; Mchaourab *et al.*, 1994; Matsuzaki *et al.*, 1995c; Mancheño *et al.*, 1996; Silvestro *et al.*, 1997; Kang *et al.*, 1998; Wang *et al.*, 1998; Mozsolits *et al.*, 2001) are by no means negligible. This partition is traced to the intrinsic hydrophobicity of the peptide (Wieprecht *et al.*, 1997), to certain segments of a sequence (Wang *et al.*, 1998), or even to specific residues (Juvvadi *et al.*, 1996a; Shin *et al.*, 1999), such as Trp. Nevertheless, this less tighter binding to neutral phospholipids may become advantageous by providing a certain degree of reversibility to peptide–membrane interaction, while not necessarily resulting in lower permeabilization abilities (Mancheño *et al.*, 1996; Wang *et al.*, 1998) (Figure 11.3).

Stoichiometry and peptide orientation

As mentioned in the previous section, permeabilization only takes place after massive peptide insertion, covering most of the membrane (Steiner *et al.*, 1988; Gazit *et al.*, 1994,

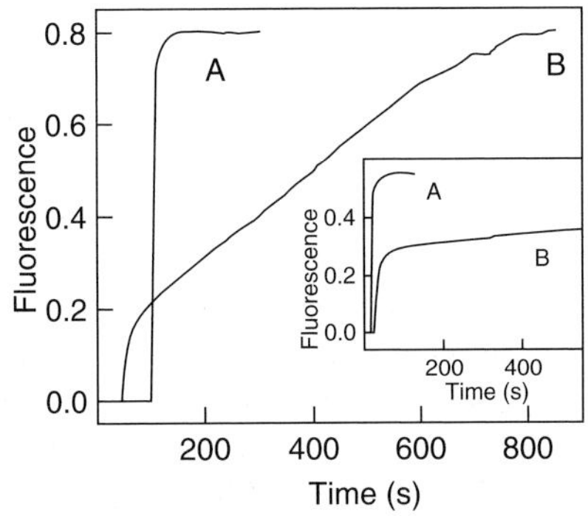

Figure 11.3 Vesicle leakage induced by CA(1–8)M(1–18). Main frame: release of aqueous contents (ANTS/DPX)(8-aminonaphthalene-1,3,6-trisulfonic acid/*N*,*N'*-*p*-xylenebispyridinium bromide) from extruded liposomes (100 nm pore diameter). Inset: Leakage of fluoresceinated dextran-20 induced by CA(1–8)M(1–18) encapsulated in liposomes extruded through 400 nm pore diameter filter. (1) bovine brain PS; (2) egg PC vesicles. In both cases the peptide:lipid ratio was 0.025. (Reprinted with permission from Mancheño J. M. *et al.* (1996), *Biochemistry*, **35**, 9892–9899. Copyright 1996. American Chemical Society.)

1995, 1996; Mancheño *et al.*, 1996; Kleinschmidt *et al.*, 1997; Silvestro *et al.*, 2000). These studies have generated some controversy, given the wide variety of techniques, vesicles, peptides, and peptide–phospholipid ratio used by different authors. Still, recent studies seem to define a certain consensus. Most peptides retain their monomeric status when bound to the membrane, at least for low-medium peptide : phospholipid ratios. This can be assessed by fluorescence energy transfer between peptides independently labeled with either donor or acceptor groups, or by spin resonance techniques (Gazit *et al.*, 1994, 1996; Matsuzaki *et al.*, 1994; Mancheño *et al.*, 1996; Hung *et al.*, 1999) at such low density, the peptide lies flat on the membrane (Bechinger, 1997, 1999; Hung *et al.*, 1999; Marassi *et al.*,1999; Silvestro and Axelxen, 2000). As the number of bound peptide molecules increases, association between monomers may take place, as found for cecropin–melittin hybrids (Fernández *et al.*, 1996), cecropin B and its self-hybrids (Wang *et al.*, 1998), cecropin AD hybrids (Mchraoubad *et al.*, 1993) or melittin (Hristova *et al.*, 2001), though not for cecropin P1 (Gazit *et al.*, 1996) which remains monomeric at high peptide concentrations. The situation for magainin is controversial (Matsuzaki *et al.*, 1994; Schümann *et al.*, 1997).

Depth of penetration has been established at the glycerol level for melittin in PC vesicles (Hristova *et al.*, 2001); for this peptide, H/D amide bond exchange experiments indicate that the hydrophobic portion of the helix faces the hydrophobic matrix (Dempsey and Butler, 1992). Magainin 2 is localized close to the phospholipid head in DPPG membranes (De Jong *et al.*, 1994), while fluorescence quenching data place the tryptophan ring 10 Å from the center of the bilayer (Matsuzaki *et al.*, 1994). Depth of penetration does not seem to depend on the nature of acidic phospholipids (Matsuzaki *et al.*, 1998), nor on the peptide–phospholipid ratio (Matsuzaki *et al.*, 1994); however, other reports seem to support such a dependence for melittin (Sui *et al.*, 1994) or cecropin A (Silvestro and Axelxen, 2000). In contrast, cecropin P1 does not penetrate too much in either acidic or zwitterionic vesicles (Gazit *et al.*, 1996). Orientation of a specific segment has been studied for cecropin A

(Marassi *et al.*, 1999) and cecropin B self-hybrids: whereas for the former peptide both helices lie at the surface of the bilayer, for the latter, penetration of the N-terminal region is more superficial than the C-terminal region, which inserts partially into the lipid bilayer (Hung *et al.*, 1999); the different orientation of this sequence or a lower flexibility of the hinge (Srisailam *et al.*, 2000) relative to cecropin A could account for this effect.

Once a threshold concentration has been achieved, a second peptide population can be detected, with the α-helix axis normal to the plane of the membrane and associated to pore formation. This population was identified by oriented circular dichroism for magainins (Ludtke *et al.*, 1995, 1996), melittin (Hristova *et al.*, 2001) and self-hybrids of cecropin B (Chen *et al.*, 2001) at membrane-permeabilizing concentrations. The same observation has been confirmed for melittin by solid state NMR in oriented lipid bilayers (Naito *et al.*, 2000) or for cecropin B and a self-hybrid by two-step kinetics of the tryptophan blue-shift fluorescence (Wang *et al.*, 1998). For magainin 2, two peptide populations have also been proposed on the basis of tryptophan quenching fluorescence by iodide (Wieprecht *et al.*, 1996); photolabeling of membrane lipids at different membrane depths also supports this dual population hypothesis (Jo *et al.*, 1998). For melittin, formation of dimers is essential to achieve this orientation (Schwarz *et al.*, 1992; Takei *et al.*, 1999; Hristova *et al.*, 2001).

Interestingly, the flat orientation of the peptide on the membrane can be "frozen" by a stronger peptide–phospholipid electrostatic interaction, either by using more acidic phospholipids or by increasing the positive charge of the peptide (Monette and Lafleur 1995; Chen *et al.*, 1997, 2001; Kleinschmidt *et al.*, 1997) this can result in non-productive peptide binding to the membrane, as found for melittin (Benachir and Lafleur, 1993), self-hybrids of cecropin B (Wang *et al.*, 1998), magainin (Matsuzaki *et al.*, 1997a) and cecropin A (Silvestro *et al.*, 1997).

Peptide–phospholipid acyl chain interaction

As expected from the above considerations, most peptides at low concentration do not produce significant changes in the lipid acyl chains. Magainin abolishes phase transition of DPPG vesicles (Matsuzaki *et al.*, 1991); a broadened transition can be observed, but to a much lower extent than other peptides that typically interact with the lipid matrix. An increase in phospholipid order has been reported for melittin, particularly in highly acidic vesicles (Kleinschmidt *et al.*, 1997); the same result was obtained for CA(1–8)M(1–18) (Mancheño *et al.*, 1996) and for a cecropin B self-hybrid (Hung *et al.*, 1999). In contrast, at higher peptide–phospholipid ratios, close to those inducing permeabilization, strong changes in the lipid phase can be observed, such as a significant broadening of the gel–liquid crystal transition for magainin 2 (Ludtke *et al.*, 1996), the higher temperature required for the POPE transition from bilayer into the H_{II} phase (Matsuzaki *et al.*, 1998), or the high perturbation of zwitterionic vesicles detected for magainin 2 by grazing X-ray incidence (Münster *et al.*, 2000). Possible aggregation of peptides within the membrane, as observed for a cecropin B self-hybrid (Hung *et al.*, 1999) and for an artificial dimer of melittin (Hristova *et al.*, 2001), may produce a perturbation much higher than the monomer, without apparent variation in the orientation, probably due to the increase of short range deformations at the membrane.

An important point related to the permeabilization process is that the peptides can induce lipid domains with different properties. Experimental evidence for this includes the observation of rich and non-rich magainin domains in POPE (Matsuzaki *et al.*, 1998), of separated phases for PGLa in DPPE/DPPG vesicles (Löhner and Prennerl, 1999), or melittin in the presence of acidic phospholipids (Lafleur *et al.*, 1989). Also, clustering of PC in a very rich PA environment can induce preferential interaction of a highly hydrophobic cecropin B self-hybrid (Wang *et al.*, 1998). This separation of domains implies that, if the peptide can achieve a preferential interaction, it can induce permeability at much lower peptide–phospholipid ratios than those expected from the general model (Shai, 1999).

Once the peptide is bound to the polar head of membrane phospholipids, steric accommodation of the peptide can induce expansion of the outer leaflet of the membrane, causing mismatches between the sectional areas of the polar and hydrophobic parts and induction of membrane thinning and a positive membrane curvature (Ludtke *et al.*, 1995); this could be compensated if the phospholipid has a polar head section smaller than that for acyl chains. This accounts for the lower permeabilization of magainin on PG vesicles when compared with other phospholipids as PS or PA (Matsuzaki *et al.*, 1998). Steric factors would also account for an easier accommodation of magainin (Hung *et al.*, 1999) and melittin (Subbarao and McDonald, 1994) into the branched diphtanoyl phospholipids, commonly used for monolayer constructions, rather than for unbranched acyl chains phospholipids; that could partially explain discrepancies between results obtained in liposomes and planar membranes (Christensen *et al.*, 1988).

Vesicles made of phospholipids with shorter chains are easier to permeabilize by melittin (Subbarao and McDonald *et al.*, 1994), and allow pore formation for shorter cecropin A–melittin analogs (Merrifield *et al.*, 1995b); in the latter case, thickness of the membrane seems to be influenced by the solvent used to build planar bilayers, by modulation of the interdigitation of acyl chains from the two leaflets (Merrifield *et al.*, 1995b).

Unsaturation of the lipid acyl chain also influences peptide permeabilization. Thus, cecropin A(1–11)D (12–37) aggregation takes place in DOPG but not in DLPG membranes (Mchaourab *et al.*, 1994), and melittin permeabilization of PC liposomes improves with the unsaturation of the lipid chain (Subbarao and McDonald, 1994). In contrast, for sarcotoxin IA no difference was found in the permeabilization of liposomes made from either *E. coli*-extracted phospholipids (saturated) or from a phospholipid mixture simulating mamalian phospholipids (unsaturated) (Nakajima *et al.*, 1987).

Cholesterol, one of the main components of eucaryotic membranes, modulates membrane fluidity and abolishes the gel to liquid crystal transition. Its incorporation into artificial membranes partially prevents liposome permeabilization by magainin 2 (Matsuzaki *et al.* 1995c; Wieprecht *et al.*, 1999a), sarcotoxin IA (Nakajima *et al.*, 1987) or melittin (Monette *et al.*, 1993), and of planar lipid bilayers by cecropin A (Christensen *et al.*, 1988). It also hampers penetration of melittin into the bilayer (Dufourcq and Faucon, 1977). A specific interaction between cholesterol and a side chain of glutamic acid in magainin was proposed, preventing penetration of the peptide into the bilayer (Tytler *et al.*, 1995). In contrast, for cecropin A, liposomes with or without cholesterol were equally permeabilized by the peptide (Steiner *et al.*, 1988). Recently, a more selective effect in liposome–cecropin A interaction has been described (Silvestro *et al.*, 1997): while cholesterol does not prevent membrane depolarization, it delays formation of lesion large enough to allow calcein release. On the other hand, cholesterol-containing liposomes are more prone to melittin permeabilization under hyposmotic shock (Benachir and Lafleur, 1995). Binding of cecropin B and its *all*-D enantiomer to cholesterol has been described, but its physiological interpretation is not easy (De Lucca *et al.*, 1998, 2000).

Pore formation

Conductivity patterns across planar lipid bilayers disclose pore formation for melittin and their stereoisomers (Dempsey 1990; Juvvadi *et al.*, 1996a), magainin (Duclohier *et al.*, 1989; Wade *et al.*, 1990), cecropins A and B (Christensen *et al.*,1988; Wade *et al.*, 1990; Merrifield *et al.*, 1995b), cecropin A–cecropin D hybrids (Christensen *et al.*, 1988), cecropin A–melittin hybrids (Wade *et al.*, 1990; Merrifield *et al.*, 1995b; Wu *et al.*, 1999; Zhang *et al.*, 1999) or cecropin P1–melittin hybrids (Juvvadi *et al.*, 1999). Additional evidence for pore formation has been obtained for magainin by crystallization of membrane–peptide complexes (Yang *et al.*, 2000) or by neutron scattering of aligned membranes (Ludtke *et al.*, 1996; Yang *et al.*, 1998). Other techniques such as calorimetry (Wenk and Seelig, 1998) or electron microscopy (Lockley and Ourth, 1996) also provide further evidence for pore existence.

In all conductivity experiments, a negative potential is needed to induce conductivity across the bilayer, whose stability can change dramatically according to the applied potential and the peptide used for experiments (Wu *et al.*, 1999). In general, no spontaneous disruption of the bilayer is observed at moderated peptide concentrations, arguing against a simple detergent-like effect (Christensen *et al.*, 1988; Wade *et al.*, 1990; Wu *et al.*, 1999; Zhang *et al.*, 1999). Nevertheless, single channels assigned to very transient structures are ubiquitous at concentrations much lower than those needed to obtain significant macroscopic conductivities, thus suggesting the existence of two kinds of membrane permeabilization (Chistensen *et al.*, 1988; Duclohier *et al.*, 1989; Silvestro *et al.*, 1997). Pore selectivity was poor, with anion/cation ratios *c*.3 for magainin and melittin (Chistensen *et al.*, 1988; Duchlohier *et al.*, 1989).

Eucaryotic antibiotic peptide induced pores do not have a fixed stoichiometry, but rather seem to be dependent on the peptide–phospholipid ratio and the kinetics of pore deactivation (Matsuzaki *et al.*, 1997b; Yang *et al.*, 1999). Since phospholipids in addition to peptide can incorporate into pores, thus reducing the number of peptide monomers required to assemble the pore, internal diameters higher than calculated from cooperativity studies have been found (Ludtke *et al.*, 1995); for magainin, 4–7 monomers per pore have been proposed, while in a modelization of a cecropin A pore 12 monomers were required (Durrell *et al.*, 1992).

A major problem with the pore formation mechanism is how to explain the bactericidal activity of peptides too short to span the membrane (*c*.30 Å). Merrifield and collaborators have found a correlation between length of α-helix and planar bilayer thickness required for conductivity (Wade *et al.*, 1990; Merrifield *et al.*, 1995b; Juvvadi *et al.*, 1999). These results need to be put into proper perspective, considering, for example, the peculiarities of the diphtanoyl phospholipids used for planar membrane formation, or the fact that thickness of biological membranes is not uniform, or that membrane thinning has been described for magainin (Ludtke *et al.*, 1995) and may likely also apply to other peptides.

Massive membrane disruption

For many peptides, induction of permeability appears to take place in a continuous fashion, that is, massive membrane disruption occurs once a high stoichiometry of peptide on the membrane is attained. This was observed for melittin (Dempsey, 1990) but also for less "harsh" peptides such as magainin (Matsuzaki *et al.*, 1989, 1991) or cecropin A (Silvestro *et al.*, 1997). Massive permeabilization also allows to explain the release of encapsulated high molecular weight markers such as IgG by cecropin A-magainin hybrids (Kang *et al.*, 1998), β-galactosidase from recombinant *Toxoplasma* Protozoa by CA(2–9)M(1–18) (Seeber, 2000) or the comparable rate of release of 4 and 20 kDa dextrans by CA(1–8)M(1–18) (Mancheño *et al.*, 1996).

Permeabilization models

During the last decade, two opposite models, the so-called barrel-stave and carpet-like mechanisms, have been proposed to interpret the different behavior of EAPs in membrane permeabilization (Shai, 1995). As new data accumulated and new intermediate models were proposed, a more continuous pattern has emerged, whereby a given peptide does not necessarily operate through a single mechanism. The following paragraphs discuss how well the EAPs and hybrids discussed in this chapter fit into the proposed models.

Barrel-stave model

This early model of EAP activity (Ehrenstein and Lecar, 1977), now fairly outdated, postulated the formation of pores made up exclusively by peptide molecules with such high self-affinity that a bunch of monomers would arrange perpendicular to the membrane plane and thus form a hydrophilic hole spanning the membrane. Melittin was for a while thought to follow a permeabilization mechanism close to this model, given the low peptide/phospholipid ratio required for permeabilization (still much higher than for bacterial toxins fitting this model). However, recent work seems to question this view too, since it has been established that melittin at moderate concentration is monomeric and lies flat on the membrane plane (Hristova *et al.*, 2001; Yang *et al.*, 2001); neither does the model explain the dependence of permeability on peptide/phospholipid rather than on peptide/vesicle ratio (Matsuzaki *et al.*, 1991; Mancheño *et al.*, 1996).

Two-state (wormhole or flip-flop) models

These have been independently proposed by Ludtke *et al.* (1996) and Matsuzaki *et al.* (1995a,b). (See Huang, 2000 for review.) Their main features are:

1 Peptide binds to the membrane and adopts α -helical conformation as monomer, lying flat on the membrane. Phospholipid order is poorly disturbed. This non-productive (S) state can at most cause some flickering conductivity due to local distorsions of the bilayer inherent to peptide insertion at the interface.
2 Progressive peptide binding perturbs the membrane creating local thinning associated to deformation energy. Once a limit is reached, peptides flip into a vertical arrangement that relieves tension and compensates for their unfavorable passage through the hydrocarbon matrix. This transition induces a pore and initiates membrane permeabilization. During this intermediate phase (I state) the peptide remains bound to phospholipid head groups. The pore is a toroidal transient structure lined with both peptide and phospholipid molecules that evolves by driving monomers toward each bilayer leaflet, creating a continuum between the two bilayer halves to which phospholipid flip-flop, ion release and peptide translocation are coupled. Pore size will thus be the result of two antagonic processes: new monomer recruitment and pore deactivation, both dependent on peptide/phospholipid ratio.

This model agrees with the observation of higher populations of peptides adopting vertical orientation at increasing peptide/phospholipid ratios, and accounts for the peptide concentration-dependent permeabilization observed for cecropin A (Silvestro *et al.*, 1997), melittin (Ladokhin *et al.*, 1997) or magainin (Matsuzaki, 1999). The fact that melittin, arguably one of the peptides with higher lytic potential, displays features inherent to this model, such as flip-flop (even in zwitterionic vesicles) (Fattal *et al.*, 1994) or peptide translocation and pore deactivation (Matsuzaki *et al.*, 1997b) suggests that the model may be valid for a wide variety of membrane-active EAPs (Yang *et al.*, 2001) and their hybrids.

Carpet-like model

In this model (Pouny *et al.*, 1992; Shai, 1999), membrane permeabilization takes place by accumulation of a huge amount of monomers, each interacting with phospholipids and producing small perturbations. Once a critical peptide concentration is reached, the membrane undergoes a catastrophic disintegration of its structure, including or not the formation of transmembrane pores.

The carpet-like model can be reconciled with the two state model, by assuming that only a very small percentage of the overall peptide population is involved in the "I" state. It has been invoked to account for the increased permeabilization of very acidic vesicles by the highly hydrophobic cecropin B self-analog CB(25–35)CB(11–24)CB(25–35) [*CB-3*], through formation of zwitterionic phospholipid clusters where the peptide acts preferentially (Wang *et al.*, 1998). It also explains how diastereomers with low α-helix percentages (Oren and Shai *et al.*, 1997) or cyclic analogs (Unger *et al.*, 2001) can permeabilize acidic vesicles, as well as the all-or-none permeabilization mechanisms observed for melittin (Schwarz *et al.*, 1992; Benachir and Lafleur, 1995) or CA(1–8)M(1–18) (Mancheño *et al.*, 1996).

In-plane diffussion model

In this model, ionic dissipation is produced at low peptide/phospholipid ratios by the membrane curvature caused by peptides inserting into the membrane and creating small temporary disturbances in phospholipid packing through which ions are released. The model agrees with the single channel conductivities observed for many EAPs. It was also assumed to account for the lethal activity of cecropin by simple bacterial depolarization without apparent membrane disruption (Silvestro *et al.*, 1997), or for the fungicidal activity of small cecropin–melittin hybrid peptides (Cavallarin *et al.*, 1998). From a formal point of view, this model may be viewed as a modification of the first step of the two-state model or the all-or-none permeabilization mechanisms observed for melittin or CA(1–8)M(1–18) (Schwarz *et al.*, 1992; Benachir and Lafleur, 1995; Mancheño *et al.*, 1996).

A somewhat frustrating conclusion to this inventory of mechanisms and putative supporting data is that, even though the main features of permeabilization processes can be analyzed by judicious choice of peptide–membrane interaction models, it remains for the moment quite difficult to unravel the complete process in detail, thereby confining the predictability of a given peptide–membrane interaction to levels not beyond a reduced set of small variations.

Cecropin–melittin hybrids as antibiotics

Bactericidal effects

Peptide interaction with bacterial structures external to the plasma membrane

For most peptides, permeabilization of the plasma membrane of the pathogen is an essential step in the killing mechanism. Still, peptide efficacy can diminish dramatically by interaction with bacterial structures and molecules other than inner membrane. For instance, embedding of bacteria in biofilm (an anionic, alginate-based, exopolysaccharide matrix; see Sutherland, 2001, for review) hampers antibiotic penetration by complexation of alginate with the cationic groups of the peptide, thereby decreasing activity against pathogens such as *Pseudomonas aeruginosa* (Friedrich *et al.*, 1999) or *Acinetobacter baumannii* (Saugar *et al.*, 2002). *Aeromonas salmonicida* deficient in the external proteinaceous paracrystalline S-layer are also more susceptible to cecropin B and P1 (Henry and Secombes, 2000).

In Gram-negative bacteria, the external leaflet of the outer membrane is almost exclusively composed by lipopolysaccharide (LPS), a complex anionic oligosaccharide acylated with fatty acids and tightly cross-linked through Mg^{2+} ions. EAPs must gain access into the periplasm by the so called self-promoted uptake mechanism (Hancock *et al.*, 1981), whereby they displace LPS-bound Mg^{2+} and cause outer membrane disorganization, allowing access of other molecules to the inner membrane. This EAPs–LPS interaction has been studied only to a limited extent. Melittin (David *et al.*, 1982; Piers and Hancock, 1994; Piers *et al.*, 1994), cecropins A (De Lucca *et al.*, 1995) and B (Vaara and Vaara, 1994), as well as cecropin A–melittin peptides and their modified analogs (Piers and Hancock, 1994; Piers *et al.*, 1994; Gough *et al.*, 1996; Scott *et al.*, 1999b; Zhang *et al.*, 1999), interact with a wide variety of bacterial LPS, even with a higher affinity than polymyxin B, the antiendotoxic molecule most commonly used in clinics. Aside from cationic character, chemical/structural requirements for EAP–LPS interaction are not clearly defined. The activity of CA(1–8)M(1–18) was improved in an analog (CEMA, Table 11.4) with a cationic C-terminus (Piers *et al.*, 1994). On the other hand, anti-LPS and bactericidal activities are not always totally parallel (Zhang *et al.*, 1999). Synergy among EAPs and other antibiotics is possible, provided a wide enough concentration gap between outer and inner membrane permeabilization exists. Several peptides capable to disrupt the outer membrane but with poor bactericidal activity have been shown to engage in synergic interactions (Friedrich *et al.*, 1999; Scott *et al.*, 1999b; Zhang *et al.*, 1999). For CA(1–8)M (1–18) and its CEMA analog, synergy with polymyxin B against *P. aeruginosa* (Piers and Hancock, 1994; Piers *et al.*, 1994) but not against *A. baumannii* (Saugar *et al.*, 2002) has been described. Since LPS is a chiral molecule, differences among peptide enantiomers are to be expected and have indeed been found: *all*-D cecropin B is only 50% as effective as its L form in displacing LPS-bound cationic dyes (De Lucca *et al.*, 2001). Chiral effects in LPS interaction have also been described for polymyxin B (Tsubery *et al.*, 2000).

In contrast with other antibiotics that promote massive release of LPS after bacterial killing, CA(1–8)M(1–18) and other analogs (Table 11.4) prevent binding of LPS to the LPS binding molecule, the first step in LPS signal transduction in the macrophage (Scott *et al.*, 2000b). These peptides are therefore very good inhibitors of LPS-induced TNF-α and IL-6 production by macrophages (Gough *et al.*, 1996; Scott *et al.*, 1999b, 2000b), can prevent endotoxic shock in mice (Gough *et al.*, 1996) or inhibit LPS- induced NOS2 expression in murine macrophages (Figure 11.4) (Velasco *et al.*, 1997). The extent of inhibition is directly related to their binding to LPS (Gough *et al.*, 1996; Scott *et al.*, 1999b).

Interaction of cecropin A–melittin peptides with lipoteichoic acid, a main component of the cell wall in Gram-positive bacteria , also abrogates most of their noxious biological activities (Scott *et al.*, 1999a).

Plasma membrane permeabilization

Susceptible organisms. Cecropin–melittin and other cecropin-derived hybrids are active on a wide variety of organisms including plant (Nordeen *et al.*, 1992; Kadono-Okuda *et al.*, 1995;

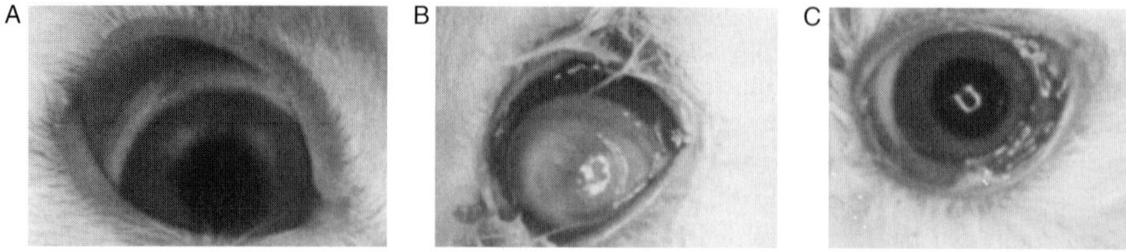

Figure 11.4 Treatment of a *Pseudomonas aeruginosa* experimental keratitis in rabbit with CA(1–7)M(2–9). Rabbits were infected with an intrastomal injection of 100 colony forming units of pathogen. After 2 h, 30 μl of a 0.1% (w/v) solution of CA(1–7)M(2–9) in PBS were applied every hour. Results were observed 24 h after initial inoculation. (A) Control rabbit eye. (B) Infected eye treated only with PBS . (C) Infected eye treated with CA(1–7)M(2–9). For more details refer to Nos-Barberá *et al.* (1997). (See Colour Plate XII.)

Osusky *et al.*, 2000) , fish (Kelly *et al.*, 1990; Jia *et al.*, 2000), and human bacteria (Oh *et al.*, 2000b). In general, however, they act preferentially on Gram-negative microorganisms (Hultmark *et al.*, 1982; Merrifield *et al.*, 1982; Andreu *et al.*, 1983, 1985, Moore *et al.*, 1996; Scott *et al.*, 1999a). The relatively poor performance of cecropins against Gram-positive bacteria was in fact one of the reasons for exploring cecropin–melittin hybrids as alternatives (Boman *et al.*, 1989; Andreu *et al.*, 1992). The structural factors underlying this differential behavior have not been conclusively elucidated. Modification at the C-terminus seems to play a certain role; thus, truncation of cecropin A [CA(1–33)] caused a sharper decrease in activity against Gram-positive than Gram-negative bacteria (Merrifield *et al.*, 1982). Also, CA(1–8)M(1–18), with a Pro–Ala–Leu–Ile–Ser–NH_2 C-terminal sequence, has better bactericidal activity than its CP26 analog (Pro–Leu–Ile–Ser–Ser C-terminus) (Scott *et al.*, 1999a). In contrast, the hinge flexibility of cecropin A–magainin peptides has a stronger effect on activity against Gram-negative than against Gram-positive bacteria (Shin *et al.*, 2000). Activity is sometimes highly dependent on a specific residue; thus, $K^6 \rightarrow E$ replacement in cecropin A preserves activity against *P. aeruginosa* but not against *E. coli* (Andreu *et al.*, 1985), and substitution or elimination of a Pro residue in CP26 can affect dramatically the susceptibility pattern of the corresponding analog (Zhang *et al.*, 1999). Another interesting finding is that, for a given set of cecropin A-melittin analogs, the spectrum of susceptible species is in general more reduced for Gram-positive than for Gram-negative bacteria (Scott *et al.*, 1999a).

Inhibition of antibacterial activity. High ionic strength, or other factors affecting the initial electrostatic interaction between peptide and bacteria, will hamper activity. This has been demonstrated for cecropin A–melittin hybrids against Gram-positive (Friedrich *et al.*, 2000), and to a lesser extent, against Gram-negative bacteria (Friedrich *et al.*, 1999; Wu *et al.*, 1999). Divalent cations are the stronger inhibitors, by competing either for negative binding sites on the membrane or with the Mg^{2+} cations that keep the LPS complex together (Piers and Hancock, 1994; Friedrich *et al.*, 1999).

Although activity is usually higher against bacteria under exponential growth conditions (Friedrich *et al.*, 1999), in some cases, (e.g. sarcotoxin IA) is independent of the growth phase (Okada and Natori, 1984). In fact, cecropins A and their analogs are able to kill depolarized bacteria: a pre-existing membrane potential is not required for a lethal hit

(Friedrich *et al.*, 1999), in agreement with permeabilization experiments on artificial membranes lacking potential (Steiner *et al.*, 1988; Mancheño *et al.*, 1996) or non-respiring (thus devoid of potential) rat liver mitochondria (Diaz-Achirica *et al.*, 1994).

Lethal hit. Membrane permeabilization triggered by massive peptide accumulation appears as the main mechanism of action for most cecropin A–melittin, cecropin A–magainin and self-hybrids from cecropin B. This affirmation is based on several observations:

(i) Fast killing kinetics. Cecropins (Hultmark *et al.*, 1980; Moore *et al.*, 1996; Silvestro *et al.*, 2000), their analogs (Nordeen *et al.*, 1992) and hybrids are capable of permeabilizing both outer (Piers and Hancock, 1994) and inner membrane (Piers *et al.*, 1994; Friedrich *et al.*, 1999; Wu *et al.*, 1999) in just a few minutes.
(ii) Correlation between concentrations required for inner membrane permeabilization and bactericidal activity (Okada and Natori, 1984; Vaara and Vaara, 1994; Silvestro *et al.*, 2000; Wu *et al.*, 1999).
(iii) Morphological evidence. Dramatic changes in bacterial membranes, such as disruption of outer membrane by sarcotoxin IA (Okada and Natori, 1984), cecropins (Moore *et al.*, 1996), cecropin B self-hybrids (Chan *et al.*, 1998a) or melittin diastereomers (Oren and Shai, 1997) have been visualized by scan and transmission electronic microscopy. Cecropin pore formation on *E. coli* membranes (Lockey and Ourth, 1994) and mesosome formation at the *S. aureus* membrane by cecropin A–melittin peptides (Friedrich *et al.*, 2000) have also been documented.
(iv) Loss of cell homeostasis. EAP-induced leakage of K^+ and membrane depolarization has been described on both Gram-negative (Okada and Natori, 1984; Wu *et al.*, 1999; Zhang *et al.*, 1999; Saugar *et al.*, 2002) and Gram-positive bacteria (Friedrich *et al.*, 2000). Also, uncoupling of respiration by sarcotoxin IA on bacteria (Okada and Natori, 1984) and by melittin, cecropins A and B (Hugosson *et al.*, 1994), magainin (Westerhoff *et al.*, 1989), and cecropin A–melittin analogs (Hugosson *et al.*, 1994; Díaz-Achirica *et al.*, 1994) on mitochondria have been reported.

There is also increasing evidence of the availability of peptides at other intracellular targets, after disruption of the membrane pore and translocation into the cytoplasm; in fact lethal hit mechanism other than a mere membrane permeabilization, such as peptide–DNA interaction was proposed for the cecropin A–melittin hybrids CP26 and CP29, as their lethal concentration was much lower than those required for membrane permeabilization (Wu *et al.*, 1999). We have also found alterations in the in vitro polymerization of FtsZ, a protein involved in septal ring during bacterial division (Díaz *et al.*, personal communication). In tune with this, alteration of *S. aureus* plasma membrane close to the divisional septum has been reported by cecropin A–melittin hybrids (Friedrich *et al.*, 2000).

Antifungal activity

Antifungal activity of EAPs has been recently reviewed (De Lucca and Walsh, 1999; De Lucca, 2000). Cecropin A kills *Candida* (Andra *et al.*, 2001) and germinating *Aspergillus* spp, whereas activity on *Fusarium* conidia is independent from germination (De Lucca *et al.*, 1998). Cecropin A degradation by secreted fungal proteinases was reported, in contrast with cecropin B and, obviously, with *all*-D cecropin B (De Lucca *et al.*, 1998, 2000). Both cholesterol and ergosterol bind to cecropin A, reducing its fungicidal activity (De Lucca *et al.*, 1998); this may be related to the inhibition of permeabilization of liposome containing either of these lipids, though not lanosterol, their immediate prebiosynthetic precursor (Feigin *et al.*, 1995). *Drosophila* cecropin A is even more active than cecropin B on most of the Insect pathogenic fungi (Ekengren and Hultmark, 1999).

Cecropin hybrids have also been used as antifungal agents. Thus, CA(1–8)M(1–12) was found to be less active than its magainin equivalent CA(1–8)MA(1–12) (Shin *et al.*, 1997a, 1998, 1999). Conversely, CA(1–8)M(1–18) and CA(1–7)M(6–9) were strongly inhibitory of a wide variety of plant fungi, with MICs in the 6.5–60 μM range (Cavallarin *et al.*, 1998) and the short hybrids being more effective than CA(1–8)M(1–18). Transgenic potato expressing MsrA1, a CEMA analog (Table 11.4), resist infection by *Phytophthora* (Osusky *et al.*, 2000).

Parasiticidal activity

The malaria-causing *Plasmodium* parasites were early tested as targets of EAPs and their hybrids. Cecropin B (Gwadz *et al.*, 1989) and analogs such as Shiva-1 and Shiva-3 (Rodriguez *et al.*, 1995; Boisbouvier *et al.*, 1998) (Table 11.2) prevented invasion of the *Anopheles* mosquito by the ookynete form of the parasite, a possible way to curtail transmission. A complementary approach has been to use the peptide to inhibit the pathology caused by the parasite in the vertebrate host, by killing the merozoite (intraerythrocytic form) or reducing its ability to invade erythrocytes. Cecropin B and its analogs (Jaynes *et al.*, 1988), as well as cecropin A–melittin hybrids (Boman *et al.*, 1989; Wade *et al.*, 1990; Andreu *et al.*, 1992) have been shown to effective in this regard. Interestingly, the *all*-D-enantiomers of the hybrids were five-fold less active than the *all*-L peptides (Wade *et al.*, 1990), pointing towards a chiral receptor either at the erythrocyte or the parasite.

Another clinically important group of protozoa are the trypanosomatids, which include *Leishmania*, and *Trypanosoma cruzi*, responsible for serious conditions such as leishmaniosis and Chagas disease, respectively. Proliferation of the *Leishmania* promastigote, the flagellated form of the parasite, is inhibited by cecropin A from either *Hyalophora* (Akkufo *et al.*, 1998; Díaz-Achirica, 1998) or *Drosophila* (Akkufo *et al.*, 1998). CA(1–8)M(1–18) displays even higher activity against this form ($LD_{50} = 1.4\,\mu M$; Díaz-Achirica *et al.*, 1998). The peptide kills by membrane permeabilization, with the *all*-D form twice as active as the L enantiomer (Figure 11.5). This increased activity of the D-form is not related to the abundant membrane-bound metalloproteinase Gp63, since a mutant with an inactive form of the proteinase is equally susceptible (Rivas and Chang, unpublished work).

The intracellular form of the parasite, that is, the amastigote, lives in a parasitophorus vacuole inside vertebrate macrophages. Despite its lack of Gp63 and the anionic lipophosphoglycan, which in the promastigote acts as a partial protecting barrier that prevents peptide access to the membrane, it is far more resistant to peptide attack (Díaz-Achirica *et al.*, 1996); a different membrane phospholipid composition, still unknown, seems to account for this difference. Interestingly, when amastigote-infected macrophages are treated with CA(1–8)M(1–18), parasites are killed at concentrations 15 times lower than those required to kill isolated amastigotes. The killing in this case has been traced to induction of nitric oxide synthase (NOS2) and subsequent production of nitric oxide (NO) by CA(1–8)M(1–18) (Figure 11.6) or its *all*-D-analog, though not by shorter peptides (Velasco *et al.*, 1997). CA(1–8)M(1–18) and its CEMA (Table 11.4) analog have been assayed for NOS2 induction without success by another group (Scott *et al.*, 2000a); high

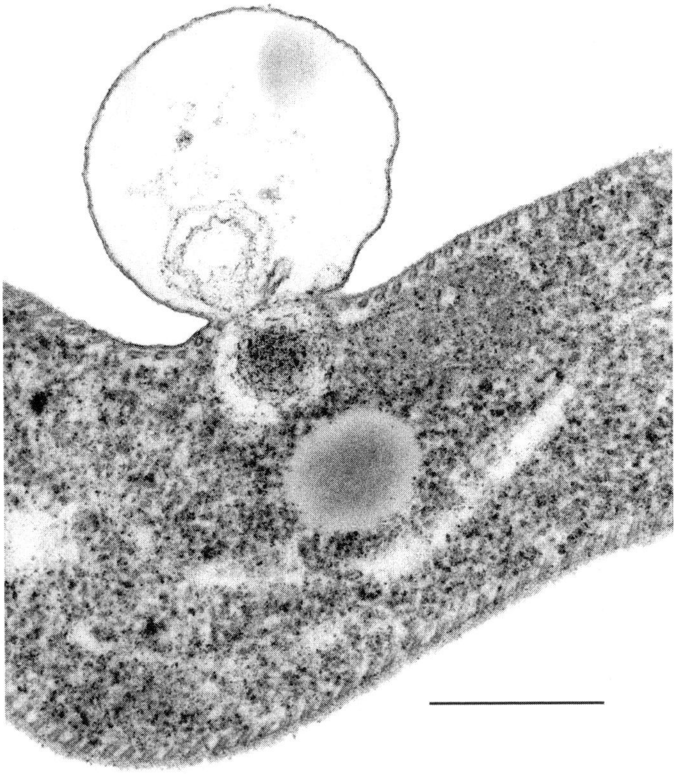

Figure 11.5 Damage to *Leishmania* promastigote by CA(1–8)M(1–18). Electron micrograph of a *Leishmania donovani* promastigote treated with 0.3 μM CA(1–8)M(1–18). Notice formation of an electron transparent bubble at the surface. Bar = 0.5 μm.

concentration of serum, a known inhibitor for these peptides, as well as differences in peptide-macrophage ratio can explain these results (Guerrero *et al.*, in preparation). As mentioned above, acylation of Lys[1] in CA(1–7)M(2–9) highly improves leishmanicidal activity against the amastigote (Chicharro *et al.*, 2001), although loss of specificity on the macrophage was also observed.

In the case of *Trypanosoma cruzi*, infection of the kissing bug (the Haemipteran transmission agent) with a cecropin A recombinant bacterial endosymbiont (*Rhodococcus rhodnii*) causes peptide production inside the gut and substantial inhibition of parasite transmission (Durvasula *et al.*, 1997; reviewed recently by Beard *et al.*, 2001). Other peptides active against *T. cruzi* inside infected cells, some causing a strong reduction in mice parasitemia have been also reported (Jaynes *et al.*, 1988; Barr *et al.*, 1995).

Other antiparasitic applications of EAPs and analogs involve *Trichomonas* (Mutwiri *et al.*, 2000), *Cryptosporidium parvum* (Arrowood *et al.*, 1991a,b; Giacometti *et al.*, 1999, 2000a,b), and the helminth (microfilariae) *Brugia pahangi* (Chalk *et al.*, 1995).

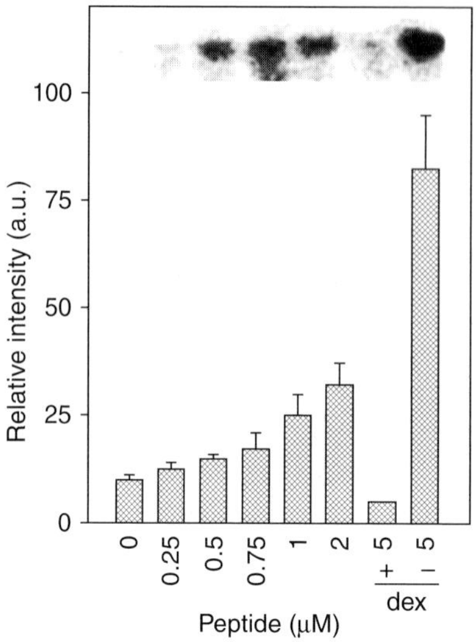

Figure 11.6 NOS2 induction in macrophages by CA(1–8)M(1–18). Northern blot of macrophage RNA (Raw 264.7) after stimulation in the absence (first lane 1) or in the presence of increasing concentrations of CA(1–8)M(1–18). A negative control with 1 μM dexamethasone (dex) at 5 μM peptide is included. The intensity of the band corresponding to NOS2 mRNA after normalization for the 18 S ribosomal RNA content is shown. (Reproduced with permission from Velasco, M. *et al.* (1997) *The Journal of Immunology*, **158**, 4437–4443. Copyright 1997. The American Association of Immunologists.)

Antiviral activity

Antiviral activities described for melittin, cecropin or their analogs are surprisingly caused by intracellular and specific intracellular events, with selective reduction of the biosynthesis of some viral proteins, as reported for the melittin analog Hecate (Table 11.2) on herpes simplex virus-1 (Baghian *et al.*, 1997) or for melittin itself on HIV-1 infected lymphoma cells (Wachinger *et al.*, 1992). Furthermore, cells transfected with melittin or cecropin also inhibited replication of HIV virus through suppression of HIV-1 gene expression (Wachinger *et al.*, 1998). This is in contrast with the antiviral activity of indolicidin, an EAP that inhibits HIV infectivity by perturbation of the viral membrane (Robinson *et al.*, 1993), or of defensins on herpes, vesicular stomatitis and influenza viruses but not on other non-enveloped viruses (Daher *et al.*, 1986). The structural similarities between the coat protein of tobacco virus and melittin have been exploited to design peptides capable to interfere with capsid formation (Marcos *et al.*, 1995). For HIV some structure-activity relationships have been derived with a set of melittin analogs: the hydrophobic N-terminal (1–20) appears to be required, while the polar section only modulates the action. Interestingly, the heterochiral hybrid M(1–20)-*all*-D-M(21–26) (Wachinger *et al.*, 1992) retained a substantial activity. To our knowledge, no cecropin A–melittin hybrid peptides have been assayed for antiviral activity.

Antitumoral activity

Tumoral cells expose anionic phospholipids, mainly PS, at the external leaflet of the plasma membrane, allowing preferential binding of the cationic peptides relative to normal cells

(Utsugi *et al.*, 1988). This fact, combined with a larger number of microvilli, can explain the increased binding of EAPs to tumor cells (Chan *et al.*, 1998b). Other factors, such as higher metabolic rates (thus higher membrane potential) or abnormal cytoskeleton and microfilament networks (Zasloff, 1992), can also contribute to peptide selectivity and may account for the synergy observed between EAPs (melittin derivatives) and cytostatic drugs (Jaynes *et al.*, 1989).

Preferential activity on tumoral cell lines has been consistently reported for magainins (Cruciani *et al.*, 1991; Baker *et al.*, 1993; Ludtke *et al.*, 1994) and for cecropins and their analogs (Jaynes *et al.*, 1989; Moore *et al.*, 1994). Cecropin-based hybrids have also been extensively studied. Thus, CA(1–8)M(1–12) and CA(1–13)M(1–13) were found to be active on several tumor cell lines; with $LD_{50} < 10\,\mu M$ (Shin *et al.*, 1997a, 1999, 2000). CEMA and their analogues were active on tumoral cells, although with a low specificity when compared with non-transformed cell lines (Johnstone *et al.*, 2000). Cecropin A–magainin hybrids have also been designed, with increased tumoricidal activity and decreased hemolytic effects (Shin *et al.*, 1997a). The relative structural similarities between the N-terminal regions of melittin and magainin, with Thr^{10}–Thr^{11} (melittin) and Lys^{10}–Lys^{11} (magainin) as the main difference, has been succesfully exploited in the design of CA(1–8)M(1–12), $Thr^{18} \rightarrow Lys$, $Thr^{19} \rightarrow Lys$, a potent analog. Other structural features such as hinge flexibility (see above), N or C-terminal size reduction, or specific replacements ($Ser^{16} \rightarrow Phe$ or $Ser^{16} \rightarrow Trp$) have also been explored with unsatisfactory results (Shin *et al.*, 1998). More recently, $Trp^2 \rightarrow Ala$ but not $Trp^2 \rightarrow Leu$ replacements in CA(1–8) MA(1–12) have been shown to decrease activity (Oh *et al.*, 2000a), again highlighting the importance of single residues.

In another group of cecropin B self-hybrids, changes in the helix character were related to tumoricidal activities. Thus, in both CB(1–25)CB(1–10) [*CB-1*] and CB(1–25)-GP-CB(1–10) [*CB-2*] (Chen *et al.*, 1997) the hydrophobic C-terminal part of cecropin B was replaced by a more amphipathic α-helix, with [*CB-2*] or without [*CB-1*] intersegmental flexibility. Both analogs were twice more effective than cecropin B in their lytic effect on several leukemia cell lines, while the additional hinge decreased somewhat the tumoricidal activity. Another self-hybrid CB(26–37)-CB(12–36) [*CB-3*], consisting of two hydrophobic α-helices, was significantly less active (Chan *et al.*, 1998a). Using scan and electron microscopy, different modes of interaction of the parental peptide, *CB-1* and *CB-3* with the tumoral cells have been proposed. (Chan *et al.*, 1998b; Srisailam *et al.*, 2001).

The low bioavailability of peptides into the deeper cellular layers of the tumor can be circumvented by in situ expression of antitumoral peptides such as preprocecropin A or prepromelittin, by infection with retroviral expression vectors with a preferential expression in cells in active division (Winder *et al.*, 1998).

Other biological activities of the EAPs

CA(1–8)M(1–18) is an apoptosis inducer on Raw 642 cells, though the peptide concentration needed to induce either apoptosis or necrosis is very similar (Velasco *et al.*, 1997); for other peptides, such as melittin, only necrosis was observed (Shaposhnikova *et al.*, 1997). Depletion of ATP will favor necrosis as a less energy-consuming process than apoptosis (Eguchi *et al.*, 1997). Furthermore, increased PS expression at the outer leaflet of the plasma membrane inherent to apoptotic cells will lead to a higher binding of peptides, finishing with necrosis. CEMA also induce a reduction in the expression of some apoptosis inhibitors (Scott *et al.*, 2000a).

Resistance against antibiotic peptides

The main drawback of antibiotic-based therapeutics, that is, the induction of resistant pathogens, is considerably less serious but not nil in the case of EAPs. Several mechanisms have been found in different organisms to decrease or annul the efficacy of EAPs. Not much is known about these mechanisms in the case of cecropin hybrids, thus inferences must be drawn from either the parental or other helical linear peptides.

Most resistance mechanisms in Gram-negative bacteria arise from an impaired interaction between EAP and LPS. Presence of the O-antigen, which confers smooth aspect to some bacterial strains, has been related to the higher resistance of *Salmonella typhimurium* against magainin (Rana *et al.*, 1991), of *Brucella* against cecropins B and P1 and melittin (Martínez de Tejada *et al.*, 1995), of *Brucella bronchiseptica* relative to *B. pertussis* against several linear EAPs (Banemann *et al.*, 1998), or of *Yersinia* against melittin differences in the outer core of LPS (Skurnik *et al.*, 1999). Absence or low expression of the O-antigen can be superseded by important LPS modifications (Bengoechea *et al.*, 1996; Guo *et al.*, 1997).

Variations in the structure of lipid A, an internal region of LPS, are also known to play a major role in resistance against EAPs (reviewed by Ernst *et al.*, 1999; Ernst and Miller, 2000; Groisman, 2001). These can involve (i) loss of negative charge by esterification by 2-aminoethyl pyrophosphate and 4-aminoarabinose at position 4′; (ii) low levels of 3-deoxyoctulosonic acid and absence of phosphate group at position 4; (iii) higher acylation of the lipid core (Guo *et al.*, 1997, 1998; Ernst *et al.*, 1999) or substitution by 2-hydroxymyristate. In all these cases, EAP resistance seems to be related to increased hydrophobicity of these structures. Decrease of anionic charge at the cell wall of the Gram-positive *S. aureus* by deletion of the *dlt* operon, increased bacterial susceptibility to EAPs (Peschel *et al.*, 1999).

Other instances of resistance have been located at the inner membrane. Thus, phospholipids with ornithine as polar head in *Brucella* (Freer *et al.*, 1996), or those formed by addition of lysine into phosphoglycerol under expression of the *mprF* gene in *S. aureus* (Peschel *et al.*, 2001) decreased the overall anionic character of the cytoplasmic membrane and hence EAPs susceptibility (Peschel *et al.*, 1999).

Efflux systems endow many microorganisms with a multiresistance pattern against antibiotics. However, implication of these systems on EAPs resistance is quite unusual. Thus, expression of the *mdr1* gene products did not confer resistance to melittin (Sharom *et al.*, 1995) or magainin (Lincke *et al.*, 1990) in different cancer cell lines. Similarly, cecropin B (Moore *et al.*, 1994) and CEMA analogs (Johnstone *et al.*, 2000) are equally active on drug-sensitive or multiresistant cell lines.

In bacterial models, several transport systems have been shown to provide resistance against linear EAPs, such as components of the *sap* operon of *S. typhimurium* against melittin and protamine (Parra-Lopéz *et al.*, 1993, 1994). In *P. aeruginosa*, expression of the *MexA-MexB* transport system did not provide protection against cecropin A-melittin analogs (Friedrich *et al.*, 1999), but the homologous *mtr* system, expressed in *Neisseria gonorrhoeae*, did induce resistance against LL-37, a helical EAP (Shafer *et al.*, 1998). On the other hand, we have tested CA(1–8)M(1–18) on a set of nosocomial isolates of *Acinetobacter baumannii* highly resistant to conventional antibiotics and found no substantial differences in activity relative to the wild type (Alarcón *et al.*, 2001) and no correlation with the resistance pattern to conventional antimicrobials.

Resistance against melittin and cecropin P1 in *Yersinia enterocolitica* is provided by the efflux system *ros A and ros B*, a pump/K^+antiporter. It is the first reported system induced by the presence of EAPs; moreover, as it is connected to a H^+-efflux glutathion dependent, a likely induced acidification of the cytoplasm would induce expression of DNA repair enzymes or DNA-binding proteins, that could protect DNA from the binding of EAPs, being the first system related to resistance against intracellular activities from EAPs (Bengoechea and Skurnik, 2000).

Proteolytic degradation is another possibility for microorganisms to foil the killing action of antimicrobial peptides (Edlund *et al.*, 1976; Sidén *et al.*, 1979). Linear, basic EAPs such as the cecropins, melittin or magainins are particularly susceptible to degradation by trypsin-like proteinases. Thus, presence of a *S. typhimurium* outer membrane proteinase (PgtE) under PhoP-PhoQ control, provides a certain degree of resistance against α-helical peptides (Guina *et al.*, 2000). Dallhammar and Steiner (1984) characterized InA, a low-specificity 70 kDa metalloproteinase from *Bacillus thuringiensis* which can degrade cecropin A and thus make this bacterium resistant. Similar systems have been described in *P. aeruginosa* (Jarosz, 1997), *Bacillus larvae*, (Jarosz and Glinski, 1990) and in insect parasitic nematodes (Götz *et al.*, 1981; Jarosz, 1998). Proteinase sensitivity is usually offset by the fast delivery of the lethal hit, which often results in the degradation process going undetected, only noticeable by the slight differences in activity between *all*-L (susceptible) and *all*-D (resistant) enantiomers. For instance, cecropin B and CA(1–7)M(2–9) are degraded in vitro by proteinases from periodontal Gram-negative pathogens such as *Porphyromonas* and *Prevotella* (Devine *et al.*, 1999). However, both peptides retain considerable activity and therefore proteolysis does not result in antibiotic resistance. The effects of proteolysis may become more perceptible when a high proteolytic activity is present at the site of action, or when the permeabilization process is somehow delayed, as found for cecropin in cholesterol-containing membranes (Silvestro *et al.*, 1997).

In fungi, degradation of cecropin A-melittin hybrids (Cavallarin *et al.*, 1998) or cecropin A (De Lucca *et al.*, 1998) was demonstrated in *Trichoderma* or *Aspergillus flavus* extracellular proteinases, respectively.

Degradation can also be carried out by the same organism that produces the peptide, either to prevent peptide accumulation into toxic levels or as a way to recover amino acids, once the stimulus for EAP production has ceased. Degradation in transgenic organisms expressing EAPs or their derivatives is also a serious concern. Plant intercellular fluids are particularly rich in proteinases, a fact that may explain why all plant EAPs are not linear but highly folded, disulfide-linked, protease-stable molecules. Up to 90% in vitro degradation of cecropin B by peach leaf extracts has been described (Mills *et al.*, 1994); even so, the remaining concentration was high enough for effective killing of *Pseudomonas syringae*. Similar results have been reported for both cecropin B and its analog SB-37 against *Erwinia amylovora* in the presence of intercellular fluid from pear leaves (Mourgues *et al.*, 1998). Failure of cecropin B to afford protection in either tobacco or potato transgenic plants has been traced to degradation by serine proteinases present in the intercellular fluids (Allefs *et al.*, 1995; Florack *et al.*, 1995). Rational study of the degradation process may allow tailoring peptide sequences for increased proteolytic resistance; this has been demonstrated for cecropin B, where $Met^{11} \rightarrow Val$ replacement increased three times its half life when challenged with leaf intercellular fluids from ten different plants (Owens and Heutte, 1997).

Production of EAPs and hybrid peptides

Chemical synthesis is the method of choice for EAP production at low scale (Andreu and Rivas, 2001) and is particularly useful for introducing specific modifications, such as non-natural amino acids, which can shed light on the mechanism of action and result in molecules with improved properties. However, the costs of peptide synthesis, even when benefiting from large scale production (e.g. in the 1$/mg range), pose still no serious challenge to those of conventional antibiotics nowadays, which may explain in part the slow progress observed in bringing EAP-derived drug candidates into clinical trials and eventually the market.

Production by recombinant technologies is an obvious alternative, given the small size and gene-encoding of EAPs and their few post-translational modifications. However, EAP recombinant production is not without its own drawbacks. Thus, the peptide can be toxic to the producing microorganism; the problem may be overcome by expression in another less susceptible microorganism or under an inducible promoter. Other inconveniences arise from secretion and degradation (see previous section) inside the organism: to achieve practical levels of peptide accumulation, a common solution is production as a fusion protein and further recovery from inclusion bodies. Secretion can also be achieved by fusing to secreted protein or by incorporation of adequate secretory sequences. The following examples serve to illustrate some of the common strategies used in recombinant EAP production.

Cecropin A from *H. cecropia* was produced as fusion protein to a L-ribulokinase gene and obtained as inclusion bodies (Callaway *et al.*, 1993). The same strategy was used for cecropin CMIV from *Bombyx mori* by fusion into TNF-α (Wang *et al.*, 1997) or *S. aureus* protein A genes (Xie *et al.*, 1996). Also, the magainin analog MSI-344 has been prepared by fusion to amidophosphoribosyltransferase from *E. coli* (Hwang *et al.*, 2001). Improved yields have been obtained by fusing the corresponding EAP sequences to proteins or peptides prone to aggregate thereby decreasing proteolytic degradation. Using this strategy, melittin or the magainin analog MSI-278 were obtained as fusion peptides to the *E. coli* polypeptide F4 (Lee *et al.*, 2000), or to a glucagon sequence (Kim *et al.*, 2000).

The Hancock group (Piers *et al.*, 1993; Zhang *et al.*, 1998) has optimized the expression of CA(1–8)M(1–18) and related analogs by fusion into the *E. coli* repA protein followed by the anionic preprodomain of human defensin 1. Incorporation of a single methionine residue before the mature sequence allowed recovery of the peptides by CNBr cleavage. A useful approach for extremely toxic or unstable peptides such as cecropin P1 is production on cell-free T7 transcription–translation systems using *E. coli* extracts (Martemyanov *et al.*, 1997, 2001).

Eucaryotic peptide production in yeast is best illustrated by preprosarcotoxin IA, expressed in *S. cerevisiae* under the constitutive phosphoglycerate kinase promoter followed by the mature peptide sequence (Aly *et al.*, 1999). Likewise, magainin 2 was expressed in *Hansenula* as a fusion product to the endogenous amine oxidase (Faber *et al.*, 1996).

The invertebrate origin of many EAPs suggests that insect cells might provide a good recognition of the DNA sequence as well as the internal machinery to carry out the postranslational modifications required for the peptide. This fact, coupled to baculovirus technology, has made this approach very attractive. EAP recombinant baculovirus can be used to infect cell cultures or moth larvae; in both cases cecropin A was successfully recovered when expressed as the preprodomain form (Hellers *et al.*, 1991), or fused into the ZZ-protein of *S. aureus* (Andersons *et al.*, 1991). On the down side, sarcotoxin IA degradation by cysteine proteinases and lack of amidation were observed when expressed in *Bombyx mori* cells (Yamada *et al.*, 1990).

Transgenic mice expressed have been prepared expressing a hybrid magainin 2 as C-terminal extension of the globin chain gene. After expression of 25% of the total hemoglobin content, mice were apparently healthy (Sharma *et al.*, 1994). Also, expression of Shiva 1a under IL-2 promoter caused a considerable reduction in the number of *Brucella abortus* cells in the liver of transgenic mice challenged with the bacteria (Reed *et al.*, 1997).

Nevertheless, expression of cecropin A_2 from *Drosophila* in mammalian cells by infection with a recombinant sarcoma virus was unsuccessful (Pore and Pal, 2000).

One of the most attractive applications envisaged for EAPs is prevention of crop loss by pathogen-resistant transgenic plants overexpressing either their own EAPs or natural or modified foreign EAP sequences. However, proteolytic degradation of foreign sequences remains a substantial problem in this respect (see section on "Other biological activities of the EAPs"). Cecropins and their analogs have been expressed in plants with contradictory results, depending on the pathogen, the gene construction or the host plant used. Thus, expression of sarcotoxin IA in tobacco (Okamoto *et al.*, 1998; Mitsuhara *et al.*, 2000) or rice (Ohshima *et al.*, 1999) promotes antibacterial and antifungal resistance. Cecropin B expression, on the other hand, has been shown to induce plant resistance in one case (Florack *et al.*, 1995; Sharma *et al.*, 2000) and to delay wilting of transgenic roses (Derks *et al.*, 1995). Expression in transgenic plants of some cecropin B analogs (Jaynes *et al.*, 1993; Huang *et al.*, 1997; Arce *et al.*, 1999; Mitsuhara *et al.*, 2000) promotes resistance against specific plant pathogens. Furthermore, expression of the cecropin A-melittin Hybrid MsrA1, a modified sequence of CEMA less harmful to the plants, in potato, induce a significant resistance against bacterial and fungal diseases of this tubercle (Osusky *et al.*, 2000).

However, as aforementioned, negative results using also cecropin B and similar protocols have been reported (Hightower *et al.*, 1994; Allefs *et al.*, 1995).

A final word in this section must be dedicated to production of EAPs as pharmaceuticals. The magainin 2 analog pexiganan is under revision for possible reformulation, once its application as a topical antibiotic cream (Locilex™) for treatment of infected diabetic foot ulcers was denied by FDA to Magainin Pharmaceuticals (now Genaera Corporation). Micrologix Biotech (http://www.mbiotech.com) is involved in development of peptide-based antimicrobials, including cecropin–melittin hybrids, for different clinical applications. At the moment of this writing, no product for human use is in the market yet, although promising results in clinical trials in phase II and III are in progress.

Conclusions and perspectives

Their mode of action on membranes has drawn attention to EAPs as a universal solution to the growing incidence of antibiotic resistance. These expectations have not been realized as quickly as initially assumed, for several reasons, the main one possibly being the still uncompetitive costs of peptide production. This situation is likely to evolve in favor of peptide antibitiotics, both by the alarming progress of multiresistance and by the technical improvements (and thus lowered costs) in industrial peptide production. Other potential drawbacks such as immunogenicity may be overcome by resorting to short peptide sequences and/or the use of self-EAPs. EAP-based gene therapy is in fact a promising scenario. Improvements in antibiotic activity or reductions in cytotoxicity are likely to be obtained from SAR studies, including cyclization (Unger *et al.*, 2001) fatty acid acylation (Chicharro *et al.*, 2001) and other strategies. The hybridization concept itself remains quite promising too, not only as applied to EAPs but also including sequences involved in intracellular transport. This may bring about substantial improvements in bioavailability, a well known hurdle in the application of peptide-based pharmaceuticals.

Acknowledgments

L. Rivas received financial support from the Comunidad de Madrid (Plan de Grupos Estratéticos, (08.2/0054/2001.2) and Grants from Fondo de Investigaciones Sanitarias, (99/0025-02), and EU (QLRT-2000-01404). Work at the University of Barcelona was supported by Generalitat de Catalunya (Centre de Referència en Biotecnologia). We thank José M. Saugar for technical advice in the preparation of the figures of the manuscript.

Notes

1 This chapter is dedicated to Profs R. B. Merrifield and H. G. Boman for their seminal contributions to the Biology and Chemistry of Animal Antibiotic Peptides, particularly the cecropins and their hybrids.

2 This chapter will follow as much as possible the nomenclature format introduced by the Boman and Merrifield groups, which we consider most informative for non-specialized readers. In this system, the segments of a natural sequence incorporated into a hybrid structure are given in parenthesis, preceded by an abbreviated name of such sequence. Thus, CA(1–11)CD(12–37) refers to a linear peptide consisting of residues 1–11 of cecropin A (CA) followed by residues 12–37 of cecropin D (CD). Other frequent abbreviations for natural sequences are cecropin B (CB), cecropin P (CP), melittin (M) and magainin (MA)(Table 11.1). Any modifications of the native sequences are indicated after the hybrid acronym, separated by a comma and in one-letter code. Thus, CA(1–8)MA(1–12), $W^2 \rightarrow L$ (Oh *et al.*, 2000a) describes a linear peptide consisting of residues 1–12 of cecropin A, with the Trp residue in position 2 replaced by Leu, and followed by residues 1–12 of magainin. Residues inserted between two hybrid segments are also given in on-letter code, for example, CA(1–8)-GP-M(1–12) denotes the insertion of Gly and Pro between residues 1–8 of cecropin A and 1–12 of melittin. Some hybrid peptides are often denoted by simple acronyms (e.g. CEME) which do not specify their constitution. See Tables (11.3–11.5).

References

Akuffo, H., Hultmark, D., Engstöm, Å., Frohlich, D. and Kimbrell, D. (1998) *Drosophila* antibacterial protein, cecropin A, differentially affects non-bacterial organisms such as *Leishmania* in a manner different from other amphipathic peptides. *International Journal of Molecular Medicine*, **1**, 77–82.

Alarcón, T., López-Hernández, S., Andreu, D., Saugar, J. M., Rivas, L. and López-Brea, M. (2001) In vitro activity of CA(1–8)M(1–18), a synthetic cecropin A-melittin hybrid peptide, against multiresistant *Acinetobacter baumannii* strains. *Revista Española de Quimioterapia*, **14**, 184–190.

Allefs, S. J. H. M., Florack, D. E. A., Hoogendoorn, C. and Stiekema, W. J. (1995) *Erwinia* soft rot resistance of potato cultivars transformed with a gene construct coding for antimicrobial peptide cecropin B is not altered. *American Journal of Potato Research*, **72**, 437–445.

Aly, R., Granot, D., Mahler-Slasky, Y., Halpern, N., Nir, D. and Galun, E. (1999) *Saccharomyces cerevisiae* cells harboring the gene encoding sarcotoxin IA secrete a peptide that is toxic to plant pathogenic bacteria. *Protein Expression and Purification*, **16**, 120–124.

Andersons, D., Engström, Å., Josephson, S., Hansson, L. and Steiner, H. (1991) Biologically active and amidated cecropin produced in a baculovirus expression system from a fusion construct containing the antibody-binding part of protein A. *Biochemical Journal*, **280**, 219–224.

Andra, J., Berninghausen, O. and Leippe, M. (2001) Cecropins, antibacterial peptides from insects and mammals, are potently fungicidal against *Candida albicans*. *Medical Microbiology and Immunology*, **189**, 169–173.

Andreu, D. and Rivas, L. (1998) Animal antimicrobial peptides: an overview. *Biopolymers*, **47**, 415–433.

Andreu, D. and Rivas, L. (2001) Synthesis of antibiotic peptides in *Peptide Antibiotics: Discovery, Modes of Action and Applications*, edited by C. J. Dutton, M. A. Haxell, H. A. I McArthur, R. G Wax, (Eds.) pp. 15–46. New York: Marcel Dekker.

Andreu, D., Merrifield, R. B., Steiner, H. and Boman, H. G. (1983) Solid-phase synthesis of cecropin A and related peptides. *Proceedings of the National Academy of Sciences USA*, **80**, 6475–6479.

Andreu, D., Merrifield, R. B., Steiner, H. and Boman, H. G. (1985) N-terminal analogues of cecropin A: synthesis, antibacterial activity and conformational properties. *Biochemistry*, **24**, 1683–1688.

Andreu, D., Ubach, J., Wade, D., Wåhlin, B., Merrifield, R. B. and Boman, H. G. (1992) Shortened cecropin A–melittin hybrids: significant size reduction retains potent antibiotic activity. *FEBS Letters*, **296**, 190–194.

Arce, P., Moreno, M., Gutiérrez, M., Gebauer, M., Dell'Orto, P., Torres, H., *et al.* (1999) Enhanced resistance to bacterial infection by *Erwinia carotovora* subsp. *atroseptica* in transgenic potato plants expressing the attacin or the cecropin SB-37 genes. *American Journal of Potato Research*, **76**, 169–177.

Arrowood, M. J., Jaynes, J. M. and Healey, M. C. (1991a) Hemolytic properties of lytic peptides active against the sporozoites of *Cryptosporidium parvum*. *Journal of Protozoology*, **38**, 161S–163S.

Arrowood, M. J., Jaynes, J. M. and Healey, M. C. (1991b) *In vitro* activities of lytic peptides against the sporozoites of *Cryptosporidium parvum*. *Antimicrobial Agents and Chemotherapy*, **35**, 224–227.

Baghian, A., Jaynes, J., Enright, F. and Kousolas, K. G. (1997) An amphipathic alpha-helical synthetic peptide analogue of melittin inhibits herpes simplex virus-1 (HSV-1)-induced cell fusion and virus spread. *Peptides*, **18**, 177–183.

Baker, M. A., Maloy, W. L., Zasloff, M. and Jacob, L. S. (1993) Anticancer efficacy of Magainin2 and analogue peptides. *Cancer Research*, **53**, 3052–3057.

Bals, R., Weiner, D. J., Moscioni, A. D., Meegalla, R. L. and Wilson, J. M. (1999) Augmentation of innate host defense by expression of a cathelicidin antimicrobial peptide. *Infection and Immunity*, **67**, 6084–6089.

Banemann, A., Deppisch, H. and Gross, R. (1998) The lipopolysaccharide of *Bordetella bronchiseptica* acts as a protective shield against antimicrobial peptides. *Infection and Immunity*, **66**, 5607–5612.

Barr, S. C., Rose, D. and Jaynes, J. M. (1995) Activity of lytic peptides against intracellular *Trypanosoma cruzi* amastigotes in vitro and parasitemias in mice. *Journal of Parasitology*, **81**, 974–978.

Bazzo, R., Tappin, M. J., Pastore, A., Harvey, T. S., Carver, J. A. and Campbell, I. D. (1998) The structure of melittin. A ^{1}H-NMR study in methanol. *European Journal of Biochemistry*, **173**, 139–146.

Beard, C. B., Dotson, E. M., Pennington, P. M., Eichler, S., Cordon-Rosales, C. and Durvasula, R. V. (2001) Bacterial symbiosis and paratransgenic control of vector-borne Chagas disease. *International Journal of Parasitology*, **31**, 620–626.

Bechinger, B. (1997) Structure and functions of channel-forming peptides: magainins, cecropins, melittin and alamethicin. *Journal of Membrane Biology*, **156**, 197–211.

Bechinger, B. (1999) The structure, dynamics and orientation of antimicrobial peptides in membranes by multidimensional solid-state NMR spectroscopy. *Biochimica et Biophysica Acta*, **1462**, 157–183.

Benachir, T. and Lafleur, M. (1995) Study of vesicle leakage induced by melittin. *Biochimica et Biophysica Acta*, **1235**, 452–460.

Bengoechea, J. A. and Skurnik, M. (2000) Temperature-regulated efflux pump/potassium antiporter system mediates resistance to cationic antimicrobial peptides in *Yersinia*. *Molecular Microbiology*, **37**, 67–80.

Bengoechea, J. A., Diaz, R. and Moriyon, I. (1996) Outer membrane differences between pathogenic and environmental *Yersinia enterocolitica* biogroups probed with hydrophobic permeants and polycationic peptides. *Infection and Immunity*, **64**, 4891–4899.

Besalle, R., Haas, H., Goria, A., Shalit, I. and Fridkin, M. (1992) Augmentation of the antibacterial activity of magainin by positive-charge chain extension. *Antimicrobial Agents and Chemotherapy*, **36**, 313–317.

Bland, J. M., De Lucca, A. J., Jacks, T. J. and Vigo, C. B. (2001) *All*-D-cecropin B: synthesis, conformation, lipopolysaccharide binding and antibacterial activity. *Molecular and Cellular Biochemistry*, **218**, 105–111.

Blondelle, S. E. and Houghten, R. A. (1991a) Hemolytic and antimicrobial activities of the twenty-four individual omission analogues of melittin. *Biochemistry*, **30**, 4671–4678.

Blondelle, S. E. and Houghten, R. A. (1991b) Probing the relationships between the structure and hemolytic activity of melittin with a complete set of leucine substitution analogs. *Peptide Research*, **4**, 12–18.

Blondelle, S. E., Simpkins, L. R., Pérez-Paya, E. and Houghten, R. A. (1993) Influence of tryptophan residues on melittin's hemolytic activity. *Biochimica et Biophysica Acta*, **1202**, 331–336.

Boisbouvier, J., Prochnicka-Chalufour, A., Nieto, A. R., Torres, J. A., Nanard, N., Rodriguez, M. H. *et al.* (1998) Structural information on a cecropin-like synthetic peptide, Shiva-3 toxic to the sporogonic development of *Plasmodium berghei*. *European Journal of Biochemistry*, **257**, 263–273.

Boman, H. G. (1995) Peptide antibiotics and their role in innate immunity. *Annual Review of Immunology*, **13**, 61–92.

Boman, H. G. (1998) Gene-encoded peptide antibiotics and the concept of innate immunity: an update review. *Scandinavian Journal of Immunology*, **48**, 15–25.

Boman, H. G. (2000) Innate immunity and the normal microflora. *Immunological Reviews*, **173**, 5–16.

Boman, H. G., Wade, D., Boman, A., Wåhlin, B. and Merrifield, R. B. (1989) Antibacterial and antimalarial properties of peptides that are cecropin–melittin hybrids. *FEBS Letters*, **259**, 103–106.

Boman, H. G., Agerberth, B. and Boman, A. (1993) Mechanisms of action on *Escherichia coli* of cecropin P1 and PR-39, two antibacterial peptides from pig intestine. *Infection and Immunity*, **61**, 2978–2984.

Callaway, J. E., Lai, J., Haselbeck, B., Baltaian, M., Bonnesen, S. P., Weickmann, J. *et al.* (1993) Modification of the C terminus of cecropin is essential for broad-spectrum antimicrobial activity. *Antimicrobial Agents and Chemotherapy*, **37**, 1614–1619.

Castle, M., Nazarian, A., Yi, S. S. and Tempst, P. (1999) Lethal effects of apidaecin on *Escherichia coli* involve sequential molecular interactions with diverse targets. *Journal of Biological Chemistry*, **274**, 32555–32564.

Cavallarin, L., Andreu, D. and San Segundo, B. (1998) Cecropin A-derived peptides are potent inhibitors of fungal plant pathogens. *Molecular Plant–Microbe Interactions*, **11**, 218–227.

Chalk, R., Townson, H. and Ham, P. J. (1995) *Brugia pahangi*: the effects of cecropins on microfilariae in vitro and in *Aedes aegypti*. *Experimental Parasitology*, **80**, 401–406.

Chan, S. C., Yau, W. L., Wang, W., Smith, D. K., Sheu, F. S. and Chen, H. M. (1998a) Microscopic observations of the different morphological changes caused by anti-bacterial peptides on *Klebsiella pneumoniae* and HL-60 leukemia cells. *Journal of Peptide Science*, **4**, 413–425.

Chan, S. C., Hui, L. and Chen, H. M. (1998b) Enhancement of the cytolytic effect of anti-bacterial cecropin by the microvilli of cancer cells. *Anticancer Research*, **18**, 4467–4474.

Chen, H. M., Wang, W., Smith, D. and Chan, S. C. (1997) Effects of the anti-bacterial peptide cecropin B and its analogs, cecropins B-1and B-2, on liposomes, bacteria and cancer cells. *Biochimica et Biophysica Acta*, **1336**, 171–179.

Chen, H. M., Clayton, A. H., Wang, W. and Sawyer, W. H. (2001) Kinetics of membrane lysis by custom lytic peptides and peptide orientations in membrane. *European Journal of Biochemistry*, **268**, 1659–1669.

Chicharro, C., Granata, C., Lozano, R., Andreu, D. and Rivas, L. (2001) N-terminal fatty acid substitution increases the leishmanicidal activity of CA(1–7)M(2–9), a cecropin–melittin hybrid peptide. *Antimicrobial Agents and Chemotherapy*, **45**, 2441–2449.

Christensen, B., Fink, J., Merrifield, R. B. and Mauzerall, D. (1988) Channel-forming properties of cecropins and related model compounds incorporated into planar lipid membranes. *Proceedings of the National Academy of Sciences USA*, **85**, 5072–5076.

Conde, R., Zamudio, F. Z., Rodríguez, M. H. and Possani, L. D. (2000) Scorpine, an anti-malaria and anti-bacterial agent purified from scorpion venom. *FEBS Letters*, **471**, 165–168.

Cruciani, R. A., Barker, J. L., Zasloff, M., Chen, H. C. and Colamonici, O. (1991) Antibiotic magainins exert cytolytic activity against transformed cell lines through channel formation. *Proceedings of the National Academy of Sciences USA*, **88**, 3792–3796.

Daher, K. A., Selsted, M. E. and Lehrer, R. I. (1986) Direct inactivation of viruses by human granulocyte defensins. *Journal of Virology*, **60**, 1068–1074.

Dalhammar, G. and Steiner, H. (1984) Characterization of inhibitor A, a protease from *Bacillus thuringiensis* which degrades attacins and cecropins, two classes of antibacterial proteins in insects. *European Journal of Biochemistry*, **139**, 247–252.

Dathe, M., Nikolenko, H., Meyer, J., Beyermann, M. and Bienert, M. (2001) Optimization of the antimicrobial activity of magainin peptides by modification of charge. *FEBS Letters*, **501**, 146–150.

David, S. A., Balasubramanian, K. A., Mathan, V. I. and Balaram, P. (1992) Analysis of the binding of polymyxin B to endotoxic lipid A and core glycolipid using a fluorescent displacement probe. *Biochimica et Biophysica Acta*, **1165**, 147–152.

Dawson, C. R., Drake, A. F., Helliwell, J. and Hider, R. C. (1978) The interaction of bee melittin with lipid bilayer membranes. *Biochimica et Biophysica Acta*, **510**, 75–86.

de Jongh, H. H., Goormaghtigh, E. and Killian, J. A. (1994) Analysis of circular dichroism spectra of oriented protein–lipid complexes: toward a general application. *Biochemistry*, **33**, 14521–14528.

De Lucca, A. J. (2000) Antifungal peptides: potential candidates for the treatment of fungal infections. *Expert Opinion in Investigational Drugs*, **9**, 273–299.

De Lucca, A. J. and Walsh, T. J. (1999) Antifungal peptides: novel therapeutic compounds against emerging pathogens. *Antimicrobial Agents and Chemotherapy*, **43**, 1–11.

De Lucca, A. J., Jacks, T. J. and Brogden, K. A. (1995) Binding between lipopolysaccharide and cecropin A. *Molecular and Cellular Biochemistry*, **151**, 141–148.

De Lucca, A. J., Bland, J. M., Jacks, T. J., Grimm, C. and Walsh, T. J. (1998). Fungicidal and binding properties of the natural peptides cecropin B and dermaseptin. *Medical Mycology*, **36**, 291–298.

De Lucca, A. J., Bland, J. M., Vigo, C. B., Jacks, T. J., Peter, J. and Walsh, T. J. (2000). D-cecropin B: proteolytic resistance, lethality for pathogenic fungi and binding properties. *Medical Mycology*, **38**, 301–308.

Dempsey, C. E. (1990) The actions of melittin on membranes. *Biochimica et Biophysica Acta*, **1031**, 143–161.

Dempsey, C. E. and Butler, G. S. (1992) Helical structure and orientation of melittin in dispersed phospholipid membranes from amide exchange analysis *in situ*. *Biochemistry*, **31**, 11973–11977.

Derks, F. H. M., van Dijk, A. J., Hanisch ten Cate, C. H., Florack, D. E. A., Dubois, L. A. M. and de Vries, D. P. (1995) Prolongation of vase life of cut roses via introduction of genes coding for antibacterial activity. Somatic embryogenesis and *Agrobacterium*-mediated transformation. *Acta horticulturae*, **405**, 205–209.

Devine, D. A., Marsh, P. D., Percival, R. S., Rangarajan, M. and Curtis, M. A. (1999) Modulation of antibacterial peptide activity by products of *Porphyromonas gingivalis* and *Prevotella spp*. *Microbiology*, **145**, 965–971.

Díaz-Achirica, P., Prieto, S., Ubach, J., Andreu, D., Rial, E. and Rivas, L. (1994) Permeabilization of the mitochondrial inner membrane by short cecropin A–melittin hybrid peptides. *European Journal of Biochemistry*, **224**, 257–263.

Díaz-Achirica, P., Guinea, A., Ubach, J., Andreu, D. and Rivas, L. (1996) Mechanism of action of cecropin A–melittin hybrid peptides on *Leishmania* parasites. In *Peptides 95, Chemistry, Structure and Biology*, edited by P. Kaumaya and R.S. Hodges, Boston: Mayflowers, pp. 183–185.

Díaz-Achirica, P., Ubach, J., Guinea, A., Andreu, D. and Rivas, L. (1998) The plasma membrane of *Leishmania donovani* promastigotes is the main target for CA(1–8)M(1–18), a synthetic cecropin A–melittin hybrid peptide. *Biochemical Journal*, **330**, 453–460.

Duclohier, H., Molle, G. and Spach, G. (1989) Antimicrobial peptide magainin I from *Xenopus* skin forms anion-permeable channels in planar lipid bilayers. *Biophysical Journal*, **56**, 1017–1021.

Dufourcq, J. and Faucon, J. F. (1977) Intrinsic fluorescence study of lipid-protein interactions in membrane models. Binding of melittin, an amphipathic peptide, to phospholipid vesicles. *Biochimica et Biophysica Acta*, **467**, 1–11.

Durell, S. R., Raghunathan, G. and Guy, H. R. (1992) Modeling the ion channel structure of cecropin. *Biophysical Journal*, **63**, 1623–1631.

Durvasula, R. V., Gumbs, A., Panackal, A., Kruglov, O., Aksoy, S., Merrifield, R. B. *et al.* (1997) Prevention of insect-borne disease: an approach using transgenic symbiotic bacteria. *Proceedings of the National Academy of Sciences USA*, **94**, 3274–3278.

Edlund, T., Sidén, I. and Boman, H. G. (1976) Evidence for two immune inhibitors from *Bacillus thuringiensis* interfering with the humoral defense system of saturniid pupae. *Infection and Immunity*, **14**, 934–941.

Eguchi, Y., Shimizu, S. and Tsujimoto, Y. (1997) Intracellular ATP levels determine cell death by apoptosis or necrosis. *Cancer Research*, **57**, 1835–1840.

Ehrenstein, G. and Lecar, H. (1977) Electrically gated ionic channels in lipid bilayers. *Quarterly Review in Biophysics*, **10**, 1–34.

Ekengren, S. and Hultmark, D. (1999) *Drosophila* cecropin as an antifungal agent. *Insect Biochemistry and Molecular Biology*, **29**, 965–972.

Ernst, R. K. and Miller, S. I. (2000) Peptides, *Pseudomonas aeruginosa*, polysaccharides and lipopolysaccharides – players in the predicament of cystic fibrosis patients: response. *Trends in Microbiology*, **8**, 250–251.

Ernst, R. K., Guina, T. and Miller, S. I. (1999) How intracellular bacteria survive: surface modifications that promote resistance to host innate immune responses. *Journal of Infectious Diseases*, **179 Suppl 2**, S326–S330.

Faber, K. N., Westra, S., Waterham, H. R., Keizer-Gunnink, I., Harder, W. and Veenhuis, G. A. (1996) Foreign gene expression in *Hansenula polymorpha*. A system for the synthesis of small functional peptides. *Applied Microbiology and Biotechnology*, **45**, 72–79.

Fattal, E., Nir, S., Parente, R. A. and Szoka, F. C. Jr (1994) Pore-forming peptides induce rapid phospholipid flip-flop in membranes. *Biochemistry*, **33**, 6721–6731.

Feigin, A. M., Teeter, J. H. and Brand, J. G. (1995) The influence of sterols on the sensitivity of lipid bilayers to melittin. *Biochemical and Biophysical Research Communications*, **211**, 312–317.

Fernández, I., Ubach, J., Reig, F., Andreu, D. and Pons, M. (1994) Effect of succinylation on the membrane activity and conformation of a short cecropin A–melittin hybrid peptide. *Biopolymers*, **34**, 1251–1258.

Fernández, I., Ubach, J., Fuxreiter, M., Andreu, J. M., Andreu, D. and Pons, M. (1996) Conformation and self-association of a hybrid peptide of cecropin A–melittin with improved antibiotic activity. *Chemistry. An European Journal*, **2**, 838–846.

Fink, J., Merrifield, R. B., Boman, A. and Boman, H. G. (1989a) The chemical synthesis of cecropin D and an analog with enhanced antibacterial activity. *Journal of Biological Chemistry*, **264**, 6260–6267.

Fink, J., Boman, A., Boman, H. G. and Merrifield, R. B. (1989b) Design, synthesis and antibacterial activity of cecropin-like model peptides. *International Journal of Peptide and Protein Research*, **33**, 412–421.

Florack, D., Allefs, S., Bollen, R., Bosch, D., Visser, B. and Stiekema, W. (1995) Expression of giant silkmoth cecropin B genes in tobacco. *Transgenic Research*, **4**, 132–141.

Freer, E., Moreno, E., Moriyon, I., Pizarro-Cerdá, J., Weintraub, A. and Gorvel, J. P. (1996) *Brucella-Salmonella* lipopolysaccharide chimeras are less permeable to hydrophobic probes and more sensitive to cationic peptides and EDTA than are their native *Brucella sp.* counterparts. *Journal of Bacteriology*, **178**, 5867–5876.

Friedrich, C., Scott, M. G., Karunaratne, N., Yan, H. and Hancock. R. E. W. (1999) Salt-resistant alpha-helical cationic antimicrobial peptides. *Antimicrobial Agents and Chemotherapy*, **43**, 1542–1548.

Friedrich, C. L., Moyles, D., Beveridge, T. J. and Hancock, R. E. W. (2000) Antibacterial action of structurally diverse cationic peptides on gram-positive bacteria. *Antimicrobial Agents and Chemotherapy*, **44**, 2086–2092.

Ganz, T. and Lehrer, R. I. (1999) Antibiotic peptides from higher eukaryotes: biology and applications. *Molecular Medicine Today*, **5**, 292–297.

Gazit, E., Lee, W. J., Brey, P. T. and Shai, Y. (1994) Mode of action of the antibacterial cecropin B2a spectrofluorometric study. *Biochemistry*, **33**, 10681–10692.

Gazit, E., Boman, A., Boman, H. G. and Shai, Y. (1995) Interaction of the mammalian antibacterial peptide cecropin P1 with phospholipid vesicles. *Biochemistry*, **34**, 11479–11488.

Gazit, E., Miller, I. R., Biggin, P. C., Sansom, M. S. and Shai, Y. (1996) Structure and orientation of the mammalian antibacterial peptide cecropin P1 within phospholipid membranes. *Journal of Molecular Biology*; **258**, 860–870.

Gesell, J., Zasloff, M. and Opella, S. J. (1997) Two-dimensional ^{1}H NMR experiments show that the 23-residue magainin antibiotic peptide is an α-helix in dodecylphosphocholine micelles, sodium dodecylsulfate micelles, and trifluoroethanol/water solution. *Journal of Biomolecular NMR*, **9**, 127–135.

Ghosh, A. K., Rukmini, R. and Chattopadhyay, A. (1997) Modulation of tryptophan environment in membrane-bound melittin by negatively charged phospholipids: implications in membrane organization and function. *Biochemistry*, **36**, 14291–14305.

Giacometti, A., Cirioni, O., Barchiesi, F., Del Prete, M. S. and Scalise, G. (1999) Antimicrobial activity of polycationic peptides. *Peptides*, **20**, 1265–1273.

Giacometti, A., Cirioni, O., Barchiesi, F., Ancarani, F. and Scalise, G. (2000a) In vitro anti-cryptosporidial activity of cationic peptides alone and in combination with inhibitors of ion transport systems. *Journal of Antimicrobial Chemotherapy*, **45**, 651–654.

Giacometti, A., Cirioni, O., Del Prete, M. S., Barchiesi, F. and Scalise, G. (2000b) Short-term exposure to membrane-active antibiotics inhibits *Cryptosporidium parvum* infection in cell culture. *Antimicrobial Agents and Chemotherapy*, **44**, 3473–3475.

Götz, P., Boman, H. G. and Boman, A. (1981) Interaction between insect immunity and insect-pathogenic nematode with symbiotic bacteria. *Proceedings of the Royal Society of London*, **B212**, 333–350.

Gough, M., Hancock, R. E. W. and Kelly, N. M. (1996) Antiendotoxin activity of cationic peptide antimicrobial agents. *Infection and Immunity*, **64**, 4922–4927.

Groisman, E. A. (2001) The pleiotropic two-component regulatory system PhoP-PhoQ. *Journal of Bacteriology*, **183**, 1835–1842.

Guina, T., Yi, E. C., Wang, H., Hackett, M. and Miller, S. I. (2000) A PhoP-regulated outer membrane protease of *Salmonella enterica* serovar *typhimurium* promotes resistance to alpha-helical antimicrobial peptides. *Journal of Bacteriology*, **182**, 4077–4086.

Guo, L., Lim, K. B., Gunn, J. S., Bainbridge, B., Darveau, R. P., Hackett, M. *et al.* (1997) Regulation of lipid A modifications by *Salmonella typhimurium* virulence genes *phoP-phoQ*. *Science*, **276**, 250–253.

Guo, L., Lim, K. B., Poduje, C. M., Daniel, M., Gunn, J. S., Hackett, M. *et al.* (1998) Lipid A acylation and bacterial resistance against vertebrate antimicrobial peptides. *Cell*, **95**, 189–198.

Gwadz, R. W., Kaslow, D., Lee, J. Y., Maloy, W. L., Zasloff, M. and Miller, L. H. (1989) Effects of magainins and cecropins on the sporogonic development of malaria parasites in mosquitoes. *Infection and Immunity*, **57**, 2628–3263.

Habermann, E. (1972) Bee and wasp venoms. *Science*, **177**, 314–322.

Habermann, E. and Kowallek, H. (1970) Modifications of amino groups and tryptophan in melittin as an aid to recognition of structure-activity relationships. *Hoppe-Seyler's Zeitschrift für Physiologische Chemie*, **351**, 884–890.

Hancock, R. E. W. (2000) Cationic antimicrobial peptides: towards clinical applications. *Expert Opinion in Investigational Drugs*, **9**, 1723–1729.

Hancock, R. E. W. and Chapple, D. S. (1999) Peptide antibiotics. *Antimicrobial Agents and Chemotherapy*, **43**, 1317–1323.

Hancock, R. E. W. and Diamond, G. (2000) The role of cationic antimicrobial peptides in innate host defences. *Trends in Microbiology*, **8**, 402–410.

Hancock, R. E. W. and Lehrer, R. (1998) Cationic peptides: a new source of antibiotics. *Trends in Biotechnology*, **16**, 82–88.

Hancock, R. E. W. and Scott, M. G. (2000) The role of antimicrobial peptides in animal defenses. *Proceedings of the National Academy of Sciences USA*, **97**, 8856–8861.

Hancock, R. E. W., Raffle, V. J. and Nicas, T. I. (1981) Involvement of the outer membrane in gentamicin and streptomycin uptake and killing in *Pseudomonas aeruginosa*. *Antimicrobial Agents and Chemotherapy*, **19**, 777–785.

Hellers, M., Gunne, H. and Steiner, H. (1991) Expression of post-translational processing of preprocecropin A using a baculovirus vector. *European Journal of Biochemistry*, **199**, 435–439.

Helmerhorst, E. J., van't Hof, W., Breeuwer, P., Veerman, E. C., Abee, T., Troxler, R. F. *et al.* (2001) Characterization of histatin 5 with respect to amphipathicity, hydrophobicity, and effects on cell and mitochondrial membrane integrity excludes a candidacidal mechanism of pore formation. *Journal of Biological Chemistry*, **276**, 5643–5649.

Henry, M. A. and Secombes, C. J. (2000) The A-layer influences the susceptibility of *Aeromonas salmonicida* to antibacterial peptides. *Fish and Shellfish Immunology*, **10**, 637–642.

Hightower, R., Baden, C., Penzes, E. and Dunsmuir, P. (1994) The expression of cecropin peptide in transgenic tobacco does not confer resistance to *Pseudomonas syringae* pv. *tabaci*. *Plant Cell Reports*, **13**, 295–299.

Holak, T. A., Engström, Å., Kraulis, P. J., Lindeberg, G., Bennich, H., Jones, T. A. *et al.* (1988) The solution conformation of the antibacterial peptide cecropin A: a nuclear magnetic resonance and dynamical simulated annealing study. *Biochemistry*, **27**, 7620–7629.

Houston, M. E. Jr, Kondejewski, L. H., Karunaratne, D. N., Gough, M., Fidai, S., Hodges, R. S. *et al.* (1998) Influence of preformed α-helix and α-helix induction on the activity of cationic antimicrobial peptides. *Journal of Peptide Research*, **5**, 81–88.

Hristova, K., Dempsey, C. E. and White, S. H. (2001) Structure, location, and lipid perturbations of melittin at the membrane interface. *Biophysical Journal*, **80**, 801–811.

Huang, H. W. (2000) Action of antimicrobial peptides: two-state model. *Biochemistry*, **39**, 8347–8352.

Huang, Y., Nordeen, R. O., Di, M., Owens, L. D. and McBeath, J. H. (1997) Expression of an engineered cecropin gene cassette in transgenic tobacco plants confers disease resistance to *Pseudomonas syringae* pv. *tabaci*. *Phytopathology*, **87**, 494–499.

Hugosson, M., Andreu, D., Boman, H. G. and Glaser, E. (1994) Antibacterial peptides and mitochondrial presequences affect mitochondrial coupling, respiration and protein import. *European Journal of Biochemistry*, **223**, 1027–1033.

Hultmark, D., Engström, Å., Bennich, H., Kapur, R. and Boman, H. G. (1982) Insect immunity: isolation and structure of cecropin D and four minor antibacterial components from *Cecropia* pupae. *European Journal of Biochemistry*, **127**, 207–217.

Hung, S. C., Wang, W., Chan, S. I. and Chen, H. M. (1999) Membrane lysis by the antibacterial peptides cecropins B1 and B3: A spin-label electron spin resonance study on phospholipid bilayers. *Biophysical Journal*, **77**, 3120–3133.

Hwang, S. W., Lee, J. H., Park, H. B., Pyo, S. H., So, J. E., Lee, H. S. *et al.* (2001) A simple method for the purification of an antimicrobial peptide in recombinant *Escherichia coli*. *Molecular Biotechnology*, **18**, 193–198.

Iwai, H., Nakajima, Y., Natori, S., Arata, Y. and Shimada, I. (1993) Solution conformation of an antibacterial peptide, sarcotoxin IA, as determined by ^{1}H-NMR. *European Journal of Biochemistry*, **217**, 639–644.

Jacobs, R. E. and White, S. H. (1989) The nature of the hydrophobic binding of small peptides at the bilayer interface: implications for the insertion of transbilayer helices. *Biochemistry*, **28**, 3421–3437.

Jarosz, J. (1997) Identification of immune inhibitor from *Pseudomonas aeruginosa* of inducible cell-free antibacterial activity in insects. *Cytobios*, **89**, 73–80.

Jarosz, J. (1998) Active resistance of entomophagous rhabditid *Heterorhabditis bacteriophora* to insect immunity. *Parasitology*, **117**, 201–208.

Jarosz, J. and Glinski, Z. (1990) Selective inhibition of cecropin-like activity of insect immune blood by protease from American foulbrood scales. *Journal of Invertebrate Pathology*, **56**, 143–149.

Jaynes, J. M., Burton, C. A., Barr, S. B., Jeffers, G. W., Julian, G. R., White, K. L. *et al.* (1988) In vitro cytocidal effect of novel lytic peptides on *Plasmodium falciparum* and *Trypanosoma cruzi*. *FASEB Journal*, **2**, 2878–2883.

Jaynes, J. M., Julian, G. R., Jeffers, G. W., White, K. L. and Enright, F. M. (1989) In vitro cytocidal effect of lytic peptides on several transformed mammalian cell lines. *Peptide Research*, **2**, 157–160.

Jaynes, J. M., Nagpala, P., Destefano-Beltran, L., Hong-Hung, J., Kim, J., Denny, T. *et al.* (1993) Expression of a cecropin B lytic peptide analogue in transgenic tobacco confers enhanced resistance to bacterial wilt caused by *Pseudomonas solanacearum*. *Plant Science*, **8**, 43–53.

Jia, X., Patrzykat, A., Devlin, R. H., Ackerman, P. A., Iwama, G. K. and Hancock, R. E. W. (2000) Antimicrobial peptides protect coho salmon from *Vibrio anguillarum* infections. *Applied and Environmental Microbiology*, **66**, 1928–1932.

Jo, E., Blazyk, J. and Boggs, J. M. (1998) Insertion of magainin into the lipid bilayer detected using lipid photolabels. *Biochemistry*, **37**, 13791–13799.

Johnstone, S. A., Gelmon, K., Mayer, L. D., Hancock, R. E. and Bally, M. B. (2000) In vitro characterization of the anticancer activity of membrane-active cationic peptides. I. Peptide-mediated cytotoxicity and peptide-enhanced cytotoxic activity of doxorubicin against wild-type and P-glycoprotein over-expressing tumor cell lines. *Anticancer Drug Design*, **15**, 151–160.

Juvvadi, P., Vunnam, S., Merrifield, E. L., Boman, H. G. and Merrifield, R. B. (1996a) Hydrophobic effects on antibacterial and channel-forming properties of cecropin A-melittin hybrids. *Journal of Peptide Science*, **2**, 223–232.

Juvvadi, P., Vunnam, S. and Merrifield, R. B. (1996b) Synthetic melittin, its enantio, retro and retroenantio isomers, and selected chimeric analogs: their antibacterial, hemolytic and lipid bilayer action. *Journal of the American Chemical Society*, **118**, 8989–8997.

Juvvadi, P., Vunnam, S., Yoo, B. and Merrifield, R. B. (1999) Structure–activity studies of normal and retro pig cecropin-melittin hybrids. *Journal of Peptide Research*, **53**, 244–251.

Kadono-Okuda, K., Taniai, K., Kato, Y., Kotani, E. and Yamakawa, M. (1995) Effects of synthetic *Bombyx mori* cecropin B on the growth of Plant pathogenic bacteria. *Journal of Invertebrate Pathology*, **65**, 309–310.

Kang, J. H., Shin, S. Y., Jang, S. Y., Lee, M. K. and Hahm, K. S. (1998) Release of aqueous contents from phospholipid vesicles induced by cecropin A (1–8)-magainin 2 (1–12) hybrid and its analogues. *Journal of Peptide Research*, **52**, 45–50.

Kelly, D., Wolters, W. R and Jaynes, J. M. (1990) Effect of lytic peptides on selected fish bacterial pathogens. *Journal of Fish Diseases*, **13**, 317–321.

Kim, D. Y., Lee, J., Saraswat, V. and Park, Y. H. (2000) Glucagon-induced self-association of recombinant proteins in *Escherichia coli* and affinity purification using a fragment of glucagon receptor. *Biotechnology and Bioengineering*, **69**, 418–428.

Kim, D. H., Lee, D. G., Kim, K. L. and Lee, Y. (2001) Internalization of tenecin 3 by a fungal cellular process is essential for its fungicidal effect on *Candida albicans*. *European Journal of Biochemistry*, **268**, 4449–4458.

Kleinschmidt, J. H., Mahaney, J. E., Thomas, D. D. and Marsh, D. (1997) Interaction of bee venom melittin with zwitterionic and negatively charged phospholipid bilayers: a spin-label electron spin resonance study. *Biophysical Journal*, **72**, 767–778.

Kragol, G., Lovas, S., Varadi, G., Condie, B. A., Hoffmann, R. and Otvos, L. Jr (2001) The antibacterial peptide pyrrhocoricin inhibits the ATPase actions of DnaK and prevents chaperone-assisted protein folding. *Biochemistry*, **40**, 3016–3026.

Ladokhin, A. S. and White, S. H. (1999) Folding of amphipathic α-helices on membranes: energetics of helix formation by melittin. *Journal of Molecular Biology*, **285**, 1363–1369.

Ladokhin, A. S., Selsted, M. E. and White, S. H. (1997) Sizing membrane pores in lipid vesicles by leakage of co-encapsulated markers: pore formation by melittin. *Biophysical Journal*, **72**, 1762–1766.

Lafleur, M., Faucon, J. F., Dufourcq, J. and Pezolet, M. (1989) Perturbation of binary phospholipid mixtures by melittin: a fluorescence and raman spectroscopy study. *Biochimica et Biophysica Acta*, **980**, 85–92.

Lee, J. Y., Boman, A., Sun, C. X., Andersson, M., Jörnvall, H., Mutt, V. *et al.* (1989) Antibacterial peptides from pig intestine: isolation of a mammalian cecropin. *Proceedings of the National Academy of Sciences USA*, **86**, 9159–9162.

Lee, J. H., Kim, J. H., Hwang, S. W., Lee, W. J., Yoon, H. K., Lee, H. S. *et al.* (2000) High-level expression of antimicrobial peptide mediated by a fusion partner reinforcing formation of inclusion bodies. *Biochemical and Biophysical Research Communications*, **277**, 575–580.

Li, Z. Q., Merrifield, R. B., Boman, I. A. and Boman, H. G. (1988) Effects on electrophoretic mobility and antibacterial spectrum of removal of two residues from synthetic sarcotoxin IA and addition of the same residues to cecropin B. *FEBS Letters*, **231**, 299–302.

Lincke, C. R., van der Bliek, A. M., Schuurhuis, G. J., van der Velde-Koerts, T., Smit, J. J. and Borst, P. (1990) Multidrug resistance phenotype of human BRO melanoma cells transfected with a wild-type human *mdr1* complementary DNA. *Cancer Research*, **50**, 1779–1785.

Lockey, T. D. and Ourth, D. D. (1996) Formation of pores in *Escherichia coli* cell membranes by a cecropin isolated from hemolymph of *Heliothis virescens* larvae. *European Journal of Biochemistry*, **236**, 263–271.

Löhner, K. and Prenner, E. J. (1999) Differential scanning calorimetry and X-ray diffraction studies of the specificity of the interaction of antimicrobial peptides with membrane-mimetic systems. *Biochimica et Biophysica Acta*, **1462**, 141–156.

Ludtke, S. J., He, K., Wu, Y. and Huang, H. W. (1994) Cooperative membrane insertion of magainin correlated with its cytolytic activity. *Biochimica et Biophysica Acta*, **1190**, 181–184.

Ludtke, S., He, K. and Huang, H. (1995) Membrane thinning caused by magainin 2. *Biochemistry*, **34**, 16764–16769.

Ludtke, S. J., He, K., Heller, W. T., Harroun, T. A., Yang, L. and Huang, H. W. (1996) Membrane pores induced by magainin. *Biochemistry*, **35**, 13723–13728.

Mancheño, J. M., Oñaderra, M., Martínez del Pozo, A., Díaz-Achirica, P., Andreu, D., Rivas, L. *et al.* (1996) Release of lipid vesicle contents by an antibacterial cecropin A–melittin hybrid peptide. *Biochemistry*, **35**, 9892–9899.

Marassi, F. M., Opella, S. J., Juvvadi, P. and Merrifield, R. B. (1999) Orientation of cecropin A helices in phospholipid bilayers determined by solid-state NMR spectroscopy. *Biophysical Journal*, **77**, 3152–3155.

Marcos, J. F., Beachy, R. N., Houghten, R. A., Blondelle, S. E. and Pérez-Payá, E. (1995) Inhibition of a plant virus infection by analogs of melittin. *Proceedings of the National Academy of Sciences USA*, **92**, 12466–12469.

Martemyanov, K. A., Spirin, A. S. and Gudkov, A. T. (1997) Direct expression of PCR products in a cell-free transcription/translation system: synthesis of antibacterial peptide cecropin. *FEBS Letters*, **414**, 268–270.

Martemyanov, K. A., Shirokov, V. A., Kurnasov, O. V., Gudkov, A. T. and Spirin, A. S. (2001) Cell-free production of biologically active polypeptides: application to the synthesis of antibacterial peptide cecropin. *Protein Expression and Purification*, **21**, 456–461.

Martínez de Tejada, G., Pizarro-Cerdá, J., Moreno, E. and Moriyón, I. (1995) The outer membranes of *Brucella spp*. are resistant to bactericidal cationic peptides. *Infection and Immunity*, **63**, 3054–3061.

Matsuzaki, K. (1998) Magainins as paradigm for the mode of action of pore forming polypeptides. *Biochimica et Biophysica Acta*, **1376**, 391–400.

Matsuzaki, K. (1999) Why and how are peptide-lipid interactions utilized for self-defense? Magainins and tachyplesins as archetypes. *Biochimica et Biophysica Acta*, **1462**, 1–10.

Matsuzaki, K., Harada, M., Handa, T., Funakoshi, S., Fujii, N., Yajima, H. *et al.* (1989) Magainin 1-induced leakage of entrapped calcein out of negatively-charged lipid vesicles. *Biochimica et Biophysica Acta*, **981**, 130–134.

Matsuzaki, K., Harada, M., Funakoshi, S., Fujii, N. and Miyajima, K. (1991) Physicochemical determinants for the interactions of magainins 1 and 2 with acidic lipid bilayers. *Biochimica et Biophysica Acta*, **1063**, 162–170.

Matsuzaki, K., Murase, O., Tokuda, H., Funakoshi, S., Fujii, N. and Miyajima, K. (1994) Orientational and aggregational states of magainin 2 in phospholipid bilayers. *Biochemistry*, **33**, 3342–3349.

Matsuzaki, K., Murase, O. and Miyajima, K. (1995a) Kinetics of pore formation by an antimicrobial peptide, magainin 2, in phospholipid bilayers. *Biochemistry*, **34**, 12553–12559.

Matsuzaki, K., Murase, O., Fujii, N. and Miyajima, K. (1995b) Translocation of a channel-forming antimicrobial peptide, magainin 2, across lipid bilayers by forming a pore. *Biochemistry*, **34**, 6521–6526.

Matsuzaki, K., Sugishita, K., Fujii, N. and Miyajima, K. (1995c) Molecular basis for membrane selectivity of an antimicrobial peptide, magainin 2. *Biochemistry*, **34**, 3423–3429.

Matsuzaki, K., Nakamura, A., Murase, O., Sugishita, K., Fujii, N. and Miyajima, K. (1997a) Modulation of magainin 2-lipid bilayer interactions by peptide charge. *Biochemistry*, **36**, 2104–2111.

Matsuzaki, K., Yoneyama, S. and Miyajima, K. (1997b) Pore formation and translocation of melittin. *Biophysical Journal*, **73**, 831–838.

Matsuzaki, K., Sugishita, K., Ishibe, N., Ueha, M., Nakata, S., Miyajima, K. *et al.* (1998) Relationship of membrane curvature to the formation of pores by magainin 2. *Biochemistry*, **37**, 11856–11863.

Matsuzaki, K., Sugishita, K. and Miyajima, K. (1999) Interactions of an antimicrobial peptide, magainin 2, with lipopolysaccharide-containing liposomes as a model for outer membranes of gram-negative bacteria. *FEBS Letters*, **449**, 221–224.

Mchaourab, H. S., Hyde, J. S. and Feix, J. B. (1993) Aggregation state of spin-labeled cecropin AD in solution. *Biochemistry*, **32**, 11895–11902.

Mchaourab, H. S., Hyde, J. S. and Feix, J. B. (1994) Binding and state of aggregation of spin-labeled cecropin AD in phospholipid bilayers: effects of surface charge and fatty acyl chain length. *Biochemistry*, **33**, 6691–6699.

Mee, R. P., Auton, T. R. and Morgan, P. J. (1997) Design of active analogues of a 15-residue peptide using D-optimal design, QSAR and a combinatorial search algorithm. *Journal of Peptide Research*, **49**, 89–102.

Merrifield, R. B., Vizioli, L. D. and Boman, H. G. (1982) Synthesis of the antibacterial peptide cecropin A (1–33). *Biochemistry*, **21**, 5020–5031.

Merrifield, R. B., Merrifield, E. L., Juvvadi, P., Andreu, D. and Boman, H. G. (1994) Design and synthesis of antimicrobial peptides. *Ciba Foundation Symposium*, **186**, 5–20.

Merrifield, R. B., Juvvadi, P., Andreu, D., Ubach, J., Boman, A. and Boman, H. G. (1995a) Retro and retroenantioanalogs of cecropin-melittin hybrids. *Proceedings of the National Academy of Sciences USA*, **92**, 3449–3453.

Merrifield, E. L., Mitchell, S. A., Ubach, J., Boman, H. G., Andreu, D. and Merrifield, R. B. (1995b) D-enantiomers of 15-residue cecropin A–melittin hybrids. *International Journal of Peptide and Protein Research*, **46**, 214–220.

Mills, D., Hammerschlag, F. A., Nordeen, R. O. and Owens, L. D. (1994) Evidence for the breakdown of cecropin B by proteinases in the intercellular fluid of peach leaves. *Plant Science*, **104**, 17–22.

Mitsuhara, I., Matsufuru, H., Ohshima, M., Kaku, H., Nakajima, Y., Murai, N. *et al.* (2000) Induced expression of sarcotoxin IA enhanced host resistance against both bacterial and fungal pathogens in transgenic tobacco. *Molecular Plant–Microbe Interactions*, **13**, 860–868.

Monette, M. and Lafleur, M. (1995) Modulation of melittin-induced lysis by surface charge density of membranes. *Biophysical Journal*, **68**, 187–195.

Monette, M., Van Calsteren, M. R. and Lafleur, M. (1993) Effect of cholesterol on the polymorphism of dipalmitoylphosphatidylcholine/melittin complexes: an NMR study. *Biochimica et Biophysica Acta*, **1149**, 319–328.

Moore, A. J., Devine, D. A. and Bibby, M. C. (1994) Preliminary experimental anticancer activity of cecropins. *Peptide Research*, **7**, 265–269.

Moore, A. J., Beazley, W. D., Bibby, M. C. and Devine, D. A. (1996) Antimicrobial activity of cecropins. *Journal of Antimicrobial Chemotherapy*, **37**, 1077–1089.

Mourgues, F., Brisset, M.-N. and Chevreau, E. (1998) Activity of different antibacterial peptides on *Erwinia amylovora* growth, and evaluation of the phytotoxicity and stability of cecropins. *Plant Science*, **139**, 83–91.

Mozsolits, H., Wirth, H. J., Werkmeister, J. and Aguilar, M. I. (2001) Analysis of antimicrobial peptide interactions with hybrid bilayer membrane systems using surface plasmon resonance. *Biochimica et Biophysica Acta*, **1512**, 64–76.

Münster, C., Lu, J., Bechinger, B. and Salditt, T. (2000) Grazing incidence X-ray diffraction of highly aligned phospholipid membranes containing the antimicrobial peptide magainin 2. *European Biophysical Journal*, **28**, 683–688.

Mutwiri, G. K., Henk, W. G., Enright, F. M. and Corbeil, L. B. (2000) Effect of the antimicrobial peptide, D-hecate, on trichomonads. *Journal of Parasitology*, **86**, 1355–1359.

Naito, A., Nagao, T., Norisada, K., Mizuno, T., Tuzi, S. and Saito, H. (2000) Conformation and dynamics of melittin bound to magnetically oriented lipid bilayers by solid-state ^{31}P and ^{13}C NMR spectroscopy. *Biophysical Journal*, **78**, 2405–2417.

Nakajima, Y., Qu, X. M. and Natori, S. (1987) Interaction between liposomes and sarcotoxin IA, a potent antibacterial protein of *Sarcophaga peregrina* (flesh fly). *Journal of Biological Chemistry*, **262**, 1665–1669.

Nordeen, R. O., Sinden, S. L., Jaynes, J. M. and Owens, L. D. (1992) Activity of cecropin SB-37 against protoplast from several plant species and their bacterial pathogens. *Plant Science*, **82**, 101–107.

Nos-Barberá, S., Portolés, M., Morilla, A., Ubach, J., Andreu, D. and Paterson, C. A. (1997) Effect of hybrid peptides of cecropin A and melittin in an experimental model of bacterial keratitis. *Cornea*, **16**, 101–106.

Oh, D., Shin, S. Y., Lee, S., Kang, J. H., Kim, S. D., Ryu, P. D. *et al.* (2000a) Role of the hinge region and the tryptophan residue in the synthetic antimicrobial peptides, cecropin A(1–8)-magainin 2(1–12) and its analogues, on their antibiotic activities and structures. *Biochemistry*, **39**, 11855–11864.

Oh, H., Hedberg, M., Wade, D. and Edlund, C. (2000b) Activities of synthetic hybrid peptides against anaerobic bacteria: aspects of methodology and stability. *Antimicrobial Agents and Chemotherapy*, **44**, 68–72.

Ohshima, M., Mitsuhara, I., Okamoto, M., Sawano, S., Nishiyama, K., Kaku, H. *et al.* (1999) Enhanced resistance to bacterial diseases of transgenic tobacco plants overexpressing sarcotoxin IA, a bactericidal peptide of insect. *Journal of Biochemistry (Tokyo)*, **125**, 431–435.

Okada, M. and Natori, S. (1984) Mode of action of a bactericidal protein induced in the haemolymph of *Sarcophaga peregrina* (flesh-fly) larvae. *Biochemical Journal*, **222**, 119–124.

Okada, M. and Natori, S. (1985) Primary structure of sarcotoxin I, an antibacterial protein induced in the hemolymph of *Sarcophaga peregrina* (flesh-fly) larvae. *Journal of Biological Chemistry*, **260**, 7147–7174.

Okamoto, M., Mitsuhara, I., Ohshima, M., Natori, S. and Ohashi, Y. (1998) Enhanced expression of an antimicrobial peptide sarcotoxin IA by GUS fusion in transgenic tobacco plants. *Plant and Cell Physiology*, **39**, 57–63.

Oren, Z. and Shai, Y. (1997) Selective lysis of bacteria but not mammalian cells by diastereomers of melittin: structure-function study. *Biochemistry*, **36**, 1826–1835.

Osusky, M., Zhou, G., Osuska, L., Hancock, R. E., Kay, W. W. and Misra, S. (2000) Transgenic plants expressing cationic peptide chimeras exhibit broad-spectrum resistance to phytopathogens. *Nature Biotechnology*, **18**, 1162–1166.

Otvos, L. Jr (2000) Antibacterial peptides isolated from insects. *Journal of Peptide Science*, **6**, 497–511.

Otvos, L. Jr, I. O., Rogers, M. E., Consolvo, P. J., Condie, B. A., Lovas, S., Bulet, P. *et al.* (2000) Interaction between heat shock proteins and antimicrobial peptides. *Biochemistry*, **39**, 14150–14159.

Owens, L. D. and Heutte, T. M. (1997) A single amino acid substitution in the antimicrobial defense protein cecropin B is associated with diminished degradation by leaf intercellular fluid. *Molecular Plant-Microbe Interactions*, **4**, 525–528.

Park, C. B., Kim, H. S. and Kim, S. C. (1998) Mechanism of action of the antimicrobial peptide buforin II: buforin II kills microorganisms by penetrating the cell membrane and inhibiting cellular functions. *Biochemical and Biophysical Research Communications*, **244**, 253–257.

Parra-López, C., Baer, M. T. and Groisman, E. A. (1993) Molecular genetic analysis of a locus required for resistance to antimicrobial peptides in *Salmonella typhimurium*. *EMBO Journal*, **12**, 4053–4062.

Parra-López, C., Lin, R., Aspedon, A. and Groisman, E. A. (1994) A *Salmonella* protein that is required for resistance to antimicrobial peptides and transport of potassium. *EMBO Journal*, **13**, 3964–3972.

Pérez-Payá, E., Houghten, R. A. and Blondelle, S. E. (1995) The role of amphipathicity in the folding, self-association and biological activity of multiple subunit small proteins. *Journal of Biological Chemistry*, **270**, 1048–1056.

Peschel, A., Otto, M., Jack, R. W., Kalbacher, H., Jung, G. and Götz, F. (1999) Inactivation of the *dlt* operon in *Staphylococcus aureus* confers sensitivity to defensins, protegrins and other antimicrobial peptides. *Journal of Biological Chemistry*, **274**, 8405–8410.

Peschel, A., Jack, R. W., Otto, M., Collins, L. V., Staubitz, P., Nicholson, G. *et al.* (2001) *Staphylococcus aureus* resistance to human defensins and evasion of neutrophil killing via the novel virulence factor MprF is based on modification of membrane lipids with L-lysine. *Journal of Experimental Medicine*, **193**, 1067–1076.

Piers, K. L. and Hancock, R. E. W. (1994) The interaction of a recombinant cecropin/melittin hybrid peptide with the outer membrane of *Pseudomonas aeruginosa*. *Molecular Microbiology*, **12**, 951–958.

Piers, K. L., Brown, M. H. and Hancock, R. E. W. (1993) Recombinant DNA procedures for producing small antimicrobial cationic peptides in bacteria. *Gene*, **134**, 7–13.

Piers, K. L., Brown, M. H. and Hancock, R. E. W. (1994) Improvement of outer membrane-permeabilizing and lipopolysaccharide-binding activities of an antimicrobial cationic peptide by C-terminal modification. *Antimicrobial Agents and Chemotherapy*, **38**, 2311–2316.

Pore, N. and Pal, S. (2000) Expression of the antibacterial peptide, cecropin, in cultured mammalian cells. *Biotechnology Letters*, **22**, 151–155.

Pouny, Y., Rapaport, D., Mor, A., Nicolas, P. and Shai, Y. (1992) Interaction of antimicrobial dermaseptin and its fluorescently labeled analogues with phospholipid membranes. *Biochemistry*, **31**, 12416–12423.

Pütsep, K., Branden, C. I., Boman, H. G. and Normark, S. (1999) Antibacterial peptide from *H. pylori*. *Nature*, **398**, 671–672.

Rana, F. R., Macias, E. A., Sultany, C. M., Modzrakowski, M. C. and Blazyk, J. (1991) Interactions between magainin 2 and *Salmonella typhimurium* outer membranes: effect of lipopolysaccharide structure. *Biochemistry*, **30**, 5858–5866.

Reed, W. A., Elzer, P. H., Enright, F. M., Jaynes, J. M., Morrey, J. D. and White, K. L. (1997) Interleukin 2 promoter/enhancer controlled expression of a synthetic cecropin-class lytic peptide in transgenic mice and subsequent resistance to *Brucella abortus*. *Transgenic Research*, **6**, 337–347.

Robinson, W. E. Jr., McDougall, B., Tran, D. and Selstedt, M. E. (1998) Anti-HIV-1 activity of indolicidin, an antimicrobial peptide from neutrophils. *Journal of Leukocyte Biology*, **63**, 91–100.

Rodríguez, M. C., Zamudio, F., Torres, J. A., González-Cerón, L., Possani, L. D. and Rodríguez, M. H. (1995) Effect of a cecropin-like synthetic peptide (Shiva-3) on the sporogonic development of *Plasmodium berghei*. *Experimental Parasitology*, **80**, 596–604.

Saugar, J. M., Alarcón, T., López-Hernández, S., López-Brea, M., Andreu, D. and Rivas, L. (2002) Activities of polymyxin B and cecropin A–melittin peptide CA(1–8)M(1–18) against a multiresistant strain of *Acinetobacter baumannii*. *Antimicrobial Agents and Chemotherapy*, **46**, 875–878.

Schümann, M., Dathe, M., Wieprecht, T., Beyermann, M. and Bienert, M. (1997) The tendency of magainin to associate upon binding to phospholipid bilayers. *Biochemistry*, **36**, 4345–4351.

Schwarz, G., Zong, R. T. and Popescu, T. (1992) Kinetics of melittin induced pore formation in the membrane of lipid vesicles. *Biochimica et Biophysica Acta*, **1110**, 97–104.

Scott, M. G. and Hancock, R. E. W. (2000) Cationic antimicrobial peptides and their multifunctional role in the immune system. *Critical Reviews in Immunology*, **20**, 407–431.

Scott, M. G., Gold, M. R. and Hancock, R. E. W. (1999a) Interaction of cationic peptides with lipoteichoic acid and gram-positive bacteria. *Infection and Immunity*, **67**, 6445–6453.

Scott, M. G., Yan, H. and Hancock, R. E. W. (1999b) Biological properties of structurally related α-helical cationic antimicrobial peptides. *Infection and Immunity*, **67**, 2005–2009.

Scott, M. G., Rosenberger, C. M., Gold, M. R., Finlay, B. B. and Hancock, R. E. W. (2000a) An α-helical cationic antimicrobial peptide selectively modulates macrophage responses to lipopolysaccharide and directly alters macrophage gene expression. *Journal of Immunology*, **165**, 3358–3365.

Scott, M. G., Vreugdenhil, A. C., Buurman, W. A., Hancock, R. E. W. and Gold, M. R. (2000b) Cationic antimicrobial peptides block the binding of lipopolysaccharide (LPS) to LPS binding protein. *Journal of Immunology*, **164**, 549–553.

Seeber, F. (2000) An enzyme-release assay for the assessment of the lytic activities of complement or antimicrobial peptides on extracellular *Toxoplasma gondii*. *Journal of Microbiological Methods*, **39**, 189–196.

Shafer, W. M., Qu, X., Waring, A. J. and Lehrer, R. I. (1998) Modulation of *Neisseria gonorrhoeae* susceptibility to vertebrate antibacterial peptides due to a member of the resistance/modulation/division efflux pump family. *Proceedings of the National Academy of Sciences USA*, **95**, 1829–1833.

Shai, Y. (1995) Molecular recognition between membrane-spanning polypeptides. *Trends in Biochemical Sciences*, **20**, 460–464.

Shai, Y. (1999) Mechanism of the binding, insertion and destabilization of phospholipid bilayer membranes by α-helical antimicrobial and cell non-selective membrane-lytic peptides. *Biochimica et Biophysica Acta*, **1462**, 55–70.

Shaposhnikova, V. V., Egorova, M. V., Kudryavtsev, A. A., Levitman, M. K. and Korystov, YuN (1997) The effect of melittin on proliferation and death of thymocytes. *FEBS Letters*, **410**, 285–288.

Sharma, A., Khoury-Christianson, A. M., White, S. P., Dhanjal, N. K., Huang, W., Paulhiac, C. *et al.* (1994) High-efficiency synthesis of human α-endorphin and magainin in the erythrocytes of transgenic mice: a production system for therapeutic peptides. *Proceedings of the National Academy of Sciences USA*, **91**, 9337–9341.

Sharma, A., Sharma, R., Imamura, M., Yamakawa, M. and Machii, H. (2000) Transgenic expression of cecropin B, an antibacterial peptide from *Bombyx mori*, confers enhanced resistance to bacterial leaf blight in rice. *FEBS Letters*, **484**, 7–11.

Sharom, F. J., DiDiodato, G., Yu, X. and Ashbourne, K. J. (1995) Interaction of the P-glycoprotein multidrug transporter with peptides and ionophores. *Journal of Biological Chemistry*, **270**, 10334–10341.

Sharon, M., Oren, Z., Shai, Y. and Anglister, J. (1999) 2D-NMR and ATR-FTIR study of the structure of a cell-selective diastereomer of melittin and its orientation in phospholipids. *Biochemistry*, **38**, 15305–15316.

Shin, S. Y., Lee, M. K., Kim, K. L. and Hahm, K. S. (1997a) Structure–antitumor and hemolytic activity relationships of synthetic peptides derived from cecropin A–magainin 2 and cecropin A–melittin hybrid peptides. *Journal of Peptide Research*, **50**, 228–279.

Shin, S. Y., Lee, D. G., Lee, S. G., Kim, K. L., Lee, M. K. and Hahm, K. S. (1997b) Structure–antifungal activity relationships of Cecropin A hybrid peptides against *Trichoderma* sp. *Journal of Microbiology (Korea)*, **35**, 21–24.

Shin, S. Y., Kang, J. H., Lee, M. K., Kim, S. Y., Kim, Y. and Hahm, K. S. (1998) Cecropin A–magainin 2 hybrid peptides having potent antimicrobial activity with low hemolytic effect. *Biochemistry and Molecular Biology International*, **44**, 1119–1126.

Shin, S. Y., Kang, J. H. and Hahm, K. S. (1999) Structure-antibacterial, antitumor and hemolytic activity relationships of cecropin A–magainin 2 and cecropin A–melittin hybrid peptides. *Journal of Peptide Research*, **53**, 82–90.

Shin, S. Y., Kang, J. H., Jang, S. Y., Kim, Y., Kim, K. L. and Hahm, K. S. (2000) Effects of the hinge region of cecropin A(1–8)-magainin 2(1–12), a synthetic antimicrobial peptide, on liposomes, bacterial and tumor cells. *Biochimica et Biophysica Acta*, **1463**, 209–218.

Sidén, I., Dalhammar, G., Telander, B., Boman, H. G. and Somerville, H. (1979) Virulence factors in *Bacillus thuringiensis*: purification and properties of a protein inhibitor of immunity in insects. *Journal of General Microbiology*, **114**, 45–52.

Silvestro, L. and Axelsen, P. H. (2000) Membrane-induced folding of cecropin A. *Biophysical Journal*, **79**, 1465–1477.

Silvestro, L., Gupta, K., Weiser, J. N. and Axelsen, P. H. (1997) The concentration-dependent membrane activity of cecropin A. *Biochemistry*, **36**, 11452–11460.

Silvestro, L., Weiser, J. N. and Axelsen, P. H. (2000) Antibacterial and antimembrane activities of cecropin A in *Escherichia coli*. *Antimicrobial Agents and Chemotherapy*, **44**, 602–607.

Simmaco, M., Barra, D., Chiarini, F., Noviello, L., Melchiorri, P., Kreil, G., *et al.* (1991) A family of bombinin-related peptides from the skin of *Bombina variegata*. *European Journal of Biochemistry*, **199**, 217–222.

Sipos, D., Chandrasekhar, K., Arvidsson, K., Engström, A. and Ehrenberg, A. (1991) Two-dimensional proton-NMR studies on a hybrid peptide between cecropin A and melittin. Resonance assignments and secondary structure. *European Journal of Biochemistry*, **199**, 285–291.

Sipos, D., Andersson, M. and Ehrenberg, A. (1992) The structure of the mammalian antibacterial peptide cecropin P1 in solution, determined by proton-NMR. *European Journal of Biochemistry*, **209**, 163–169.

Skurnik, M., Venho, R., Bengoechea, J. A. and Moriyón, I. (1999) The lipopolysaccharide outer core of *Yersinia enterocolitica* serotype O : 3 is required for virulence and plays a role in outer membrane integrity. *Molecular Microbiology*, **31**, 1443–1462.

Srisailam, S., Arunkumar, A. I., Wang, W., Yu, C. and Chen, H. M. (2000) Conformational study of a custom antibacterial peptide cecropin B1: implications of the lytic activity. *Biochimica et Biophysica Acta*, **1479**, 275–285.

Srisailam, S., Kumar, T. K., Arunkumar, A. I., Leung, K. W., Yu, C. and Chen, H. M. (2001) Crumpled structure of the custom hydrophobic lytic peptide cecropin B3. *European Journal of Biochemistry*, **268**, 4278–4284.

Steiner, H. (1982) Secondary structure of the cecropins: antibacterial peptides from the moth *Hyalophora cecropia*. *FEBS Letters*, **137**, 283–287.

Steiner, H., Hultmark, D., Engström, A., Bennich, H. and Boman, H. G. (1981) Sequence and specificity of two antibacterial proteins involved in insect immunity. *Nature*, **292**, 246–248.

Steiner, H., Andreu, D. and Merrifield, R. B. (1988) Binding and action of cecropin and cecropin analogues: antibacterial peptides from insects. *Biochimica et Biophysica Acta*, **939**, 260–266.

Subbarao, N. K. and MacDonald, R. C. (1994) Lipid unsaturation influences melittin-induced leakage of vesicles. *Biochimica et Biophysica Acta*, **1189**, 101–107.

Subbalakshmi, C., Nagaraj, R. and Sitaram, N. (1999) Biological activities of C-terminal 15-residue synthetic fragment of melittin: design of an analog with improved antibacterial activity. *FEBS Letters*, **448**, 62–66.

Sui, S. F., Wu, H., Guo, Y. and Chen, K. S. (1994) Conformational changes of melittin upon insertion into phospholipid monolayer and vesicle. *Journal of Biochemistry (Tokyo)*, **116**, 482–487.

Sutherland, I. W. (2001) The biofilm matrix – an immobilized but dynamic microbial environment. *Trends in Microbiology*, **9**, 222–227.

Takei, J., Remenyi, A. and Dempsey, C. E. (1999) Generalised bilayer perturbation from peptide helix dimerisation at membrane surfaces: vesicle lysis induced by disulphide-dimerised melittin analogues. *FEBS Letters*, **442**, 11–14.

Tang, Y. Q., Yuan, J., Osapay, G., Osapay, K., Tran, D., Miller, C. J. *et al.* (1999) A cyclic antimicrobial peptide produced in primate leukocytes by the ligation of two truncated α-defensins. *Science*, **286**, 498–502.

Terwilliger, T. C. and Eisenberg, D. (1982a) The structure of melittin. I. Structure determination and partial refinement. *Journal of Biological Chemistry*, **257**, 6010–6015.

Terwilliger, T. C. and Eisenberg, D. (1982b) The structure of melittin. II. Interpretation of the structure. *Journal of Biological Chemistry*, **257**, 6016–6022.

Tsubery, H., Ofek, I., Cohen, S. and Fridkin, M. (2000) The functional association of polymyxin B with bacterial lipopolysaccharide is stereospecific: studies on polymyxin B nonapeptide. *Biochemistry*, **39**, 11837–11844.

Tytler, E. M., Anantharamaiah, G. M., Walker, D. E., Mishra, V. K., Palgunachari, M. N. and Segrest, J. P. (1995) Molecular basis for prokaryotic specificity of magainin-induced lysis. *Biochemistry*, **34**, 4393–4401.

Unger, T., Oren, Z. and Shai, Y. (2001) The effect of cyclization of magainin 2 and melittin analogues on structure, function and model membrane interactions: implication to their mode of action. *Biochemistry*, **40**, 6388–6397.

Utsugi, T., Schroit, A. J., Connor, J., Bucana, C. D. and Fidler, I. J. (1991) Elevated expression of phosphatidylserine in the outer membrane leaflet of human tumor cells and recognition by activated human blood monocytes. *Cancer Research*, **51**, 3062–3066.

Vaara, M. and Vaara, T. (1994) Ability of cecropin B to penetrate the enterobacterial outer membrane. *Antimicrobial Agents and Chemotherapy*, **38**, 2498–2501.

Velasco, M., Díaz-Guerra, M. J., Díaz-Achirica, P., Andreu, D., Rivas, L. and Boscá, L. (1997) Macrophage triggering with cecropin A and melittin-derived peptides induces type II nitric oxide synthase expression. *Journal of Immunology*, **158**, 4437–4443.

Vunnam, S., Juvvadi, P., Rotondi, K. S. and Merrifield, R. B. (1998) Synthesis and study of normal, enantio, retro and retroenantio isomers of cecropin A–melittin hybrids, their end group effects and selective enzyme inactivation. *Journal of Peptide Research*, **51**, 38–44.

Wachinger, M., Saermark, T. and Erfle, V. (1992) Influence of amphipathic peptides on the HIV-1 production in persistently infected T lymphoma cells. *FEBS Letters*, **309**, 235–241.

Wachinger, M., Kleinschmidt, A., Winder, D., von Pechmann, N., Ludvigsen, A., Neumann, M. *et al.* (1998) Antimicrobial peptides melittin and cecropin inhibit replication of human immunodeficiency virus 1 by suppressing viral gene expression. *Journal of General Virology*, **79**, 731–740.

Wade, D., Boman, A., Wåhlin, B., Drain, C. M., Andreu, D., Boman, H. G. *et al.* (1990) *All* D-amino acid containing channel-forming antibiotic peptides. *Proceedings of the National Academy of Sciences USA*, **87**, 4761–4765.

Wade, D., Andreu, D., Mitchell, S. A., Silveria, A. M., Boman, A., Boman, H. G. *et al.* (1992) Antibacterial peptides designed as analogs on hybrids of cecropins and melittin. *International Journal of Peptide and Protein Research*, **40**, 429–436.

Wang, L., Wu, H., Dou, F., Xie, W. and Xu, X. (1997) High-level expression of cecropin CMIV in *E. coli* from a fusion construct containing the human tumor necrosis factor. *Biochemistry and Molecular Biology International*, **41**, 1051–1056.

Wang, W., Smith, D. K., Moulding, K. and Chen, H. M. (1998) The dependence of membrane permeability by the antibacterial peptide cecropin B and its analogs, CB-1 and CB-3, on liposomes of different composition. *Journal of Biological Chemistry*, **273**, 27438–27448.

Wenk, M. R. and Seelig, J. (1998) Magainin 2 amide interaction with lipid membranes: calorimetric detection of peptide binding and pore formation. *Biochemistry*, **37**, 3909–3916.

Westerhoff, H. V., Juretic, D., Hendler, R. W. and Zasloff, M. (1989) Magainins and the disruption of membrane-linked free-energy transduction. *Proceedings of the National Academy of Sciences USA*, **86**, 6597–6601.

White, S. H. and Wimley, W. C. (1998) Hydrophobic interactions of peptides with membrane interfaces. *Biochimica et Biophysica Acta*, **1376**, 339–352.

Wieprecht, T., Dathe, M., Schümann, M., Krause, E., Beyermann, M. and Bienert, M. (1996) Conformational and functional study of magainin 2 in model membrane environments using the new approach of systematic double-D-amino acid replacement. *Biochemistry*, **35**, 10844–10853.

Wieprecht, T., Dathe, M., Beyermann, M., Krause, E., Maloy, W. L., MacDonald, D. L. *et al.* (1997) Peptide hydrophobicity controls the activity and selectivity of magainin 2 amide in interaction with membranes. *Biochemistry*, **36**, 6124–6132.

Wieprecht, T., Beyermann, M. and Seelig, J. (1999a) Binding of antibacterial magainin peptides to electrically neutral membranes: thermodynamics and structure. *Biochemistry*, **38**, 10377–10387.

Wieprecht, T., Apostolov, O., Beyermann, M. and Seelig, J. (1999b) Thermodynamics of the α-helix-coil transition of amphipathic peptides in a membrane environment: implications for the peptide-membrane binding equilibrium. *Journal of Molecular Biology*, **294**, 785–794.

Winder, D., Gunzburg, W. H., Erfle, W. and Salmons, B. (1998) Expression of antimicrobial peptides has an antitumour effect in human cells. *Biochemical and Biophysical Research Communications*, **242**, 608–612.

Wu, M., Maier, E., Benz, R. and Hancock, R. E. W. (1999) Mechanism of interaction of different classes of cationic antimicrobial peptides with planar bilayers and with the cytoplasmic membrane of *Escherichia coli*. *Biochemistry*, **38**, 7235–7242.

Xia, W., Liu, Q., Wu, J., Xia, Y., Shi, Y. and Qu, X. (1998) Secondary structure of an antibacterial peptide Abp3 studied by two-dimensional proton-NMR. *Biochimica et Biophysica Acta*, **1384**, 299–305.

Xie, W., Qiu, Q., Wu, H. and Xu, X. (1996) Expression of cecropin CMIV fusion protein in *E. coli* under T7 promoter. *Biochemistry and Molecular Biology International*, **39**, 487–492.

Yamada, K., Nakajima, Y. and Natori, S. (1990) Production of recombinant sarcotoxin IA in *Bombyx mori* cells. *Biochemical Journal*, **272**, 633–636.

Yang, L., Harroun, T. A., Heller, W. T., Weiss, T. M. and Huang, H. W. (1998) Neutron off-plane scattering of aligned membranes. I. Method of measurement. *Biophysical Journal*, **75**, 641–645.

Yang, L., Weiss, T. M., Harroun, T. A., Heller, W. T. and Huang, H. W. (1999) Supramolecular structures of peptide assemblies in membranes by neutron off-plane scattering: method of analysis. *Biophysical Journal*, **77**, 2648–2656.

Yang, L., Weiss, T. M., Lehrer, R. I. and Huang, H. W. (2000) Crystallization of antimicrobial pores in membranes: magainin and protegrin. *Biophysical Journal*, **79**, 2002–2009.

Yang, L., Harroun, T. A., Weiss, T. M., Ding, L. and Huang, H. W. (2001) Barrel-stave model or toroidal model? A case study on melittin pores. *Biophysical Journal*, **81**, 1475–1485.

Yau, W. M., Wimley, W. C., Gawrisch, K. and White, S. H. (1998) The preference of tryptophan for membrane interfaces. *Biochemistry*, **37**, 14713–14718.

Zasloff, M. (1987) Magainins, a class of antimicrobial peptides from *Xenopus* skin: isolation, characterization of two active forms, and partial cDNA sequence of a precursor. *Proceedings of the National Academy of Sciences USA*, **84**, 5449–5453.

Zasloff, M. (1992) Antibiotic peptides as mediators of innate immunity. *Current Opinion in Immunology*, **4**, 3–7.

Zhang, L., Falla, T., Wu, M., Fidai, S., Burian, J., Kay, W. and Hancock, R. E. W. (1998) Determinants of recombinant production of antimicrobial cationic peptides and creation of peptide variants in bacteria. *Biochemical and Biophysical Research Communications*, **247**, 674–680.

Zhang, L., Benz, R. and Hancock, R. E. W. (1999) Influence of proline residues on the antimicrobial and synergistic activities of α-helical peptides. *Biochemistry*, **38**, 8102–8111.

12 The syringomycins

Lipodepsipeptide pore formers from plant bacterium *Pseudomonas syringae*

Jon Y. Takemoto, Joseph G. Brand, Yuri A. Kaulin, Valery V. Malev, Ludmila V. Schagina and Katalin Blasko

The syringomycins are a class of cyclic lipodepsipeptides produced by the plant bacterium *Pseudomonas syringae* pv. *syringae* (Pss). They possess antifungal activities and target the fungal plasma membrane. Studies with lipid bilayers, unilamellar lipid vesicles, and erythrocytes show that syringomycin E forms voltage-dependent ion conducting pores of about 1 nm average diameter. Pore formation is influenced by lipid charge, and the pores cluster into structures with higher conductances but similar 1 nm pore diameters. Molecular genetic studies with yeast reveal that sphingolipids and sterols promote susceptibility to syringomycin E. Mutants defective in sphingolipid biosynthetic enzyme sphinganine C4-hydroxylase are resistant to syringomycin E, and they produce sphingolipids that lack the C4-OH group. When yeast non C4-OH sphingolipids are included in lipid bilayers, pore formation is impeded. These results suggest that pore formation is the mechanism that underlies the antifungal activity of syringomycin E. It is speculated that hydrogen bonding as well as lipid charge strongly influence pore formation. Future research will be directed toward a molecular description of the syringomycin E pore structure and determination of the mechanisms by which the pores cluster.

Introduction

The plant bacterium (Pss) leads a dual lifestyle (Takemoto, 1992). As an epiphyte on plant surfaces, Pss defends itself against other microbes and competes for nutrients and space, and its prevalence on many plants suggests that it is a successful competitor. But, Pss is also an opportunistic pathogen that can exhibit marked virulence, and it has been implicated in a number of devastating necrotic plant diseases (DeVay *et al.*, 1968). The two modes of Pss' existence appear to be promoted, respectively, by the production of two classes of bioactive cyclic lipodepsipeptides (Iacobellis *et al.*, 1992; Takemoto, 1992; Bender *et al.*, 1999). These are the small cyclic lipodepsipeptides (MW 1,100–1,300) that possess potent antifungal activities, and the large cyclic lipodepsipeptides (MW 2,200–2,500) that cause plant tissue necroses. Members of the former group are often termed the syringomycins (for the prototype syringomycin and convenience) and the latter are commonly called the syringopeptins. It must be emphasized that the syringomycins and the syringopeptins, respectively, are not exclusively antifungal and plant toxic. At relatively higher concentrations, each is inhibitory to both fungi and plants (Iacobellis *et al.*, 1992; Scholz-Schroeder *et al.*, 2001). In addition, low concentrations of syringopeptins were shown to be inhibitory to the Gram-positive bacterium *Bacillus megaterium* (Iacobellis *et al.*, 1992). The antibacterial properties of the syringopeptins, however, have not been adequately explored.

Table 12.1 Fungicidal activities of syringomycin E

Fungus	MIC (μg/ml)
Aspergillus fumigatus	3.12
Candida albicans	6.25
Candida kefyr	3.12
Cryptococcus neoformans	1.56
Saccharomyces cerevisiae	1.56

In addition to effects against plant-associated fungi, the syringomycins are fungicidal against human fungal pathogens (Sorensen *et al.*, 1996; De Lucca *et al.*, 1999). For example, syringomycin E (described in the next section) inhibits certain clinically important fungi at low minimum inhibitory concentrations (Table 12.1). A fruit postharvest biological control system based on syringomycin-producing Pss strains is practiced (Janisiewicz *et al.*, 1999), and genetic studies indicate that syringomycin production is at least partly responsible for the suppression of fungal proliferation in these crops (Bull *et al.*, 1998).

This chapter focuses on the syringomycins – emphasizing their structures and pore-forming mechanism of action. A review that deals in addition with the biology and biosynthesis of the Pss cyclic lipodepsipeptides including the syringopeptins was recently published (Bender *et al.*, 1999).

The structure of syringomycin E and structural variations

Among the syringomycins, the best studied in terms of mechanism of action is syringomycin E. Its structure is shown in Figure 12.1. It is a cyclic lipodepsipeptide composed of nine amino acids with the C-terminus 4-chlorothreonine ester linked to the N-terminus serine (Segre *et al.*, 1989; Fukuchi *et al.*, 1992). The N-terminus serine is N-acylated to 3-OH dodecanoic acid. The closest analogs, syringomycin A_1 and G, vary in the length of the 3-OH acyl chain with 10 and 14 carbons, respectively. The structures of more distant analogs are shown in Figure 12.2. These are syringostatin, syringotoxin, and pseudomycin (Ballio *et al.*, 1990, 1994; Fukuchi *et al.*, 1992). Each of these latter analogs also occurs with variations in the length of the 3-OH acyl chain as in the case of syringomycins E, A1 and G. All the syringomycins possess the lipopeptide motif: dehydroaminobutanoic acid-β-OH aspartic acid-4-chlorothreonine-serine-3-OH acyl chain. This conserved motif may thus serve as the structural signature for the identification of additional members of the syringomycin family.

The kind of syringomycin analog produced is often related to the specificity of the bacterium for its plant host. For example, syringomycins E, A_1 and G are typically produced by Pss strains isolated from stone fruits and grasses (Segre *et al.*, 1989; Fukuchi *et al.*, 1992), and Pss strains from citrus produce the syringotoxins (Ballio *et al.*, 1990). The syringostatins were extracted from a Pss isolate of lilac (Fukuchi *et al.*, 1992), and the pseudomycins from a barley strain (Ballio *et al.*, 1994). However, production of the syringomycins is not confined to Pss. Recently, it was shown that syringotoxin is produced by a strain of *P. fuscovaginae* – the causative agent of bacterial sheath brown rot of rice (Flamand *et al.*, 1996). Also, syringomycins E and G are produced by *P. syringae* pv. *atrofaciens*, the causative agent of basal glume rot of cereals (Vassilev *et al.*, 1996). The plant-dependent variability in syringomycin structure evolved likely because of differences in plant host biochemistry and

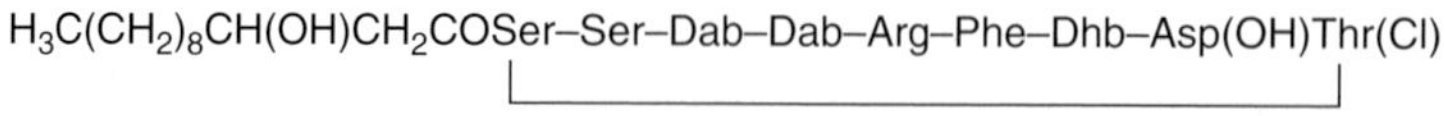

Figure 12.1 The structure of syringomycin E. Analogs syringomycin A_1 and syringomycin G have C_{10} and C_{14} 3-OH acyl chains, respectively. Dab=2,4 diaminobutyric acid; Dhb=dehydroaminobutanoic acid; Asp(OH)=β-OH aspartic acid; Thr (Cl)= 4-chlorothreonine.

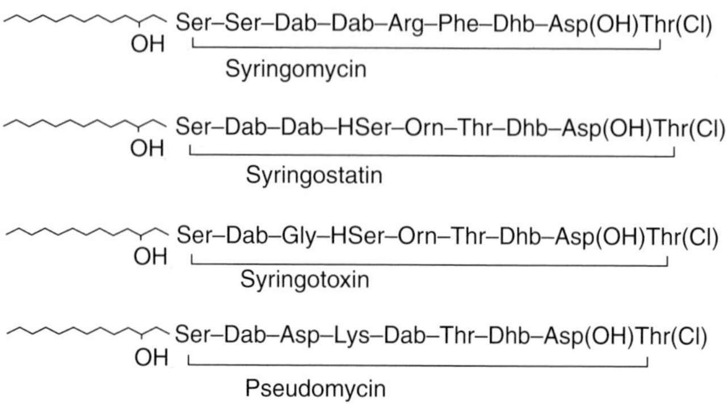

Figure 12.2 Cyclic lipodepsipeptide analogs of syringomycin. Shown are the forms with 3-OH dodecanoyl acyl chains. Other forms exist with acyl chains of varying lengths. All analogs possess the lipopeptide motif: Dhb–Asp(OH)–Thr(Cl)–Ser–3-OH acyl chain.

the fungal flora confronted by the pseudomonads. As a consequence, it is conceivable that more – and as yet undiscovered – structural variations of the syringomycins exist in nature.

Syringomycin E mechanism of action – physiological studies

Syringomycin E is an amphiphilic molecule, and it is easy to predict that it would interact with membranes of the target organism. Earlier physiological studies with plant tissues and with yeasts strongly suggested this to be the case (Takemoto, 1992). The observations particularly implicate the plasma membrane as the cellular target. With exposure to syringomycin E at μg per ml concentrations, plant cells, tissues and yeasts all show marked changes in ion fluxes across the cell surface. For example, red beet slices and yeast cells in glucose medium all displayed rapid and pronounced Ca^{2+} fluxes across their cell surfaces when treated with syringomycin E (Takemoto, 1992). In yeast, a dramatic K^+ efflux is also observed (Zhang and Takemoto, 1989). In yeast, red beet, maize and mung bean, syringomycin E caused pronounced changes in H^+ fluxes and alterations in H^+-ATPase activities (Di Giorgio *et al.*, 1996; Bidwai *et al.*, 1987; Reidl and Takemoto, 1987; Che *et al.*, 1992; Batoko *et al.*, 1998). The variety of flux changes coincide with Ca^{2+}-dependent phosphorylation and dephosphorylation of plasma membrane proteins that may either cause or result from ion movements in response to syringomycin E (Bidwai and Takemoto, 1987). Altogether, these effects point to the target cell plasma membrane as the primary site of action for syringomycin E. Also, when the nature of the various physiological effects are considered together, it is reasonable to speculate that pore formation at the plasma membrane is the underlying basis for the inhibitory effects (Takemoto, 1992; Hutchison *et al.*, 1995).

Evidence for membrane pore formation by syringomycin E

Membrane pore formation by syringotoxin was first reported in 1984 by Ziegler *et al.* (1984) based on lipid bilayer studies. Although the structures of syringotoxin and the syringomycins were unknown at the time, a similar pore-forming activity of syringomycin was conceivable. However, it was not until a decade later that Feigin *et al.* (1996) and Hutchison *et al.* (1995) independently reported formation of voltage-dependent ion channels by syringomycin E. Syringomycin E (1–5 μg per ml) addition on the *cis* side of planar lipid bilayers composed of DOPS and a *cis* side positive transmembrane potential in 0.1 M NaCl (pH 6) bathing solution caused pronounced increases in macroscopic ion currents (Feigin *et al.*, 1996). A typical current–time profile is shown in Figure 12.3. A reversal of the voltage polarity (*cis* side negative) after channel formation led to channel inactivation that occurred over a time period of several minutes. During the inactivation phase, small conducting single channels were evident leading to the notion of the existence of more than one kind of syringomycin E channel. The membrane conductance induced by syringomycin E increased with the sixth power of syringomycin E concentration, suggesting that at least six monomers are required for channel formation. Similar asymmetric voltage-dependent properties were observed with syringopeptin 25A incorporated into planar lipid bilayers (Dalla Serra *et al.*, 1999a). Further studies (Kaulin *et al.*, 1998) were aimed at

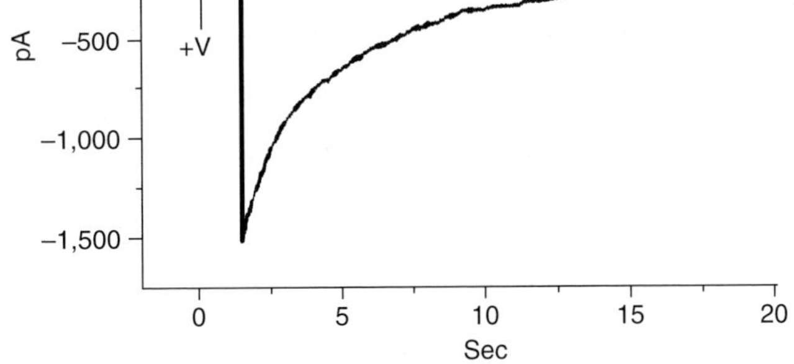

Figure 12.3 Time course of a DOPS bilayer current in a field reversal experiment in the presence of syringomycin E. Syringomycin E (2 μg ml) was added on the *cis* side before application of transmembrane potential. At zero time, 120 mV was applied, *cis* side positive (+V). The voltage polarity was then reversed (–V). The bathing solution was 100 mM NaCl and 5 mM MOPS, pH 6.0.

characterizing the nature of different types of syringomycin E channels. It was shown that syringomycin E forms two major types of channels (small and large) that are different in conductance by a factor of six. Both shared the same properties for ion selectivity and voltage dependence. Pore size estimates based upon single channel conductance measurements in the presence of hydrophilic polyethylene glycol polymers in the aqueous phase indicated that both kinds of channels have pores with similar dimensions (about 1 nm diameter averages). These observations lead to the suggestion that the large channels are clusters of small channels that exhibit synchronous opening and closing.

The lipid composition of the bilayers strongly influences the formation and properties of the syringomycin E channels. For example, cholesterol impeded the macroscopic conductance although ergosterol and stigmasterol had little affect (Feigin *et al.*, 1997). With bilayers containing neutral phospholipids (e.g. DPhPC) instead of negatively charged phospholipids (e.g. DOPS), channel opening was driven by a negative potential (*cis* side) and a subsequent shift to a positive potential caused channel inactivation. In other words, an opposite response to voltage polarity was achieved as compared to the case with charged lipids. Interestingly, this opposite response was mimicked by the addition of 1 M NaCl to the charged membrane bathing solution or decreasing the pH to 2. The addition of the dipole reagent phloretin to the electrolyte bathing the uncharged bilayers reduced the voltage gating to that observed with charged lipids bathed in dilute electrolyte solutions (⩾0.1 M NaCl, pH 6) (Malev *et al.*, 2001; Schagina *et al.*, 2001). In sum, these observations suggest that lipids are part of the syringomycin E channel structure. They also suggest that Coulombic forces and the lipid dipole orientation in relation to the interchannel electric field contribute to the properties of voltage-dependent channel opening.

The pore-forming capabilities of syringomycin E were also shown with liposomes (Dalla Serra *et al.*, 1999b). Large unilamellar vesicles made from phosphatidylcholine were permeabilized by syringomycin E as well as by syringotoxin and syringopeptins in a dose-dependent manner. Interestingly, syringomycin E pore formation was more effective when the vesicles also included cholesterol – an effect that is inconsistent with observations of sterol effects on pore formation in planar lipid membranes (Feigin *et al.*, 1997).

Finally, the properties of syringomycin E pores in erythrocytes have been described (Blasko *et al.*, 1998; Dalla Serra *et al.*, 1999b). Concentration-dependent hemolysis was induced in a minor fraction (24–37%) of the erythrocytes with the remainder showing syringomycin E enhanced permeability for ^{86}Rb and monomeric hemoglobin with kinetics that indicated formation of pores (Blasko *et al.*, 1998). As with planar lipid bilayers, the erythrocyte membrane pores displayed discrete lifetimes and eventual inactivation. The pore dimensions were estimated to range between 0.7 and 1.7 nm with increasing syringomycin E concentration (Dalla Serra *et al.*, 1999b) – large enough to accommodate monomeric hemoglobin (Blasko *et al.*, 1998). Cholesterol depletion or ergosterol substitution had the effect of promoting pore-forming activity. At face value, this effect appears to be consistent with the sterol effects in planar lipid membranes but not in liposomes. Although the differences lack an explanation, the observations support the general notion that sterols influence syringomycin E's pore-forming capabilities or properties of the pore.

In summary, it is clear that syringomycin E is capable of forming pores in planar lipid bilayers, liposomes and erythrocyte membranes. It is less clear that pore formation is responsible for syringomycin E's antifungal and plant virulence activities. Recent findings described in the next section, however, provide support for such a mechanism as the basis for the growth inhibitory effects against yeast.

Syringomycin E action against yeast requires sterols and sphingolipids

Molecular genetic studies with yeast were initiated to define more precisely the antifungal mechanism of action of syringomycin E. Syringomycin E-resistant mutants were generated to permit identification of the mutated genes by complementation (Takemoto *et al.*, 1993). To date, 12 syringomycin E-resistant complementation groups corresponding to gene functions needed for syringomycin E action have been characterized. Mutants defective in genes of these groups were also growth resistant to syringotoxin and syringostatin (Takemoto *et al.*, 1993). Four complementation groups represent functions related to sterol biosynthesis. For example, *SYR1/ERG3* encodes the sterol C5,6 desaturase for the biosynthesis of ergosterol, the primary sterol in the yeast plasma membrane (Taguchi *et al.*, 1994). The remaining seven complementing groups represent sphingolipid biosynthetic genes (Figure 12.4). *SYR2* is required for sphinganine C4-hydroxylation, revealing that C4-OH phytoceramide-based sphingolipids are required for syringomycin E action (Grilley *et al.*, 1998). *ELO2* and *ELO3* are responsible for the conversion of C16 and C18 fatty acids to the very long chain (C20–C26) fatty acids that are N-acylated to the ceramide moieties of sphingolipid molecules (Stock *et al.*, 2000). *IPT1* encodes the enzyme that catalyzes the terminal yeast sphingolipid biosynthetic step – the transfer of phosphoinositol from phosphatidylinositol to MIPC to form $M(IP)_2C$ and diacylglycerol, and *CSG1* and *CSG2* are necessary for mannosylation of IPC to $M(IP)_2C$ (Stock *et al.*, 2000). Finally, *FAH1* encodes the enzyme that α-hydroxylates the N-acylated very long fatty acid chain (Hama *et al.*, 2000). Although sphingolipids are required for cell viability (Dickson and Lester, 1999), each of the above sphingolipid biosynthetic functions is not essential for growth on rich, glucose-containing medium. Therefore, their expression imparts structural features to sphingolipids that allow susceptibility to syringomycin E but that are dispensable for growth. Altogether, the identification of these genes reveals that sterols and sphingolipids are important for yeast's susceptibility to syringomycin E. Both of these lipid classes occur predominantly in the plasma membrane, and therefore, recognition of their importance

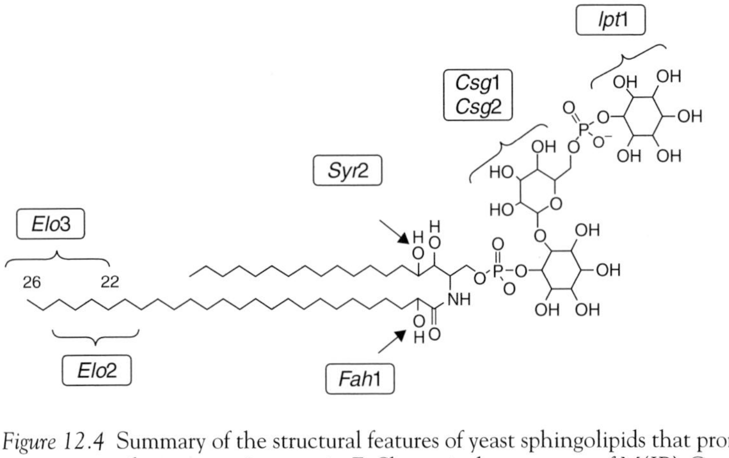

Figure 12.4 Summary of the structural features of yeast sphingolipids that promote susceptibility of yeast to syringomycin E. Shown is the structure of $M(IP)_2C$ – the terminal product of the yeast sphingolipid biosynthetic pathway. The indicated yeast sphingolipid biosynthetic enzymes responsible for these structural features are described in the text.

for syringomycin E action is consistent with the notion that the plasma membrane is the cellular target.

C4 hydroxylated sphingolipids promote pore formation in planar lipid bilayers

As defects in the above sphingolipid biosynthetic genes do not prevent yeast growth, the respective yeast mutants may be propagated and the various defective sphingolipids may be extracted. Such defective sphingolipids can then be studied for their influences on syringomycin E pore formation in planar lipid bilayers or liposomes. As an example, studies were recently conducted with non C4-OH sphingolipids isolated from a syr2 deletion mutant to determine the influence of the C4-OH group on pore structure and formation (Kaulin *et al.*, 1999).

The influence of sphinganine C4 hydroxylation on pore formation was studied using planar lipid bilayers containing either wild-type C4-OH sphingolipids or syr2 mutant non C4-OH sphingolipids, each at 20 mol% with 80 mol% of dioleylphosphatidylserine (DOPS) : dioleylphosphatidylethanolamine (DOPE) : ergosterol (1 : 1 : 1). Figure 12.5 shows macroscopic bilayer currents for the two cases with exposure to syringomycin E (added on the *cis* side of the bilayer). Both showed qualitatively the same current properties (an increase with positive (*cis* side) voltage and reversal with negative voltage followed by

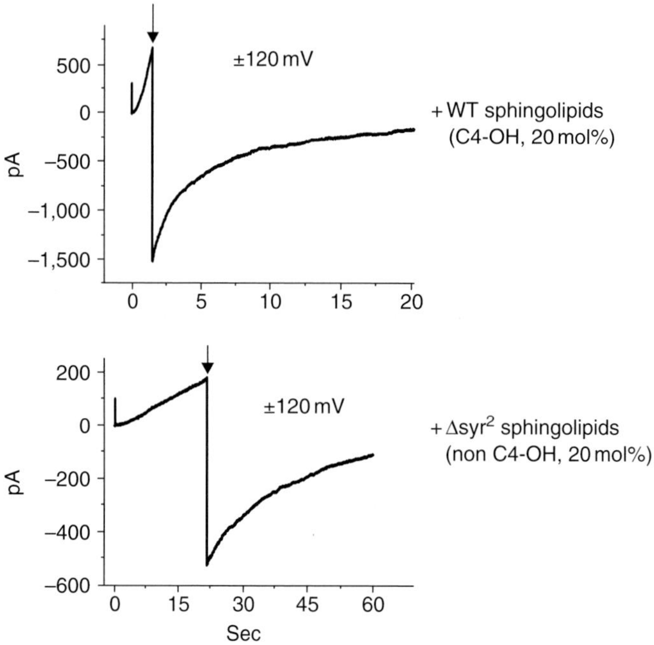

Figure 12.5 Time course of planar bilayer currents with field reversal in the presence of syringomycin E (2 μg ml, *cis* side) and wild type C4-OH (upper trace) or mutant non C4-OH yeast sphingolipids. Bilayers were composed of DOPS/DOPE/ergosterol (1 : 1 : 1) and 20 mol% sphingolipids. Bathing solution and voltages (±120 mV) were as described in the Figure 12.3 legend. The arrows indicate when field reversals to negative voltages were imposed (*cis* side).

inactivation). However, the rate of current increase with positive voltage was significantly lower with the mutant non C4-OH sphingolipids as compared to the case with wild-type sphingolipids. At 120 mV, 220 and 5 s were required to achieve 150 pA with non C4-OH and wild-type sphingolipids, respectively. The current responses with syringomycin E addition to bilayers containing no sphingolipids were nearly identical to those seen with wild-type sphingolipids. In contrast to the difference in macroscopic currents, the mutant and wild-type sphingolipids showed no difference in their influences on the properties of low conductance single channels. For example, the current vs voltage curves of the single unit currents measured between −50 mV and −200 mV (Feigin *et al.*, 1996) were the same for both cases. Overall, these results show that the non C4-OH sphingolipids do not affect the syringomycin E single channel properties, but they instead impede its pore-forming activity. When combined with the observation that syr2 mutants are resistant to syringomycin E, it is suggested that the diminished ability to form pores is the basis for the resistance to this metabolite. In other words, pore formation is suggested to be the mechanism by which syringomycin E inhibits the growth of yeast.

Role of sphingolipid hydrogen bonding in pore formation

As shown in Figure 12.4, several structural features of the yeast sphingolipids – including sphingoid base C4-hydroxylation – promote syringomycin E's antifungal action. The majority of these features contribute to the hydrogen bonding capacities of the sphingolipids – both between lipids and between lipids and syringomycin E. More specifically, the sphingoid base C4-OH and very long fatty acid α-OH groups, and the carbonyl and hydroxyl groups of the mannosyl and phosphatidylinositol sugars are all excellent hydrogen bond donors and acceptors. While elongation of the very long fatty acid chain itself does not directly contribute to hydrogen bonding, it will do so indirectly. Acyl chain length will influence the position of the lipid, and this will, in turn, influence the accessibility of the sphingolipid head groups for hydrogen bonding. Such hydrogen bonding is invoked as a major driving force for the occurrence of predominantly liquid ordered state sphingolipid and sterol-rich domains in biomembranes (Brown and London, 1998). Analogous domains have been reported to occur in yeast (Kubler *et al.*, 1996; Bagnat *et al.*, 2000), and it is tempting to speculate that the basis for the sphingolipid structural and sterol requirements for syringomycin E pore formation lie in the specificities and degrees of hydrogen bonding that occur in these domains. However, it should be noted that syringomycin E pore formation in planar lipid bilayers occurs without sphingolipids and sterols, for example, in DOPS/DOPE (1 : 1) bilayers. Although the DOPS head group is an excellent hydrogen bond donor, the presence of sphingolipids lacking the C4-OH impedes pore formation (Figure 12.5). Therefore, hydrogen bond potential alone is not a requirement for pore formation, and other interactions that promote pore formation must also be considered. Primary among these are interactions of charged channel components (e.g. syringomycin and lipids) and of interactions between lipid dipoles and the channel electric field as suggested by the effects of salt, pH and phloretin on the voltage-dependency properties (discussed above).

Summary and future directions

It is clear that syringomycin E is capable of forming pores in membranes. Furthermore, the correlation between yeast growth resistance and reduced lipid bilayer pore formation imparted by the lack of sphingolipid C4-hydroxylation suggests that syringomycin E inhibition of yeast growth is due to pore formation. This sphingolipid defect (as well as other lipid biosynthetic defects (Figure 12.4)) also confers resistance to syringotoxin and syringostatin (Takemoto *et al.*, 1993). Therefore, lipid modulated pore formation is likely the antifungal mechanism of action for the *P. syringae* bacterial cyclic lipodepsispeptides in general. Less certain is the mechanism of syringomycin E action on plant tissues which is less pronounced compared to its antifungal effects.

Major questions remain regarding the molecular details of syringomycin E pore formation and structure. Although hydrogen bonding and charge interactions almost certainly play roles, their relative importance and precise contributions to channel formation and structure remain to be determined. The occurrence of lipids in the channel structure has been suggested, but obviously lacking is an adequate molecular view of the channel structure. A molecular description of the syringomycin E channel represents a major goal in current research on cyclic lipodepsipeptide mechanism of action. Finally, the structure of the syringomycin E channel cluster and its formation are important topics that have broad implications in biological pore-forming phenomena. Membrane-mediated clustering occurs with many pore formers that include anthrax protective antigen (Sellman *et al.*, 2001), and cytolytic toxins such as aerolysin (Abrami *et al.*, 2000) and perfringolysin (Shatursky *et al.*, 1999). The mechanisms by which syringomycin E forms clustered pores could well serve as paradigms for learning how clustering occurs with these other pore formers to facilitate their toxic actions.

Acknowledgments

Our research was supported by the National Science Foundation, Utah Agricultural Experiment Station, Eli Lilly and Co., National Institutes of Health, Department of Veterans Affairs, Russian Fund for Basic Research and Hungarian Ministry of Welfare.

References

Abrami, L., Fivaz, M. and van der Goot, F. G. (2000) Adventures of a pore-forming toxin at the target cell surface. *Trends Microbiol.*, **8**, 168–172.

Bagnat, M., Keranen, S., Shevchenko, A. and Simons, K. (2000) Lipid rafts function in biosynthetic delivery of proteins to the cell surface in yeast. *Proc. Natl. Acad. Sci. USA*, **97**, 3254–3259.

Ballio, A., Bossa, F., Collina, A., Gallo, M., Iacobellis, N. S., Paci, M. *et al.* (1990) Structure of syringotoxin, a bioactive metabolite of *Pseudomonas syringae* pv. *syringae*. *FEBS Lett.*, **269**, 377–380.

Ballio, A., Bossa, F., Di Giorgio, D., Ferranti, P., Paci, M., Scaloni, A. *et al.* (1994) Novel bioactive lipodepsipeptides from *Pseudomonas syringae*: the pseudomycins. *FEBS Lett.*, **355**, 96–100.

Batoko, H., de Kerchove d'Exaerde, A., Kinet, J. M., Bouharmont, J., Gage, R. A., Maraite, H. *et al.* (1998) Modulation of plant plasma membrane H+-ATPase by phytotoxic lipodepsipeptides produced by the plant pathogen *Pseudomonas fuscovaginae*. *Biochim. Biophys. Acta.*, **1372**, 216–226.

Bender, C. L., Alarcon-Chaidez, F. and Gross, D. C. (1999) *Pseudomonas syringae* phytotoxins: mode of action, regulation, and biosynthesis by peptide and polyketide synthetases. *Microbiol. Mol. Biol. Rev.*, **63**, 266–292.

Bidwai, A. P. and Takemoto, J. Y. (1987) Bacterial phytotoxin, syringomycin, induces a protein kinase-mediated phosphorylation of red beet plasma membrane polypeptides. *Proc. Natl. Acad. Sci. USA*, **84**, 6755–6759.

Bidwai, A. P., Zhang, L., Bachmann, R. C. and Takemoto, J. Y. (1987) Mechanism of action of *Pseudomonas syringae* phytotoxin, syringomycin. Stimulation of red beet plasma membrane ATPase activity. *Plant Physiol.*, **83**, 39–43.

Blasko, K., Schagina, L. V., Agner, G., Kaulin, Y. A. and Takemoto, J. Y. (1998) Membrane sterol composition modulates the pore forming activity of syringomycin E in human red blood cells. *Biochim. Biophys. Acta*, **1373**, 163–169.

Brown, D. A. and London, E. (1998) Structure and origin of ordered lipid domains in biological membranes. *J. Mem. Biol.*, **164**, 103–114.

Bull, C., Wadsworth, M., Sorensen, K., Takemoto, J., Austin, R. and Smilanick, J. (1998) Syringomycin E produced by biological control agents controls green mold on lemons. *Biol. Control*, **12**, 89–95.

Che, F. S., Kasamo, K., Fukuchi, N., Isogai, A. and Suzuki, A. (1992) Effects of syringostatin on plasma membrane ATPase of mung beans (Vigna radiata L.): putative mechanism of action of syringostatin. *Physiol. Plant.*, **86**, 518–524.

Dalla Serra, M., Bernhart, I., Nordera, P., Di Giorgio, D., Ballio, A. and Menestrina, G. (1999a) Conductive properties and gating of channels formed by syringopeptin 25A, a bioactive lipodepsipeptide from *Pseudomonas syringae* pv. *syringae*, in planar lipid membranes. *Mol. Plant–Microbe Interact.*, **12**, 401–409.

Dalla Serra, M., Fagiuoli, G., Nordera, P., Bernhart, I., Della Volpe, C., Di Giorgio, D. *et al.* (1999b) The interaction of lipodepsipeptide toxins from *Pseudomonas syringae* pv. *syringae* with biological and model membranes: a comparison of syringotoxin, syringomycin, and two syringopeptins. *Mol. Plant–Microbe Interact.*, **12**, 391–400.

De Lucca, A. J., Jacks, T. J., Takemoto, J., Vinyard, B., Peter, J., Navarro, E. *et al.* (1999) Fungal lethality, binding, and cytotoxicity of syringomycin-E. *Antimicrob. Agents Chemother.*, **43**, 371–373.

DeVay, J. E., Lukezic, F. L., Sinden, S. L., English, H. and Coplin, D. L. (1968) A biocide produced by pathogenic isolates of *Pseudomonas syringae* and its possible role in the bacterial canker disease of peach trees. *Phytopathology*, **58**, 95–101.

Di Giorgio, D., Camoni, L., Mott, K. A., Takemoto, J. Y. and Ballio, A. (1996) Syringopeptins, *Pseudomonas syringae* pv. *syringae* phytotoxins, resemble syringomycin in closing stomata. *Plant Pathol.* **45**, 564–571.

Dickson, R. C. and Lester, R. L. (1999) Yeast sphingolipids. *Biochim. Biophys. Acta*, **1426**, 347–357.

Feigin, A. M., Schagina, L. V., Takemoto, J. Y., Teeter, J. H. and Brand, J. G. (1997) The effect of sterols on the sensitivity of membranes to the channel-forming antifungal antibiotic, syringomycin E. *Biochim. Biophys. Acta*, **1324**, 102–110.

Feigin, A. M., Takemoto, J. Y., Wangspa, R., Teeter, J. H. and Brand, J. G. (1996) Properties of voltage-gated ion channels formed by syringomycin E in planar lipid bilayers. *J. Membrane Biol.*, **149**, 41–47.

Flamand, M. C., Pelser, S., Ewbank, E. and Maraite, H. (1996) Production of syringotoxin and other bioactive peptides by Pseudomonas fuscovaginae. *Physiol. Mol. Plant Pathol.*, **48**, 217–231.

Fukuchi, N., Isogai, A., Nakayama, J., Takayama, S., Yamashita, S., Suyama, K. *et al.* (1992) Structures and stereochemistry of three phytotoxins, syringomycin, syringotoxin and syringostatin, produced by *Pseudomonas syringae* pv. *syringae*. *J. Chem. Soc. Perkin Trans. 1*, **1992**, 1149–1157.

Grilley, M. M., Stock, S. D., Dickson, R. C., Lester, R. L. and Takemoto, J. Y. (1998) Syringomycin action gene *SYR2* is essential for sphingolipid 4-hydroxylation in *Saccharomyces cerevisiae*. *J. Biol. Chem.*, **273**, 11062–11068.

Hama, H., Young, D. A., Radding, J. A., Ma, D., Tang, J., Stock, S. D. *et al.* (2000) Requirement of sphingolipid alpha-hydroxylation for fungicidal action of syringomycin E. *FEBS Lett.*, **478**, 26–28.

Hutchison, M. L., Tester, M. A. and Gross, D. C. (1995) Role of biosurfactant and ion channel-forming activities of syringomycin in transmembrane ion flux: a model for the mechanism of action in the plant–pathogen interaction. *Mol. Plant–Microbe Interact.*, **8**, 610–620.

Iacobellis, N. S., Lavermicocca, P., Grgurina, I., Simmaco, M. and Ballio, A. (1992) Phytotoxic properties of *Pseudomonas syringae* pv. *syringae* toxins. *Physiol. Mol. Plant Pathol.*, **40**, 107–116.

Janisiewicz, W. J., Conway, W. S. and Leverentz, B. (1999) Biological control of postharvest decays of apple can prevent growth of *Escherichia coli* O157 : H7 in apple wounds. *J. Food Prot.*, **62**, 1372–1375.

Kaulin, Y. A., Schagina, L. V., Feigin, A. M., Wangspa, R., Takemoto, J. Y., Teeter, J. H. *et al.* (1999) Sphingolipid hydroxylation influences sensitivity of lipid bilayers and yeast to syringomycin E. *Biophys. J.*, **76**, 182a.

Kaulin, Y. A., Schagina, L. V., Bezrukov, S. M., Malev, V. V., Feigin, A. M., Takemoto, J. Y. *et al.* (1998) Cluster organization of ion channels formed by the antibiotic syringomycin E in bilayer lipid membranes. *Biophys. J.*, **74**, 2918–2925.

Kubler, E., Dohlman, H. G. and Lisanti, M. P. (1996) Identification of Triton X-100 insoluble membrane domains in the yeast *Saccharomyces cerevisiae*. Lipid requirements for targeting of heterotrimeric G-protein subunits. *J. Biol. Chem.*, **271**, 32975–32980.

Malev, V. V., Schagina, L. V., Takemoto, J. Y., Nestorovich, E. M. and Bezrukov, S. M. (2001) Gating and conductance of syringomycin E channels as manifestations of an asymmetrical channel structure. *Biophys. J.*, **80**, 132a.

Reidl, H. H. and Takemoto, J. Y. (1987) Mechanism of action of bacterial phytotoxin, syringomycin. Simultaneous measurement of early responses in yeast and maize. *Biochim. Biophys. Acta*, **898**, 59–69.

Schagina, L. V., Gurnev, P. A., Kaulin, Y. A., Takemoto, J. Y., Brand, J. G. and Malev, V. V. (2001) Phloretin affects properties of syringomycin E channels formed in lipid bilayer membranes. *Biophys. J.*, **80**, 133a.

Scholz-Schroeder, B. K., Hutchison, M. L., Grgurina, I. and Gross, D. C. (2001) The contribution of syringopeptin and syringomycin to virulence of *Pseudomonas syringae* pv. *syringae* strain B301D on the basis of sypA and syrB1 biosynthesis mutant analysis. *Mol. Plant–Microbe Interact.*, **14**, 336–348.

Segre, A., Bachmann, R. C., Ballio, A., Bossa, F., Grgurina, I., Iacobellis, N. S. *et al.* (1989) The structure of syringomycins A1, E and G. *FEBS Lett.*, **255**, 27–31.

Sellman, B. R., Mourez, M. and Collier, R. J. (2001) Dominant-negative mutants of a toxin subunit: an approach to therapy of anthrax. *Science*, **292**, 695–697.

Shatursky, O., Heuck, A. P., Shepard, L. A., Rossjohn, J., Parker, M. W., Johnson, A. E. *et al.* (1999) The mechanism of membrane insertion for a cholesterol-dependent cytolysin: a novel paradigm for pore-forming toxins. *Cell*, **99**, 293–299.

Sorensen, K. N., Kim, K. H. and Takemoto, J. Y. (1996) In vitro antifungal and fungicidal activities and erythrocyte toxicities of cyclic lipodepsinonapeptides produced by *Pseudomonas syringae* pv. *syringae*. *Antimicrob. Agents Chemother.*, **40**, 2710–2713.

Stock, S. D., Hama, H., Radding, J. A., Young, D. A. and Takemoto, J. Y. (2000) Syringomycin E inhibition of *Saccharomyces cerevisiae*: requirement for biosynthesis of sphingolipids with very-long-chain fatty acids and mannose- and phosphoinositol-containing head groups. *Antimicrob. Agents Chemother.*, **44**, 1174–1180.

Taguchi, N., Takano, Y., Julmanop, C., Wang, Y., Stock, S., Takemoto, J. *et al.* (1994) Identification and analyses of the *Saccharomyces cerevisiae SYR1* reveals that ergosterol is involved in the action of syringomycin. *Microbiology*, **140**, 353–359.

Takemoto, J. Y. (1992) Bacterial phytotoxin syringomycin and its interaction with host membranes. In: *Molecular Signals in Plant–Microbe Communications*, edited by D. P. S. Verma, pp. 247–260. Boca Raton, CRC Press, Inc.

Takemoto, J. Y., Yu, Y., Stock, S. D. and Miyakawa, T. (1993) Yeast genes involved in growth inhibition by *Pseudomonas syringae* pv. *syringae* syringomycin family lipodepsipeptides. *FEMS Microbiol. Lett.*, **114**, 339–342.

Vassilev, V., Lavermicocca, P., Di Giorgio, C. and Iacobellis, N. (1996) Production of syringomycins and syringopeptins by *Pseudomonas syringae* pv. *atrofaciens*. *Plant Pathol.*, **45**, 316–322.

Zhang, L. and Takemoto, J. Y. (1989) Syringomycin stimulation of potassium efflux by yeast cells. *Biochim. Biophys. Acta*, **987**, 171–175.

Ziegler, W., Pavlovkin, J. and Pokorny, J. (1984) Effect of syringotoxin on the permeability of bilayer lipid membranes. *Biologia*, **39**, 693–699.

13 Molecular mechanisms of action of syringopeptins, antifungal peptides from *Pseudomonas syringae* pv. *syringae*

Mauro Dalla Serra, Gianfranco Menestrina, Armando Carpaneto, Franco Gambale, Vincenzo Fogliano and Alessandro Ballio

Introduction

The syringopeptins (SPs) are cyclic lipodepsipeptides (LDP) produced by the Gram-negative bacterium *Pseudomonas syringae* pv. *syringae* (Ballio *et al.*, 1991), a plant pathogen with a wide host range which includes also a number of plants of economical importance (Bradbury, 1986). Although similar to the lipodepsinonapeptides produced by the same bacterium (described in another chapter), SPs have peculiar structure (Figure 13.1) and biological activity. Due to their phytotoxic activity (Iacobellis *et al.*, 1992; Di Giorgio *et al.*, 1994, 1996a,b; Lavermicocca *et al.*, 1997) they are considered additional virulence factors for *P. syringae* pv. *syringae*, and probably contribute to a variety of plant diseases. They have also significant antimicrobial activity, particularly evident in the presence of cell wall degrading enzymes (Fogliano *et al.*, 2002).

In analogy with the other lipopeptides produced by *Pseudomonas* (Nutkins *et al.*, 1991) and *Bacillus* spp. (Peypoux *et al.*, 1986; Maget-Dana and Peypoux, 1994; Eshita *et al.*, 1995, Yakimov *et al.*, 1996), SPs are amphipathic molecules whose primary site of action is probably the cell membrane. A wealth of studies has indicated that a common motif in the mode of action of all bacterial lipopeptides is their capacity to form transmembrane pores in a lipid membrane (Maget-Dana and Ptak, 1990; Brodey *et al.*, 1991; Sheppard *et al.*, 1991; Hutchison *et al.*, 1995; Feigin *et al.*, 1996). Some simplified systems, such as artificial lipid membranes (e.g. monolayers, liposomes and planar bilayers) and mammalian RBC or plant tonoplasts, have proven very useful to clarify the SPs–membrane interactions leading to the opening of ion channels. The results of these studies can shed some light upon the mechanism of plant–pathogen interaction, suggesting also possible biotechnological developments.

Structure of syringopeptins

Primary structure and composition

The SPs share some structural features with the syringomycins (Ballio *et al.*, 1991). They contain a peptide moiety and a long chain 3-hydroxy fatty acyl. The peptide moiety is composed of either 22 (SP_{22}) or 25 (SP_{25}) amino acid residues, mostly of a hydrophobic nature and with D-stereochemistry. Its N-terminal Z-2,3-dehydro-2-aminobutyric acid residue is acylated by an unbranched C_{10} or C_{12} *D*-3-hydroxy fatty acid, and its C-terminal tyrosine

SYRINGOPEPTIN 25 (SP_{25})

R–*z*Dhb–Pro–Val–Ala–Ala–Val Leu–Ala–Ala–*z*Dhb–Val–*z*Dhb–Ala–Val–Ala–Ala–*z*Dhb–*a*Thr–Ser–Ala–Val–Ala–*z*Dhb–*z*Dhb–Tyr (residues 1–25; O of *a*Thr18 linked to CO of Tyr25)

R: 3-hydroxydecanoyl ($SP_{25}A$); 3-hydroxydodecanoyl ($SP_{25}B$)

[Phe[25]]–SYRINGOPEPTIN 25 ([Phe[25]]–$SP_{25}A$)

R–*z*Dhb–Pro–Val–Ala–Ala–Val Leu–Ala–Ala–*z*Dhb–Val–*z*Dhb–Ala–Val–Ala–Ala–*z*Dhb–*a*Thr–Ser–Ala–Val–Ala–*z*Dhb–*z*Dhb–Phe (residues 1–25; O of *a*Thr18 linked to CO of Phe25)

R: 3-hydroxydecanoyl

SYRINGOPEPTIN 22 (SP_{22})

R–*z*Dhb–Pro–Val–Val–Ala–Ala–Val–Val–*z*Dhb–Ala–Val–Ala–Ala–*z*Dhb–*a*Thr–Ser–Ala–*z*Dhb–Ala–Dab–Dab–Tyr (residues 1–22; O of *a*Thr15 linked to CO of Tyr22)

R: 3-hydroxydecanoyl($SP_{22}A$); 3-hydroxydodecanoyl ($SP_{22}B$)

SYRINGOPEPTIN SC (SP[SC])

R–*z*Dhb–Pro–Val–Leu–Ala–Ala–Leu–Val–Dhp–Ala–Val–Ala–Ala–*z*Dhb–*a*Thr–Ser–Ala–*z*Dhb–Ala–Dab–Dab–Tyr (residues 1–22; O of *a*Thr15 linked to CO of Tyr22)

R: 3-hydroxydecanoyl (SP[SC]-1); 3-hydroxydodecanoyl (SP[SC]-2)

SYRINGOPEPTIN 22 Phv($SP_{22}Phv$)

R–*z*Dhb–Pro–Val–Leu–Ala–Ala–Ala–Val–*z*Dhb–Ala–Val–Ala–Ala–*z*Dhb–*a*Thr–Ser–Ala–*z*Dhb–Ala–Dab–Dab–Tyr (residues 1–22; O of *a*Thr15 linked to CO of Tyr22)

R: 3-hydroxydecanoyl ($SP_{22}PhvA$); 3-hydroxydodecanoyl ($SP_{22}PhvB$)

Figure 13.1 Primary structure of all *P. syringae* syringopeptins. Non standard amino acids are: Dab, 2,4-diaminobutanoic acid; *z*Dhb, 2,3-dehydro-2-aminobutyric acid; *a*Thr, *allo*threonine; Dhp dehydro-2-aminopropanoic acid. The 3-hydroxy fatty acid composition is indicated for each compound. The amino acid chirality of syringopeptins 22 and 25 is reported in Ballio *et al.*, 1991.

residue closes a 25-membered lactone ring on the hydroxy group of an *allo*threonine residue.

The primary structure of SPs has been determined by a combination of chemical methods, ^{1}H and ^{13}C NMR spectroscopy, and FAB mass spectrometry.

The first members of the SP group, described by Ballio *et al.* (1991), were $SP_{22}A$, $SP_{22}B$, $SP_{25}A$ and $SP_{25}B$; the suffix A indicates the forms containing D-3-hydroxydecanoic acid, while the suffix B indicates those with D-3-hydroxydodecanoic acid (Figure 13.1). All four metabolites were made by syringomycin-producing isolates (the two SP_{22} from a peach and a pear isolate, and the two SP_{25} from a millet isolate), and the two SP_{25} also by a syringotoxin-producing citrus isolate. Their occurence in previously reported partially purified preparations of SR and ST (Surico and DeVay, 1982) was evidenced by reverse-phase HPLC, the technique of choice for the fractionation, isolation and purification of pseudomonad LDPs.

$SP_{25}A$ and $SP_{25}B$ are also present in cultures of the pseudomycin-producer *P. syringae* MSU 16H (Ballio *et al.*, 1994); furthermore $SP_{25}A$ was found in cultures of the SR-producers *P. syringae* pv. *atrofaciens* (Vassilev *et al.*, 1996), and *P. syringae* pv. *lachrymans*, (Monti *et al.*, 2001) as well as in those of the saprophytic SR-E-producer *P. syringae* M1 isolated from wheat (Adetuy *et al.*, 1995).

TOLAASIN(Tol)

1 2 3 4 5 6 7 8 9 10 11 12 13 14 15 16 17 18
3 OH–C_8–*z*Dhb–Pro–Ser–Leu–Val–Ser–Leu–Val–Val–Gln–Leu–Val–*z*Dhb–*a*Thr–Ile–Hse–Dab–Lys
(lactone: O of *a*Thr14 — CO of Lys18)

FUSCOPEPTIN(FP)

1 2 3 4 5 6 7 8 9 10 11 12 13 14 15 16 17 18 19
R–*z*Dhb–Pro–Leu–Ala–Ala–Ala–Ala–Val–Gly–Ala–Val–Ala–Val–*z*Dhb–*a*Thr–Ala–Dab–Dab–Phe
(lactone: O of *a*Thr15 — CO of Phe19)

R: 3-hydroxyoctanoyl(FP A), 3-hydroxydecanoyl(FP B)

CORPEPTIN(CP)

1 2 3 4 5 6 7 8 9 10 11 12 13 14 15 16 17 18 19 20 21 22
R–*z*Dhb–Pro–Ala–Ala–Ala–Val–Val–*z*Dhb–Hse–Val–*a*Ile–Dhp–Ala–Ala–Ala–Val–*z*Dhb–*a*Thr–Ala–Dab–Ser–Ile
(lactone: O of *a*Thr18 — CO of Ile22)

R: 3-hydroxydecanoyl(CPA); *cis*-3-hydroxy-5-dodecanoyl(CP B)

SYRINGOMYCIN-E(SRE)

1 2 3 4 5 6 7 8 9
R–Ser–Ser–Dab–Dab–Arg–Phe–Dhb–(3–OH)Asp–(4–Cl)Thr
(lactone: Ser1 — O — (4–Cl)Thr9)

R: 3-hydroxydodecanoyl R: CH_3–$(CH_2)_8$–CH(OH)–CH_2–CO

SYRINGOTOXIN(ST)

1 2 3 4 5 6 7 8 9
R–Ser–Dab–Gly–Hse–Orn–Thr–Dhb–(3–OH)Asp–(4–Cl)Thr
(lactone: Ser1 — O — (4–Cl)Thr9)

R: 3-hydroxytetradecanoyl R: CH_3–$(CH_2)_{10}$–CH(OH)–CH_2–CO

Figure 13.2 Primary structure of all *P. syringae* LDPs. Non standard amino acids are as in Figure 13.1 and: Hse, homoserine; Orn, ornithine; *a*Ile, *allo*Isoleucine (amino acid chirality is not specified). The 3-hydroxy fatty acid composition is indicated for each compound. With the exception of corpeptin the amino acid chirality of other toxins is reported in the paper concerning structure elucidation.

More recently other isoforms of SP_{22} have been reported (Figure 13.1). Isogai *et al.* (1995) found in cultures of a sugar cane isolate of *P. syringae* pv. *syringae* two LDPs, called SP(SC)-1 and SP(SC)-2, that differ from the two above described SP_{22} only for three amino acid residues. Grgurina *et al.* (2002) have demonstrated the formation of a new SP_{22} by a *P. syringae* pv. *syringae* isolated from bean and Scaloni *et al.* (1997) have shown that a new peptin made by a laurel isolate is [Phe^{25}]-$SP_{25}A$ (Figure 13.1). Other pseudomonad metabolites sharing structural features and biological properties with the SPs are (Figure 13.2): the tolaasins produced by *P. tolaasii* (Nutkins *et al.*, 1991), the fuscopeptins produced by *P. fuscovaginae* (Ballio *et al.*, 1996), and the corpeptins produced by *P. corrugata* (Emanuele *et al.*, 1998).

Tertiary structure and folding

Detailed information on the three-dimensional structure of $SP_{25}A$ in aqueous solution was obtained by the interpretation of its two-dimensional NMR spectra resulting in the extensive collection of NOE data (Ballio *et al.*, 1995; Baré *et al.*, 1999). First, the chirality of all its amino acid residues was elucidated. Thereafter, computer simulation was performed, applying both distance geometry and molecular dynamics procedures. Figure 13.3 illustrates the solution conformation emerged as the one with lower number of violations, lower potential energy and better agreement with experimental NMR data. This conformation is characterized by three structural regions: a loop including residues from Pro2 to Val6,

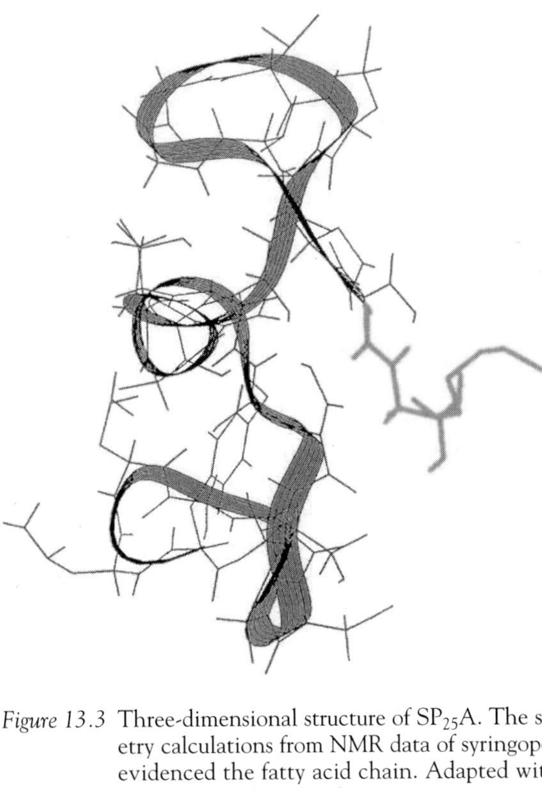

Figure 13.3 Three-dimensional structure of $SP_{25}A$. The structure is obtained by distance geometry calculations from NMR data of syringopeptin in D_2O at pH 3.6. In yellow it is evidenced the fatty acid chain. Adapted with permission from Ballio *et al.*, 1995. (See Colour Plate XVI.)

a helicoidal zone (close to a left-handed alpha-helix) including residues from Ala8 to Ala15, and a large lactone ring including residues from aThr18 to Tyr25. Interestingly the type of ring conformation resembles the seam of a tennis ball, the same previously found in two bioactive lipodepsinonapeptides, the white line inducing principle (WLIP) produced by *P. reactans* (Mortishire-Smith *et al.*, 1991; Han *et al.*, 1992) and siringotoxin produced by *P. syringae* pv. *syringae* (Ballio *et al.*, 1990). The structure found for $SP_{25}A$ probably changes in a membrane environment, as suggested by the CD spectrum in a 4 : 1 trifluoroethanol–water solution. In fact, at variance with the CD spectrum in aqueous solution, which does not contain readily identifiable elements of secondary structure, the former has a calculated amount of 15% α-helix and of 11% of β-turn (Ballio *et al.*, 1995).

Biological role of syringopeptins (SPs)

Toxin production in vitro and in planta

All peptin-producing strains of *P. syringae* so far investigated produce nonapeptides. The amount of toxins produced in static culture is in the range of 1–100 mg l^{-1} per each compounds. MALDI–TOF and electrospray ionization (ESI) mass spectrometry proved to be a powerful tool for detecting and assessing the amount of bacterial LDPs in extracts from cultures of various strains of *P. syringae* (Monti *et al.*, 2001). MALDI–TOF MS gives a snapshot of the LDPs present in raw samples without any previous chromatographic separation, while HPLC-coupled ESI MS makes possible to achieve precise quantification of individual LDPs.

The quantitative determination of peptins in the culture filtrate of *P. syringae* can also be achieved by an immunological assay (Fogliano *et al.*, 1999). This approach gives a simple and reliable tool for the quantitative assessment of SPs in culture filtrates or in aqueous extracts from plant tissues infected by plant pathogenic pseudomonads. The competitive ELISA is about 100 times more sensitive than the chromatographic methods (HPLC) and does not require organic extraction of the toxin from the culture filtrate. With this method, and the SRs immuno-assay developed later (Gallo *et al.*, 2000), the ability of nine pathovars and strains of *P. syringae* to produce LDP in culture was assessed. Under the conditions of the experiment, the maximum yield of SPs was about 50 mg l^{-1}. All strains, but three, produced more SRs than SPs.

Using the immunological approach it was established that in the course of the infection of zucchini cotyledons by *P. syringae* pv. *lachrymans*, the estimated concentration of SPs in the plant tissue was 370 mg kg^{-1}. Production of LDP in planta seems to have occurred relatively early. At the initial stages of the disease the concentration of the toxins was well beyond those known for their activity on biological systems. However, the values obtained for aqueous extracts of infected plants appear to be overestimated and are relatively high if compared to the figures of the "in vitro" experiments reported above. This can be correlated with previous observations (Iacobellis *et al.*, 1989; Fogliano *et al.*, 1999) indicating that immune-related compounds or high molecular weight-complexes are formed with LDP in the infected plant tissue. The antibodies recognition of these complexes seems to be much higher than that exhibited with the pure toxins.

The SPs immuno-assay might be useful for studies on the synthesis, biological activity and role in pathogenesis of the bacterial LDP toxins. Apart from this potential application, the immuno-assay could also be used for diagnostic purposes. Cultural methods and antimicrobial assays have been adopted for differentiating *P. syringae* pathovars and LDP-producing strains. However, these tests are time-consuming and sometimes give imprecise or unreliable results. Recently, a PCR test has been proposed to screen pseudomonads for their potential ability to produce LDP (Bultreys and Gheysen, 1999). The test does not discriminate between SR- and SP-producing bacteria and does not provide an estimate of the amount of toxin produced.

Biological activities

Early studies on the LDPs biological activities were performed using partially purified preparations of pseudomonad LDPs, which contained both nonapeptides and peptins. These studies demonstrated that LDPs induce necrosis in plant tissue and that their target is the plasma membrane. They have shown that LDPs are able to impair the growth of a wide range of microorganisms including fungi, and have a marked phytotoxic activity (De Vay *et al.*, 1968; Backman and DeVay, 1971). Actually only the availability of the purified components of LDP mixtures (Ballio *et al.*, 1988) has allowed the evaluation of the specific properties of individual metabolites belonging to the different groups of LDPs.

The investigations on the biological activities of SPs begun immediately after elucidation of their structure (Ballio *et al.*, 1991), and results are often part of comparative studies between peptins and nonapeptides. The antimicrobial activity of the SPs has been investigated both with fungi and with bacteria, although the number of detailed investigation is still limited. SPs have a spectrum of activity distinct from that of SRs and related nonapeptides. A first comparative evaluation of some biological activities of SRs and SPs demonstrated that, while SRE and ST are more effective than $SP_{25}A$ in inhibiting the growth of *Rhodotorula pilimanae*, nonapeptides are much more active in inhibiting conidial germination and germ tube elongation of *Botrytis cinerea*. $SP_{25}A$ is much more phytotoxic than the two nonapeptides, as shown by the electrolyte leakage of carrot tissues and viability of potato tuber tissue (Iacobellis *et al.*, 1992; Lavermicocca *et al.*, 1997). Data by Hutchison and Gross (1997) on tobacco tissue also confirmed that SPs are more phytotoxic than SRs.

During the following years, a more extended comparison of the effects of the two groups on several physiological processes in higher plants, has shown that SRs are more active inhibitors of the fusicoccin-promoted proton excretion from root segments of maize than SPs. Yet, they are much less effective in impairing the H^+-ATPase activity and the ATP-dependent proton transport associated with inside-out plasma membrane vesicles from the same tissue (Di Giorgio *et al.*, 1994). These apparently contrasting results have suggested that the H^+-pump, which is located at the cytoplasmic side of the plasma membrane, is more accessible by the nonapeptides than by the more hydrophobic and larger peptins. Furthermore, the SPs are much stronger reversible-inhibitors of the hydrolytic activity of the H^+-ATPase purified from plasma membranes of maize roots (Camoni *et al.*, 1995), better uncouplers of oxidative phosphorylation in isolated plant mitochondria (Di Giorgio *et al.*, 1996a) and better promoters of stomatal closure in detached leaves of *Xanthium strumarium* and in epidermal strips of *Vicia faba* (Di Giorgio *et al.*, 1996b).

Lavermicocca *et al.* (1997) observed a week activity of $SP_{22}A$ on *Rhodotorula pilimanae* and a negligible activity of both SPs on *Geotrichum candidum*. The same authors have instead reported that $SP_{25}A$ is nearly ten times more active than $SP_{22}A$, an unexpected result in consideration of the structural and biophysical similarity of the two SPs (Ballio *et al.*, 1991; Dalla Serra *et al.*, 1999a). Some data are also available for other peptins: Ballio *et al.* (1996) reported that FPA is nearly as active as $SP_{22}A$ on *B. cinerea*, moderately active on *R. pilimane*, and inactive on *G. candidum*. Also the last discovered peptin, that is, the

Table 13.1 Fifty percent effective dose (ED_{50}) for inhibition of spore germination or bacteria colony growth, *Pseudomonas* LDPs (SRE and $SP_{25}A$). Similar results were obtained for the inhibition of hyphal elongation in the filamentous fungi tested

Peptide	*SRE*	*$SP_{25}A$*
Microorganism tested	*ED_{50}*	*(μM)*
Fusarium oxysporum	0.44	0.33
Verticillium dahliae	0.48	0.39
Botrytis cinerea	0.73	>1
Penicillium expansum	0.38	0.47
Phytophthora infestans	0.48	0.31
Trichoderma atroviride	0.30	>1

corpeptines (CPs) isolated from *P. corrugata*, have a negligible activity against *R. pilimane* (Emanuele *et al.*, 1998).

Gram-negative bacteria resistant to SRE and ST are also resistant to $SP_{22}A$ and $SP_{25}A$. Instead, the two SPs inhibit the growth of three Gram-positives that are weakly affected by SRE and ST (Lavermicocca *et al.*, 1997). In particular, *B. megaterium* is highly inhibited by the two SPs, as well as by the CPs (Emanuele *et al.*, 1998), thus becoming the preferred organism for the bioassay of peptins. Previous results of Rainey *et al.* (1991, 1992) with tolaasin, the peptin produced by *P. tolaasi*, had evidenced a similar selectivity in antibacterial tests.

Recently, Fogliano *et al.* (2002) have assayed SRE and $SP_{25}A$ against different fungi for the ability of inhibiting fungal spore germination and reducing hyphal elongation. The inhibition of fungal spore germination is roughly similar for the two toxins (Table 13.1). Appreciable differences were only detected with *Trichoderma atroviride* and *B. cynerea* where SRE was more active.

Antagonism and biocontrol

Lipodepsipeptides are considered virulence factors for plant pathogenic *Pseudomonas*, in fact their production leads to an increased disease severity, though some disease can occur even in their absence (Xu and Gross, 1988; Takemoto, 1992). Knock out mutants producing only one LDP have reduced virulence, although some saprophitic strains also produce LDPs (Adetuy *et al.*, 1995; Grgurina *et al.*, 1996; Scholz-Schroeder *et al.*, 2001). It seems that the contribution of LDPs to virulence may vary with the plant host parasitized by *P. syringae* (Bender *et al.*, 1999).

On the other hand, the antifungal properties of LDPs may help LDPs-producing *P. syringae* strains to survive in nature. In fact, these bacteria are epiphytes on the surface of plant leaves where they have to compete with other microbes for scarce nutrients. Conceivably, LDPs play a role in controlling fungal populations to allow bacterial survival, thus they can be responsible for the biocontrol action exerted by some *Pseudomonas* strains which are marketed as bio-pesticides.

In this respect, it is of interest to note that *Pseudomonas* strains are also able to produce enzymes such as chitinases and glucanases which degrade fungal cell walls (CWDEs)

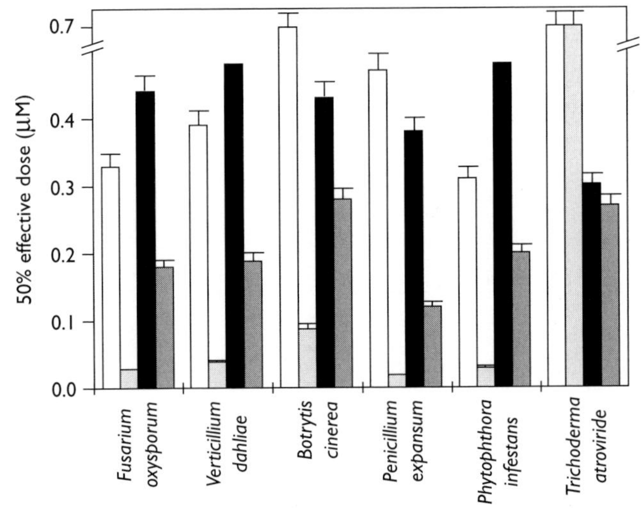

Figure 13.4 Growth inhibition activity of LDPs. Fifty percent effective dose (ED_{50}) for inhibition of spore germination of mixtures containing varying concentrations of *Pseudomonas* LDPs and given concentrations of cell wall degrading enzymes (CWDEs) from various sources. $SP_{25}A$ alone (white bars); $SP_{25}A$ plus CWDEs (light gray bars); SRE alone (black bars), SRE plus CWDEs (dark gray bars).

(Whipps, 1997). The synergistic action of the enzymes acting on the cell wall with peptides targeting the plasma membrane has been already demonstrated for the fungal biocontrol agent *Trichoderma* (Lorito *et al.*, 1996). Recently, Fogliano *et al.* (2002) have shown that a strain of *P. syringae* pv. *syringae* simultaneously produces both LDPs and CWDEs, suggesting that CWDEs are used by *P. syringae* to synergize the fungicidal action of its toxins during the interaction with other microbes. As shown in Figure 13.4, CWDEs at a concentration causing a limited inhibition of the target fungi, dramatically increase the efficacy of both LDPs, with $SP_{25}A$ becoming 4 to 10 times more active than SRE on a molar basis. The higher activity levels of $SP_{25}A$, compared to SRE, is in agreement with the fact that $SP_{25}A$ is much more active than SRE when tested on unilamellar phospholipid liposomes (Camoni *et al.*, 1995), and that $SP_{25}A$ has higher affinity than SRE for a preformed lipid layer (Dalla Serra *et al.*, 1999a).

Pore-formation by syringopeptins

Binding and permeabilisation of model membranes

Surface activity

When injected into an aqueous phase, all LDPs have the tendency to migrate to the water–air interface where they form a monolayer and increase the surface pressure π. This behavior is known as surface-activity and is typical of amphipathic molecules. The increase induced by peptins, $SP_{25}A$ and $SP_{22}A$, is larger than that induced by SRE (typically in the range of 20 mN/m for the peptins and 10 mN/m for the nonapeptide at a concentration of

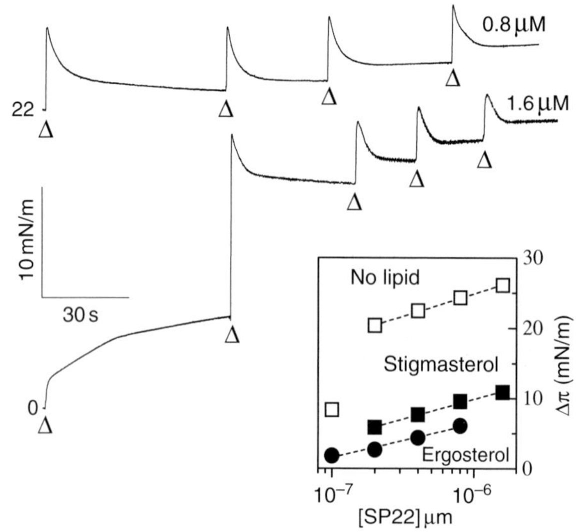

Figure 13.5 Surface pressure increase induced by $SP_{22}A$ with or without a preformed lipid layer. (A) Effects on the interfacial surface pressure of increasing doses of $SP_{22}A$ in the absence (lower trace) or in the presence (upper trace) of an equimolar PC/ergosterol lipid monolayer of initial pressure 22 mN/m. Injections in the subphase are marked by arrow-heads. Each addition induced a fast increase in the surface pressure followed by a slower decrease. The final concentration of $SP_{22}A$ is given next to each trace. (B) Half-logarithmic dose dependence of the steady-state surface pressure increase ($\Delta\pi$) induced by $SP_{22}A$. Open symbols are for the peptide alone, closed symbols for the peptide in the presence of an equimolar monolayer of PC/stigmasterol (squares) or PC/ergosterol (circles) with initial pressure 22 mN/m. Using Gibbs equation, the slope of the regression lines gave the average interface area occupied by each molecule reported in Figure 13.6 (Dalla Serra *et al.*, 1999a).

0.4 μM). Notably, the average surface area (A_m) occupied at the interface by each LDPs molecule is instead substantially similar: 1.2 nm^2 for SRE, 1.3 nm^2 for $SP_{25}A$ and 1.5 nm^2 for $SP_{22}A$ (Figure 13.6). All LDPs displayed a peculiar behavior, which was observed also with some other surface active cyclic LPs of bacterial origin (Maget-Dana and Ptak, 1995). After injection, the surface pressure undergoes a fast growth followed by a slower relaxation towards a steady state increase of lower entity (Figure 13.5). This transient behavior could indicate a rapid absorption of the LDPs to the interface in the water-soluble configuration, followed by a reorganization of the molecule in the new environment to a more stable, and more compact, conformation or aggregation state.

A complementary approach is represented by contact angle measurements. In these experiments, the influence of LDPs inclusion on the angle of contact formed by water drops on a surface is examined. LDPs decrease the contact angle. The long chain $SP_{25}A$ was more effective than the nonapeptide SRE, since a decrease of 10° in contact angle required 0.7 mM of SRE, but only 0.25 mM of $SP_{25}A$. This result is in qualitative agreement with the monolayer experiment, though the concentration required for observing a significant variation of contact angle was definitely higher than for surface pressure. It was proposed that, by decreasing the contact angle of water drops, LDPs may favor the spread on the

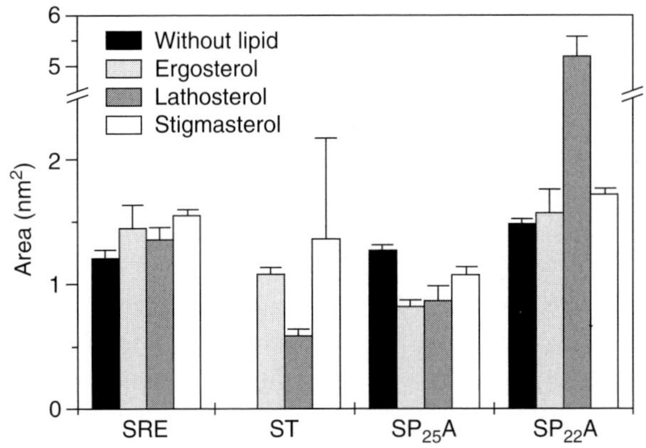

Figure 13.6 Average area occupied by LDP molecules at a water–air interface with or without a pre-formed lipid monolayer of variable composition. A_m was calculated as shown in Figure 13.5. Error is ± SD of two to three experiments. The LDPs used were: $SP_{22}A$, syringopeptin 22 A; $SP_{25}A$, syringopeptin 25 A; SRE, syringomycin E; ST, syringotoxin. Black bars are for the LDPs in the absence of lipid. Monolayers were comprised of an equimolar mixture of PC and ergosterol (light gray bars), lathosterol (dark gray bars) or stigmasterol (white bars). The initial surface pressure was $\pi_i \sim 22$ mN/m. *This large value indicates that practically no interaction occurred.

foliar surface of the bacteria inside these drops (Hutchison and Johnstone, 1993; Hutchison *et al.*, 1995, Hutchison and Gross, 1997). However, the concentration of LDPs necessary to achieve a significant change of contact angle appears to be too large to be relevant on the field.

Lipid monolayers

Useful information on the tendency of the LDPs to partition from the aqueous to the lipid phase can be obtained by repeating the surface pressure experiments in the presence of a pre-formed lipid monolayer. Also in this case the LDPs migrate to the interface, inserting into the monolayer and further increasing its surface pressure. For a monolayer of initial pressure 22 mN/m the increase induced by 0.4 μM LDP was around 5 mN/m, independently of the LDP (Dalla Serra *et al.*, 1999a). The adsorption kinetics was again biphasic with a fast increase in π followed by a slower decrease (Figure 13.5), independently of the LDPs and the lipid composition. When equimolar mixtures of egg PC with one of three different sterols were used, the pressure increments with the different LDPs were rather uniform with just a few exceptions. ST was significantly more effective with lathosterol containing monolayers, whereas $SP_{22}A$ was more active in the presence of stigmasterol and less with lathosterol. Values of A_m for four LDPs, in the presence or absence of lipid monolayers are shown in Figure 13.6. In the presence of lipids they ranged from 0.6 nm^2 (ST with lathosterol) to 1.7 nm^2 ($SP_{22}A$ with stigmasterol). For the sake of comparison, the molecular area occupied by a phospholipid is 0.7 nm^2, that occupied by a sterol is 0.55 nm^2, and that occupied by the structurally similar lipopeptide surfactin was 1.7 nm^2 (Maget-Dana and Ptak, 1995).

Lipid vesicles

Like many cytolytic toxins (Menestrina *et al.*, 1994) *P. syringae* LDPs can induce permeabilization of pure lipid vesicles (Camoni *et al.*, 1995). An extensive investigation of the role of different lipid compositions in this interaction was recently undertaken (Dalla Serra *et al.*, 1999a). It was observed that ST and SR are poorly active on lipid bilayers comprised of phospholipids alone, but become active when sterols are included. SPs instead are fully active also on phospholipid LUVs and do not specifically require sterols. The parameter C_{50}, that is, the concentration necessary for 50% release of vesicle content, can be evaluated from dose dependence curves (Figure 13.7). Values obtained with different LDPs and lipid compositions are reported in Table 13.2. As far as the sterols are considered, it was noted that ST had a preference for lathosterol (mainly present in animal cell membranes), SR was very active on ergosterol (a typical component of fungi membranes) whereas both SPs were more active with stigmasterol (a peculiar component of plant cell membranes).

The time course and dose dependence of LDPs pore-formation in lipid vesicles was analysed in terms of a phenomenological model envisaging three main steps (Parente *et al.*, 1990; Rapaport *et al.*, 1996): (i) binding and incorporation of peptide monomers into the lipid bilayer; (ii) reversible aggregation of bound monomers to form oligomers of increasing size; (iii) channel formation by oligomers that have reached a critical size. The model allowed to derive three parameters: K_1, a partition coefficient which expresses the tendency of the LDPs to insert into the lipid bilayer; K_2, the equilibrium constant for the two-dimensional aggregation of LDPs in the membrane, and m, the minimal number of monomers necessary to form a pore. The average value of m found was 5 ± 1 for the peptins and 6 ± 1 for SRE and ST (Dalla Serra *et al.*, 1999a). Comparing the derived parameters it was observed that the higher, or lower, activity was due to a combination of higher, or lower, partition constant (K_1) and aggregation equilibrium (K_2).

Table 13.2 Permeabilizing and cell-lytic effects of LDPs[a]

Peptide		C_{50} (μM)[b]			
		SR	*ST*	*SP$_{25}$A*	*SP$_{22}$A*
LUVs	Cholesterol[c]	0.2 ± 0.1	4.3 ± 0.5	3.2 ± 0.5	n.d.[d]
	Ergosterol	0.8 ± 0.3	4.9 ± 0.8	1.7 ± 0.5	1.2 ± 0.5
	Lathosterol	1.0 ± 0.7	0.8 ± 0.3	2.8 ± 0.2	9.1 ± 0.5
	Stigmasterol	1.2 ± 0.4	≈40	1.2 ± 0.3	0.4 ± 0.2
	PC:PE:PS	∞[e]	∞[e]	0.7 ± 0.3	0.7 ± 0.3
RBC	Human	0.41 ± 0.05	1.9 ± 0.3	3.1 ± 0.4	2.4 ± 0.3
	Rabbit	0.08 ± 0.01	2.0 ± 0.3	0.38 ± 0.04	n.d.

Notes

a Data were taken from Dalla Serra *et al.*, 1999a,b, except for the lysis of Rabbit RBC, which are unpublished results.

b C_{50} indicates the LDP concentration causing release from 50% of vesicles or hemolysis of 50% of red blood cells (RBC). Values are mean ± SD of two to three experiments.

c LUV were comprised of an equimolar mixture of PC with the indicated sterol.

d n.d.: not determined.

e Up to the higher concentration tested (≈20 μM) there was no indication of any release.

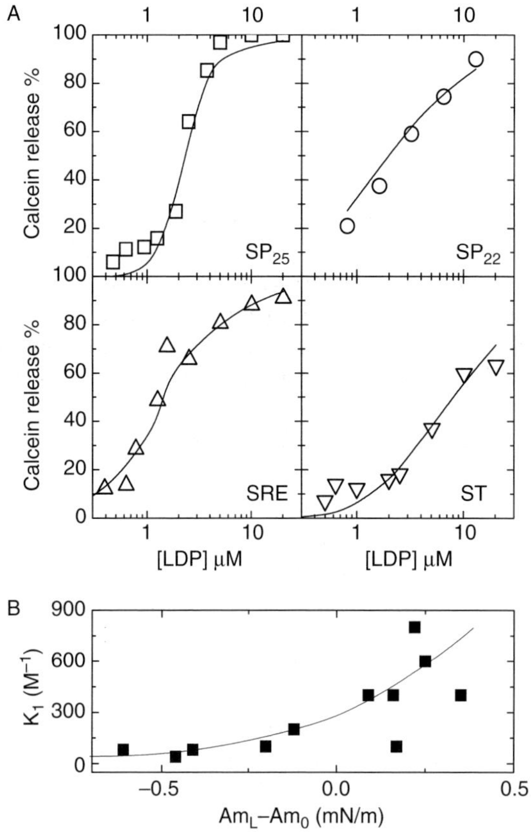

Figure 13.7 Dose dependence of the permeabilizing activity of different LDPs on lipid vesicles. (A) Percent of calcein released by LUVs comprised of an equimolar mixture of PC with ergosterol exposed to different LDPs for 45 min. The lines drawn through the points are best fit according to a model described in (Anzlovar *et al.*, 1998; Dalla Serra *et al.*, 1999a). Similar results were obtained also with vesicles containing other sterols. Lipid concentration was 6.25 μM. (B) Correlation between the parameter K_1 (partition coefficient), obtained from best fit of LUV permeabilization data, and the change in molecular area of the same peptide when it inserts into a monolayer of the same composition (with respect to the area it occupies when it is alone). Values are taken from (Dalla Serra *et al.*, 1999a) and replotted. The line through the points is drawn by eye.

It is interesting to compare the permeabilization of lipid vesicles to the adsorption onto lipid monolayers, for various LDPs and lipid compositions. It appears that the partition coefficient K_1 is large when the molecular area occupied by the LDP in the mixed lipid–peptide monolayer (Am_L) is larger than in the monolayer of peptide alone (Am_0), and vice versa (Figure 13.7B). This suggests that the necessity to stay in a more compact conformation (lower Am_L) decreases the partitioning into the monolayer, whereas the possibility to stay in a more relaxed form (larger Am_L) improves it. Conversely, the tendency to aggregate (equilibrium constant K_2) is inversely related to Am_L suggesting that more compact LDPs conformations are more prone to aggregate (not shown).

Determination of critical micellar concentration (CMC)

Some of the properties of LDPs that we have described, for example, surface activity, partitioning into lipid monolayers and permeabilisation of lipid vesicles are shared also by detergent molecules (Simone and Henkart, 1982; Ruiz *et al.*, 1988; Lasch, 1995). Since detergents are often characterized by their CMC, the CMC of LDPs was also determined. By contact angle measurements, Dalla Serra *et al.* (1999a) found values of 1.3 mM for SRE and 0.9 mM for SP_{25}, whereas by the drop weight method values of 1 mM for SRE and 0.4 mM for $SP_{22}A$ were reported (Hutchison *et al.*, 1995; Hutchison and Gross, 1997). This means that, being in the mM range, the CMC of the LDPs, are two to three orders of magnitude larger than the C_{50} observed for monolayer penetration and vesicle permeabilisation (Figures. 13.4–13.6). In the case of detergents instead, the C_{50}/CMC ratio is between 0.3 and 0.7 (Dalla Serra *et al.*, 1999a). In addition, the parameter m, that is, the molecularity of the unit permeabilising the vesicles, was in the range 8–15 with the detergents but much smaller, between 4 and 7, with the LDPs. This clearly indicates a different mode of action.

Planar lipid bilayers

The system of choice for studying the formation of ion channels is the planar lipid bilayer membrane (PLM) (Menestrina, 1991; Kagan and Sokolov, 1997). Some of the pseudomonad nonapeptidic LDPs have been extensively studied with this system, including SRE (Hutchison *et al.*, 1995; Feigin *et al.*, 1996, 1997; Kaulin *et al.*, 1998; Malev *et al.*, 2000,

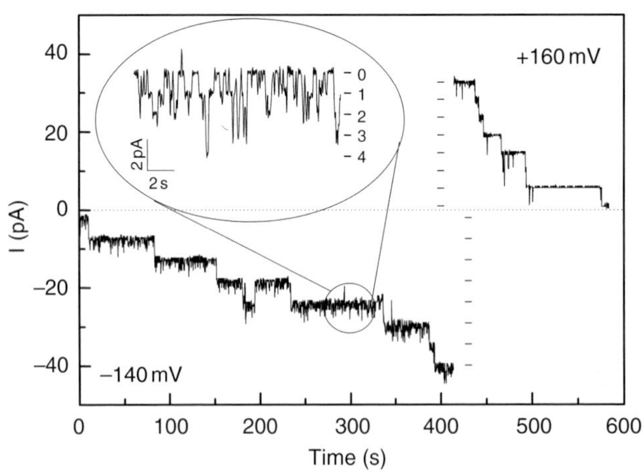

Figure 13.8 Formation of ion channels by SP_{25}. Discrete fluctuations of the ionic current flowing through the membrane after addition into the aqueous phase of $SP_{25}A$ to a final concentration of 4 nM. Buffer was NaCl 100 mM, MES 10 mM, EDTA 1 mM, pH 6.0 and the toxin was added to the *cis* side only (which was maintained at virtual ground). The membrane was comprised of a mixture of PC, PE and PS in a 2 : 2 : 1 molar ratio. Applied voltage was − 140 mV and + 160 mV as indicated. Application of the negative voltage stimulated the opening of the channels whereas switching to the positive voltage produced their closure. The conductance of the pore was 38.5 pS at − 140 mV and 28.5 pS at + 160 mV. The thin dotted line indicates zero of current. At negative voltages, smaller, short-lived, fluctuations of ≈9.6 pS were observed (inset) which represent substates of the open channel.

2001 and see other chapters) and an incompletely purified preparation of ST (Ziegler *et al.*, 1984, 1986). Also some of the related lipopeptides from *Bacillus* have been studied in this way (Maget-Dana *et al.*, 1985a,b; Maget-Dana and Ptak, 1990; Sheppard *et al.*, 1991). For what concerns the peptins, demonstration of SP_{22} channels in PLM was early reported (Hutchison and Gross, 1997; Agner *et al.*, 2000), but a thorough characterization of the ion channels formed by $SP_{25}A$ appeared only recently (Dalla Serra *et al.*, 1999b). The opening of $SP_{25}A$ channels in a voltage-clamped PLM, indicated by uniform step-increases of the current (Figure 13.8) was strictly voltage-dependent. Pores opened when negative voltages were applied, but quickly closed again when the voltage was switched positive. A similar voltage gating (Hille, 1984), was observed also with SRE (Feigin *et al.*, 1996) and with unpurifed ST (Ziegler *et al.*, 1984). Gating was dependent on lipid composition, since larger negative voltages were required to open pores in PLM comprised of purified phospholipids, as compared to PLM comprised of asolectin, a natural mixture of phospholipids of plant origin containing approximately 20% of acidic lipids (Carpaneto *et al.*, 1999, 2002; Dalla Serra *et al.*, 1999b).

CHANNEL CONDUCTANCE

Channel conductance (in 0.1 M NaCl) was about 40 pS at negative voltages and 30 pS at positive voltages. These values are in line with those reported for SRE (Feigin *et al.*, 1996), unpurified ST (Ziegler *et al.*, 1984) and SP_{22} (Hutchison and Gross, 1997), all in the range of 20–40 pS under similar conditions. They are also similar to a conductance of some 20 pS found for tolaasin (Brodey *et al.*, 1991). Different values and quite heterogeneous, were observed instead with the branched-fatty acid lipopeptides of *Bacillus*, for example, iturin A (Maget-Dana *et al.*, 1985a,b), mycosubtilin (Maget-Dana and Ptak, 1990), bacillomycin (Maget-Dana *et al.*, 1985a) and surfactin (Sheppard *et al.*, 1991).

In addition to the main step, with $SP_{25}A$ fluctuations of conductance almost exactly 1/4 that of the stable 40 pS pore were observed at negative potentials. Their lifetime (typically tens of milliseconds) was much shorter than that of the larger channel (which normally lasts for several seconds). In analogy with what demonstrated for SRE (Kaulin *et al.*, 1998), it appears that large pores are actually formed by clusters of the small pores which open independently and close simultaneously. The formation of a tetrameric cluster provides a substantial stabilization to the pore, since the lifetime of the tetramer increases by at least three orders of magnitude compared to that of the components and no other kind of clusters, larger or smaller, was observed. Cluster size of SRE was 6 and its appearance was dose dependent (Kaulin *et al.*, 1998). With $SP_{25}A$ instead, only the large channels were observed at positive voltages, independently of peptide concentration.

CHANNEL SELECTIVITY

The selectivity of $SP_{25}A$ channels was determined by measuring V_{rev}, the reversal voltage (i.e. the voltage at which no current flows) in the presence of a salt gradient (Schultz, 1980; Hille, 1984). It was observed that, independently of membrane composition and gradient magnitude, the channels were anion selective, since the permeability of Cl^- was some three times that of Na^+ (Dalla Serra *et al.*, 1999b). It should be emphasized that such selectivity is rather poor when compared for example to the exquisite selectivity of endogenous ion channels found in animal or plant cells (Hille, 1984), but coherent with a toxin pore that is supposed to inflict an unselective damage to a target cell (Menestrina, 1991; Kagan and Sokolov, 1997). The selectivity properties of these pores correlated nicely with the

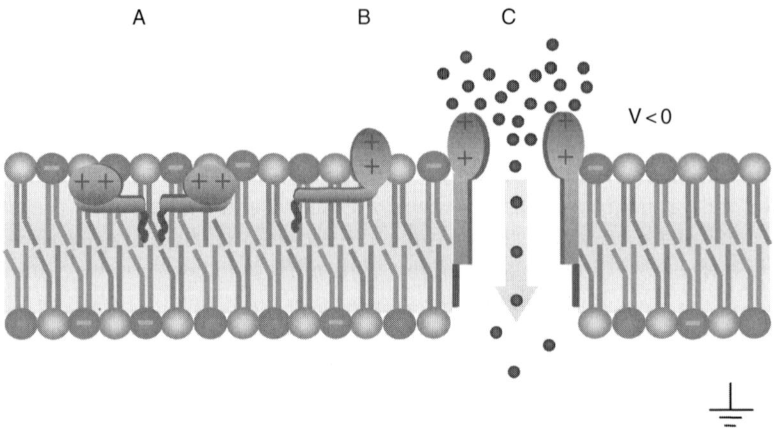

Figure 13.9 Schematic model of the mechanism of pore formation by $SP_{25}A$. (A) $SP_{25}A$ molecules adsorb onto the membrane with the hydrophobic acyl chain inserted between those of the lipid, whereby the hydrophobic linear peptide portion (rectangle) and the hydrophilic cyclic moiety (ellipse) remain folded together as they are in solution. The two positively charged Dab residues of the ring are partially embedded into the membrane at the level of the glycerol backbone, in favorable interaction with the negative charges of the lipid head-groups. Adsorbed monomers form small aggregates. (B) Application of a negative voltage at the *cis* side might open the monomer by attracting the positive charges out of the membrane and causing the unfolding of the hydrophobic part of the peptide which is anchored to the apolar core of the membrane. (C) The hydrophobic part of the peptide aligns with the lipid tails spanning the membrane and forming the staves of a barrel pore. The movement of the charges provides the basis for the voltage-dependent rate of opening and closing of the channels (see Figure 13.8). The presence of positive charges at the *cis* pore entrance accumulates counter ions (anions, red spots) and repels co-ions (cations, blue spots) originating: (i) the anion selectivity of the pore; (ii) a larger current when negative voltages are applied (i.e. a non-linear current/voltage curve, see text). (See Colour Plate XVII.)

molecular structure of the component LDP in view of the presence, at the pore entrance, of the positive charge of the cyclic peptide moiety that would attract anions and repel cations (see Figure 13.9). Also ST and SRE pores were found to be anion selective (Ziegler *et al.*, 1986; Feigin *et al.*, 1996). Among the pores formed by *Bacillus* lipopeptides those opened by uncharged iturin A and mycosubtilin were slightly anion selective (Maget-Dana *et al.*, 1985a; Maget-Dana and Ptak, 1990), whereas those formed by surfactin, which bears two negative charges, were cation selective (Sheppard *et al.*, 1991).

CURRENT VOLTAGE CHARACTERISTIC

As already stated, the single pore conductance was larger at negative than at positive voltages (see Figure 13.8). Also for membranes containing many channels the I/V curve was markedly non-linear, in the sense that the conductance (slope of the I/V curve) was larger at negative than at positive potentials. The ratio of the two limiting conductances observed at positive and negative voltages (called G_+ and G_-) was 0.5. The origin of this non-linearity is conceivably an asymmetric distribution of the charges on the pore (see the model in Figure 13.9). In fact, the presence of the positive charges of the lactone ring near the pore entrance, would accumulate locally the anions and generate a larger anion current when negative voltages push these anions through the pore.

SALT DEPENDENCE

When the concentration of NaCl present was changed, both G_+ and G_- varied almost linearly with the conductivity of the solution, while keeping constant the ratio $G_+/G_- = 0.5$. This linear dependence suggests that $SP_{25}A$ pores are water-filled. Assuming that the pore is simply a cylindrical hole filled with water, that the mobility of ions is similar to the bulk aqueous solution and that its length is at least 5 nm (the average thickness of a lipid bilayer), a pore radius of 0.28 nm could be evaluated from G_-. A similar value, 0.25 nm, can be estimated from the conductance of the pores observed with unpurified ST (Ziegler *et al.*, 1984). Considering the selectivity for Cl^-, the value of $SP_{25}A$ pore radius should be increased to 0.4 nm. However, this value remains smaller than that necessary to release calcein from vesicles (0.6 nm), an effect that was also observed (Dalla Serra *et al.*, 1999a). One possible reason for this discrepancy could be that, except for the polar groups at its entrance, all the rest of the $SP_{25}A$ pore is lined by hydrophobic residues or groups (see Figure 13.9). This hydrophobic portion, while not preventing water from entering the pore to permit ion flux, would certainly restrict the number of ions crossing the channel per unit of time. It should be noted that, based on experiments with uncharged polymers, it was suggested that the smaller unit of the SRE channel (with a conductance of only 4 pS in 0.1 M NaCl) has already a radius of about 1 nm (Kaulin *et al.*, 1998).

VOLTAGE GATING

The initial rate of opening of the channels at negative voltages, and the rate at which the channels decayed to the closed state with large positive voltages, were both exponentially regulated by the applied voltage. Increasing the voltage, the rate of opening decreased *e*-fold every 12 mV, while that of closing increased *e*-fold every 45 mV. A qualitatively similar behavior was reported also for SRE (Feigin *et al.*, 1996), however in that case an *e*-fold change every 13 mV was observed for the opening and every 24–27 mV for the closing.

It is possible to explain the voltage-dependent gating of the pore with a standard two-state model assuming that the transition between the open and the closed state corresponds to a major conformational change triggered by the movement of a charge through the electric field (Neher and Stevens, 1977; Schwarz, 1978; Hille, 1984). The net value of this charge, in elementary units, is derived from the sum of the exponential slopes of the closing and opening rates. In the case of $SP_{25}A$, it was found that the charge moving during the opening of one pore was 2.6 elementary units. Considering that each $SP_{25}A$ carries two positive charges, provided by the two Dab residues located at the cyclic peptide moiety and that pores may be pentamers (Dalla Serra *et al.*, 1999a), the gating should be due to the displacement of the ring through approximately one-quarter of the applied electric field. Because a negative voltage on the same side of toxin addition opens the pore, the involved events were suggested to be the following. (1) Partitioning of the folded LDP monomer into the membrane, with partial embedding of the cyclic peptide moiety into the lipid at the level of the glycerol backbone and insertion of the acyl chain between those of the lipid. (2) Aggregation of the embedded molecules into the lipid film. (3) Upon application of a negative voltage, extrusion of the positive charges from the membrane causing a major unfolding of the adsorbed LDP, because of the anchoring of the hydrophobic part of the peptide to the hydrophobic core of the membrane. (4) Alignment of the unfolded hydrophobic part of the peptide with the lipid tails. (5) Formation of a barrel-stave pore. This model, outlined in Figure 13.9, implies that the cycled peptide moiety remains always on the side of addition and never crosses the bilayer even if the channels open and close (Dalla Serra *et al.*, 1999b).

Effects on biological membranes

Plant vacuoles

The vacuole is the dominant intracellular organelle in mature plant cells and can constitute up to 90% of the cell volume. Compared to artificial membranes composed of pure lipids, the vacuole is a fully natural membrane which contains a variety of lipids, phospholipids, glycolipids and neutral lipids (Tavernier *et al.*, 1993; Behzadipour *et al.*, 1998), as well as proteins that may act as regulators; or even receptors, for $SP_{25}A$. Like the planar lipid bilayer, the tonoplast can be used as a model system for the characterization of channel forming compounds. The technical reasons supporting the idea of incorporating ion channels into a plant tonoplast are: (i) easy isolation from plant-cell, (ii) fast and durable seal, (iii) reliable control/inhibition of endogenous vacuolar channels. Moreover, although the main target of $SP_{25}A$ is the plasma membrane of plant cells, several lines of evidence indicate that intracellular membranes may be also affected: (a) the capacity of destabilizing and crossing the plasma membrane (Camoni *et al.*, 1995), (b) the closure of leaf stomata induced by the toxin (Di Giorgio *et al.*, 1996a), which involves a redistribution of vacuolar ions and a decrease of the vacuolar volume (MacRobbie, 2000), (c) the effects induced by sub-micromolar doses of the toxin on isolated mitochondria (Di Giorgio *et al.*, 1996b). Therefore tonoplasts (vacuolar membranes) could be a biological target for *Pseudomonas* peptins as a source of nutrients for bacterial growth.

Having this in mind, the permeabilizing activity of $SP_{25}A$ on sugar beet tonoplasts was studied with the patch-clamp technique (Carpaneto *et al.*, 2002). Application of 1 μM $SP_{25}A$ induces a significant increase of the basal membrane current (Figure 13.10A). After ceasing toxin administration, a slow decay of the toxin-induced current is observed. Interestingly, as with asolectin (see the Section on "Determination of critical micellar concentration (CMC)"), small applied voltages (− 50 mV) are sufficient to gate toxin insertion, confirming that lipid composition may modulate the gating mechanism possibly by direct participation of the lipid molecules in pore opening. The macroscopic currents induced by the toxin, elicited by different membrane potentials and recorded in symmetrical solutions, are shown in Figure 13.10B (left); the filled symbols to the right represent the corresponding IV-characteristic. When the salt content (KCl) in the extravacuolar solution was decreased from 150 mM to 50 mM the IV-curve (empty symbols) shifts toward negative potential indicating that the $SP_{25}A$ channels are more selective for chloride than for potassium ($P_{Cl}/P_K = 6.7 \pm 1.3$). Substitution of chloride with different anion results in the following selectivity sequence: $NO_3^- \approx Cl^- > F^- >$ gluconate$^-$. The permeabilization of the tonoplast is due to the formation of single channels that can be recorded by incorporating few toxin molecules in small membrane patches. Representative single channel transitions at $V = -80$ mV are shown in Figure 13.10C. The analysis of the single channel values at various membrane potentials reveals a different conductance at large positive and large negative potentials (data not shown); this asymmetric conductance is in agreement with

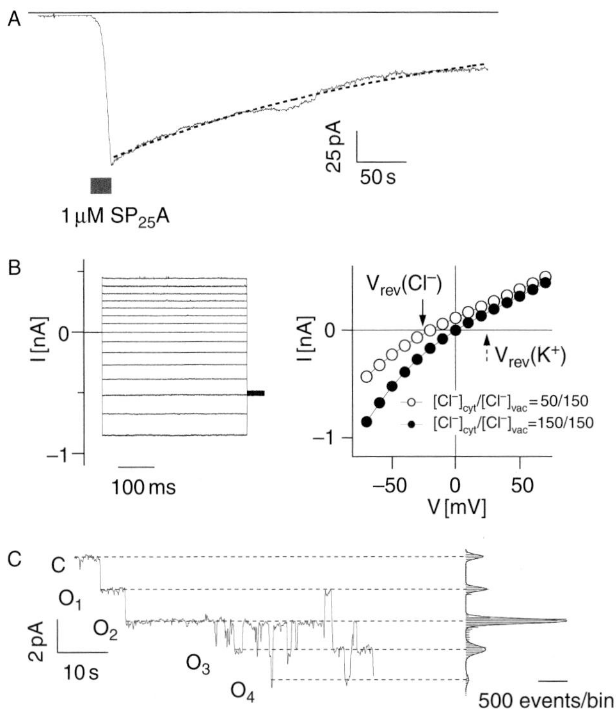

Figure 13.10 $SP_{25}A$ permeabilization of the sugar beet vacuolar membrane. (A) Short perfusion of sugar beet tonoplast with 1 μM $SP_{25}A$ (grey bar). After stopping the perfusion, a slow decrease of the ionic current is evident, which is fitted by a single exponential function with relaxation time 332±8 (dashed line). Internal solution (in mM): KCl 200, $MgCl_2$ 2, $CaCl_2$ 1, Mes 10, pH 5.6. Bath solution: KCl 150, Hepes 15, pH 7.2. Solutions were supplemented with sorbitol up to the sugar beet osmolarity (665 mOsm in that case). Applied potential −50 mV. (B) Left: after the application of 1 μM $SP_{25}A$ in the bath solution macroscopic currents were elicited by a series of voltage pulses ranging from −70 to +70 mV in 10 mV steps. Holding and tail potential were 0 and −50 mV respectively. Symmetrical solutions: KCl 150, EDTA 1, Mes 10, pH 6.0. Right: I–V characteristic in symmetrical (filled symbols) and asymmetrical (empty symbols) conditions. The asymmetrical bath solution is KCl 50, EDTA 1, Mes 10, pH 6.0. The arrows indicate the reversal voltages for chloride and potassium. (C) Single channels transitions of ion channels formed by $SP_{25}A$ at +80 mV. The closed state (C) and the different open levels (O_i, $i = 1, 2, 3$ and 4) are indicated by dashed lines. Histograms to the right show the occupation time of the corresponding trace. Symmetrical solutions as in B.

the inward rectification shown in Figure 13.10 B (right panel, filled symbols) and can be due to an asymmetric distribution of the positive charges that conceivably remain located at the cytoplasmic entrance of the pore (see Figure 13.9). If gluconate permeates through the pore with a single hydration shell, the pore radius of the $SP_{25}A$ channel is estimated to be larger than 0.66 nm, consistent with the results obtained from the osmotic protection experiments and from calcein release (see the section on "Determination of critical micellar concentration (CMC)").

Red blood cells

The ability of LDPs to open pores in lipid membranes and the fact that they do not require a specific receptor to do so, suggest they may also attack non-specifically other cells. One example is their capacity to lyse RBC, which are not normal targets (Hutchison *et al.*, 1995). Earlier findings indicated that SR, ST (Hutchison *et al.*, 1995; Sorensen *et al.*, 1996) and SP_{22} (Hutchison and Gross, 1997) were all hemolytic. More recently it was shown (Dalla Serra *et al.*, 1999a) that the C_{50} (the concentration required for 50% hemolysis) ranked in the order $SP_{25}A > SP_{22}A > ST > SRE$ both on human and rabbit RBC (Table 13.2). The C_{50}s of the peptins were roughly ten times larger than that of SRE, being therefore considerably less hemolytic. Hemolysis was observed also with other cyclic LPs produced by *Pseudomonas* spp. (e.g. tolaasin (Brodey *et al.*, 1991)) or *Bacillus* spp. (e.g. the iturin group (Quentin *et al.*, 1982; Latoud *et al.*, 1986)). Hemolysis was of the colloid-osmotic nature, caused by the opening of peptide-induced pores (Menestrina *et al.*, 1994). This was demonstrated by the fact that the addition in the external solution of osmotic protectants, too large to permeate through the channel, could prevent cell breakdown. From the size of protective sugars, it was estimated that syringopeptins had a radius around 0.9 nm for $SP_{22}A$ and 1 nm for $SP_{25}A$. This value is virtually independent of their concentration, indicating a fixed conformation. It is larger than that estimated with PLM (for the reasons already discussed), but compatible with the release of calcein (ionic radius 0.6 nm) from lipid vesicles. In the case of SRE, instead, the apparent radius determined by osmotic protection ranged from 0.7 to 1.7 nm when the amount of toxin was increased from 0.6 to 3.5 μM (Dalla Serra *et al.*, 1999a), suggesting that the pore is less defined in size and can grow by addition of further monomers when the LDP concentration is increased. The formation in planar lipid bilayers of SRE channels of 1 nm radius was recently described (Kaulin *et al.*, 1998).

Perspectives

Soil-borne, non-pathogenic bacteria with the ability to antagonize fungal phytopathogens and prevent plant diseases represent a realistic alternative to chemical fungicides (Walsh *et al.*, 2001). The potentiality of *P. syringae* strains in this respect deserves attention. Certain strains of *P. syringae* have been successfully used in the biocontrol of post-harvest fungal diseases of citrus and small fruits. It has been observed that the antifungal effect of these strains is associated with their ability to produce LDPs, particularly SRE (Bull *et al.*, 1998). However the role of pseudomonad peptins should be reevaluated in view of their synergic action with CWDEs which dramatically increased the efficacy of $SP_{25}A$ to inhibit target fungi. The use of CWDEs more potent than those from *Pseudomonas*, such as those isolated from the biocontrol fungus *Trichoderma*, further increases the synergism and the inhibitory effect of the mixture, whereas chitinase inhibition reduces the biocontrol effect of *P. syringae* in vivo.

Many fundamental challenges still remain. First of all the development of formulates containing *Pseudomonas* strains combinations having broad antifungal spectrum. This goal can be achieved taking advantage from the interaction between *Pseudomonas* and *Trichoderma*, which involves a synergism between the antimicrobial compounds produced by the two microorganisms, possibly in addition to other biocontrol mechanisms (i.e. exclusion of the pathogen from the infection site or competition for the substrate). In addition, the availability of biosafety data, necessary for the registration of biocontrol agents, are a required prerequisite to produce *Pseudomonas* products marketable as environmentally friendly alternative to chemical fungicides.

Literature cited

Adetuy, F. C., Isogai, A., Di Giorgio, D., Ballio, A. and Takemoto, J. Y. (1995) Saprophytic *Pseudomonas syringae* strain M1 of wheat produces cyclic lipodepsipeptides. *FEMS Microbiology letters*, **131**, 63–67.

Agner, G., Kaulin, Y. A., Gurnev, P. A., Szabo, Z., Schagina, L. V., Takemoto, J. Y., *et al.* (2000) Membrane-permeabilizing activities of cyclic lipodepsipeptides, syringopeptin 22A and syringomycin E from *Pseudomonas syringae* pv. *syringae* in human red blood cells and in bilayer lipid membranes. *Bioelectrochemistry*, **52**, 161–167.

Anzlovar, S., Dalla Serra, M., Dermastia, M. and Menestrina, G. (1998) Membrane permeabilizing activity of linusitin from flax seed. *Molecular Plant – Microbe Interactions*, **11**, 610–617.

Backman, P. A. and DeVay, J. E. (1971) Studies on the mode of action and biogenesis of the phytotoxin syringomycin. *Physiology of Plant Pathology*, **1**, 215–233.

Ballio, A., Barra, D., Bossa, F., DeVay, J. E., Grgurina, I., Iacobellis, N. S., *et al.* (1988) Multiple forms of syringomycin. *Plant Pathology*, **33**, 493–496.

Ballio, A., Bossa, F., Collina, A., Gallo, M., Iacobellis, N. S., Paci, M., *et al.* (1990) Structure of syringotoxin, a bioactive metabolite of *Pseudomonas syringae* pv. *syringae*. *FEBS Letters*, **269**, 377–380.

Ballio, A., Barra, D., Bossa, F., Collina, A., Grgurina, I., Marino, G., *et al.* (1991) Syringopeptins, new phytotoxic lipodepsipeptides of *Pseudomonas syringae* pv. *syringae*. *FEBS Letters*, **291**, 109–112.

Ballio, A., Bossa, F., Di Giorgio, D., Ferranti, P., Paci, M., Pucci, P., *et al.* (1994) Novel bioactive lipodepsipeptides from *Pseudomonas syringae*: the pseudomycins. *FEBS Letters*, **355**, 96–100.

Ballio, A., Bossa, F., Di Giorgio, D., Di Nola, A., Manetti, C., Paci, M., *et al.* (1995) Solution conformation of the *Pseudomonas syringae* pv. *syringae* phytotoxic lipodepsipeptide syringopeptin 25A: two-dimensional NMR, distance geometry and molecular dynamics. *European Journal of Biochemistry*, **234**, 747–758.

Ballio, A., Bossa, F., Camoni, L., Di Giorgio, D., Flamand, M. C., Marcite, H., *et al.* (1996) Structure of fuscopeptins, phytotoxic metabolites from *Pseudomonas fuscovaginae*. *FEBS Letters*, **381**, 213–216.

Baré, S., Coiro, V. M., Scaloni, A., DiNola, A., Paci, M., Segre, A. L., *et al.* (1999) Conformations in solution of the fuscopeptins – phytotoxic metabolites of *Pseudomonas fuscovaginae*. *European Journal of Biochemistry*, **266**, 484–492.

Behzadipour, M., Ratajczak, R., Faist, K., Pawlitschek, P., Trémolières, A. and Kluge, M. (1998) Phenotypic adaptation of tonoplast fluidity to growth temperature in the CAM plant *Kalanchoe daigremontiana* Ham. et Per. is accompanied by changes in the membrane phospholipid and protein composition. *Journal of Membrane Biology*, **166**, 61–70.

Bender, C. L., Alarcon Chaidez, F. and Gross, D. C. (1999) *Pseudomonas syringae* phytotoxins: Mode of action, regulation, and biosynthesis by peptide and polyketide synthetases. *Microbiology and Molecular Biology Reviews*, **63**, 266–292.

Bradbury, J. F. (1986) *Guide to Plant Pathogenic Bacteria*. Farnham Royal, Slough, England: CAB International Mycological

Brodey, C. L., Rainey, P. B., Tester, M. and Johnstone, K. (1991) Bacterial blotch disease of the cultivated mushroom is caused by an ion channel forming lipodepsipeptide toxin. *Molecular Plant–Microbe Interactions*, **4**, 407–411.

Bull, C. T., Wadsworth, M. L., Sorensen, K. N., Takemoto, J. Y., Austin, R. K. and Smilanick, J. L. (1998) Syringomycin E produced by biological control agent controls green mold on citrus. *Biocontrol*, **12**, 89–95.

Bultreys, A. and Gheysen, I. (1999) Biological and molecular detection of toxic lipodepsipeptide-producing *Pseudomonas syringae* strains and PCR identification in plants. *Applied Environmental Microbiology*, **65**, 1904–1909.

Camoni, L., Di Giorgio, D., Marra, M., Aducci, P. and Ballio, A. (1995) *Pseudomonas syringae* pv. *syringae* phytotoxins reversibly inhibit the plasma membrane H^+-ATPase and disrupt unilamellar liposomes. *Biochemical and Biophysical Research Communications*, **214**, 118–124.

Carpaneto, A., Dalla Serra, M., Menestrina, G. and Gambale, F. (1999) Incorporation of channel forming phytotoxic peptides in vacuoles from higher plant cells. *Biophysical Journal*, **76**, A442.

Carpaneto, A., Dalla Serra, M., Menestrina, G., Fogliano, V. and Gambale, F. (2002) The phytotoxic lipodepsipeptide syringopeptin 25A from *Pseudomonas syringae* pv *syringae* forms ion channels in sugar beet vacuoles. *Journal of Membrane Biology*, **188**(3), 237–248

Dalla Serra, M., Fagiuoli, G., Nordera, P., Bernhart, I., Della Volpe, C., Di Giorgio, D., *et al.* (1999a) The interaction of lipodepsipeptide toxins from *Pseudomonas syringae* pv. *syringae* with biological and model membranes: a comparison of syringotoxin, syringomycin and syringopeptins. *Molecular Plant–Microbe Interactions*, **12**, 391–400.

Dalla Serra, M., Nordera, P., Bernhart, I., Di Giorgio, D., Ballio, A. and Menestrina, G. (1999b) Conductive properties and gating of channels formed by syringopeptin 25-A, a bioactive lipodepsipeptide from *Pseudomonas syringae* pv. *syringae*, in planar lipid membranes. *Molecular Plant–Microbe Interactions*, **12**, 401–409.

De Vay, J. E., Lukezic, F. L., Sinden, S. L., English, H. and Coplin, D. L. (1968) A biocide produced by pathogenic isolate of *Pseudomonas syringae* and its possible role in the bacterial canker disease of peach trees. *Phytopathology*, **58**, 95–101.

Di Giorgio, D., Camoni, L. and Ballio, A. (1994) Toxins of *Pseudomonas syringae* pv. *syringae* affect H^+-transport across the plasma membrane of maize. *Physiologia Plantarum*, **91**, 741–746.

Di Giorgio, D., Camoni, L., Mott, K. A., Takemoto, J. Y. and Ballio, A. (1996a) Syringopeptins, *Pseudomonas syringae* pv. *syringae* phytotoxins, resemble syringomycin in closing stomata. *Plant Pathology*, **45**, 564–571.

Di Giorgio, D., Lavermicocca, P., Marchiafava, C., Camoni, L., Surico, G. and Ballio, A. (1996b) Effect of syringomycin-E and syringopeptins on isolated plant mitochondria. *Physiological and Molecular Plant Pathology*, **48**, 325–334.

Emanuele, M. C., Scaloni, A., Lavermicocca, P., Iacobellis, N. S., Camoni, L., Di Giorgio, D., *et al.* (1998) Corpeptins, new bioactive lipodepsipeptide from culture of *Pseudomonas corrugata*. *FEBS Letters*, **433**, 317–320.

Eshita, S. M., Roberto, N. H., Beale, J. M., Mamiya, B. M. and Workman, R. F. (1995) Bacillomycin L_c, a new antibiotic of the iturin group: isolation, structures and antifungal activities of the congeners. *Journal of Antibiotics*, **48**, 1240–1247.

Feigin, A. M., Takemoto, J. Y., Wangspa, R., Teeter, J. H. and Brand, J. G. (1996) Properties of voltage-gated ion channels formed by syringomycin E in planar lipid bilayers. *Journal of Membrane Biology*, **149**, 41–47.

Feigin, A. M., Schagina, L. V., Takemoto, J. Y., Teeter, J. H. and Brand, J. G. (1997) The effect of sterols on the sensitivity of membranes to the channel-forming antifungal antibiotic, syringomycin E. *Biochimica et Biophysica Acta*, **1324**, 102–110.

Fogliano, V., Gallo, M., Vinale, F., Ritieni, A., Randazzo, G., Greco, M., *et al.* (1999) Immunological detection of syringopeptins produced by *Pseudomonas syringae* pv. *lachrymans*. *Physiological and Molecular Plant Pathology*, **55**, 255–261.

Fogliano, V., Ballio, A., Gallo, M., Woo, S., Scala, F. and Lorito, M. (2002) *Pseudomonas* lipodepsipeptides and fungal cell wall degrading enzymes act synergistically in biocontrol. *Molecular Plant–Microbe Interactions*, **15**(4), 323–333.

Gallo, M., Fogliano, V., Ritieni, A., Peluso, L., Greco, M. L., Lops, R., *et al.* (2000) Immunoassessment of *Pseudomonas syringae* lipodepsipeptides (syringomycins and syringopeptins). *Phytopathologia Mediterranea*, **39**, 410–416.

Grgurina, I., Gross, D. C., Iacobellis, N. S., Lavermicocca, P., Takemoto, J. Y. and Benincasa, M. (1996) Phytotoxin production by *syr* mutants effective in biosynthesis or secretion of syringomycin. *FEMS Microbiology Letters*, **138**, 35–41.

Grgurina, I., Mariotti, F., Fogliano, V., Gallo, M., Scaloni, A., Iacobellis, N. S., *et al.* (2002) A new syringopeptin produced by bean strains of *Pseudomonas syringae* pv. *syringae*. *Biochimica et Biophysica Acta*, **1597**(1), 81–89.

Han, F., Mortishire-Smith, R. J., Rainey, P. B. and Williams, D. H. (1992) Structure of the White-Line-Inducing Principle isolated from *Pseudomonas reactans*. *Acta Crystallographica*, **C48**, 1965–1968.

Hille, B. (1984) *Ionic Channels of Excitable Membranes*. Sinauer Associates Publishers, Sunderland Massachussets.

Hutchison, M. L. and Gross, D. C. (1997) Lipopeptide phytotoxins produced by *Pseudomonas syringae* pv. *syringae*: Comparison of the biosurfactant and ion channel-forming activities of syringopeptin and syringomycin. *Molecular Plant–Microbe Interactions*, **10**, 347–354.

Hutchison, M. L. and Johnstone, K. (1993) Evidence for the involvement of the surface active properties of the extracellular toxin tolaasin in the manifestation of the brown blotch disease symptoms by *Pseudomonas tolaasii* on *Agaricus bisporus*. *Physiological and Molecular Plant Pathology*, **42**, 373–384.

Hutchison, M. L., Tester, M. A. and Gross, D. C. (1995) Role of biosurfactant and ion channel-forming activities of syringomycin in transmembrane ion flux: a model for the mechanism of action in the plant–pathogen interaction. *Molecular Plant–Microbe Interactions*, **8**, 610–620.

Iacobellis, N. S., Lavermicocca, P., Surico, G. and Durbin, R. D. (1989) Occurrence of a syringomycin-high molecular weight complex in *Pseudomonas syringae* pv. *syringae*. In *Phytotoxins and Plant Pathogenesis*, edited by A. Graniti, R. D. Durbin and A. Ballio, pp. 429–431. Berlin, Germany: Springer Verlag.

Iacobellis, N. S., Lavermicocca, P., Grgurina, I., Simmaco, M. and Ballio, A. (1992) Phytotoxic properties of *Pseudomonas syringae* pv. *syringae* toxins. *Physiological and Molecular Plant Pathology*, **40**, 107–116.

Isogai, A., Iguchi, H., Nakayama, J., Kusai, A., Takemoto, J. Y. and Suzuki, A. (1995) Structural analysis of new syringopeptins by tandem mass spectrometry. *Bioscience Biotechnology and Biochemistry*, **59**, 1374–1376.

Kagan, B. L. and Sokolov, Y. (1997) Use of lipid bilayer membranes to detect pore formation by toxins. In *Bacterial Pathogenesis*, edited by V. L. Clark and P. M. Bavoil, pp. 395–409. San Diego, USA: Academic Press.

Kaulin, Y. A., Schagina, L. V., Bezrukov, S. M., Malev, V. V., Feigin, A. M., Takemoto, J. Y., *et al.* (1998) Cluster organization of ion channels formed by the antibiotic syringomycin E in bilayer lipid membranes. *Biophysical Journal*, **74**, 2918–2925.

Lasch, J. (1995) Interaction of detergents with lipid vesicles. *Biochimica et Biophysica Acta*, **1241**, 269–292.

Latoud, C., Peypoux, F., Michel, G., Genet, R. and Morgat, J. L. (1986) Interactions of antibiotics of the iturin group with human erythrocytes. *Biochimica et Biophysica Acta*, **856**, 526–535.

Lavermicocca, P., Iacobellis, N. S., Simmaco, M. and Graniti, A. (1997) Biological properties and spectrum of activity of *Pseudomonas syringae* pv. *syringae* toxins. *Physiological and Molecular Plant Pathology*, **50**, 129–140.

Lorito, M., Woo, S. L., D'Ambrosio, M., Harman, G. E., Hayes, C. K., Kubicek, C. P., *et al.* (1996) Synergistic interaction between cell wall degrading enzymes and membrane affecting compounds. *Molecular Plant–Microbe Interactions*, **9**, 206–213.

MacRobbie, E. A. (2000) ABA activates multiple Ca(2+) fluxes in stomatal guard cells, triggering vacuolar K(+)(Rb(+)) release. *Proceedings of the National Academy of Sciences of the United States of America*, **97**, 12361–12368.

Maget-Dana, R. and Ptak, M. (1990) Iturin lipopeptides: interaction of mycosubtilin with lipids in planar membranes and mixed monolayers. *Biochimica et Biophysica Acta*, **1023**, 34–40.

Maget-Dana, R. and Peypoux, F. (1994) Iturins, a special class of pore-forming lipopeptides: biological and physicochemical properties. *Toxicology*, **87**, 151–174.

Maget-Dana, R. and Ptak, M. (1995) Interactions of surfactins with membrane models. *Biophysical Journal*, **68**, 1937–1943.

Maget-Dana, R., Heitz, F., Ptak, M., Peypoux, F. and Guinand, M. (1985a) Bacterial lipopeptides induce ion-conducting pores in planar bilayers. *Biochemical and Biophysical Research Communications*, **129**, 965–971.

Maget-Dana, R., Ptak, M., Peypoux, F. and Michel, G. (1985b) Pore-forming properties of iturin A, a lipopeptide antibiotic. *Biochimica et Biophysica Acta*, **815**, 405–409.

Malev, V. V., Kaulin, Y. A., Bezrukov, S. M., Gurnev, P. A., Takemoto, J. Y. and Schagina, L. V. (2000) Kinetics of opening-closure of syringomycin E channels formed in lipid bilayers. *Biologicheskie Membrany*, **17**, 653–665.

Malev, V. V., Kaulin, Y. A., Gurnev, P. A., Bezrukov, S. M., Takemoto, J. and Shchagina, L. V. (2001) Spatial charge distribution effects in the conductance of syringomycin E ion channels formed in lipid bilayers. *Biologicheskie Membrany*, **18**, 145–153.

Menestrina, G. (1991) Electrophysiological methods for the study of toxin-membrane interaction. In *Sourcebook of Bacterial Protein Toxins*, edited by J. E. Alouf and J. H. Freer, pp. 215–241. London: Academic Press.

Menestrina, G., Schiavo, G. and Montecucco, C. (1994) Molecular mechanism of action of bacterial protein toxins. *Molecular Aspects of Medicine*, **15**, 79–193.

Monti, S. M., Gallo, M., Ferracane, R., Borrelli, R. C., Ritieni, A., Greco, M. L., *et al.* (2001) Analysis of bacterial lipodepsipeptides by MALDI-TOF and electrospray mass spectrometry. *Rapid Communications in Mass Spectrometry*, **15**, 623–628.

Mortishire-Smith, R. J., Nutkins, J. C., Packman, L. C., Brodey, C. L., Rainey, P. B., Johnstone, K., *et al.* (1991) Determination of the structure of an extracellular peptide produced by the mushroom saprophytic *Pseudomonas reactans*. *Tetrahedron*, **47**, 3645–3654.

Neher, E. and Stevens, C. F. (1977) Conductance fluctuations and ionic pores in membranes. *Annual Review of Biophysics and Bioengineering*, **6**, 345–381.

Nutkins, J. C., Mortishire-Smith, R. J., Packman, L. C., Brodey, C. L., Rainey, P. B., Johnstone, K., *et al.* (1991) Structure determination of tolaasin, an extracellular lipodepsipeptide produced by the mashroom pathogen *Pseudomonas tolaasi* Paine. *Journal of the American Chemical Society*, **113**, 2621–2627.

Parente, R. A., Nir, S. and Szoka, F. C., Jr. (1990) Mechanism of leakage of phospholipid vesicle contents induced by the peptide GALA. *Biochemistry*, **29**, 8720–8728.

Peypoux, F., Pommier, M. T., Marion, D., Ptak, M., Das, C. and Michel, G. (1986) Revised structure of mycosubtilin, a peptidolipid antibiotic from *Bacillus subtilis*. *Journal of Antibiotics*, **39**, 636–641.

Quentin, M. J., Besson, F., Peypoux, F. and Michel, G. (1982) Action of peptidolipidic antibiotics of the iturin group on erythrocytes. Effect of some lipids on haemolysis. *Biochimica et Biophysica Acta*, **684**, 207–211.

Rainey, P. B., Brodey, C. L. and Johnston, K. (1991) Biological properties and spectrum of activity of tolaasin a lipodepsipeptide toxin produced by the mushroom pathogen *Pseudomonas tolaasii*. *Physiological and Molecular Plant Pathology*, **39**, 57–70.

Rainey, P. B., Brodey, C. L. and Johnston, K. (1992) Biology of *Pseudomonas tolaasii*, cause of brown blotch disease of the cultivated mushroom. *Advances in Plant Pathology*, **8**, 105–106.

Rapaport, D., Peled, R., Nir, S. and Shai, Y. (1996) Reversible surface aggregation in pore formation by pardaxin. *Biophysical Journal*, **70**, 2502–2512.

Ruiz, J., Goñi, F. M. and Alonso, A. (1988) Surfactant-induced release of liposomal contents. A survey of methods and results. *Biochimica et Biophysica Acta*, **937**, 127–134.

Scaloni, A., Camoni, L., DiGiorgio, D., Scortichini, M., Cozzolino, R. and Ballio, A. (1997) A new syringopeptin produced by a *Pseudomonas syringae* pv. *syringae* strain isolated from diseased twigs of laurel. *Physiological and Molecular Plant Pathology*, **51**, 259–264.

Scholz-Schroeder, B. K., Hutchison, M. L., Grgurina, I. and Gross, D. C. (2001) The contribution of syringopeptin and syringomycin to virulence of *Pseudomonas syringae* pv. *syringae* strain B301D on the basis of sypA and syrB1 biosynthesis mutant analysis. *Molecular Plant–Microbe Interactions*, **14**, 336–348.

Schultz, S. G. (1980) *Basic Principles of Membrane Transport*, New York: Cambridge University Press.

Schwarz, G. (1978) On the physico-chemical basis of voltage-dependent molecular gating mechanisms in biological membranes. *Journal of Membrane Biology*, **43**, 127–148.

Sheppard, J. D., Jumarie, C., Cooper, D. G. and Laprade, R. (1991) Ionic channels induced by surfactin in planar lipid bilayer membranes. *Biochimica et Biophysica Acta*, **1064**, 13–23.

Simone, C. B. and Henkart, P. A. (1982) Inhibition of marker influx into complement-treated resealed erythrocytes ghosts by anti-C5. *Journal of Immunology*, **128**, 1168–1175.

Sorensen, K. N., Kim, K. H. and Takemoto, J. Y. (1996) In vitro antifungal and fungicidal activities and erythrocyte toxicities of cyclic lipodepsinonapeptides produced by *Pseudomonas syringae* pv. *syringae*. *Antimicrobial Agents Chemotherapy*, **40**, 2710–2713.

Surico, G. and DeVay, J. E. (1982) Effect of syiringomicin and syringotoxin by *Pseudomonas syringae* pv. *syringae* on structure and function of mitochondria isolated from holcus spot resistant and susceptible maize lines. *Physiology of Plant Pathology*, **21**, 39–45.

Takemoto, J. Y. (1992) Bacterial phytotoxins syringomycin and its interaction with host membranes. In *Molecular Signals in Plant–Microbe Communications*, edited by D. P. S. Verma, pp. 247–260. Boca Raton: CRC Press.

Tavernier, E., Le Quoc, D. and Le Quoc, K. (1993) Lipid composition of vacuolar membrane of *Acer pseudoplatanus* cultured cells. *Biochimica et Biophysica Acta*, **167**, 242–247.

Vassilev, V., Lavermicocca, P., Di Giorgio, D. and Iacobellis, N. S. (1996) Production of syiringomicin and syringopeptin by *Pseudomonas syringae* pv. *atrofaciens*. *Plant Pathology*, **45**, 316–322.

Walsh, U. F., Morrisey, J. P. and O'Gara, F. (2001) *Pseudomonas* for biocontrol of phytopathogens: from functional genomics to commercial exploitation. *Current Opinion in Biotechnology*, **12**, 289–295.

Whipps, J. M. (1997) Developments in the biological control of soil-borne plant pathogens. *Advances in Botanical Research Incorporating Advances in Plant Pathology*, **26**, 1–134.

Xu, G. W. and Gross, D. C. (1988) Evaluation of the role of syringomycin in plant pathogenesis by using Tn5 mutants of *Pseudomonas syringae* pv. *Syringae* defective in syringomycin production. *Applied Environmental Microbiology*, **54**, 1345–1353.

Yakimov, M. M., Fredrickson, H. L. and Timmis, K. N. (1996) Effect of heterogeneity of hydrophobic moieties on surface activity of lychenysin A, a lipopeptide biosurfactant form *Bacillus licheniformis* BAS50. *Biotechnology and Applied Biochemistry*, **23**, 13–18.

Ziegler, W., Pavlovkin, J. and Pokornj, J. (1984) Effect of syringotoxin on the permeability of bilayer lipid membranes. *Biologia*, **39**, 693–699.

Ziegler, W., Pavlovkin, J., Remis, D. and Pokornj, J. (1986) The anion/cation selectivity of the syringotoxin channel. *Biologia*, **41**, 1091–1096.

14 The role of amyloid peptide channels in amyloid disease

Yutaka Hirakura, Rustam Azimov, Rushana Azimova, Meng-Chin Lin and Bruce L. Kagan

The amyloidoses comprise a polymorphous group of chronic, progressive and sometimes fatal disorders which have in common the deposition of insoluble, proteinaceous fibrils known as amyloid. Over 16 biochemically unrelated polypeptides and proteins have been reported to form amyloid fibrils and cause disease. Characteristics of amyloid include β-pleated sheet conformation, aggregation into rigid fibrils of 6–10 nanometers, cross beta X-ray diffraction pattern and staining with Congo red. The pathophysiologic mechanisms of the amyloidoses remain unknown. Several amyloidogenic peptides have been shown to be cytotoxic, especially for susceptible cells in the target tissue. Seven different amyloidogenic peptides can form highly similar, relatively nonselective, long-lived ion channels in lipid bilayer membranes. Some of these peptides have also been shown to alter Ca^{++} homeostasis in cells and to associate tightly with the plasma membrane. These channels could allow the influx of toxic ions such as calcium as well as the efflux of important cellular species such as potassium and magnesium, and place a severe burden on cellular energy metabolism due to the discharging of ionic gradients. We propose that amyloid channels mediate the pathologic mechanism of action of amyloidogenic peptides. Compounds which inhibit channel formation or block amyloid channels could have a potential role in treatment of the amyloidoses.

Introduction

Amyloid disease

The amyloid diseases include a broad range of chronic and often fatal illnesses characterized by amyloid fibril deposition. Systematic amyloidosis exhibits deposits throughout the body. Disorders of this type can be associated with neoplasia, genetic variations, inflammatory conditions or iatrogenic causes (e.g. hemodialysis). Amyloid deposits may also be limited to specific tissues (see Table 14.1). Over 16 polypeptides sharing little or no sequence homology have been reported as significant contributors to amyloid deposits (Sipe and Cohen, 2000 for review). Because these proteins adopt non-native conformations before aggregating and depositing as amyloid, these diseases have been called "Protein folding diseases." Amyloid deposits can also occur in the absence of clinical illness. For example, while Alzheimer's disease (AD) is characterized by the amyloid deposits occurring in senile plaques, neurofibrillary tangles and vascular locations, healthy elderly subjects also exhibit significant, if lesser, amounts of amyloid deposits in their brains. Amyloidogenic peptides and proteins all adopt a β-pleated sheet conformation, and this appears to enable them to form homomeric aggregates which eventually become extended fibrils with a characteristic

Table 14.1 Amyloid peptides and associated diseases. The peptides till ANF have been shown to form channels

Peptide	*Disease*
Aβ – amyloid –beta protein	Alzheimer's disease, Down's syndrome (trisomy 21), hereditary cerebral angiopathy with bleeding (Dutch)
PrP^{sc} – prion protein	Creutzfeldt-Jakob disease, Gerstmann-Straussler Syndrome, Kuru, fatal familial insomnia, scrapie, mad cow disease
IAPP – Islet amyloid polypeptide-amylin	Type II diabetes mellitus (adult onset)
AA – Serum amyloid A	Reactive systemic amyloidosis, familial Mediterranean fever, Muckle-Wells syndrome
β2-M – β_2-microglobulin	Chronic hemodialysis arthropathy
TTR – Transthyretin (Prealbumin)	Senile systemic cardiac amyloidosis, familial (Portugese, Japanese, Swedish)
ANF – Atrial natriuretic factor	Senile cardiac amyloidosis
Calcitonin	Endocrine amyloidosis (c-cell thyroid tumors)
Immunoglobulin light chains	Idiopathic amyloidosis, multiple myeloma, local modular amyloidosis
Apolipoprotein AI	Familial amyloid polyneuropathy (FAP)-Iowa (Irish), Aortic
Gelsolin	Familial amyloid polyneuropathy (FA)-Finnish
Cystatin C	Hereditary cerebral angiopathy with bleeding (Iceland)
Lysozyme	Nonneuropathic hereditary amyloid with renal disease
Fibrinogen A *a*	Nonneuropathic hereditary amyloid with renal disease, familial
y1 heavy chain	Macroglobulinemia
Lactoferrin	Corneal deposits
Prolactin	Prolactinomas, aging pituitary
Insulin	Iatrogenic

appearance on light and electron microscopy. Amyloid fibrils are rigid, nonbranching, have a width of 6–10 nanometers (nm), and a length that is variable, but often quite extended. X-ray diffraction studies show that they produce a cross beta-pattern (Serpell *et al.*, 1997). The defining feature of amyloid deposits is their characteristic staining with Congo red, and their exhibition of green birefringence under polarized microscopy. The amyloidogenic peptides tend to be small, polyanionic proteins with molecular weights between 3 and 30 kDa. The peptides usually contain notable amounts of glutamic acid, aspartic acid and serine residues. Amyloid fibrils are curiously insoluble and unusually resistant to proteolysis. Serum amyloid P, a pentraxin (pentamer) which facilitates the aggregation of amyloid is found in virtually all amyloid deposits. Amyloid deposits are also associated with a carbohydrate moiety, glycosaminoglycans (GAG). Many amyloid deposits have been reported in association with extracellular matrix and basement membrane components (Kisilevsky, 1996).

Recently, the amyloidoses have received increasing attention as it has become clear that a number of important medical illnesses such as Alzheimer's disease, prion diseases (spongiform encephalopathies), type II diabetes mellitus and rheumatoid arthritis may result partially or wholly from the buildup of amyloid fibrils and/or deposits. In Alzheimer's disease, for example, it has been demonstrated that the number of amyloid deposits correlates highly with cognitive decline (Cummings *et al.*, 1996).

Evidence for a pathogenic role of amyloid

Alzheimer's disease

A full understanding of the role amyloid deposits play in the pathophysiology of amyloid disease has remained elusive. The natural history of amyloidosis suggests that deposits accumulate progressively with time and bear at least some relationship (not necessarily causal) to disease progression. Lacking definite evidence, some observers suggested that amyloid fibrils might contribute to disease in a purely physical manner by disrupting normal tissue architecture (Gilmore, 1997). A growing number of investigations has now pointed to amyloid in the pathogenesis of disease. This mounting evidence supersedes an older theory which hypothesized that amyloid deposits were only a "tombstone"; a remnant of past destructive injury to tissue. In the place of the destroyed tissue only amyloid remained. Amyloid was considered to be a scar that bore a non-causal relationship to illness. This theory has recently been overthrown in AD research due to substantial evidence linking the A beta-amyloid peptide (Aβ) to cellular, tissue and system damage in AD (Cotman, 1997). This evidence includes: (1) Mutations in the gene for the amyloid precursor protein (APP) or its processing appear to be involved in many cases of familial AD (Mullan and Crawford, 1993). (2) Excessive gene dose of APP (trisomy 21, Down's Syndrome) leads to lifelong buildup of A beta deposits and early onset of an Alzheimer's-like dementia (Hardy and Allsop, 1991). (3) Recent studies have suggested a quantitative relationship between the amount of amyloid deposits in the brain and cognitive/intellectual deficits (Cummings *et al.*, 1996), although other studies have failed to find such a relationship. (4) Certain alleles of apolipoprotein E, a component of virtually all amyloid deposits, can increase the risk for sporadic AD several fold (Small *et al.*, 1995). (5) Numerous studies to be described below have demonstrated that A beta, the 39–43 amino acid long peptide product of APP can be toxic to neurons in culture as well as in vivo and can inhibit processes involved in memory such as LTP (Mattson, 1997). (6) Transgenic mice with an added APP or Aβ gene exhibit amyloid deposits, synapse loss, electrophysiologic and behavioral deficits (Moran *et al.*, 1995; Hsiao *et al.*, 1996; Nalbantoglu *et al.*, 1997). and (7) This syndrome can be reversed by vaccination with A beta (Schenk *et al.*, 1999).

Although a large number of studies have now shown that Aβ peptides can cause toxicity to neurons, initial investigations of this effect were not reproducible. Variations of cell culture, ionic solutions, laboratories and even lots of the same peptide were reported to cause different levels of toxicity. Pike *et al.* (1993) helped resolve the paradox when they demonstrated that neurotoxicity was related to the aggregation state of Aβ. Thus, Aβ monomers were non-toxic while aggregated A beta was highly toxic. Other reports have refined this view and showed that it may be possible for Aβ to overaggregate to the point where toxicity decreased. An intermediate level of aggregation may be critical for the neurotoxicity. These data are consistent with the notion that insoluble amyloid fibrils may be non-toxic, while a moderately aggregated soluble form of amyloid could be responsible for toxicity (Harper and Lansbury, 1997).

Numerous investigations of the mechanism of neurotoxicity of Aβ peptides have been reported (see Mattson, 1997 for review). Three main hypotheses appear to have garnered substantial experimental support. First, the regulation of intercellular calcium homeostasis is profoundly altered by Aβ peptide action on cells. Second, generation of reactive oxygen species (ROS) has been implicated in several studies of Aβ cellular neurotoxicity. Antioxidants and antioxidant enzymes appear to be able to protect cells from

A beta-induced toxicity. A substantial body of evidence has built suggesting that Aβ peptides induce toxic oxygen species which then go on to kill neurons in AD. A third line of evidence has demonstrated that Aβ peptides induce permeability alterations in neuronal plasma membranes and disturb ionic balance in the cells. This work has produced evidence both that Aβ can interact with ion channels present on neurons, and that Aβ can form ion channels itself, permeable to cations, calcium and possibly other physiologic ions which might lead to depolarization of cells and alterations in calcium influx (Kourie, 2001a). Calcium channel blockers can sometimes ameliorate the toxicity associated with A beta peptides. The three hypotheses are not exclusive since it is possible that ion channel formation and disruption of calcium homeostasis could lead to oxygen radical production.

Diabetes mellitus

Type II (adult onset) diabetes mellitus has long been associated with the accumulation of amyloid deposits in the Islets of Langerhans of patients. The severity of the illness as measured by the patient insulin requirements was shown to be directly related to the size of the amyloid lesions (Schneider *et al.*, 1980; Ludvik *et al.*, 1997 for review). This suggests a critical, causal relationship between amyloid and clinical illness. Over 90% of patients with type II diabetes mellitus exhibit amyloid deposits. The major component of these deposits is the islet amyloid polypeptide (IAPP or amylin). IAPP is a 37 amino acid hormone which is cosecreted with insulin from the beta cells of the Islets of Langerhans. Studies suggest it may possess a regulatory function opposite that of insulin. Human IAPP kills islet beta cells, whereas the nonamyloidogenic rat IAPP is nontoxic (Lorenzo *et al.*, 1994). Intriguingly, rats are among the few mammalian species that do not form amyloid and do not develop type II diabetes. Other pathological evidence has suggested that IAPP may interact with cell membranes of the beta cells of the Islets of Langerhans (Westermark and Wilander, 1978). Our group has shown that human but not rat IAPP can form ion channels at cytotoxic concentrations in planar phospholipid bilayer membranes. Channel formation is strongly influenced by factors such as lipid membrane composition, ionic strength and membrane potential. The channels are relatively nonselective amongst physiologic ions and are homogenous, exhibiting a single channel conductance of 7.5 picoSiemens in 10 mM KCl. These channels, like the Aβ channels observed by our group and others, would be capable of producing a significant leak in cell membranes due to their lack of ion selectivity, long lifetimes and irreversibility (Mirzabekov *et al.*, 1996). Janson *et al.*, (1999) have shown that small IAPP aggregates are responsible for beta cell toxicity. They employed planar membranes and light scattering to show that aggregates containing 25–6,000 IAPP molecules could cause toxicity and membrane disruption. Once these aggregates exceed 10^6 molecules, they are no longer toxic or capable of damaging membranes. These results are in remarkable accord with the findings for Aβ peptides (Cotman, 1997; Hirakura *et al.*, 1999b and see below).

Prion diseases

PrP106–126 is a cytotoxic, amyloidogenic peptide fragment from the prion protein. The prion diseases (e.g. mad cow disease, scrapie, Creutzfeld–Jakob disease) are caused by a novel protein called the prion (PrPsc), which appears to consist of a single polypeptide chain of 244 amino acids (see Prusiner, 1998 for review). It possesses an amino acid sequence identical to that of a cellular protein (PrPc), but seems to have an altered secondary and tertiary conformation perhaps involving the conversion of alpha-helices to β-sheets. Animals and humans with prion diseases often show amyloid deposits, lesions containing mostly PrPsc, which differs from PrPc in being resistant to proteolytic degradation, relatively insoluble and containing more beta sheet and less alpha-helix. Several peptides from this protein have been reported to be highly amyloidogenic and one of them, PrP106–126, has been shown to be neurotoxic in cell culture (Forloni *et al.*, 1993). Lin *et al.* (1997) reported that PrP106–126 formed relatively nonselective, long lifetime ion channels of highly variable conductances in planar phospholipid bilayer membranes. Treatments which enhanced aggregation and neurotoxicity, such as peptide aging or acidity, resulted in marked increases in channel forming ability of the peptide and in single channel conductance (Lin *et al.*, 1997). The increase of single channel conductance could have indicated that channels were formed by progressively larger peptide aggregates. The channel forming ability of PrP106–126 was recently confirmed and extended to describe several distinct subtypes of channel (Kourie and Culverson, 2000), perhaps reflecting different aggregation states. Channel formation can be inhibited by Congo red (Hirakura *et al.*, 2000a), a treatment which can slow the progression of prion disease (Caspi *et al.*, 1998). Channels can also be blocked by Zn^{+2}, similar to Aβ channels (Hirakura *et al.*, 2000a). One group has failed to find evidence of PrP106–126 neurotoxicity or channel formation (Manunta *et al.*, 2000), but this may reflect the variable aggregation state of the peptide. Similar discrepancies were widely reported for Aβ peptides (see above and Hirakura *et al.*, 1999a). Evidence for prion channels as a toxic mechanism has been recently reviewed (Kourie, 2001b). It is noteworthy that the mutation in PrP which results in the neurogenerative illness Gerstmann–Scheinker–Straussler (GSS) syndrome causes PrP to become a transmembrane protein and this mutation (A116V) is located in the middle of PrP106–126 (Hegde *et al.*, 1998).

An oxidative mechanism for toxicity has been demonstrated in amyloid peptides such as Aβ, IAPP, calcitonin and atrial natriuretic peptide (ANF) (Schubert *et al.*, 1995). These authors extended the toxicity studies done previously to the novel amyloid peptides, calcitonin and ANF. They showed that toxicity was associated with increased hydrogen peroxide production and could be prevented by antioxidants. They were unable to show a correlation between β-sheet formation and toxicity and went on to suggest that amphiphilicity might be the relevant feature in toxicity. This would be consistent with a channel forming mechanism since pore-forming peptides typically have an amphiphilic structure. It is well known that peptides may kill cells without adopting β-sheet formation, however, it remains to be determined if that is critical for amyloid peptides. Current evidence suggests that amyloid peptides must have β-sheet conformation in order to induce amyloid formation and for them to induce toxicity. Hirakura *et al.* (1999a, 2000a) showed that Congo red could inhibit channel formation of Aβ, PrP106–126, and IAPP, suggesting that beta-sheet is necessary.

Channel formation

A large body of evidence has now shown that various amyloid peptides can form ion channels. Arispe *et al.* (1993a), used a lipid bilayer (tip dip) system to observe ion channels induced by Aβ1–40. They found ideally cation selective, calcium permeable, voltage independent channels at concentrations of 0.1 μM, well below that required for toxicity and suggested that these channels might mediate the neurotoxicity of Aβ. They went on to show that Aβ1–40 could form very large conductance channels (Arispe *et al.*, 1993b). A later report suggested that Zn^{+2} blocked Aβ channels through a direct interaction with the pore (Arispe *et al.*, 1996). Durell *et al.* (1995) used molecular modeling to propose several pore structures for Aβ1–40 channels, indicating how a change in conformation could lead to a change in channel conductance.

Kawahara *et al.* (1997) showed that Aβ1–40 could form channels in GnRH secreting neurons that had properties similar to those seen in lipid bilayers. Kawahara *et al.* (2000) have recently shown that Aβ1–40, PrP106–126 and IAPP can all elevate intracellular Ca^{+2} levels in GT1-7 neurons when added to the exterior of these cells at cytotoxic concentrations. Aβ1–40 and Aβ1–42 have been shown to kill fibroblasts and increase intracellular Ca^{+2} levels in a manner inhibitable by Zn^{+2} and Congo red, but not by antioxidants (Rhee *et al.*, 1998; Lin *et al.*, 1999; Bhatia *et al.*, 2000; Zhu *et al.*, 2000). Hirakura *et al.* (1999a) demonstrated that Aβ1–42 and Aβ1–40 could form channels in planar bilayers and that channel activity and conductance depended on solvent history (aggregation state) and pH and could be inhibited by Congo red and blocked by Zn^{+2}. Taken together, these studies provide a powerful impetus for the channel hypothesis of AD – the idea that channel formation is the molecular mechanism of neurotoxicity in AD.

Other groups have reported channel formation by Aβ peptides in a variety of preparations (Fraser *et al.*, 1997; Kawahara and Kuroda, 2000, for review). These include channel formation in neuroblastoma cells, human fibroblasts, acutely dissociated hippocampal neurons, HNT cells and channel formation or induction of conductances by Aβ 25–35 in bullfrog sympathetic neurons, cortical neurons and artificial lipid bilayers. A carboxy terminal fragment of the APP (CT 105) which actually is found in vivo and is neurotoxic has also been found to produce nonselective ionic conductances in oocytes.

Other amyloid channels

These findings have now been extended to several other amyloid peptides (Kourie and Shorthouse, 2000). This evidence argues for a common channel forming mechanism of toxicity, although it does not conclusively demonstrate it.

It has been recently shown that ANF related peptides can form ion channels similar to those of other amyloid peptides (Kourie, 1999a,b). These channels are relatively large and long-lived. They exhibit a variety of conductance states similar to that observed for other amyloid peptide channels.

Serum amyloid A (SAA) is a 104 amino acid peptide which is the precursor protein of a 75 amino acid peptide, amyloid A (AA), which deposits in "reactive amyloidosis." SAA is an acute phase reactant whose serum concentration increases as much as 1000-fold during inflammation or infection and may play a significant role in atherosclerosis, rheumatoid arthritis and amyloidoses associated with other inflammatory conditions (Steel *et al.*, 1994). SAA forms channels in planar lipid bilayers at concentrations well within the physiologic range (Hirakura *et al.*, 2001b). The N-terminal fragment of SAA can kill bacteria, suggesting a possible physiologic role for channel formation. SAA is also found in AD amyloid deposits (Chung *et al.*, 2000).

Beta 2 microglobulin (B2M) is an immune system protein whose serum concentration increases dramatically in dialysis patients. B2M forms amyloid deposits in virtually all such patients. Channels formed in planar lipid bilayers by B2M have physiologic properties similar to those of other amyloid channels. They are also inhibited by Congo red and blocked by Zn^{+2} (Hirakura and Kagan, 2001).

Transthyretin (TTR) is a serum protein which deposits in familial amyloid polyneuropathy (FAP) and mutations linked to the illness have been shown to destabilize the mature form of the protein, resulting in amyloid formation. Channels formed by TTR show a strong physiologic similarity to other amyloid channels in that they have little selectivity amongst physiologic ions, exhibit variable conductances and have long lifetimes. These channels are also inhibited by Congo red and blocked by Zn^{+2} (Hirakura *et al.*, 2001a).

Table 14.2 shows the comparative properties of amyloid peptide channels. Most notable is that with the exception of IAPP, all the channel formers exhibit a variety of single channel conductances, a property not usually associated with ion channels, but particularly associated with channels formed by oligomers of varying sizes. Furthermore, channel formation in all of these peptides occurs at cytotoxic or neurotoxic concentrations, and conditions which favor aggregation and amyloid fibril formation also appear to favor channel formation, as well as toxicity. Properties of these channels differ substantially from classical channels involved in nerve conduction and excitation. Single channel conductances are substantially larger, ionic selectively is considerably weaker, lifetime is dramatically longer and there is little dependence on voltage of opening and closing of the channel. Although some blockers have been described of these channels, they appear to be relatively nonspecific ones so far and do not appear to hold significant therapeutic potential at this point. However, the fact that these channels can be blocked suggests that searching for and/or designing blockers of these channels might significantly reduce cellular toxicity associated with these channels.

Polyglutamine

Although Huntington's Disease (HD) has not been classically considered as an amyloidosis, it is now clear that amyloid-like deposits accumulate in HD, and that these are related to the mutant protein *huntingtin* which contains large tracts of polyglutamine. Hirakura *et al.* (2000b) and Monoi *et al.* (2000) have demonstrated that polyglutamine can form ion channels in planar lipid bilayers. These channels also exhibit variable conductances and relatively low selectivity. Since other evidence indicates a toxic "gain of function" in HD, the polyglutamine channels may be the toxic element in the mutant *huntingtin* protein (Kim and Tanzi, 1998).

Cellular target

Current evidence suggests that the plasma membrane is the likely target of amyloid channels. Leakage pathways in this membrane would cause cellular dysfunction and death. However, another possibility is that the mitochondial membrane might be the target of Aβ, and that Aβ may interact with the permeability transition pore (PTP) and induce apoptosis. This intriguing idea is worthy of exploration.

Table 14.2 Properties of channels formed by amyloid peptides

	Voltage dependence	*Single channel conductance*	*Ion selectivity (Permeability ratio)*	*Blockade by zinc*	*Inhibition by Congo red*	*References*
Aβ25–35	Dependent[a]	10–400 pS	Cation ($P_K/P_{Cl} = 1.6$)	+	+	Mirzabekov *et al.* (1994)
Aβ1–40	Independent	10–2000 pS	Cation ($P_K/P_{Cl} = 1.8$)	+	n.d.	Hirakura *et al.* (1999a)
Aβ1–40	Independent	50–4000 pS	Cation ($P_K/P_{Cl} = 11.1$)	+	n.d.	Arispe *et al.* (1993a,b, 1994, 1996)
Aβ1–42	Independent	10–2000 pS	Cation ($P_K/P_{Cl} = 1.6$)	+	+	Hirakura *et al.* (1999a)
Islet amyloid polypeptide (amylin)	Dependent[b]	7.5 pS	Cation ($P_K/P_{Cl} = 1.9$)	+	+	Mirzabekov *et al.* (1996)
PrP106–126	Independent	10–400 pS	Cation ($P_K/P_{Cl} = 2.5$)	+	+	Lin *et al.* (1997)
PrP106–126	Independent	Various	Cation variable	n.d.	n.d.	Kourie and Culverson (2000)
Serum amyloid A	Independent	10–1000 pS	Cation ($P_K/P_{Cl} = 2.9$)	+	+	Hirakura *et al.* (2001b)
CT105	Independent	120 pS	Cation ($P_K/P_{Cl} = 1.6$)	+	+	Kim *et al.* (1999)
C-type natiuretic peptide	Independent		Cation ($P_K/P_{Cl} = 1.6$)	+	+	Kourie (1999a,b)
B2-microglobulin	Independent		Cation ($P_K/P_{Cl} = 1.6$)	+	+	Hirakura and Kagan (2001)
Transthyretin	Independent		Cation (variable)	+	+	Hirakura *et al.* (2001a)

Notes
a Channels open at negative voltages and close at positive voltages.
b Channels close at positive voltages.
n.d. not determined.

Conclusions

We have reviewed the broad range of clinical syndromes of amyloid diseases, as well as the common pathology observed in them due to β-sheet formation, aggregation and insoluble fibril deposition. Physical chemical properties of these peptides appear to unite them and may predispose them to formation of β-barrels which can form relatively nonselective pores in cell membranes, thus exposing these cells to toxic leaks in their cellular integrity. These leaks may prove injurious or even fatal to cells and may be responsible for the pathogenesis of cellular and organ damage in these various syndromes. More recent evidence suggests that cellular dysfunction such as neuronal dysfunction and inhibition of processes such as LTP might be responsible for early clinical symptoms such as memory deficits in Alzheimer's disease and/or normal aging due to the demonstration of deficits induced by Aβ peptides in the absence of cellular toxicity (Chen *et al.*, 2000). Other biochemical classes of peptides need to be tested for their channel formation potential. This work will likely determine whether channel formation is an important universal mechanism of toxicity in amyloid diseases.

Acknowledgments

It is a pleasure to acknowledge the assistance of Doris N. Finck. Support was provided to B. L. K. by the National Institute of Health (MH01174), the UCLA AIDS Institute and the UCLA Stein-Oppenheimer Fund. Y. H. was supported by a fellowship from the Japanese Society for Promotion of Science.

References

Arispe, N., Rojas, E. and Pollard, H. B. (1993a) Alzheimer disease amyloid beta protein forms calcium channels in bilayer membranes: blockade by tromethamine and alminum. *Proc. Natl. Acad. Sci.*, **90**, 567–571.

Arispe, N., Pollard, H. B. and Rojas, E. (1993b) Giant multilevel cation channels formed by Alzheimer disease amyloid beta-protein [A beta P-(1–40)] in bilayer membranes. *Proc. Natl Acad. Sci.*, **90**, 10573–10577.

Arispe, N., Pollard, H. B. and Rojas, E. (1994) Beta-amyloid Ca(2+)-channel hypothesis for neuronal death in Alzheimer disease. *Mol. Cell. Biochem.*, **140**, 119–125.

Arispe N., Pollard, H. B. and Rojas, E. (1996) Zn2+ interaction with Alzheimer amyloid beta protein calcium channels. *Proc. Natl. Acad. Sci., USA*, **93**, 1710–1715.

Bhatia R., Lin, H. and Lal, R. (2000) Fresh and globular amyloid beta protein (1–42) induces rapid cellular degeneration: evidence for A beta P channel-mediated cellular toxicity. *FASEB J.*, **14**, 1233–1243.

Caspi, S., Halimi, M., Yani, A., Sasson, S. B., Taraboulous, A. and Gabizon, R. (1998) The antiprion activity of Congo Red: putative mechanism. *J. Biol. Chem.*, **273**, 3484–3489.

Chen, Q. S., Kagan, B. L. and Hirakura, Y. (2000) Impairment of hippocampal long-term potentiation by Alzheimer amyloid beta-peptides. *J. Neurosci. Res.*, **60**(1), 65–72.

Chung, T. F., Sipe, J. D., McKee, A., Fine, R. E., Schreiber, B. M., Liangg, J. S. and Johnson, R. J. (2000) Serum amyloid A in Alzheimer's brain is predominantly localized to myelin sheaths and axonal membrane. *Amyloid*, **7**, 105–110.

Cotman, C. W. (1997) The beta-amyloid peptide, peptide self-assembly, and the emergence of biological activities. A new principle in peptide function and the induction of neuropathology. *Ann. NY Acad. Sci.*, **814**, 1–16.

Cummings, B. J., Pike, C. J., Shankle, R. and Cotman, C. W. (1996) Beta-amyloid deposition and other measures of neuropathology predict cognitive status in Alzheimer's disease. *Neurobiol. Aging*, **17**, 921–933.

Durell, S. R., Guy, H. R., Arispe, N., Rojas, E. and Pollard, H. B. (1994) Theoretical models of the ion channel structure of amyloid beta-protein. *Biophys. J.*, **67**(6), 2137–2145.

Forloni, G., Angeretti, N., Chiesa, R., Monzani, E., Salmona, M., Bugiani, O. and Tagliavini, F. (1993) Neurotoxicity of a prion protein fragment. *Nature*, **362**, 543–546.

Fraser, S. P., Suh, Y. H. and Djamgoz, M. B. (1997) Ionic effects of the Alzheimer's disease beta-amyloid precursor protein and its metabolic fragments. *Trends Neurosci.*, **20**, 67–72.

Hardy, J. and Allsop, D. (1991) The amyloid cascade hypothesis trends. *Pharm. Sci.*, **12**, 383–388.

Harper, J. D. and Lansbury, P. T. Jr. (1997) Models of amyloid seeding in Alzheimer's disease and scrapie: mechanistic truths and physiological consequences of the time-dependent solubility of amyloid proteins. *Ann. Rev. Biochem.*, **66**, 385–407.

Hegde, R. S., Mastrianni, J. A., Scott, M. R., De Fea, K. A., Trenblay, P., Torchia, M., De Armond, S. J., Prusiner, S. B. and Lingappa, R. V. (1998) A transmembrane form of the prion protein in neurodegenerative disease. *Science*, **279**, 827–834.

Hirakura, Y., Lin, M. C. and Kagan, B. L. (1999a) Alzheimer amyloid beta1–42 channels: effects of solvent, pH, and Congo Red. *J. Neurosci. Res.*, **57**, 458–466 and **58**, 726.

Hirakura, Y., Satoh, Y., Hirashima, N., Suzuki, T., Kagan, B. L. and Kirino, Y. (1999b) Membrane perturbation by the neurotoxic Alzheimer's disease amyloid fragment β 25–35 requires aggregation and β-sheet formation. *Biochem. and Molec. Biol. Interntl.*, **46**, 787–794.

Hirakura, Y., Yiu, W. W., Yamamoto, A. and Kagan, B. L. (2000a) Amyloid peptide channels: blockade by zinc and inhibition by Congo red (amyloid channel block). *Amyloid*, **7**, 194–199.

Hirakura, Y., Azimov, R., Azimova, R. and Kagan, B. L. (2000b) Polyglutamine-induced ion channels. A possible mechanism for the neurotoxicity of Huntington and other CAG repeat diseases. *J. Neurosci. Res.*, **60**(4), 490–494.

Hirakura, Y. and Kagan B. L. (2001) Channel formation by beta-2-microglobulin: a mechanism for the pathogenesis of dialysis associated amyloidosis. *Amyloid*, **8**, 94–100.

Hirakura, Y., Azimova, R., Azimov, R. and Kagan, B. L. (2001a) Ion channels with different selectivity formed by TTR. *Biophys. J.*, **80**, 129a.

Hirakura, Y., Carrera, I., Sipe, J. D. Kagan, B. L. (2001b) Channel formation by serum amyloid A: A potential mechanism for host defense and amyloid pathogenesis. *Amyloid*, **9**, 13–23.

Hsiao, K., Chapman, P., Nilsen, S., Eckman, C., Harigaya, Y., Younkin, S., Yang, F. and Cole, G. (1996) Correlative memory deficits, A beta elevation and amyloid plaques in transgenic mice. *Science*, **274**, 99–102.

Janson, J., Ashley, R. H., Harrison, D., McIntyre, S. and Butler, P. C. (1999) The mechanism of islet amyloid polypeptide toxicity is membrane disruption by intermediate-sized toxic amyloid particles. *Diabetes*, **48**(3), 491–498.

Kawahara, M. and Kuroda Y. (2000) Molecular mechanism of neurodegeneration induced by Alzheimer's beta-amyloid protein: channel formation and disruption of calcium homeostaisis. *Brain Res. Bull.*, **53**(4), 389–397.

Kawahara, M., Arispe, N., Kuroda, Y. and Rojas, E. (1997) Alzheimer's disease amyloid beta-protein forms Zn(2+)-sensitive, cation-selective channels across excised membrane patches from hypothalamic neurons. *Biophys J.*, **73**, 67–75.

Kawahara, M., Kuroda, Y., Arispe, N. and Rojas, E. (2000) Alzheimer's beta-amyloid, human islet amylin, and prion protein fragment evoke intracellular free calcium elevations by a common mechanism in a hypothalamic GnRH neuronal cell line. *J. Biol. Chem.*, **275**, 14077–14083.

Kim, T. W. and Tanzi, R. E. (1998) Neuronal intranuclear inclusions in polyglutamine diseases: nuclear weapons or nuclear fallout? *Neuron*, **21**, 657–659.

Kim, H. J., Suh, Y. H., Lee, M. H. and Ryu, P. D. (1999) Cation selective channels formed by a C-terminal fragment of beta-amyloid precursor protein. *Neuroreport*, **10**(7), 1427–1431.

Kourie, J. I. (1999a) Synthetic mammalian C-type natriuretic peptide forms large cation channels. *FEBS. Lett.*, **445**(1), 57–62.

Kourie, J. I. (1999b) Calcium dependence of C-type natriuretic peptide-formed fast K(+) channel. *Am. J. Physiol.*, **277**, C43–C50.

Kourie, J. I. (2001a) Mechanism of amyloid beta protein-induced modification in ion transport systems: implications for neurodegenerative diseases. *Cell. Mol. Neurobiol.*, **21**, 173–213.

Kourie, J. I. (2001b) Mechanism of prion-induced modification in membraine transport properties: implications for signal transduction and neurotoxicity. *Chem. Biol. Interactions*, **138**, 1–26.

Kourie, J. I. and Culverson A. (2000) Prion peptide fragment PrP[106–126] forms distinct cation channel types. *J. Neurosci. Res.*, **62**, 120–133.

Kourie, J. I. and Shorthouse, A. A. (2000) Properties of cytotoxic peptide-formed ion channels. *Am. J. Physiol-Cell Ph.*, **278**(6), C1063–C1087.

Lin, M. C., Mirzabekov, T. and Kagan, B. L. (1997) Channel formation by a neurotoxic prion protein fragment. *J. Biol. Chem.*, **272**, 44–47.

Lin, H., Zhu, Y. W. J. and Lal, R. (1999) Amyloid beta protein (1–40) forms calcium-permeable, Zn2+-sensitive channel in reconstituted lipid vesicles. *Biochemistry-US.*, **38**(34), 11189–11196.

Lorenzo, A., Razzaboni, B., Weir, G. C. and Yankner, B. A. (1994) Pancreatic islet cell toxicity of amylin associated with type-2 diabetes mellitus. *Nature*, **368**, 756–760.

Ludvik, B., Kautzky-Willer, A., Prager, R., Thomaseth, K. and Pacini, G. (1997) Amylin: history and overview. *Diabetic Med.*, **14**(Suppl 2), S9–S13.

Manunta, M., Kunz, B., Sandmeier, E., (2000) Reported channel formation by prion protein fragment 106–126 in planar lipid bilayers cannot be reproduced. *FEBS Lett.*, **474**(2–3), 255–256.

Mattson, M. P. (1997) Cellular actions of beta-amyloid precursor protein and its soluble and fibrillogenic derivatives. *Physiol. Rev.*, **77**, 1081–1132.

Mirzabekov, T., Lin, M. C., Yuan, W. L., Marshall, P. J., Carman, M., Tomaselli, K., Lieberburg, I. and Kagan, B. L. (1994) Channel formation in planar lipid bilayers by a neurotoxic fragment of the beta-amyloid peptide. *Biochem. Biophys. Research. Com.*, **202**, 1142–1148.

Mirzabekov, T. A., Lin, M. C. and Kagan, B. L. (1996) Pore formation by the cytotoxic islet amyloid peptide amylin. *J. Biol. Chem.*, **271**, 1988–1992.

Monoi, H., Futaki, S. and Kugimiya, S. (2000) Poly-(L)-glutamine forms cation channels: relevance to the pathogenesis of the polyglutamine diseases. *Biophys. J.*, **78**(6), 2892–2899.

Moran, P. M., Higgins, L. S., Cordell, B. and Moser, P. C. (1995) Age-related learning deficits in transgenic mice expressing the 751-aminoacid isoform of human beta-amyloid precursor protein. *Proc. Natl. Acad. Sci.*, **92**, 5341–5345.

Mullan, M. and Crawford, F. (1993) Genetic and molecular advances in Alzheimer's disease. *Trends Neurosci.*, **16**, 398–403.

Nalbantoglu, J., Tirado-Santiago, G., Lahsaini, A., Poirier, J., Goncalves, O., Verge, G., Momoli, F., Welner, S. A., Massicotte, G. and Julien, J. P. (1997) Impaired learning and LTP in mice expressing the carboxy terminus of the Alzheimer amyloid precursor protein. *Nature*, **387**, 500–505.

Pike, C. J., Burdick, D., Walencewicz, A. J., Glabe, C. G. and Cotman, C. W. (1993) Neurodegeneration induced by beta-amyloid peptides in vitro: the role of peptide assembly state. *J. Neurosci.*, **13**, 1676–1687.

Prusiner, S. B. (1998) The prion diseases. *Brain Path.*, **8**, 499–513.

Rhee, S. K., Quist, A. P. and Lal, R. (1998) Amyloid beta protein-(1–42) forms calcium-permeable, Zn2+-sensitive channel. *J. Biol. Chem.*, **273**, 13379–13382.

Schenk, D., Barbour, R., Dunn, W., Gorden, G., Grajeden, M. *et al.* (1999) Immunization with amyloid–beta attenuates Alzheimer's-disease-like pathology. In the PDAPP Mouse. *Nature*, **400**, 173–177.

Schneider, H. M., Storkel, S. and Will, W. (1980) Amyloid of islets of Langerhans and its relation to diabetes mellitus (author's transl). *Deutsche Medizinische Wochenschrift*, **105**, 1143–1147.

Schubert, D., Behl, C., Lesley, R., Brack, A., Dargusch, R., Sagara, Y. and Kimura, H. (1995) Amyloid peptides are toxic via a common oxidative mechanism. *Proc. Natl. Acad. Sci.*, **92**, 1989–1993.

Serpell, L. C., Sunde, M. and Blake, C. C. (1997) The molecular basis of amyloidosis. *Cell. Mol. Life. Sci.*, **53**, 871–887.

Sipe, J. D. and Cohen, A. S. (2000) Review: history of the amyloid fibril. *J. Struct. Biol.*, **130**, 88–98.

Steel, D. M. and Whitehead, A. S. (1994) The major acute phase reactants: C-reactive protein, serum amyloid P component and serum amyloid A protein. *Immunol. Today*, **15**, 81.

Small, G. W., Mazziotta, J. C., Collins, M. T., Baxter, L. R., Phelps, M. E., Mandekern, M. A., Kaplan, A., La Rue, A., Adamson, C. F., Chang, L. *et al.* (1995) Apolipoprotein E type 4 allele and cerebral glucose metabolism in relatives at risk for familial Alzheimer's disease. *JAMA*, **273**(12), 942–947.

Westermark, P. and Wilander, E. (1978) The influence of amyloid deposits on the islet volume in maturity onset diabetes mellitus. *Diabetologia*, **15**, 417–421.

Zhu, Y. J., Lin, H. and Lal, R. (2000) Fresh and nonfibrillar amyloid beta protein(1–40) induces rapid cellular degeneration in aged human fibroblasts: evidence for A beta P-channel-mediated cellular toxicity. *FASEB. J.*, **14**(9), 1244–1254.

Index